PROGRESS IN COLLOID & POLYMER SCIENCE

Editors: F. Kremer (Leipzig) and G. Lagaly (Kiel)

Volume 98 (1995)

Trends in Colloid and Interface Science IX

Guest Editors:

J. Appell and
G. Porte (Montpellier)

SPRINGER-VERLAG BERLIN
HEIDELBERG GMBH

IV

ISBN 978-3-662-15709-1
ISSN 0340-255 X

Die Deutsche Bibliothek –
CIP-Einheitsaufnahme

Trends in colloid and interface science.

Früher begrenztes Werk in verschie-
denen Ausg.
9 (1995)
(Progress in colloid & polymer science ;
Vol. 98)
ISBN 978-3-662-15709-1
ISBN 978-3-7985-1667-0 (eBook)
DOI 10.1007/978-3-7985-1667-0
NE: GT

© 1995 by Springer-Verlag Berlin Heidelberg
Originally published by Dr. Dietrich Steinkopff Verlag GmbH & Co. KG, Darmstadt in 1995
Softcover reprint of the hardcover 1st edition 1995

Chemistry editor: Dr. Maria Magdalene Nabbe; English editor: James C. Willis; Production: Holger Frey, Bärbel Flauaus.

Type-Setting: Macmillan Ltd., Bangalore, India

Progr Colloid Polym Sci (1995) V
© Steinkopff Verlag 1995

The VIIIth annual meeting of the European Colloid and Interface Society took place in Montpellier-France, September 25-30, 1994. Nearly 300 scientists from nearly every country of Europe attended this meeting. About 50 oral contributions and over 200 poster presentations were given during the meeting. These contributions illustrated almost all fields of modern colloid science in their technical, experimental and theoretical aspects. The present volume reflects only a part of the wealth of contributions and exchanges between participants. But due to lack of space, we were obliged to restrict the number of published contributions. The written contributions all pertain to the topics we chose to highlight during this meeting. They are published here under the corresponding headings.

We gratefully acknowledge financial support from the Euroconference Action of the Human Capital and Mobility Program of the Commission of the European Communities, the Ministère de l'Enseignement Supérieur et de la Recherche, the Centre National de la Recherche Scientifique, University of Montpellier II, the Région Languedoc-Roussillon and the District of Montpellier, Institut Français du Pétrole, the companies L'Oréal and Rhône-Poulenc.

On behalf of ECIS, we would like to thank all participants for their contributions. The ever increasing number of participants attending this annual meeting and the very lively and stimulating discussions which took place during the oral and poster sessions illustrate the good health of ECIS and meet the aims of ECIS in bringing together young and senior colloid scientists each year at an unformal meeting. We also thank our colleagues from Groupe de Dynamique des Phases Condensées (URA233- C.N.R.S.) at University Montpellier II for their collaboration in the organization of the meeting.

Jacqueline Appell
Grégoire Porte

CONTENTS

VIII   Contents

Supramolecular structures under flow

Suspensions and emulsions

Biocolloids

Monolayers, bilayers and interfaces

Progr Colloid Polym Sci (1995) 98:1–5
© Steinkopff Verlag 1995

D. Bastos-González
F.J. de las Nieves

Colloidal stability of model polymer colloids with different functional groups

D. Bastos-González
F.J. de las Nieves-Lopez (✉)
Biocolloids and Fluid Physics Group
Department of Applied Physics
University of Granada
Fueutenveva av.
18071 Granada, Spain

Abstract In this work we describe the colloidal stability of different functionalized latexes: four monodisperse latexes were prepared by surfactant-free emulsion polymerization. The particle sizes of the samples varied between 276 and 350 nm, when measured by TEM. The surface charge densities were determined by conductimetric and potentiometric titrations, the lowest and highest values being 1.3 and 19.2 μC/cm^2. The relative hydrophobicity of latexes was estimated by adsorption of a non-ionic surfactant onto their surfaces. The critical coagulation concentration (ccc) was determined in the presence of 1:1 electrolyte (KBr) at several different pH. By means of the DLVO theory the characteristic parameters A and Ψ_d were calculated. We have employed the Eversole and Boardman equation for calculating Δ and, using this factor, we can find the best agreement between the theoretical and experimental Hamaker Constant values for latexes with the most hydrophobic character.

Key words Colloidal stability – polymer colloids – functional surface groups

Introduction

Monodisperse latexes are widely used in many practical medical applications as immunodiagnostic tests [1, 2]. The size, together with the chemistry of the surface groups and surface charge densities of the particles are important parameters that we need to know in order to control the final properties of the test. Furthermore, looking for this type of application one of the most important factors is the colloid stability, i.e., the critical coagulation concentration of the latexes. To improve the stability we can prepare particles of a high surface charge density. Also, the covalent bonding of the protein to the microsphere surface can improve the reactivity and stability of immunoassays. For that reason it is important to prepare monodisperse particles of specific functionalities on their surface. In particular, latexes with aldehyde and carboxyl functionalities on their surface can be used for covalent bonding of amino-group-carrying biological material such as proteins. In this work four monodisperse latexes with different functionalities were prepared by surfactant-free emulsion polymerization. Sulfate (DBG-0) [3] and carboxyl (DJL-6) [4] functionalities were obtained by convectional emulsion polymerization of styrene using potassium persulfate and ACPA as initiators, respectively. To obtain sulfonate (DBG-2) [5] and aldehyde (AD-1) [6,7] functionalities, the copolymerization of styrene/sodium styrene sulfonate was used in the first case, and styrene/acrolein in the latter. The particle sizes of the samples varied between 276 and 350 nm, when measured by TEM. The surface charge densities were determined by conductimetric and potentiometric titrations with the lowest and highest values being 1.3 and 19.2 μC/cm^2.

The critical coagulation concentration (ccc) of each latex was obtained by turbidity measurements in presence of 1:1 electrolyte (KBr) at several pH. The ccc values were always higher than the physiological ionic strength, which means that they are potentially applicable in immunodiagnostic tests. By means of the DLVO theory the characteristic parameters A and Ψ_d were calculated. We found that both parameters were very low in comparison with the theoretical values. In order to explain these discrepancies with the theory, we adsorbed a nonionic surfactant onto the surface of the particles to determine the relative hydrophobicity of the latexes. The latexes AD-1 and DBG-2 that were synthesized with hydrophilic comonomers present the most hydrophilic character which brings about an additional stabilization that could be due to a liquid layer of water or to the presence of oligomers. We have employed the Eversole and Boardman equation for calculating Δ and, using this factor, we can find the best agreement between the theoretical and experimental Hamaker Constant values.

Experimental

Four surfactant-free latexes with different chemical surface groups (sulfate, sulfonate, aldehyde and carboxilic) have been used in this work. The sulfate latex (DBG-0) was synthesized according to the method of Goodwin et al. [3]. The sulfonate latex (DBG-2) was prepared following the shot-injection method described previously [5,8] without a second injection. To obtain carboxilated latex (DJL-6) we have followed the recipe developed by Guthrie [4]. And the latex aldehyde (AD-1) was synthesized following the indication reported by Yan et al. [6], but with some modifications [7]. Styrene was obtained from Merck and was distilled under reduced nitrogen pressure at 40 °C. The rest of the chemicals used in this study were of analytical grade and were used without further purification. Water used in all experiments was double distilled and deionized with a Milli-Q Water Purification System (Millipore).

Prior to the surface characterization, the latexes were cleaned using several procedures. All the latexes were previously filtered through glass wool to remove the coagula formed during the synthesis. After this, the latexes DBG-0, DJL-6 and AD-1 were cleaned by repeated cycles of centrifugation/decantation/redispersion in a centrifuge of Kontron Instruments, followed by serum replacement for several days until the conductivity of the supernatant was constant and similar to the conductivity of the DDI water [8]. The cleaning process of latex DBG-2 was done first by serum replacement and then the purified latex was treated with mixed ion exchange resins (Amberlite).

Table 1 Particle size, polydispersity index, and surface charge density of latexes.

Latex	Diameter (nm)	P.D.I.	σ_0 (μC/cm^2)
DBG-0	352 ± 10	1.0038	10.4 ± 0.5
DBG-2	287 ± 12	1.0051	1.3 ± 0.1
DJL-6	281 ± 13	1.0060	19.2 ± 0.8
AD-1	324 ± 16	1.0064	2.9 ± 0.2

Transmission electron microscopy was used to determine the mean particle diameter and the polydispersity index (PDI). The particle size was determined by direct measurement of at least 500 particles for each sample using a Calcomp Drawing Board Digitizer and taking two different points on the sectional view for each particle. With the use of a computer program the weight-average (D_w) diameters and number average (D_n) diameters and the standard deviation for each sample were obtained. The latex was considered to be monodisperse if the PDI (defined as D_w/D_n) was less than 1.05.

Surface charge densities of the latexes were determined by conductimetric and potentiometric titrations of the cleaned latexes. The method and procedure for carrying out the automatic titrations have been described previously [5, 7]. As a summary, Table 1 shows the particle size, polydispersity index and surface charge density of the latexes.

The methods and procedure to measure the stability ratios was described in a previous paper [9].

The electrophoretic mobilities of these latexes were measured at 25 °C using a Zeta-Sizer 4 instrument from Malvern Instruments. The reported mobilities are the average of six measurements at stationary level. The experimental error was taken as the standard deviation in these measurements. For the data shown throughout this paper, the standard deviation was always lower than 0.2 10^{-8} m^2/Vs.

Results and discussion

As we know from the DLVO theory, the relation among critical coagulation concentration (ccc), Hamaker constant (A), and the surface potential (Ψ_0) can be found by the following equation [10].

$$ccc = const \frac{\gamma^4}{A^2 z^6}, \tag{1}$$

where the constant depends on the properties of the medium, z is the absolute value of the valence of the coagulat-

Progr Colloid Polym Sci (1995) 98:1-5
© Steinkopff Verlag 1995

ing electrolytes, and γ is:

$$\gamma = \frac{\exp(\frac{ze\psi_0}{2kT}) - 1}{\exp(\frac{ze\psi_0}{2kT}) + 1} . \qquad (2)$$

In Eq. (2), e is the elemental charge, k is the Boltzmann constant and T is the absolute temperature.

Equation (1) was derived taking into account the attraction energy between two parallel plates and the repulsion energy described as a pure Gouy–Chapman double-layer model.

Further analysis of the relation between colloid stability and electrostatic and van der Waals forces [11] modified Eq. (1) by introducing the attraction energy between two equal spheres of radius "a" given by Hamaker [12] and the fact that ions have a finite size. The latter led to the choice of the Gouy–Stern double layer as a model. This model implies that the van der Waals attraction acts over a distance H and the repulsion over a distance $H - 2\Delta$, where Δ is the radius of the ions, and therefore the surface potential must be replaced by diffuse potential (Ψ_d).

All these modifications lead to the relation among ccc, A, and Ψ_d given by:

$$\text{ccc} = \text{const}\,\frac{\gamma^4}{A^2 z^6}\left(1 - \frac{3}{2\kappa a} + \frac{4}{\kappa a}\ln\kappa a\right)\left(1 + \frac{2\Delta}{a}\right)e^{4\Delta\kappa} \qquad (3)$$

The experimental value of ccc and Ψ_d can be obtained from:

$$\frac{d\log W}{d\log C} = -2.15\,10^9\left(\frac{a\gamma^4}{z^2}\right), \qquad (4)$$

where C is the concentration of coagulating electrolyte, a is the radius of the particle, W is the stability ratio (inverse of the collision efficiency factor) and the numerical constant applies to aqueous systems at 25 °C.

Equation (4) applies to concentrations up to the ccc and predicts a linear dependence of $\log W$ on $\log C$ (assuming a constant value of γ). At higher concentrations, the stability ratio remains constant ($W = 1$) since aggregation then occurs at the quickest rate. For this reason, it is usual for stability data to be plotted in $\log W$ versus $\log C$ form. The value of ccc is obtained at the intersection of the two lines and the diffuse potential is taken from the slope of the straight line.

Figures 1 and 2 show $\log W$ versus $\log[\text{KBr}]$ for latexes DBG-2 and DJL-6 at two pH; the correlation coefficients of the fitted straight lines were higher than 0.98. Similar results and correlations were found for latexes DBG-0 and AD-1. The ccc values were calculated and are shown in Table 2 for four latexes.

The most striking result is that the value of the latexes is higher than the physiological ionic strength (approximately 150 mM) and therefore, all of them could be poten-

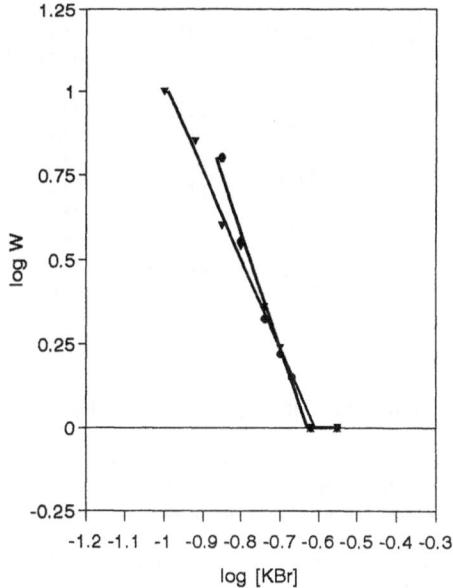

Fig. 1 Log W versus log [KBr] for latex DBG-2 at two pH (● pH 7, ▼ pH 5)

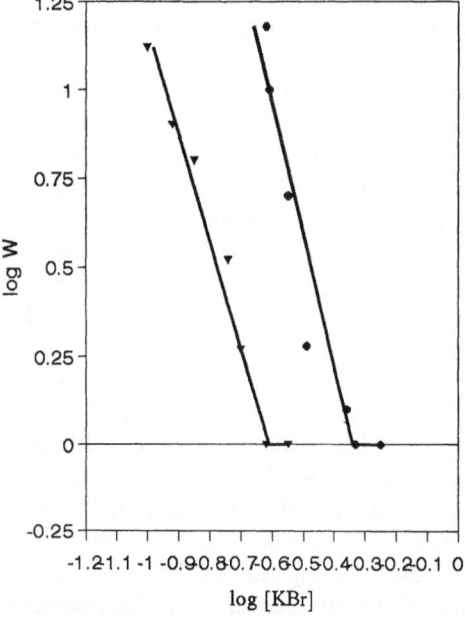

Fig. 2 Log W versus log [KBr] for latex DJL-6 at two pH (● pH 7, ▼ pH 5)

tially applicable to the development of immunodiagnostic tests, where the monodispersity and colloid stability of the latexes are important conditions. The latex DJL-6 with only weak charges on its surface presents a clear difference in the ccc values at both pH, due to the fact that carboxyl groups are more protonated at pH 5 and therefore the effective surface charge is lower. However, it is appropriate

Table 2 ccc values for four latexes at several different pH.

Latex	pH	c.c.c. (mM)
DBG-0	pH = 5	437
	pH = 7	474
DBG-2	pH = 5	246
	pH = 7	234
DJL-6	pH = 5	253
	pH = 7	453
AD-1	pH = 6	345
	pH = 9	371

Table 3 Experimental values for diffuse potentials and Hamaker constants.

Latex	A (10^{-20} J)	ψ_D (mV)
DBG-0	0.3	4.1 ± 0.1
DBG-2	0.4	5.1 ± 0.3
DJL-6	0.5	5.4 ± 0.3
AD-1	0.3	4.8 ± 0.1

Table 4 Occupied area per surfactant molecule and Δ values (from Eq. (6)) for each latex.

Latex	Å^2/molecule	Δ (nm)
DBG-0	120 ± 4	0.65 ± 0.01
DBG-2	129 ± 5	1.07 ± 0.04
DJL-6	111 ± 5	0.78 ± 0.02
AD-1	150 ± 6	1.43 ± 0.03

stabilize latex particles by means of a steric hindrance mechanism. These conclusions could be acceptable for latexes with a relatively strong hydrophilic character or with significant amounts of oligomers on their surface.

In order to determine the relative hydrophobicity of the latexes, we have adsorbed a nonionic surfactant (Triton X-100) onto their surfaces. Table 4 shows the results expressed as occupied area per surfactant molecule and, therefore, a low value implies a more pronunciated hydrophobic character. We can observe that the latexes AD-1 and DBG-2 show the most hydrophilic character. The result for DBG-1 can be explained since it has been synthesized using the comonomer hydrophilic, sodium styrene sulphonate [5], while the DBG-0 and DJL-5 have the least hydrophilic character.

The thickness of a liquid layer of water or the presence of oligomers onto the surface of the particle can also be estimated by means of the Eversole and Boardman equation [17, 18].

$$\ln \tanh \left(\frac{ze\psi}{4kT} \right) = \ln \tanh \left(\frac{ze\psi_d}{4kT} \right) - \kappa\Delta \ , \qquad (5)$$

where Δ is the thickness of this layer and it represents the displacement of the shear plane due to the existence of such a layer.

For an electrolyte 1:1 (KBr) and 25 °C, Eq. (5) is transformed into:

$$\ln \tanh(9.727 \, 10^{-3}\zeta) = \ln \tanh(9.727 \, 10^{-3}\psi_d) - 32.85\Delta c^{1/2} \ . \qquad (6)$$

The value of Δ is obtained from the resultant slope when the first part of Eq. (6) is plotted versus 32.85 $c^{1/2}$. The ζ-potentials have been obtained transforming the electrophoretic mobility values by means of the Dukhin and Semenikhin theory. Figure 3 shows the ζ-potentials versus the electrokinetic radius κa (κ is the reciprocal e.d.l. thickness and a is the radius of the spherical particles) and Fig. 4 shows the experimental points used and the straight lines fixed for four latexes and KBr as electrolyte. In Table 4 the Δ values appear for each latex. We can see that the highest values for Δ can be found in latexes DBG-2 and AD-1. These results match those obtained from the adsorption of the nonionic surfactant. Therefore, the low Hamaker constant obtained from both latexes can be explained, as we

to emphasize that latexes DBG-2 (1.3 μC/cm^2) and AD-1 (2.9 μC/cm^2) present a high ccc in relation to their surface charge densities.

Using Eqs. (2) and (4), we have obtained experimental values for the diffuse potentials and the Hamaker constants. The results are shown in Table 3 with a standard deviation of $0.1 \, 10^{-20}$ J for A and with a value of $\Delta = 0.363$ nm for the mean radius of hydration of the ions used in this work [13]. The A and Ψ_d values are the average obtained from the results carried out at pH 5 and pH 7 for DBG-0, DJL-6, and DBG-2 latexes and, at pH 6 and pH 9 for AD-1 one.

The theoretical value of the Hamaker constant for a polystyrene/water system is $1.37 \, 10^{-20}$ J [14]. Our experimental values are lower than this one, although Hamaker constants experimentally obtained for the same system can vary by even one order of magnitude [15, 16].

In a previous paper [7], the Hamaker constant for this latex had been calculated using Eq. (1). The value of Ψ_d had been obtained taking into account that this potential should be similar to the ζ-potential at such a high electrolyte concentration. In this way, we calculated the ζ-potential estimated by the Dukhin and Semenikhin theory, at ccc conditions. The average values of A and ζ (Ψ_d) in that paper were $0.4 \, 10^{-20}$ J and 16.3 mV, respectively. We concluded that the low value of A might be due to the more hydrophilic character of the aldehyde latexes. This additional stabilization of the aldehyde latexes might have been caused by the surface hydrated layers. Nevertheless, the copolymerization of acrolein and styrene should also have given rise to a large amount of oligomers which could

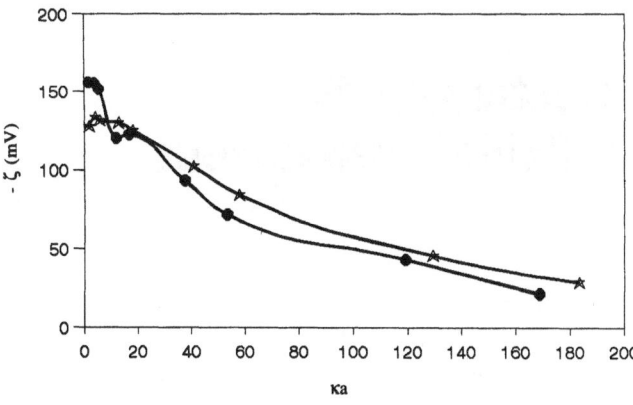

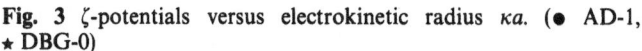

Fig. 3 ζ-potentials versus electrokinetic radius κa. (● AD-1, ★ DBG-0)

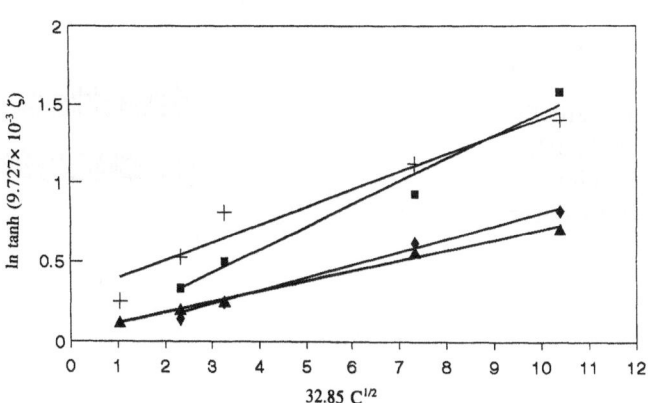

Fig. 4 Experimental values of $[\ln \tanh (9.727 \times 10^{-3}\zeta)]$ versus $[32.85\ C^{1/2}]$ and fixed straight lines for four latexes (▼ DBG-0, + DBG-2, ■ AD-1, ◆ DJL-6)

discussed previously, due to the existence of a relatively strong hydrophilic character, or due to a layer of oligomers on the surface of these latexes.

The hydrophobic character of the DBG-0 and DJL-6 latexes is confirmed by the nonionic surfactant adsorption and the low values of the hydration layer obtained by the Eversole and Boardman equation.

However, a better agreement between the theoretical and experimental Hamaker values is found when in Eq. (3)

Δ-values are used. Thus, the Hamaker constant values for these hydrophobic latexes are $0.9\ 10^{-20}$ J and $1.5\ 10^{-20}$ J, respectively.

Acknowledgements This work is supported by the Comisión Interministerial de Ciencia y Tecnología (CICYT), project MAT 93-0530-C02-02. We would like to express our sincere thanks to Dr. R. Hidalgo-Alvarez for his help with manuscript preparation and Ms. Gráinne Bryan for her assistance in proof-reading the English version.

References

1. Millan JL, Nustad K, Norgaad-Pedersen B (1985) Clin Chem 31:54
2. Masuzawa S, Itoh Y, Kimura H, Kobayashi R, Miyauchi Ch (1983) J Immunol Meth 60:189
3. Goodwin JW, Hearn J, Ho CC, Ottewill RH (1974) Colloid Polym Sci 252:464
4. Guthrie WH (1985) Ph D Thesis, Lehigh University (USA)
5. Bastos D, de las Nieves FJ (1993) Colloid Polym Sci 271:860
6. Yan Ch, Zhang X, Sun Z, Kitano H, Ise N (1990) J Appl Polym Sci 40:89
7. Bastos D, Santos R, Forcada J, Hidalgo R, de las Nieves FJ (1994) Colloids Surfaces A, 92:137
8. de las Nieves FJ, Daniels ES, El-Aasser MS (1991) Colloids Surfaces 60:107
9. Bastos D, de las Nieves FJ (1994) Coll Polym Sci 272:592
10. Verwey EJW, Overbeek JThG (1948) in "Theory of the stability of lyophobic colloids" Elsevier, Amsterdam
11. Overbeek JThG (1980) Pure Appl Chem 52:1181
12. Hamaker HC (1937) Physica 4:1058
13. Robinson RA, Stokes RH (1970) in "Electrolyte solutions" Butterworth, London
14. Prieve DC, Russel WB (1988) J Colloid Interf Sci 115:463
15. Tsaur SL, Fitch RM (1987) J Colloid Interf Sci 115:463
16. Ottewill RH, Shaw JN (1966) Disc Faraday Soc 42:154
17. Eversole WG, Boardman VW (1941) J Chem Phys 9:978
18. Moleón-Baca JA, Rubio-Hernández FJ, de las Nieves-López FJ, Hidalgo-Alvarez R (1991) J Non-Equilib Thermodyn 16:187

Progr Colloid Polym Sci (1995) 98:6–11
© Steinkopff Verlag 1995

T. Palberg
M. Würth
J. Schwarz
P. Leiderer

Kinetics of crystal growth in charged colloidal suspensions

Dr. T. Palberg (✉) · M. Würth
J. Schwarz · P. Leiderer
Faculty of Physics
University of Konstanz
P. O. Box 55 60 M 676
78434 Konstanz, FRG

Abstract We report on the
solidification of a metastable colloidal
melt of monodisperse, highly charged
latex spheres. Light-scattering and
video microscopy are used to study
the growth velocities of crystals
nucleated at the walls of the
observation cell and in the bulk melt.
The velocity observed for the planar
(110) face of the body-centered cubic
wall crystals v_{110} is found to be
significantly smaller than the radial
growth velocity v_R of the
homogeneously nucleated crystals of
rounded polyhedric shape. Under
isothermal conditions the interaction
determining suspension parameters
packing fraction Φ salt concentration
c, and surface charge Z were
systematically varied with high
accuracy using advanced preparation
methods. Growth velocities v_{110} in
the $\langle 110 \rangle$ direction increase over
more than three orders of magnitude
with increasing Φ and decreasing c.
All data collapse on a single curve if
plotted against a reduced energy
density Π^* between melt and fluid at
melting. This master curve shows an
initially linear increase and saturates
at large Π^* with $v_\infty = 9.1~\mu\mathrm{ms}^{-1}$ as
the limiting velocity. It can be
excellently fitted with
a Wilson–Frenkel growth law
yielding a conversion factor of
$B = 6.7~k_B T$ between Π^* and the
chemical potential difference $\Delta\mu$
between melt and solid. Detailed
analysis of the saturation value v_∞
provides evidence for two different
growth mechanisms operative in the
solidification of colloidal crystals

Key words Charged colloids –
colloidal crystals – crystal growth –
Wilson–Frenkel-growth –
time-resolved static light scattering

Introduction

There are numerous examples of colloid and interface systems forming supramolecular order ranging from equilibrium lamellar and cubic phases in microemulsion to shear-induced ordering of latex spheres, from nematic to cholesteric and from ringing gels to opals. Many of these systems offer extremely rich phase diagrams with a large number of parameters. Moreover, they often reveal complicated coupled dynamics on a multitude of time scales

[1]. Investigations of the phase transition kinetics prove extremely difficult due to the great complexity of the systems, yet adding another facet to the fascinating and useful interdisciplinary field of colloid science. The strategy to solve this problem in the long run will lie in resorting to comparably simple systems and to concepts from condensed matter physics first and then gradually introducing larger numbers of system specific boundary conditions.

Solidification of materials in general is of great importance in the processing of materials and has a long history of investigation. A number of semiempirical theories of

Progr Colloid Polym Sci (1995) 98:6–11
© Steinkopff Verlag 1995

nucleation and growth have already been developed in the first part of this century which capture the basic features of the first order freezing transition [2, 3]. Their quantitative experimental verification and the development of corresponding microscopic models for the relevant kinetic processes face two major problems even in the case of atomic or molecular systems with only a small number of internal degrees of freedom. The relevant length scales are accessible only with high technical effort. Moreover direct observation is difficult due to the high rates of nucleation and high velocities of growth. As a consequence experimental data of both high spatial and temporal resolution are rare. In addition, poorly defined or changing experimental conditions often restrict quantitative comparisons to the above-mentioned theories. Limited thermal diffusion, for example, often prevents the development of sufficiently stationary conditions during and after an applied temperature quench or the release of latent heat.

Colloidal systems, on the other hand, may be isothermally shear molten due to the low particle density of some 10^{12} to 10^{14} cm^{-3} and the presence of the suspending fluid acting as a heat bath [4]. This allows for the formation of sufficiently stationary states after quasi instantaneous quenches. Solidifying samples may be prepared and observed by optical techniques under conditions very close to those defined by the assumptions of the above mentioned theories [5]. Rigid spherical particles suspended in a homogeneous medium would provide a particularly simple example of solidifying colloidal systems. Freezing and melting are located at packing fractions of $\Phi_f = 0.494$ and $\Phi_m = 0.521$, respectively [6]. Suspensions with electrostatic stabilization solidify at considerably smaller packing fraction depending on the particle charge and the amount of screening electrolyte c [7–10]. The presence of the viscous suspending medium shifts the relevant time scales to the range between milliseconds and hours. This facilitates observations in real and in reciprocal space with high temporal and spatial resolution. Finally, as a point important for systematic and quantitative measurements, one can precisely adjust the interaction between the suspended particles while it is well described by theoretical expressions [11].

These colloid-specific advantages have led to a strong interest in the solidification dynamics of colloidal crystals and a considerable number of papers exist both on the nucleation and the growth from the metastable shear melt [4, 11–18]. Still, the body of systematic data is yet too small for rigorous and comprehensive comparisons to the aforementioned theories or to recent computer simulations [19].

We here summarize some of our recent results on the growth kinetics observed during the solidification of metastable melts of highly charged latex spheres. We report

systematic studies under variation of the packing fraction, the strength and the range of the Yukawa interaction. We focus on both the presentation of high precision data and their detailed quantitative analysis. The paper is organized as follows. The second section gives a short description of the experimental system and the methods used in preparing well-defined shear melts. We then concentrate on the growth velocities to give unequivocal evidence for Wilson–Frenkel growth in highly charged colloidal suspensions. Finally, the structure of the solid-melt interface is discussed in some detail in the fourth section.

Experimental

Two kinds of particles were used in this study which have been characterized extensively by light scattering and other methods [20]. In particular, we used commercially available polystyrene latex spheres (Seradyn, U.S.A. Lot Nos. 2010M9R and 2011M9R) of equal hydrodynamic radius $a_h = 51$ nm (dynamic light scattering), but different surface group numbers N. At completely deionized conditions the renormalized or effective charge numbers differ by some 15%. The pair interaction potential is well described using either the renormalization procedures or the modified DLVO approximation (MDA), as has been tested by shear modulus titration and also by dynamical criteria [20–22]. A compilation of particle and suspension data is given in Table 1.

The monodisperse particles crystallize into a body-centered cubic (bcc) lattice if the packing fraction is above $\Phi = 0.002$ and the concentration of added salt c is in the sub-micromolar range [20]. To adjust the suspension parameters Φ and c with errors below 1% and 2%, respectively, a recently developed advanced preparation procedure is used [23]. The suspension flows through the observation cell under a hydrostatic pressure difference between two reservoirs of adjustable height difference. It is peristaltically pumped back to the upper reservoir through an inert Teflon tubing system connecting a separate ion exchange chamber, a conductivity measurement to control the salt concentration, and a Debye–Scherrer-like setup to determine the packing fraction via static light scattering. The ion exchange chamber may be bypassed, and salt or further suspension added under inert gas atmosphere to the reservoirs. This technique allows for fast and reproducible preparation of gradient free samples with well-characterized suspension parameters.

During preparation the suspension is in a shear molten state which readily solidifies after termination of flow by closing the two electromagnetic valves at the cell in- and outlet. Using flat shear cells with a ratio between height

Table 1 Compilation of experimental data for the two samples. The range of variation of suspension parameter is indicated where appropriate.

Sample	Diameter σ [nm]	Packing fraction Φ	Salt concentration c_s [μmol l^{-1}]	Surface group number N	Charge number Z	Renormalized charge Z_{PBC}
A	102	≤ 0.0024	0–2	1200	—	450
B	102	0.003	0–2	950	580	395

and depth of $K > 10$ the growth of wall nucleated crystals is usually completed before significant disturbance by homogeneous nucleation processes occur. Furthermore, a high shear rate during preparation significantly increases the induction period for the formation of critical nuclei. On the other hand, homogeneous nucleation and radial growth are the predominant mechanisms of solidification in cells with small K and after lower preparation shear rates. In both cases, however, the variation of shear parameters during the preparation did not lead to an observable change in the growth velocities.

We measured the propagation velocity of a planar (110) face of a bcc crystal nucleated at the cell wall against the metastable melt. The image of a laser beam crossing the cell is monitored with a long distance telescopic microscope (QM1, Questar, NL) under an angle of observation not fulfilling the Bragg condition. This angle is chosen such that the structure factor $S(k)$ of the fluid phase has a significantly larger value than that of the crystalline phase. In the resulting video frames regions of largely different intensities are clearly visible. Each of these frames is further analyzed using image processing (Optimas, Stemmer, Germany) directly yielding the temporal evolution of the spatial intensity distribution and thus the velocity of growth.

It is instructive to qualitatively compare this growth velocity in $\langle 110 \rangle$ direction to the radial growth velocity of homogeneously nucleated crystals. For the latter measurements the cell is mounted on the stage of a polarization microscope (Laborlux 12, Leitz, Germany) and the shape and size of crystals is directly observed. In Fig. 1, we show two data sets measured in different cells, but on the same suspension. Growth of the wall crystal starts immediately after stopping the shear, but is significantly slower. In both cases the growth is strictly linear in time. This was observed for all experimental conditions resulting in complete solidification. In agreement with recently reported studies on growth instabilities [14] nonlinear growth was observed only in the coexistence region of the phase diagram.

Fig. 1 Comparison of the velocities of radial growth (○) to those measured for a planar (110) interface (□) in sample A. We show the evolution of the crystal radii, respective the wall crystal thickness, as a function of time. Growth is strictly linear in both cases. At $\Phi = 0.0022$ and $c = 0.5\ \mu$mol l^{-1} the radial growth velocity $v_R = 9.6\ \mu$ms^{-1} is considerably larger than for the planar interface $v_{110} = 8.4\ \mu$ms^{-1}. Note that the wall crystal thickness quasi instantaneously jumps to a finite value. This might be caused by the registering of shear stabilized layers adjacent to the cell wall during preparation. Note also the the lag time observed for homogeneous nucleation which also is observed to depend on the shearing conditions during preparation. The growth velocities, however, are found to be independent of the sample history in both cases

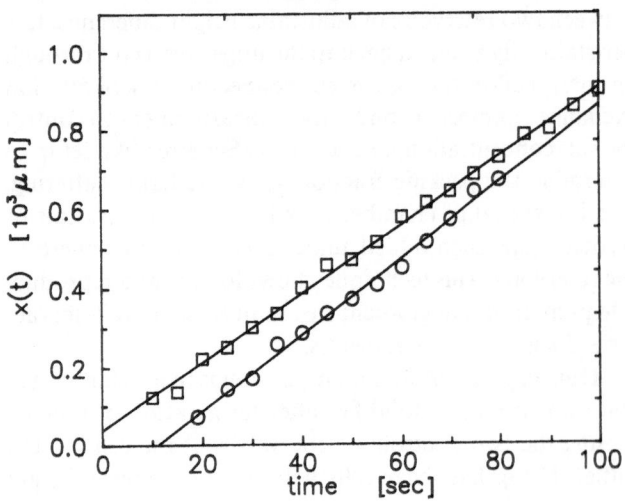

Wilson–Frenkel growth

In their pioneering work, Aastuen et al. were the first to observe a monotonous increase in the radial growth velocities v_R as a function of the packing fraction Φ in a system of highly charged colloidal spheres [12, 13]. In weakly charged systems the growth rate of the linear dimension of crystallites was reported to be fairly independent of Φ [15], whereas in hard sphere systems it shows a pronounced maximum in the upper coexistence range and decays to practically zero as the glass transition is approached [17]. In the latter case density fluctuations were observed to give rise to a $t^{1/2}$ behavior as a function of time t.

In this study, we vary the particle charge and the concentration of screening electrolyte in addition to the packing fraction. In Figs. 2 and 3, we show the results for the propagation velocities of the planar bcc (110) faces obtained for our system of wall nucleated crystals. The growth velocity v_{110} increases over more than two orders of magnitude, if Φ is increased above melting, and saturates at a value of approximately 9 μms^{-1}. Similar results are observed for the salt-dependent growth velocities of the two differently charged samples A and B. The phase boundary for the higher charged sample A ($Z_{PBC} = 450$) is situated at significantly higher salt concentrations even though the packing fraction of that sample ($\Phi = 0.0022$) is somewhat lower than for sample B ($Z_{PBC} = 395$; $\Phi = 0.003$). The limiting velocities, however, are practically identical. Data measured at equilibrium coexistence are given by the stars and will not be included in further analysis.

We further proceed much in the spirit of Aastuen et al. [12]. They suggested a Wilson–Frenkel growth law to interpret their data:

$$v = v_{\infty} \left[1 - \exp\left(-\frac{\Delta\mu}{k_B T} \right) \right] \qquad (1)$$

Fig. 2 Propagation velocity v_{110} of the planar (110) interface for a completely deionized sample A ($Z_{PBC} = 450$) as a function of increasing packing fraction Φ. The stars denote data recorded in the coexistence region of the phase diagram

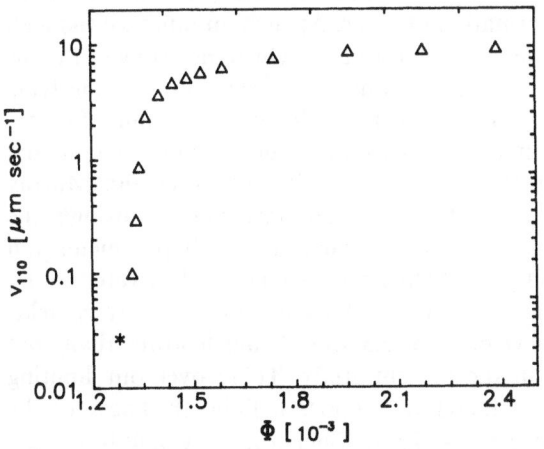

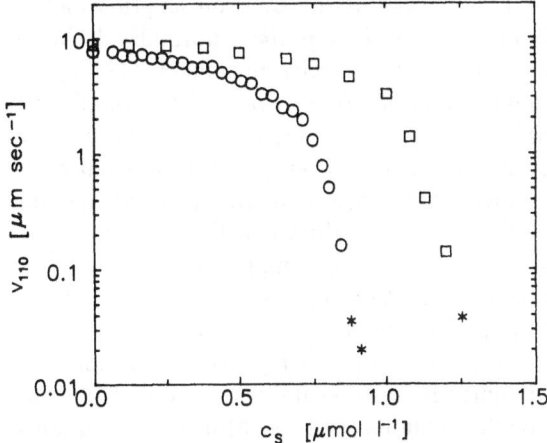

Fig. 3 Propagation velocity v_{110} of the planar (110) interface for (□) sample A ($Z_{PBC} = 450$; $\Phi = 0.0022$) and (○) sample B ($Z_{PBC} = 395$; $\Phi = 0.003$) as a function of increasing salt concentration. Stars again denote data recorded in the coexistence region of the phase diagram. Note the coincidence of the velocities at deionized conditions for the two differently charged samples of equal diameter $\sigma = 102$ nm

Here, v_{∞} is a limiting velocity determined by the self-diffusion of particles of the melt to their target place in the crystal lattice. $k_B T$ denotes the thermal energy and the difference in chemical potential $\Delta\mu$ between metastable melt and solid was approximated by the reduced difference in particle densities $\Delta\mu = B(\rho - \rho_m)/\rho_m$, where m denotes the value at melting. A fit of Eq. (1) was compatible with their experimental data, a quantitative verification of the initial linear increase and the saturation at large $\Delta\mu$ was, however, still missing. Though appropriate for their study in dependence on $\Phi = 3\rho/4\pi a^3$, this approximation does not capture variations in the other suspension parameters. We will therefore derive a reduced energy density difference suited for particles interacting via a Yukawa-type potential formulated within the recently tested modified DLVO approximation [21, 22]:

$$V(r) = Z_{PBC}^2 \lambda_B \left(\frac{e^{+\kappa a}}{1 + \kappa a} \right)^2 \frac{e^{-\kappa r}}{r} \qquad (2)$$

with the screening parameter κ given as:

$$\kappa^2 = \lambda_B \left(|Z_{PBC}|\rho + 2000 N_A c\kappa^2 \right), \qquad (3)$$

We note, that our present analysis neglects several points, needed for a comprehensive understanding. Most importantly the choice of Π_m as reference point may seem somewhat arbitrary. However, if we use $\Pi^{**} = \Pi - \Pi_f/\Pi_f$, the correct value at freezing, and fit Eq. (1) to the data points taken above melting only, the extrapolation of the Wilson–Frenkel curve to zero growth velocity yields $\Pi^{**} = \Pi_m$. This is equivalent to a horizontal shift of the master curve by Π_m in a plot of v_{110} vs. Π^{**}. Correspondingly in our Fig. 4 the data points at coexistence are at negative Π^*. An improved description therefore should consider two here, which are important at low Π^*, i.e. where energetic differences are small. Moreover, at coexistence the crystallization scenario might be significantly different, since the development of a density difference between fluid and coexisting crystal should actually be governed by collective diffusion. Consequently, the Wilson Frenkel prefactor will not remain constant. The data points taken at coexistence are therefore not described by Eq. (1) with $v_{\infty} = const.$, and Π^* defined by energetic criteria alone. Further work in this direction is currently in progress.

where $\lambda_B = e^2/4\pi\varepsilon\varepsilon_0 k_B T$ is the Bjerrum length which is about 0.7 nm in water. Z_{PBC} is the renormalized charge number derived from the bare charge number Z [9, 22, 23].

We define a quantity $\Pi = \rho V(r)$ which formally has the dimension of an osmotic pressure. The data are then plotted against the reduced energy density difference between the metastable melt and the fluid at melting $\Pi^* = (\Pi - \Pi_m)/\Pi_m$. As is shown in Fig. 4, all three experimental series collapse on a single curve.

We fit the master curve with Eq. (1) and obtain a value of $v_\infty = 9.1 \ \mu ms^{-1}$ for the limiting velocity. The fit to the data is extremely good for $\dot{\Pi} > \Pi_m$ and we consider this the first unambigous experimental verification of Wilson–Frenkel growth in colloidal systems. Moreover, the quality of the data allows the derivation of a proportionality factor between Π^* and the chemical potential difference as $\Delta\mu = B\ \Pi^*$ with $B = 6.7\ k_B T$. It provides a quick but reasonably accurate estimate for $\Delta\mu$ given the suspension parameters and one point of the melting line.

The prefactor v_∞ describes a limiting velocity. It is considerably lower than the limiting radial growth velocity observed by Aastuen et al. for only slightly smaller particles [12]. This is, however, consistent with the differences between v_{110} and v_R shown in Fig. 1. To explain this difference, we suggest the presence of alternative growth mechanisms.

In recent computer simulations on the growth of planar (100) and (111) faces of face-centered cubic (fcc) Lennard–Jones crystals it was observed that two different mechanisms were operative [19]. For the (100) face growth was limited by the thermal velocity of the particles. It has

been suggested to use the diffusional transport of particles towards the rough interface of homogeneously nucleated crystals as the corresponding mechanism in colloidal systems [12]. Assuming bulk diffusion, we set $v_\infty = 6D_L l/(d_{110}\lambda)^2$, where $D_L = 0.1\ D_S$ is the long time self-diffusion coefficient which is about of one-tenth of the short-time self-diffusion coefficient D_S throughout the metastable melt. l is the interfacial thickness and $\lambda = 0.39$ is the average relative distance of homogeneously distributed points from the center of a sphere of diameter d_{110}. Equating l to one lattice spacing d_{110} leads to a limiting velocity of $v_\infty = 33\ \mu ms^{-1}$, much larger than the values measured here on v_{110} or v_R. Increasing l yields even larger values. Unfortunately, we do not have enough data on the radial growth velocities of our system to safely extract a limiting *radial* velocity. We note, however, that using our analysis on the data of Aastuen their limiting radial velocity is quantitatively reproduced. This strongly supports bulk long-time self-diffusion of particles towards their target places within a rough interface of one layer thickness to be the microscopic mechanism of growth limitation in the case of rounded polyhedric crystals formed after homogeneous nucleation in the bulk.

Since this mechanism has to be excluded for the growth of our bcc (110) face, we have to check for alternatives. For the (111) face of the Lennard Jones crystals the simulations [19] shows that growth is limited by a reaction-like process operative on interfaces with more than one possibility of arranging the next layer. In the fcc system this is the stacking sequence. For our bcc (110) face it is the possible formation of twins. At such an interface the melt first acquires a layered strcture, which then shows in-plane crystalline ordering and finally registers on the underlying solid layer. This last step is velocity determining. In a recent experimental study on the nucleation of fcc colloidal crystals in the vicinity of a cell wall, Grier and Murray observed a similar three-step mechanism through the evolution of a sixfold coordination order parameter and the self-diffusion coefficients [16]. Using their values of D_L during registering in the above-mentioned analysis yields a limiting velocity of $3.3\ \mu ms^{-1}$, much lower than that observed in the present study. To recover our limiting velocity, we therefore propose a finite thickness of the interface of $l \approx 2 - 5\ d_{110}$. Such a result would be consistent also with both recent computer simulations and density functional calculations [19, 24].

This analysis for the first time provides strong evidence for the presence of two limiting processes in the growth of colloidal crystals depending on the respective crystal lattice plane. One is due to diffusional transport in the fluid-like ordered melt which is slowed as compared to short-time self-diffusion. The other involves a complex three-step mechanism of layer formation, in-layer crystal-

Fig. 4 Propagation velocity v_{110} for all three series plotted versus the reduced energy density difference Π^*. Symbols are as in Figs. 2 and 3. All data collapse on a single curve. The solid line is the best fit of Eq. (1) to the experimental data using $v_{110} = 9.1\ \mu ms^{-1}$ and $B = 6.7\ k_B T$

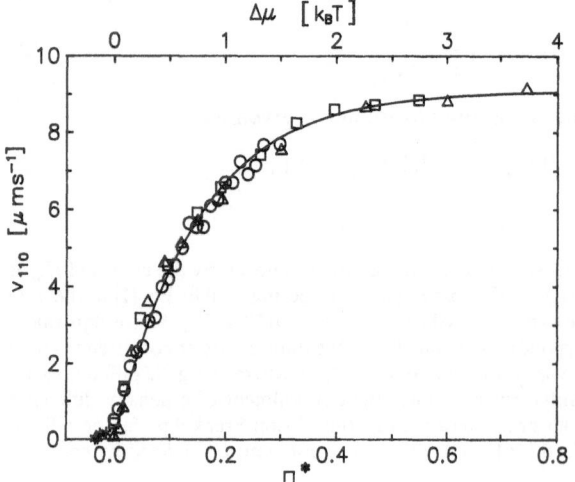

Progr Colloid Polym Sci (1995) 98:6–11
© Steinkopff Verlag 1995

line ordering and subsequent registering. The first one is operative at interfaces with only one choice of target places, while the second one occurs whenever alternative registrations are possible.

Conclusions

We are still far from a comprehensive understanding of the kinetics of the first-order freezing transition repeatedly observed also in colloid and interface systems. Performed on the selected case of highly charged latex spheres suspended in aqueous electrolyte, however, this study reported some results of broader relevance.

The observed phase transition is indeed driven by the potential energy difference between solid and melt and it is limited by interfacial dynamics. Thus, our data are excellently described by a Wilson–Frenkel law. A detailed analysis identified two distinct microscopic processes. While the limiting mechanisms may be of different nature in other complex fluids, the general frame is expected to hold even for systems of complicated internal dynamics.

The suggested derivation of the energy density proved successful in our case. This provides a quick but reasonably accurate estimation procedure for the potential energy difference between colloidal melts and the corresponding solids. While the conversion factor B is now only known for bcc systems it is, in principle, also accessible for other crystal structures.

Future work will therefore involve gradual inclusion of further structural changes and more sophisticated kinetics. As a first step the growth velocities of systems forming depletion layers could be studied.

Acknowledgements The authors thank W. Dieterich, A. Majhofer, and H. Löwen for critical discussions on the issue of the growth mechanisms and the interfacial thickness. R. Klein is thanked for his critical remarks on the issue of entropic contributions. Financial support from the DFG is gratefully acknowledged.

References

1. Chen SH, Huang JS, Tartaglia P (1992) Structure and Dynamics of strongly interacting Colloids and Supramolecular Aggregate Kluwer, Doordrecht, NATO-ASI 369:39
2. Wilson HA (1990) Philos Mag 50:238
3. Frenkel J (1932) Phys Z Sowjetunion 1:498
4. Ackerson BJ, Clark NA (1981) Phys Rev Lett 46:123
5. Ackerson BJ (ed.) (1990) Phase Transitions 21:(2–4)
6. Pusey PN, van Megen W (1989) Nature 320:340
7. Voegtli LP, Zukoski CF, IV (1991) J Colloid Interface Sci 141:79
8. Robbins MO, Kremer K, Grest GS (1988) J Chem Phys 88:3286
9. Palberg T, Mönch W, Bitzer F, Leiderer P, Bellini L, Belloni T, Piazza R (1994) Helvetica Physica Acta 67:225
10. Sirota EB, Ou-Yang HD, Sinha SK, Chaikin PM, Axe JD, Fujii Y (1989) Phys Rev Lett 62:1524
11. Pusey PN in Hansen JP, Levesque D, Zinn-Justin J (eds.) (1989): "Liquids, freezing and glass transition", 51st summer school in theoretical physics, Les Houches (F) Elsevier Amsterdam 1991, pp 763
12. Aastuen DJW, Clark NA, Kotter LK (1986) Phys Rev Lett 57:1733
13. Aastuen DJW, Clark NA, Swindal JC, Muzny CD in [4], pp 139
14. Gast AP, Monovoukas Y (1991) Nature 351:552
15. Dhont JKG, Smits C, Lekkerkerker HNW (1992) J Colloid Interface Sci 152:386
16. Grier DA, Murray CA (1994) J Chem Phys 100:9088
17. Schätzel K, Ackerson BJ (1993) Phys Rev E 48:3766
18. Davis KE, Russel WB (1987) Adv Ceram 21:573
19. Burke E, Broughton JQ, Glimer GH (1988) J Chem Phys 89:1030
20. Palberg T, Würth M, Simon R, Leiderer P (1994) Prog Colloid Polym Sci 96:62
21. Palberg T, Kottal J, Bitzer F, Simon R, Würth M, Leiderer P (1995) J Colloid Interf Sci 168:85
22. Bitzer F, Palberg T, Löwen H, Simon R, Leiderer P (1994) Phys Rev E 50:2821
23. Palberg T, Härtl W, Wittig U, Versmold H, Würth M, Simnacher E (1992) J Phys Chem 96:8180
24. Nieswand M, Majhofer A, Dieterich W (1993) Phys Rev E 48:2521

Progr Colloid Polym Sci (1995) 98:12–17
© Steinkopff Verlag 1995

Lateral capillary forces between colloidal particles incorporated in liquid films or lipid bilayers

P.A. Kralchevsky
C.D. Dushkin
V.N. Paunov
N.D. Denkov
K. Nagayama

P.A. Kralchevsky (✉) · C.D. Dushkin
V.N. Paunov · N.D. Denkov
Laboratory of Thermodynamics and
Physicochemical Hydrodynamics
University of Sofia
Faculty of Chemistry
James Boucher Ave.
Sofia 1126, Bulgaria

K. Nagayama
Protein Array Project
Program ERATO
Research and Development
Corporation of Japan
5–9–1 Tokodai
Tsukuba 300–26, Japan

Abstract The lateral capillary force appears between particles protruding from a liquid film and its physical origin is the overlap of the menisci formed around the separate particles. In turn, the size and shape of a separate meniscus is determined by the wetting properties of the particle surface, i.e., by the intermolecular forces. Therefore, the lateral capillary interaction is operative between particles of size from 1 mm down to 10 nm. Here, we first report data for direct measurement of capillary forces between particles of millimeter and submillimeter size by means of a torsional balance. Next, we present theoretical calculations of the capillary interaction between much smaller particles confined in spherical thin liquid films (vesicles, liposomes, etc.). As a result, we obtain a strong interparticle attraction, which can bring about aggregation and ordering of the particles in consonance with the experimental observations.

Key words Capillary forces
– interaction between floating
particles – particle aggregation
– Pickering emulsions –
two-dimensional crystallization

Introduction

Capillary forces have been found to play an important role in aggregation and ordering of colloidal particles and protein macromolecules captured in liquid films [1–5]. These forces appear between particles, which are partially immersed in a liquid film [6]; it can be a wetting film (Fig. 1), free foam or emulsion film, and even a lipid bilayer [7].

The cause of the lateral capillary forces is the deformation of the liquid surface(s) which is supposed to be flat (or spherical, see below) in the absence of particles. The deformation of the menisci formed around the particles (Fig. 1) gives rise to a force of lateral capillary attraction between similar particles, or repulsion between dissimilar particles, say, hydrophilic and hydrophobic [7, 8].

One can distinguish two kinds of lateral capillary forces. As known, two similar particles floating on a liquid interface attract each other. This attraction appears because the liquid surface deforms in such a way that the gravitational potential energy of the two particles decreases when they approach each other. Hence, the origin of this force is the particle weight (including the upthrust) [9, 10].

On the other hand, the gravity effect on the capillary interaction between the particles depicted in Fig. 1 is negligible. The deformation of the liquid surface in this case is related to the wetting properties of the particles surface, rather than to gravity.

To distinguish the capillary forces in the case of floating particles from ones in the case of partially immersed particles in liquid films, the former are called lateral *flotation* forces and the latter lateral *immersion* forces [7, 10, 11].

Progr Colloid Polym Sci (1995) 98:12–17
© Steinkopff Verlag 1995

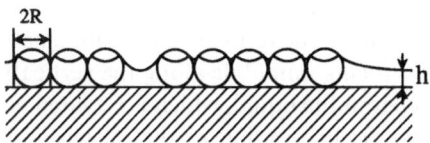

Fig. 1 Sketch of particles protruding from a liquid layer, which have aggregated under the action of lateral capillary forces

The theory of lateral capillary forces is based on the solution of the Laplace equation of capillarity for the shape of the meniscus deformed by the particles. For small meniscus slope the Laplace equation can be linearized. The *linearized* theory provides the following expression for calculating the lateral capillary force between two praticles of radii R_1 and R_2 separated by a center-to-center distance L [7–10].

$$F = 2\pi\gamma Q_1 Q_2 q K_1(qL)[1 + O(q^2 R_k^2)], \quad r_k \ll L, \qquad (1)$$

where K_1 is modified Bessel function, γ is the interfacial tension, r_1 and r_2 are the radii of the two contact lines; $Q_k = r_k \sin \psi_k (k = 1, 2)$ is the "capillary charge" of the particle with ψ_k being the meniscus slope angle at the contact line of the k-th particle. The range of capillary force (cf. Eq. (1)) is determined by the parameter

$$q^2 = (\Delta\rho g - \Pi')/\gamma, \qquad (2)$$

where $\Delta\rho$ is the difference between the mass densities of the two adjacent fluids, g is the acceleration due to gravity, and Π' is the derivative of the disjoining pressure with respect to the film thickness; for a thick film $\Pi' \equiv 0$. The asymptotic form of Eq. (1) for $qL \ll 1$ ($q^{-1} = 2.7$ mm for water/air interface)

$$F = 2\pi\gamma Q_1 Q_2 / L, \quad r_k \ll L \ll q^{-1} \qquad (3)$$

looks like a two-dimensional analogue of Coulomb's law in electrostatics. This explains the name "capillary charge" of Q_1 and Q_2. The immersion and flotation capillary forces exhibit the same functional dependence on the interparticle separation, L (see Eqs. (1) and (3)). Their different physical origins are manifested in the different magnitudes of the "capillary charges" of these two kinds of capillary force. In particular, when $R_1 = R_2 = R$; $r_k \ll L \ll q^{-1}$, one can derive [10]

$$\begin{aligned}
F &\propto (R^6/\gamma) K_1(qL) \quad \text{for flotation force} \\
F &\propto \gamma R^2 K_1(qL) \quad \text{for immersion force.}
\end{aligned} \qquad (4)$$

One sees that the flotation force decreases, while the immersion force increases, when the interfacial tension γ increases. Moreover, the flotation force decreases much more strongly with the decrease of R than the immersion force. The numerical calculations show that $F_{\text{flotation}}$ is

negligible for $R < 10 \, \mu$m, whereas $F_{\text{immersion}}$ can be significant even when $R = 10$ nm [7, 10] and is able to bring about aggregation and ordering of small colloidal particles and protein globules in the film [1–5].

Direct measurement of the immersion force was carried out by Camoin et al. [12], who established an exponential decay for large distances, in agreement with Eq. (1). Detailed experiments with vertical cylinders [13] showed that Eq. (1) (in the range of its validity) compares very well with the experiment without using any adjustable parameter. Good agreement between theory and experiment was established with the so-called "capillary image forces," expressing the capillary interaction of floating particles with a vertical wall [14].

The present article is devoted to two new studies of capillary immersion forces. First, we describe a new torsion balance for measuring capillary forces between two spherical particles, sphere and vertical cylinder or wall, two vertical cylinders, etc. Second, we investigate theoretically the lateral capillary forces between spherical colloidal particles confined in *spherical* liquid films; this is a problem related to the stability of Pickering emulsions and to the interaction between inclusions in lipid vesicles.

Torsion balance for capillary forces

As known, the torsion balances find application in various measurements of forces and torques, including the classical measurements of the gravitational and electrical forces. Our torsion balance (Fig. 2) is a precise instrument designed for measurement of capillary forces of magnitude from 5×10^{-8} to 5×10^{-5} N. The force moment due to two identical couples of interacting particles is counterbalanced by the torsional moment of a metal wire. (Experiments with one couple of particles are also possible.) One of the particles in each couple is attached to the anchor hanging on a wire (Fig. 2). The second member of each couple is attached to an appropriate holder. The particles are partially immersed in liquid contained in a Petri dish of diameter 9 cm, i.e., wide enough to make the interaction of the particles with the walls of the dish negligible. The length of the anchor is 10 mm and its weight is 0.135 g. We used platinum wire of diameter 25 μm or 100 μm (in the latter case larger anchor was required). Details about the materials used and the procedure of measurements can be found elsewhere [15].

When one of the two interacting particles is a sphere, it is attached to the anchor (or the holder) from below, so only the tops of the particles protrude from the liquid (Fig. 3). In all experiments the liquid was 8×10^{-2} mol/l aqueous solution of sodium dodecyl sulfate. This relatively

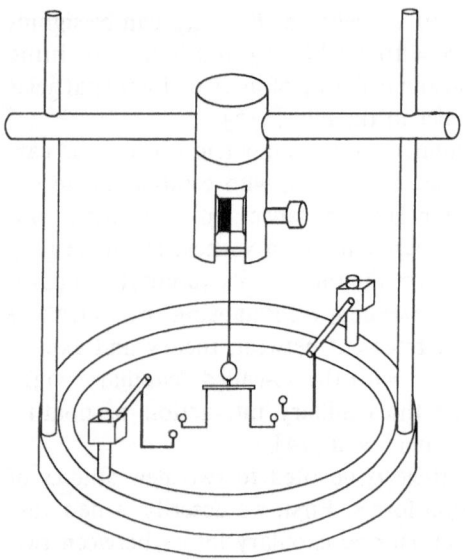

Fig. 2 Schematic view of torsion microbalance used to measure the lateral capillary force between two couples of particles immersed in a liquid. The anchor with two identical particles attached to it is suspended on a platinum wire whose length can be varied by rotating the reel. The second particle of each pair is attached to the respective holder. The torsion angle is detected by reflection of a laser beam from a small mirror above the anchor

high surfactant concentration ensures constancy of the surface tension. Since all particles used in the experiments (spheres, capillaries, plates) were made of glass, the solid-liquid-gas contact angle was equal to zero.

The measured capillary force between two spheres and/or vertical cylinders of radii in the range between 100 and 1000 μm agree qualitatively with Eq. (1). Quantitative agreement is present at large interparticle separations, while at smaller separations the measured force is greater than that calculated from the linearized theory, Eq. (1). This is not surprising in view of the relatively large meniscus slope in the close vicinity of the hydrophilic particles. On the other hand, at large interparticle separations the menisci overlap in regions of small interfacial slope, where the linearized theory is valid. A large collection of data with the respective interpretation and discussion can be found in ref. [15].

The interaction between a sphere and a vertical wall was also measured. The "wall" is a microcover glass plate of dimensions $18 \times 18 \times 0.15$ mm attached to the holder. The spherical particle of diameter 1.2 mm is attached to the anchor from below by means of a special glue. The sphere approaches the wall perpendicularly to the central part of its surface. In principle, there might be some effect of the wall edges since the distance sphere-edge is not always much larger than the distance sphere-wall. The

experimental data (Fig. 4) exhibit a maximum for small values of the distance. This is due to the fact that, for small separations, the sphere (being immobile along the vertical) is being immersed deeper in the liquid, which is elevated at the wall. Thus, the contact line radius, r_c, diminishes and the particle capillary charge decreases. By using theoretical considerations like those in refs. [13, 14], we derive the expression

$$F(L) \approx - \pi\gamma \left[2qQ^2K_1(qL) + 2QDe^{-\frac{qL}{2}} + q(r_c De^{-\frac{qL}{2}})^2 \right], \tag{5}$$

where L is the distance between the particle and its mirror image with respect to the wall, i.e., two times the particle-wall separation;

$$D = 4 \tan\frac{\psi_w}{4} \exp\left(-4\sin^2\frac{\psi_w}{4} \right), \quad qL \geq 1 \,,$$

with ψ_w being the meniscus slope at the wall. The first term in the brackets in Eq. (5) is the capillary image force, i.e., the interaction of the particle with its mirror image (cf. Eq. (1)); the second term originates from the horizontal projection of the buoyancy force at an inclined meniscus; the last term stems from the pressure jump at the liquid surface due to the curvature of the meniscus on the wall. Equation (5) well describes the data for large separations (Fig. 4) without use of any adjustable parameter. The difference between theory and experiment for small L (where the non-linear effects become important) is due to the fact that Eq. (5) is derived by means of the superposition approximation [13, 14], which is based on linearization.

Forces between particles trapped in spherical films

The spherical geometry implies some specific conditions not present in planar films. For example, the volume of the liquid layer is finite. In addition, the capillary force between two diametrically opposed particles is always zero irrespective of the range of the capillary interaction determined by the capillary length, q^{-1}, see Eq. (7) below.

Examples for systems containing spherical layers are shown in Fig. 5. A solid particle (substrate) covered with a liquid film intervening between the substrate and the outer fluid phase is sketched in Fig. 5a; such configurations can appear in some suspensions. Smaller particles incorporated in the liquid layer interact through the perturbations in the shape of the liquid-fluid interface caused by them.

Figure 5b shows a similar system, but with two liquid interfaces: the spherical film intervenes between an emulsion droplet and the outer fluid phase. In this case the particles immersed in the film deform both fluid interfaces.

Progr Colloid Polym Sci (1995) 98:12–17
© Steinkopff Verlag 1995

Fig. 3 Two interacting glass spheres of diameters 1.2 mm with protrusion heights 0.8 mm (distance from the top of the sphere to the flat nondisturbed surface). The left sphere is attached to the anchor, whereas the right sphere is attached to the holder. The separation distance is $L = 0.406$ cm; the measured lateral capillary force is $F = 0.491$ dyn

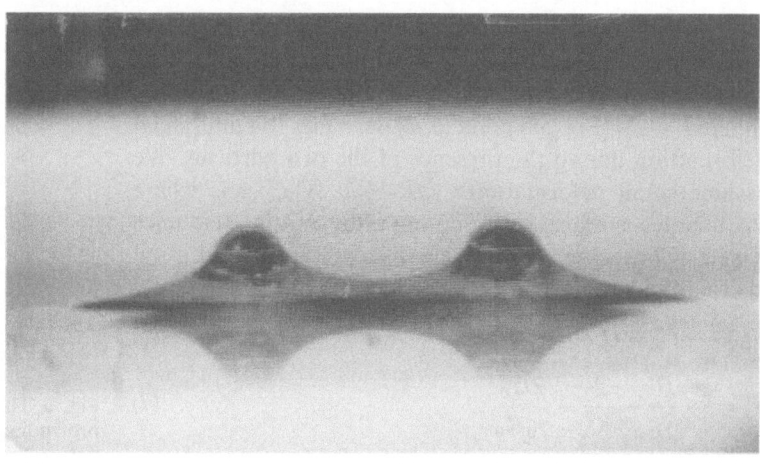

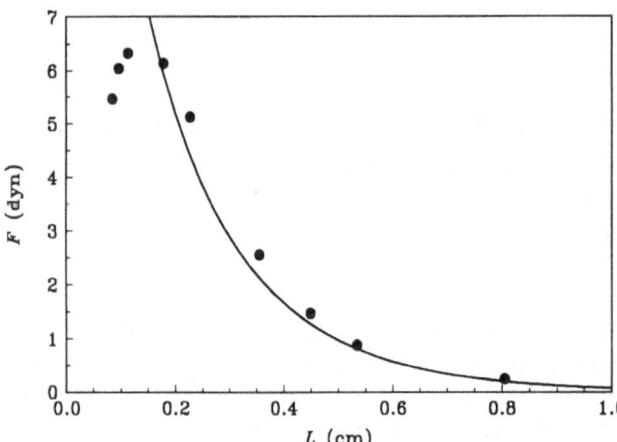

Fig. 4 Capillary immersion force F versus distance L between glass sphere of diameter 1.2 mm and glass wall. The length of platinum wire is 5 cm; the height of sphere protrusion from the solution is 1.05 mm. The solid curve is theoretical fit of the data by means of Eq. (5)

Fig. 5 Sketch of three possible configurations of particles protruding from a spherical liquid film: a) two particles immersed in a thin liquid film on a spherical solid substrate; b) two particles included in a vesicle; c) two membrane proteins incorporated into a lipid bilayer

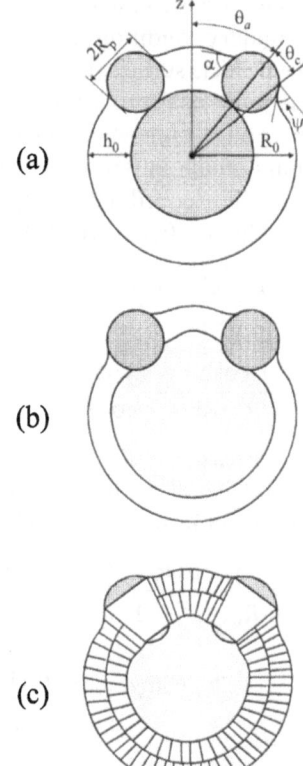

A spherical lipid bilayer (vesicle) containing incorporated membrane proteins is depicted in Fig. 5c. The difference in the thickness of the hydrophobic zones of the protein and the bilayer gives rise to interfacial deformations and protein-protein interaction [16]. The specificity of this system is that the hydrocarbon core of the lipid bilayer exhibits some elastic behavior and cannot be treated as a simple fluid [17].

In summary, we deal with relatively small particles contained in a liquid phase; hence, we can assume that the effect of gravity on the interfacial shape is negligible. At these conditions the (non-disturbed) spherical liquid film may have stable uniform thickness only due to the action of some repulsive surface forces inside the film [18]. That is the reason why below we will consider only *thin* liquid films, i.e., films in which the effect of the surface force (the disjoining pressure) is not negligible.

Here, we restrict our considerations to the simpler case depicted in Fig. 5a, where there is only one deformable surface and the two particles are identical. The results can be further extended to the more complicated systems, viz. two deformable surfaces (Fig. 5b) or film of elastic behavior (Fig. 5c).

16

P.A. Kralchevsky et al.
Lateral capillary forces between colloidal particles

The radial coordinate of a point of the deformed film surface can be presented in the form $r = R_0 + \zeta(\theta, \varphi)$, where θ and φ are polar coordinates on the reference sphere $r = R_0$ (Fig. 5a) and $\zeta(\theta, \varphi)$ describes the interfacial deformation due to the presence of the two particles. We assume small deformations: $|\zeta/R_0| \ll 1$, $|\nabla_{II}\zeta|^2 \ll 1$, where ∇_{II} denotes surface gradient operator in the reference sphere. Consequently, the generalized Laplace equation for the film surface can be linearized [19]:

$$\nabla_{II}^2\zeta - q^2\zeta = Q/R_0^2 \tag{6}$$

$$q^2 = -\frac{\Pi'}{\gamma_0} - \frac{2}{R_0^2} - \frac{2\Pi_0}{\gamma_0 R_0}, \tag{7}$$

where $Q = R_0 \sin\theta_c \sin\psi_c$ is particle capillary charge; γ_0 and Π_0 are the surface tension and disjoining pressure of the non-disturbed spherical film of thickness h_0. To determine ζ, we integrated Eq. (6) numerically [19] by using the boundary condition for constancy of the contact angle α at the particle surface (Fig. 5a). We simplified and accelerated the numerical procedure by utilizing bipolar coordinates (σ, τ) on the reference sphere, which transform the integration domain into a rectangle. The connection between the Cartesian coordinates (x, y, z) and the spherical bipolar coordinates (r, σ, τ) reads [19]:

$$x = \frac{r\sqrt{\lambda^2 - 1}\sinh\tau}{\lambda\cosh\tau - \cos\sigma}, \qquad y = \frac{r\sqrt{\lambda^2 - 1}\sin\sigma}{\lambda\cosh\tau - \cos\sigma},$$

$$z = \frac{r(\cosh\tau - \lambda\cos\sigma)}{\lambda\cosh\tau - \cos\sigma}.$$

In our case $r = R_0$ and $\lambda = \cos\theta_c/\cos\theta_a$ (see Fig. 5a). The lateral capillary force exerted on each particle can be expressed in the form [19]:

$$F = \frac{\gamma_0\cos\theta_c}{R_0\sqrt{\lambda^2 - 1}} \int_0^\pi d\sigma \left(\frac{d\zeta_c}{d\sigma}\right)^2 (\lambda\cosh\tau_c\cos\sigma - 1)$$

$$- 2\gamma_0(\sin\psi_c - \sin\theta_c)\int_0^\pi d\sigma \frac{d\zeta_c}{d\sigma}\frac{\sqrt{\lambda^2 - 1}\sin\sigma\cot\theta_a}{\lambda\cosh\tau_c - \cos\sigma}, \tag{8}$$

where $\zeta_c = \zeta_c(\sigma)$ is the value of ζ at the contact line, whose equation in bipolar coordinates reads $\tau = \tau_c = \text{const}$;

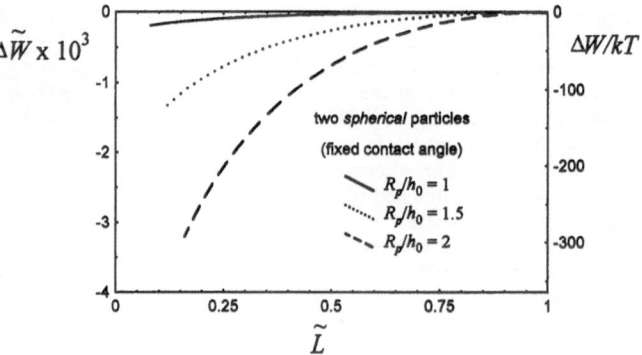

Fig. 6 The capillary interaction energy of two *spherical* particles protruding from a spherical liquid layer as a function of the interparticle separation, $\tilde{L}$. The different curves correspond to different values of the ratio R_p/h_0 at fixed $R_p/R_0 = 0.05$. The other parameters are: $qR_0 = 1$ and $\alpha = 60°$. The right-hand side scale shows the values of the capillary interaction energy for the special case $R_0 = 1\,\mu m$, $T = 298\,K$ and $\gamma_0 = 30\,mN/m$

$\tau_c \equiv \text{arctanh}\left(\sqrt{\lambda^2 - 1}\cot\theta_a\right)$ – see ref. [19] for details. From the calculated values of F, one can then determine the capillary interaction energy:

$$\Delta W(L) = \int_L^{\pi R_0} F(L)dL, \tag{9}$$

where L is the length of the (shortest) arc on the reference sphere connecting the two particles. ΔW is set zero for two diametrically opposed particles, i.e., for $L = \pi R_0$. In Fig. 6 the calculated dimensionless interaction energy, $\Delta\tilde{W} = \Delta W/(\gamma R_0^2)$ is plotted against the dimensionless distance, $\tilde{L} = L/(\pi R_0)$. One sees that the energy is negative, i.e., corresponds to attraction between the two particles. The right-hand side scale in Fig. 6 shows the values of $\Delta W/kT$ calculated by substituting typical parameters values. One sees that ΔW can be much larger than kT. Consequently, the lateral capillary forces can be strong enough to bring about particle aggregation and ordering in the spherical film.

Acknowledgement This work was supported by the Japanese Research and Development Corporation (JRDC) under the program Exploratory Research for Advanced Technology (ERATO).

References

1. Yoshimura H, Matsumoto M, Endo S, Nagayama K (1990) Ultramicroscopy 32:265–271
2. Denkov ND, Velev OD, Kralchevsky PA, Ivanov IB, Yoshimura H, Nagayama K (1992) Langmuir 8:3183–3190
3. Denkov ND, Velev OD, Kralchevsky PA, Ivanov IB, Yoshimura H, Nagayama K (1993) Nature (London) 361:26
4. Dushkin CD, Yoshimura H, Nagayama K (1993) Chem Phys Lett 204:455–460
5. Lazarov GS, Denkov ND, Velev OD, Kralchevsky PA, Nagayama K (1994) J Chem Soc Faraday Trans 90:2077–2083
6. Kralchevsky PA, Paunov VN, Ivanov IB, Nagayama K (1992) J Colloid Interface Sci 151:79–94

7. Kralchevsky PA, Nagayama K (1994) Langmuir 10:23–36
8. Kralchevsky PA, Paunov VN, Ivanov IB, Nagayama K (1993) J Colloid Interface Sci 155:420–437
9. Chan DYC, Henry JD, White LR (1981) J Colloid Interface Sci 79:410–418
10. Paunov VN, Kralchevsky PA, Denkov ND, Nagayama K (1993) J Colloid Interface Sci 157:100–112
11. Kralchevsky PA, Denkov ND, Paunov VN, Velev OD, Ivanov IB, Yoshimura H, Nagayama K (1994) J Phys: Condens Matter 6:A395–A402
12. Camoin C, Roussel JF, Faure R, Blanc R (1987) Europhys Lett 3:449–457
13. Velev OD, Denkov ND, Kralchevsky PA, Paunov PN, Nagayama K (1993) Langmuir 9:3702–3709
14. Kralchevsky PA, Paunov VN, Denkov ND, Nagayama K (1994) J Colloid Interface Sci 167:47–65
15. Dushkin CD, Kralchevsky PA, Yoshimura H, Nagayama K, Paunov VN (1995) Langmuir – submitted
16. Israelachvili JN (1977) Biochim Biophys Acta 469:221–225
17. Petrov AG, Bivas I (1984) Prog Surface Sci 16:389–511
18. Derjaguin BV, Churaev, NV, Muller VM (1987) Surface Forces, Plenum Press, Concultants Bureau, New York
19. Kralchevsky PA, Paunov VN, Nagayama K (1995) J Fluid Mech at press

Progr Colloid Polym Sci (1995) 98:18–22
© Steinkopff Verlag 1995

N.D. Denkov
P.A. Kralchevsky

Colloid structural forces in thin liquid films

N.D. Denkov · Prof. P.A. Kralchevsky (✉)
Laboratory of Thermodynamics and
Physico-chemical Hydrodynamics
University of Sofia
Faculty of Chemistry
1 James Boucher Ave.
1126 Sofia, Bulgaria

Abstract In recent years the interest in non-DLVO forces has continuously increased because of their tremendous importance for dispersion stability. Colloid structural forces appear when thin liquid films are formed from concentrated suspension of monodisperse particles (micelles, latex particles, proteins). At low particle volume fraction these oscillatory forces degenerate into the depletion attraction. We propose simple semiempirical formulas describing the oscillatory structural force and energy. Their comparison with available experimental data and theoretical calculations shows a good agreement. Depending on the conditions (particle size, concentration, charge, etc.) the oscillatory structural forces may stabilize or destabilize the thin films and dispersions (emulsions, suspensions). The role of different factors such as temperature, electrolyte, and micelle concentrations, etc., are treated in a unified way and can be explained by the change of the effective particle volume fraction.

Key words Depletion attraction – micellar structures – non-DLVO surface forces – oscillatory structural forces – stratifying films

Introduction

The oscillatory structural force, which is the subject of the present article, appears in two cases: i) in thin liquid films between two smooth solid surfaces, and ii) in liquid films containing colloidal particles, e.g., surfactant micelles or macromolecules. In the first case the oscillatory forces are called the "solvation forces" [1, 2] as the period of oscillations is of the order of the diameter of a solvent molecule; they contribute to the short-range interaction between molecularly smooth solid surfaces. In the second case the structural forces affect the stability of foam and emulsion films containing colloidal particles as well as the particle interactions in various colloids. At higher particle concentrations these colloid structural forces *stabilize* the liquid films and dispersion [3, 4]. At lower particle concentra-

tions the structural force degenerates into the so-called depletion force, which is found to *destabilize* various dispersions and to bring about coagulation [1].

In summary, the interplay of two physical factors gives rise to the oscillatory structural forces: i) volume exclusion effect due to the finite particle size, and ii) particle structuring induced by the two films surfaces.

The oscillatory structural forces can be directly measured by means of the surface force apparatus [2, 5]. Another experimental tool for studying the colloid structural forces is provided by the phenomenon stratification, i.e., step-wise thinning of a foam or emulsion type liquid film containing colloidal particles [3, 4, 6]. This is a universal phenomenon observed with particle diameters varying between 5 nm and 0.2 μm [7].

Theoretically the oscillatory structural forces can be accounted for by adding an extra term Π_{os}, in the DLVO

Progr Colloid Polym Sci (1995) 98:18–22
© Steinkopff Verlag 1995

expression for the disjoining pressure [3]

$$\Pi(h) = \Pi_{vw}(h) + \Pi_{el}(h) + \Pi_{os}(h) \,, \qquad (1)$$

where h denotes film thickness. Simple expressions are available for estimating the van der Waals disjoining pressure, Π_{vw}, and the electrostatic disjoining pressure, Π_{el}. An analogous convenient expression for Π_{os} is still missing. The integral equations of statistical mechanics lead to sophisticated computational procedures for Π_{os} [8–11]. Our aim in the present study is to construct a simpler expression for Π_{os} which is to correctly represent the dependence of Π_{os} on both film thickness (slit width), h, and particle volume fraction, φ. Below, we propose a semiempirical formula for Π_{os}, which is tested against the predictions of integral equations, computer simulations, and experimental data for stratifying films. This article is a brief communication; a detailed description will be published elsewhere [12].

Period and decay length of the oscillatory force

For the sake of estimates Israelachvili [1] proposed an approximated expression, in which both the oscillatory period and the characteristic decay length are set equal to the particle diameter, d, in accordance with the theoretical predictions for high volume fractions. On the other hand, the works on stratifying films quoted above indicate that the period can depend on the particle volume fraction. To elucidate this point, we examined the predictions of the theory by Henderson [9]. The latter theory is based on an inversion of the Laplace transforms of the radial correlation functions determined for a mixture of larger and smaller hard spheres in the framework of the Percus–Yevick closure. At a certain point of the derivation the radius of the larger spheres is set infinitely large and thus the interaction between two walls is determined.

By using Henderson theory [9], we calculated Π_{os} for various volume fractions of the (smaller) hard spheres, φ. The results confirm than in a first approximation Π_{os} is a periodical function of h with an exponential decay (for $h > d$). The oscillatory period, d_1, and the decay length, d_2, are determined by averaging over many consecutive maxima and minima. The data show that $d_1 \approx d_2 \approx d$ only around $\varphi \approx 0.4$. For smaller volume fractions ($\varphi \approx 0.1$) d_1 increases with c.a. 20%, whereas d_2 decreases by more than three times. The latter fact, coupled with the circumstance that d_2 takes place under the sign of an exponent, implies a strong concentration dependence of Π_{os}. (Roughly speaking, for a given h the oscillatory disjoining pressure Π_{os} increases five times when φ is increased with 10%-see Fig. 2 below.)

We obtained interpolation formulas for the data about d_1 and d_2 by using the following heuristic considerations. When φ tends to the concentration of close packing. $\varphi_{max} = \pi/(3\sqrt{2})$, the dimensionless period, d_1/d, should tend to $\sqrt{2/3}$, whereas the decay length d_2 should tend to infinity in so far as we deal with a densely packed fcc lattice. Then we seek d_1 and d_2 in the form of truncated series expansions with respect to $\Delta\varphi = \varphi_{max} - \varphi$:

$$\frac{d_1}{d} = \sqrt{\frac{2}{3}} + a_1\Delta\varphi + a_2(\Delta\varphi)^2, \quad \frac{d_2}{d} = \frac{b}{\Delta\varphi} - b_2 \qquad (2)$$

In Fig. 1 the data for d_1 and d_2 vs φ, obtained from the Henderson theory, are compared with Eq. (2) and the coefficients a_1, a_2, b_1 and b_2 are determined.

For volume fraction $\varphi = 0.419$ and $d = 0.74$ nm Attard and Parker [11] calculated theoretically $d_1 = 0.74$ nm and $d_2 = 0.78$ nm, which are to be compared with the predictions of Eq. (2), viz. $d_1 = 0.71$ nm and $d_2 = 0.81$ nm. For volume fraction $\varphi = 0.357$ Kjellander

Fig. 1 The data for the period d_1 and the decay length d_2 fitted by means of Eq. (2). The coefficients a_1, a_2, b_1 and b_2 are determined from the slopes and intercepts of the corresponding lines

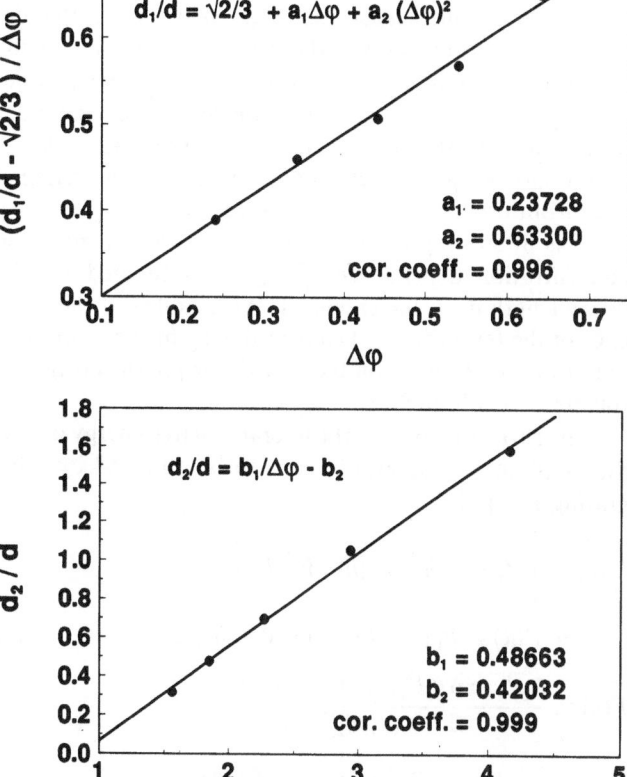

and Sarman [10] determined $d_1/d = 1$, $d_2/d = 0.89$, which compares well with the outputs of Eq. (2): $d_1/d = 1.00$ and $d_2/d = 0.85$. The lack of perfect numerical agreement between the different theories is not surprising in view of the different approximations and model assumptions used.

Expressions for disjoining pressure and interaction free energy

Next, we analyzed numerical data for the positions and magnitudes of the minima and maxima of Π_{os} calculated by means of Henderson theory [9]. We concluded that the data for various h and φ are fitted well by means of the following formula

$$\Pi_{os} = P_0 \cos\left(\frac{2\pi h}{d_1}\right) \exp\left(\frac{d^3}{d_1^2 d_2} - \frac{h}{d_2}\right) \quad \text{for } h > d,$$

$$= -P_0 \quad \text{for } 0 < h < d, \tag{3}$$

where d_1 and d_2 are given by Eq. (2), and P_0 is the particle osmotic pressure determined by means of Carnahan–Starling formula [13]:

$$P_0 = \rho kT \frac{1 + \varphi + \varphi^2 - \varphi^3}{(1 - \varphi)^3}, \quad \rho = \frac{6\varphi}{\pi d^3}, \tag{4}$$

where ρ is the particle's number density, k is the Boltzmann constant, and T is temperature. It is clear that for $h < d$, when the particles are expelled from the slit into the neighboring bulk suspension, Eq. (3) describes the depletion attraction [1]. On the other hand, for $h > d$ the structural disjoining pressure oscillates around P_0 as defined by Eq. (4) in agreement with the finding of Kjellander and Sarman [10]. The finite discontinuity of Π_{os} at $h = d$ is not surprising as, at this point, interaction is switched over from oscillatory to depletion regime.

It is interesting to note that in oscillatory regime the concentration dependence of Π_{os} is dominated by the decay length d_2 in the exponent. We established the presence of the term $d^3/(d_1^2 d_2)$ empirically, by fitting numerical data from the Henderson theory [9]; the physical origin of this term is still unclear.

The surface density of the interaction free energy due to the oscillatory structural forces can be obtained by integrating Π_{os} [1]:

$$f_{os}(h) = \int_h^\infty \Pi_{os}(h')dh' = F(h) \quad \text{for } h \geq d$$

$$= F(d) - P_0(d - h) \quad \text{for } 0 \leq h \leq d \tag{5}$$

$$F(h) \equiv \frac{P_0 d_1 \exp\left((d^3/d_1^2 d_2) - (h/d_2)\right)}{4\pi^2 + (d_1/d_2)^2}$$

$$\times \left[\frac{d_1}{d_2} \cos\left(\frac{2\pi h}{d_1}\right) - 2\pi \sin\left(\frac{2\pi h}{d_2}\right)\right]$$

It should be noted that Eqs. (3) and (5) refer to hard spheres of diameter d. In practice, however, the interparticle potential can be "soft" because of the action of some long-range forces. If such is the case, one can obtain an estimation of the structural force by introducing an effective hard core diameter [4]

$$d(T) = \left[\frac{3}{2\pi} B_2(T)\right]^{1/3}, \tag{6}$$

where B_2 is the second virial coefficient in the virial expansion of the particle osmotic pressure: $P_{osm}/(\rho kT) = 1 + B_2\rho + \cdots$

Numerical test and discussion

In Figs. 2a, b, c the effect of particle volume fraction, φ, on the oscillatory disjoining pressure is examined. The curves calculated from Eq. (3) compare well with the predictions of Henderson theory [9], except in the region around $h = d$, where the Henderson theory predicts about three times larger values of Π_{os}. In Fig. 2b the theoretical curve calculated by Kjellander and Sarman [10] for $\varphi = 0.357$ and $h > 2$ by using the anisotropic Percus–Yevick approximation is shown by the dashed line; the crosses represent grand canonical Monte Carlo simulation results due to Karlström [14]. (To avoid misunderstandings, we note that the slit width used in ref. [10] corresponds to $h - d$ in our notation.) Figures 2a, b, c show that the oscillatory character is more pronounced for the higher volume fractions, which is correlated with the increase of the decay length, d_2, cf. Eq. (2).

One can check [12] that Eq. (5) for the oscillatory free energy, f_{os}, also compares well with the theories of Henderson [9] and Mitchell et al. [8]. An exclusion is the region of depletion attraction ($0 < h < d$), where f_{os} should be a linear function of h (Eq. (5)), whereas nonlinear behavior is obtained in refs. [8,9]. This nonlinearity (as well as the large value of Π_{os} around $h = d$ mentioned above) is probably an artefact due to some approximations used in refs. [8,9]; for more details – see ref. [12].

In the experiments with stratifying liquid films [3,4,6,7] the plane-parallel film is encircled by a Plateau border. The capillary pressure, P_c, in the Plateau border can be controlled and measured. Equilibrium films can be formed only when the condition for mechanical equilibrium of the film surfaces, $\Pi(h) = P_c$, is satisfied [3]. If P_c is kept constant during an experiment, the states of *stable* equilibrium (with $d\Pi/dh < 0$) are given by the intersection points of the disjoining pressure isotherm Π vs. h with the horizontal line $\Pi = P_c = $ const, see Fig. 3. Below, we compare the number of the metastable states observed

Progr Colloid Polym Sci (1995) 98:18-22
© Steinkopff Verlag 1995

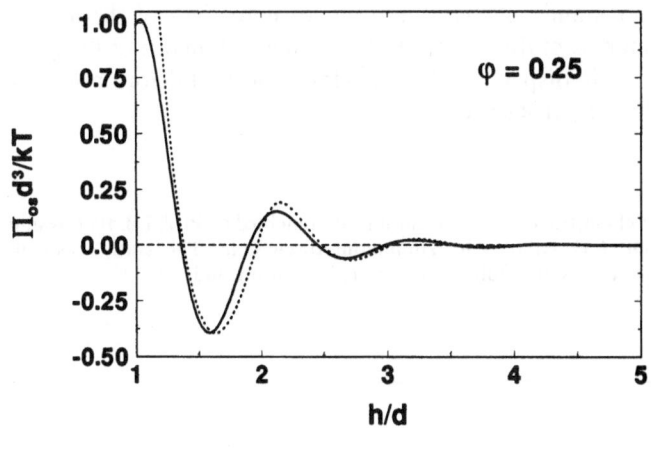

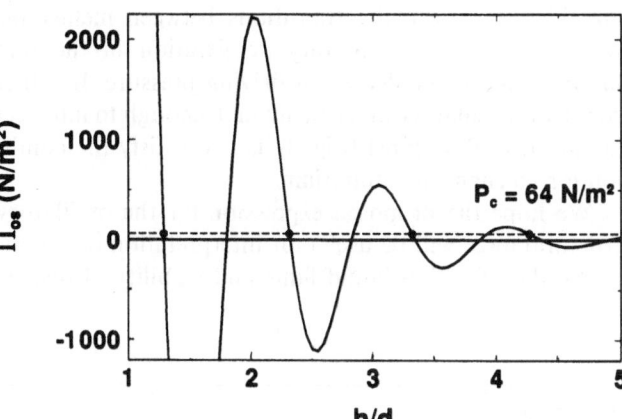

Fig. 3 Plot of the oscillatory disjoining pressure vs. film thickness for stratifying foam film from 0.06 M aqueous solution of SDS

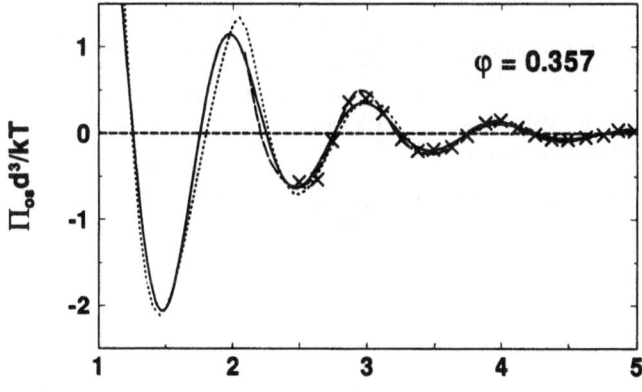

Table 1 Test of the predictions of Eq. (3) against data from ref. [3] for stratifying films from aqueous solutions of SDS.

$C(M)$	φ	$d_0 + 2/\kappa$ (nm)	Δh (nm)	d_1 (nm)	n_{obs}	n_{pred}
0.03	0.20	11.2	15.3	12.7	3	2
0.06	0.32	10.4	11.8	10.8	4	4
0.08	0.39	10.1	10.8	9.9	4	6
0.10	0.43	9.7	10.4	9.2	5	8

refs. [3, 6]. The diameter of the micelles is $d_0 = 4.8$ nm and the capillary pressure is $P_c \approx 50$ Pa. We used $d = d_0 + 2/\kappa$ (κ is Debye screening parameter) as an effective diameter of the micelles taking into account the thickness of the counterion atmosphere. The first column in Table 1 presents the total surfactant concentration, C; the second column gives the volume fraction of the micelles calculated by using the effective diameter, d, listed in the third column. The next two columns compare the experimentally measured period characterized by the magnitude of the stepwise changes in the film thickness, Δh (see ref. [3]), with the period of oscillations d_1 calculated from Eq. (2). In spite of the fact that d_1 is slightly smaller than Δh, these two quantities exhibit the same tendency of decreasing with the increase of φ. The last two columns in Table 1 compare the observed and predicted numbers of the film metastable states, n_{obs} and n_{pred}, the final stable state being excluded. The agreement between n_{obs} and n_{pred} is satisfactory in view of the fact that the softness of the double-layer repulsion between the micelles and the possible role of dynamic factors are not accounted for when calculating n_{pred}. Figure 3 illustrates the $\Pi(h)$ isotherm calculated with the data for $C = 0.06$ M SDS in Table 1 by using Eq. (3). A general conclusion is that the observed step-wise changes in the

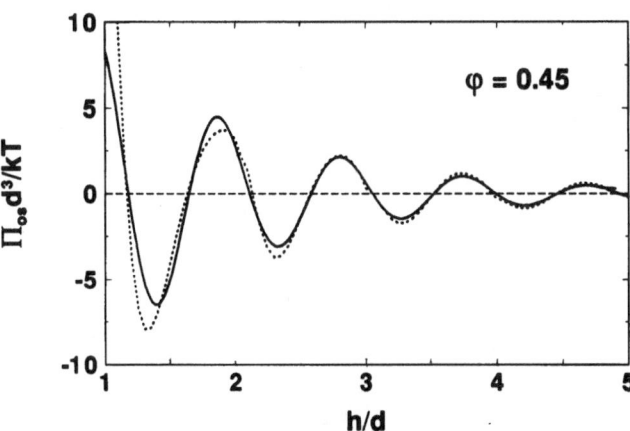

Fig. 2 Comparison of the oscillatory disjoining pressure, Eq. (3), the solid curves, with the results from Henderson theory [9], the dotted curves, for different particle volume fractions, φ. Numerical results from refs. [10] and [14] are shown in Fig. 2b with dashed curve and crosses, respectively

experimentally with the number of the intersection points predicted by Eqs. (1) and (3).

Table 1 contains data for stratifying films from aqueous solutions of sodium dodecyl sulfate (SDS), studied in

film thickness are really transitions between metastable states, rather than temporary deceleration in the film thinning due to the decreased driving pressure. In other words, the oscillatory maxima are high enough to intersect the line $\Pi = P_c = \text{const}$ (Fig. 3), i.e., to satisfy the condition for mechanical equilibrium.

We hope the proposed expression for the oscillatory structural force will be useful for interpretation of experimental data for thin liquid films and stability of dispersions and will be utilized as an ingredient of theoretical models of dynamics of film drainage, kinetics of flocculation in dispersions of nondeformable and deformable colloidal particles, etc.

Acknowledgement The authors are indebted to Prof. I.B. Ivanov and Dr. A.D. Nikolov for stimulating discussions. This study was supported by the Bulgarian National Science Fund.

References

1. Israelachvili JN (1992) Intermolecular and surface forces. Academic Press, London
2. Horn RG, Israelachvili JN (1980) J Chem Phys 75:1400–1411
3. Nikolov AD, Wasan DT, Kralchevsky PA, Ivanov IB (1989) J Colloid Interface Sci 133:1–22
4. Nikolov AD, Wasan DT, Denkov ND, Kralchevsky PA, Ivanov IB (1990) Prog Colloid Polym Sci 82:87–98
5. Richetti P, Kekicheff P (1992) Phys Rev Lett 68:1951–1954
6. Bergeron V, Radke CJ (1992) Langmuir 8:3020–3026
7. Wasan DT, Nikolov AD, Kralchevsky PA, Ivanov IB (1992) Colloids Surfaces 67:139–145
8. Mitchell DT, Ninham BW, Pailthorpe BA (1978) J Chem Soc Faraday Trans II 74:1116–1125
9. Henderson D (1988) J Colloid Interface Sci 121:486–490
10. Kjellander R, Sarman S (1988) Chem Phys Lett 149:102–108
11. Attard P, Parker JL (1992) J Phys Chem 96:5086–5093
12. Kralchevsky PA, Denkov ND (1994) Chem Phys Lett – submitted
13. Carnahan NF, Starling KE (1969) J Chem Phys 51:635–644
14. Karlström G (1985) Chem Scripta 25:89–96

Progr Colloid Polym Sci (1995) 98:23–29
© Steinkopff Verlag 1995

V. Cabuil
E. Dubois
S. Neveu
J.-C. Bacri
E. Hasmonay
R. Perzynski

Phase separation in aqueous magnetic colloidal solutions

Dr. V. Cabuil (✉) · E. Dubois · S. Neveu
Laboratoire de Physicochimie Inorganique
Université Pierre et Marie Curie
Casier 63
4, Place Jussieu
75252 Paris Cedex 05, France

J.-C. Bacri · E. Hasmonay · R. Perzynski
Laboratoire d'Acoustique et Optique de la
Matiére Condensée
Université Pierre et Marie Curie
URA CNRS 800

Abstract Ionic magnetic fluids are aqueous colloidal solutions of magnetic nanoparticles which wear surface charges. Stability of the solution is ensured by screened electrostatic repulsions, and phase transitions "gas-liquid" like are observed when an electrolyte is added to the colloidal solution. The nature and the onset of the transitions are determined as functions of the nature of the particles (maghemite or cobalt ferrite), of the surface charges (anionic oxide particles without or with specifically adsorbed ions), and of the added electrolyte. The onset of the transitions is also related to the particle size distribution parameters.

Key words Magnetic fluids – phase transitions – colloidal solutions

Introduction

Ionic magnetic fluids are colloidal solutions of magnetic nanoparticles dispersed in water without tensioactive molecules [1]. The magnetic particles wear surface charges and the stability of the solution is ensured by screened electrostatic repulsions between grains. A stable colloidal solution is usually compared to a gas of particles, and theories state that a phase separation "gas-liquid" like between a phase dense in particles (liquid phase) and a phase poor in particles (gas phase) may occur [2–8], when some parameters (for example, temperature) are modified. In such ionic solutions, the interparticle interactions are Van der Waals ones, to which screened electrostatic repulsions and magnetic dipolar interactions have to be added. As in any electrostatically stabilized colloid, an addition of electrolyte, which increases the ionic strength, decreases the repulsions and may induce phase transitions. Experimentally, we have found in the case of anionic maghemite (γ-Fe$_2$O$_3$) particles dispersed in alkaline aqueous solutions, that addition of tetramethylammonium hy-

droxide induces such a phase separation: a liquid phase rich in magnetic particles appears as droplets in another liquid phase poor in particles [9, 10]. An experimental phase diagram has been constructed describing this phase separation in alkaline magnetic fluids when extra base was added [10].

It fact, such results raise many questions. i) The systems are polydisperse and the first problem is to compare the theories, generally dealing with monodisperse systems and our experimental results; another question is if obtaining a liquid phase rich in particles is or is not an artefact due to the polydispersity and then if it really corresponds to a thermodynamic phenomenon. ii) The second problem is to check if this behavior is a universal one, i.e., to study how it depends on the nature of the particles, the parameters of their size distribution, the nature of their surface charges, and the nature of the added electrolyte.

Thus, we present here an experimental work concerning **anionic** magnetic particles of different nature (maghemite γ-Fe$_2$O$_3$, and cobalt ferrite CoFe$_2$O$_4$), with different size distributions, dispersed in water (alkaline and pH 7 solutions according to the nature of the adsorbed species), and to which various electrolytes will be added. The aim of this work is to determine as a function of these parameters, the nature of the phase transition observed by addition of electrolyte ("gas-liquid" like or precipitation) and the concentration of electrolyte corresponding to the onset of the transitions.

Samples

The ionic magnetic fluids (ferrofluids) that can be obtained are schematized in Fig. 1.

The magnetic nanoparticles are grains of spinel-type ferrimagnetic ferric oxide (MFe_2O_4) and are synthesized as anionic ones through a chemical process i.e., coprecipitation of mixtures of the metallic salts in alkaline medium as described by Massart [11].

They are dispersable in water as soon as the counterion associated to particles is $N(CH_3)_4^+$ (Na^+, NH_3^+ induce flocculation of the solution [12]) and stability is a function of the pH as the surface charges are due to surface acid-base equilibria. The pH of the point of zero charge is located for maghemite, as it is for cobalt ferrite particles, around pH 7.5. To modify the pH of this point in case of zero charge, it is necessary to adsorb on the surface small ions, such as for example citrate anions (LNa_3, see Fig. 1), which lower the pH of the point of zero charge to 3, allowing to obtain magnetic fluids of pH 7 [13].

The characteristics that allow to describe a magnetic fluid are firstly the **parameters of the size distribution**. The maghemite and cobalt ferrite particles obtained are always polydisperse, the size distribution usually being well described by a log-normal law characterized by the two parameters D_0 and σ [14]:

$$P(D) = \frac{1}{\sqrt{2\pi}D\sigma} \exp\left(-\frac{\ln^2(D/D_0)}{2\sigma^2}\right)$$

These parameters are determined either by **electron microscopy**, either through the **analysis of the magnetization curve** of the fluid [15]. Usual magnetic fluids obtained in the laboratory have a D_0 of the order of 7 nm and a σ of the order of 0.4 (Fig. 2).

Another size characteristic useful for the present study, but specific of magnetic fluid is the **average diameter D_{bir} determined by relaxation of birefringence**. Its measure is based on the fact that a magnetic fluid, submitted to a magnetic field becomes birefringent because of the alignment of the ferromagnetic grains [16]. This diameter is of hydrodynamical type and, because of the measure, is an average which gives an important weight to the biggest particles and thus is a characterization of the tail of the distribution [17]. It is thus always found to be much greater than D_0, the difference between these diameters being more important when σ value is great (Table 1). In the case of the phase separation observed in alkaline sols of maghemite particules, we have shown that this diameter was the relevant parameter which controls the onset of the phase transitions.

The average **number of structural surface charges per particle** has been estimated in the absence of specific adsorption of citrate ions, that means for acid or alkaline fluids, and was found equal to about 200. For these particles, the pH of the point of zero charge is about 7.5, but it is shifted to about 3 by adsorption of citrate [18].

Fig. 1 Schematic description of the ionic magnetic fluids synthesized and studied

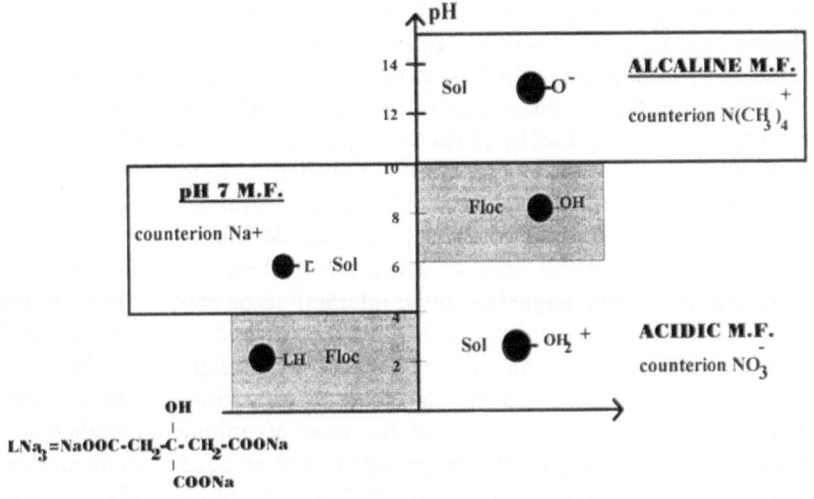

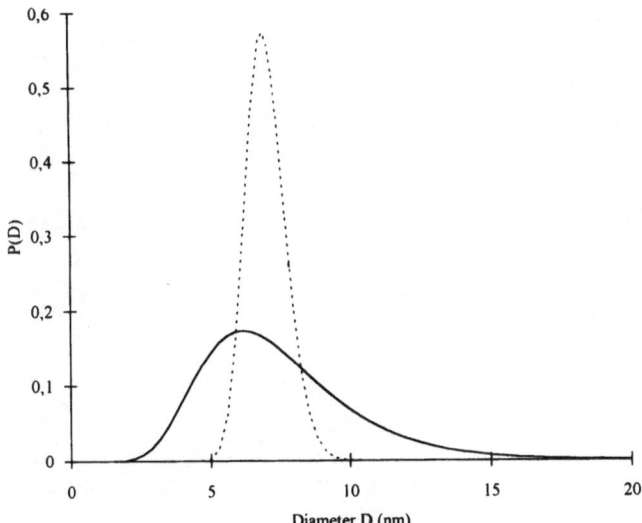

Fig. 2 Calculated lognormal distributions corresponding to different values of typical parameters D_0 and σ. **Plain line:** $D_0 = 7$ nm, $\sigma = 0.4$; **dotted line:** $D_0 = 7$ nm, $\sigma = 0.1$

Table I Experimental relationship found between the average diameter D_{bir} (determined from relaxation of birefringence) and the parameters D_0 et σ of the lognormal size distribution (determined from magnetization measurements).

σ	0.1	0.15	0.2	0.25	0.3	0.4
D_{bir}/D_0	1.9	2.1	2.3	2.4	3.8	7.2

Finally, a typical magnetic fluid is constituted by ionized particles associated to their counterions, with a **volume fraction of particles** that typically ranges between 0 and 18%. This volume fraction is measured either by **chemical titration of iron** [19] either from the value of the **saturation magnetization** [15]. But ionic magnetic fluids inevitably contain some residual ionic strength that has to be determined through, for example, acid-base titrations.

Experiments

Nature and onset of the transitions

All these experiments are based on observations with an optical microscope (enlargement ×900) of the solutions entrapped in spectroscopic cells. The stability of the initial sample is related to homogeneity of the preparation (Fig. 3a). Then, an electrolyte is added to the sample, and

this latter is again checked by microscopic observation. When a phase transition occurs, droplets of a liquid phase, concentrated in particles appear inside a less concentrated phase (Fig. 3b). These droplets present a net interface with the dilute phase and they elongate easily if a magnetic field is applied (Fig. 3c). The surface tension between the two phases is very low (σ ranging between 10^{-6} and 10^{-8} Jm^{-2}) [19]. In that case, we speak of a "gas-liquid"-like transition. When the phase which appears inside the dilute solution is rocklike (Fig. 3d) and keeps its shape under the action of the magnetic field (it eventually moves inside the cell), we say that the transition is a precipitation. It is clear that these criteria for determination of transition onset are related to the degree of magnification: the smallest drops that can be detected have a size of the order of 1 μm.

Construction of diagrams

To a given volume of stable magnetic fluid is added the extra electrolyte; after stirring, the sample is observed with an optical microscope and allowed to rest for 2 days to let the droplets of concentrated phase settle; concentrated and dilute phases are then separated and iron is titrated by a chemical procedure [20] in both phases.

Results

The aim of this work is to determine the influence of the nature of particles, of surface charges and of added electrolyte on the nature and onset of the transitions observed in magnetic colloid, here in absence of an external magnetic field. The size distribution of the particles is an important parameter, and experiments have been performed on more or less polydisperse samples.

Role of the nature of particles

This latter has been checked with alkaline solutions of maghemite and cobalt ferrite anionic particles. In these samples, the extra electrolyte added was a solution of tetramethylammonium hydroxide. For maghemite particles solutions, we have shown in ref. [10] that addition of tetramethylammonium hydroxide induces phase "gas-liquid"-like transitions, whose onset is strongly dependent on the particles size distribution, and we have proposed a phase diagram taking into account this effect of size through the diameter D_{bir} (Fig. 4). The experiments performed with alkaline solutions of cobalt ferrite showed the same dependence on size and the points corresponding to

Fig. 3 Pictures of a magnetic
fluid entrapped in
a spectroscopy cell; optical
microscopy, enlargement
×900. **a**) a stable magnetic
fluid; **b**) "gas-liquid"-like phase
separation: the concentrated
phase forms dark droplets
dispersed inside the diluted
phase; **c**) application of
a magnetic field, which
elongates the droplets, proving
that the latter are liquid;
d) precipitation 0.9 cm
corresponds to 10 μm

a

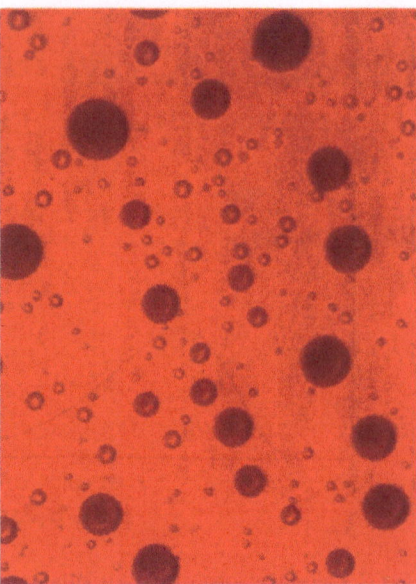

b

c

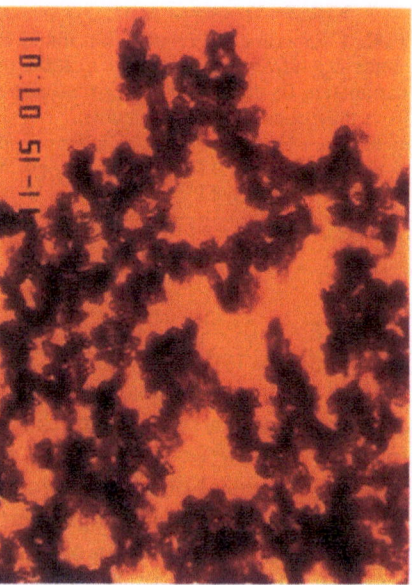

d

cobalt ferrite have been plotted on the phase diagram
concerning maghemite particles. It is noticeable that they
integrate fairly well in this diagram (Fig. 4).

Role of the nature of surface charges

First experiments have been carried out with **anionic**
maghemite particles dispersed **in alkaline medium**. Figure

5a shows that a "gas-liquid"-like phase separation was
observed in these samples, and followed by precipitation;
the domain of electrolyte concentration in the range of
which such a phase transition may be observed is increasingly less important when the average diameter D_{bir}
decreases, and when D_{bir} is lower than about 30 nm, precipitation is directly observed.

In the case of **anionic citrate coated particles**, "gas-
liquid"-like phase transitions have been observed for any

Progr Colloid Polym Sci (1995) 98:23–29
© Steinkopff Verlag 1995

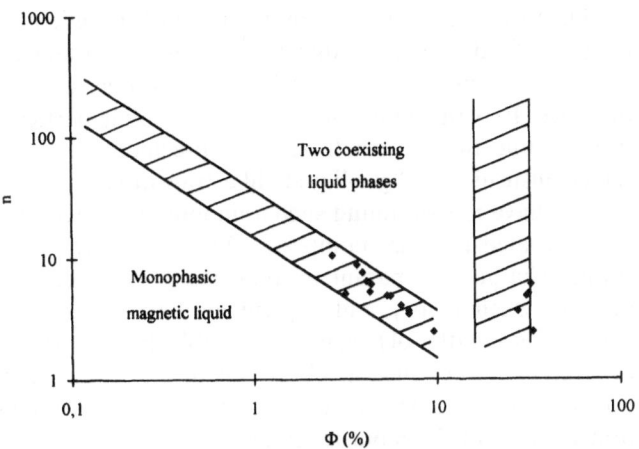

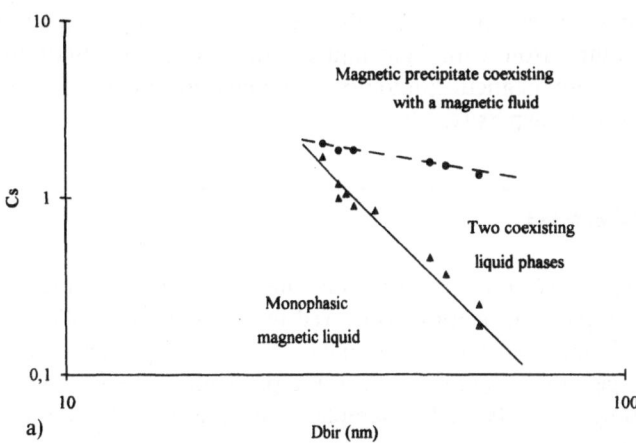

Fig. 4 General normalized phase diagram when tetramethyl-ammonium hydroxide is added to an alkaline magnetic fluid. The **hachured area** corresponds to the location of the points describing the behavior of maghemite particles (from ref. [10]); the **points** correspond to cobalt ferrite magnetic fluids. Φ = volume fraction of particles; n is a reduced parameter which represents a number of free counterions per particle: $n = \mathrm{Na} \cdot c_s \cdot D_{bir}^3 / \Phi$, where Na is the Avogadro number and c_s, the concentration of added electrolyte

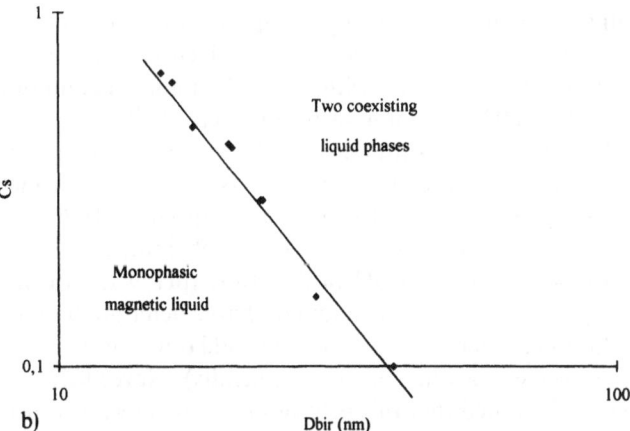

sample synthesized, even when D_{bir} is as small as 15 nm, and this for several electrolytes.

Role of the nature of electrolyte

The first "gas-liquid"-like transition was observed in **alkaline magnetic fluids** when **tetramethylammonium hydroxide** was added. In fact, tetramethylammonium hydroxide is the only electrolyte that induces such a transition. The other electrolytes tested here ($N(CH_3)_4Cl$, ($N(CH_3)_4NO_3$, NaCl, NaOH, ...) induce precipitation of the solution which, moreover, is non-reversible.

On the other hand, in pH 7 magnetic fluids created by dispersion of **anionic citrate coated particles**, all the salts added to the solution induced a "gas-liquid"-like phase transition with the same onset concentration.

Role of the size distribution parameters (Fig. 5)

We noted in [9] that in alkaline magnetic fluids phase transitions induced by addition of tetramethylammonium hydroxide allow to perform size sorting of the particles, as the largest particles are more numerous in the concentrated phase, although the smallest particles stay in the dilute one. Bibette also used this kind of result to perform size sorting in emulsions [21].

Fig. 5 Concentration of electrolyte at the onset of phase separations as a function of the average diameter D_{bir} of the particles; **a)** case of alkaline magnetic fluids and addition of tetramethylammonium hydroxide (from ref. [10]); *plain lines*: onset of "gas-liquid" like-transitions; *dotted line*: onset of precipitation (in the case of alkaline magnetic fluids). **b)** case of pH 7 magnetic fluids and addition of NaCl: the line corresponds to the onset of "gas-liquid" like-transitions; precipitation appears when the concentration of NaCl increases, but it has still not been studied

The same phenomenon has been observed for citrate-coated particles dispersed in water at pH 7 when NaCl is added as extra electrolyte. Size distribution parameters are again monitoring parameters for the onset of transitions, as is shown in Fig. 5b where the concentration of NaCl at the onset of the transition is plotted as a function of the diameter D_{bir}.

Using this result, it was possible to perform an efficient size sorting of particles and to get monodisperse fractions of particles ($D_0 = 7.8$ nm, 9.4 nm; $\sigma = 0.1$), which corresponds to a ratio M_w/M_n equal to 1.1) from a fairly polydisperse initial sample ($D_0 = 7$ mm, $\sigma = 0.4$). It is noticeable that even in such monodisperse samples, a "gas-liquid"-

28

V. Cabuil et al.
Phase separation in aqueous magnetic colloidal solutions

like transition may be observed. This result justifies the comparison with "gas-liquid" transitions, as obtaining a dense magnetic liquid does not constitute an artefact due to polydispersity.

Discussion

This study is an experimental one which aim is to check if the phase transitions observed in the past with colloidal alkaline solutions of anionic magnetic nanoparticles were a general behavior, or if it was a particular case due to the polydispersity of the samples, the nature of the surface charges or the nature of the electrolyte.

The present results show that, in the case of such alkaline solutions, "gas-liquid"-like phase transitions are observed only for one electrolyte, which is rather original as it is a base with an voluminous counterion $(N(CH_3)_4OH)$. It is not surprising that $N(CH_3)_4^+$ cation, which allows the synthesis of stable magnetic fluids contrary to smaller cations such as Na^+, K^+, ..., which strongly associate to particles, plays a specific role. Nevertheless, the role of the co-ion is less obvious; even if the particles at the high pH of the study (pH > 12) are not supposed to adsorb any additive charge determining ions, addition of such ions, as OH^-, may add other effects to the pure screening phenomenon (appendix). Nevertheless, it has to be noted that this behaviour is general with respect to the nature of the particles, as the diagram corresponding to cobalt ferrite magnetic fluids is the same as the diagram of the points corresponding to maghemite particles. This result is in good accordance with theoretical considerations on the role played by magnetic interactions in transitions, these interactions being considered as negligible in zero magnetic field [22].

In the case of anionic citrate coated particles dispersed in water, the behavior seems very general as "gas-liquid"-like transitions are observed for each ionic strength tested and for any value of σ, even for monodisperse samples. The onset of the phase transition has been found correlated to D_{bir}^{-2}, which is in good accordance with theoretical calculations of Victor and Hansen in [6], but different to what was found in the case of alkaline solutions in which the transition was induced by addition of tetramethylammoniumhydroxide (dependence to D_{bir}^{-3}[10]); the specificity of this electrolyte may be at the origin of this distortion).

The theoretical works in refs. [22] and [23] concerning magnetic fluids predicted that there is an onset diameter under which phase "gas-liquid"-like transitions are not observed. Experimentally, in the case of alkaline particles, it was found that for small particles, precipitation is observed instead of a "gas-liquid"-like transition. The fact that we have not yet found such a minimum diameter for citrate-coated particles points out that the nature of the surface charges is a parameter that has to be taken into account when theory and experiment have to be considered. As a matter of fact, it is reasonable that adsorption of large molecules such as citrate introduces some additional steric repulsions and thus modification of the quantitative value of the repulsive potential.

Conclusion

This experimental study has shown that in colloidal solutions of magnetic nanoparticles stabilized through screened electrostatic repulsions, the existence of "gas-liquid"-like transitions is a general feature, even if the nature of the particles surface charges may introduce some quantitative modifications. In any case, the onset of the transition depends clearly on the parameters of the size distribution, allowing to use these transitions in order to produce monodisperse samples.

These samples will now be used to study phase transitions when temperature decreases or when a magnetic field is applied, and comparison with theories [22–25] will then be easier.

Acknowledgements The authors are greatly indebted to M. Carpentier for technical assistance and M. Lavergne for electron micrographs.

Appendix

The impossibility to observe a "gas-liquid" like transition with other co-ions than OH^- ions is still not explained. Nevertheless, specificity of OH^- anions has to be related to the fact that they are charge determining ions and to the possible existence of adsorption equilibria of counterions on particles. This possibility is discussed on the theoretical point of view in reference [25].

References

1. Bacri J-C, Perzynski R, Salin D, Cabuil V, Massart R (1990) J Magn Magn Mater 85:27–32
2. Jansen JW, Kruif CG, Vrij A (1986) J Colloid Interface Sci 114(2):471–481
3. Vrij A, Penders, Rouw PW, Kruif CG, Dhont JKG, Smits C, Lekerkerker HNW (19) Faraday Discuss Chem Soc 90:31–40

Progr Colloid Polym Sci (1995) 98:23–29
© Steinkopff Verlag 1995

4. Cowell C, Vincent B (1982) J Colloid Interface Sci 31:267–298
5. Vincent B, Edwards J, Emmett S, Croot R (1988) Colloids & Surfaces 31:267–298
6. Victor J-M, Hansen J-P (1985) J Chem Soc Faraday Trans 81:43–61
7. Biben T, Hansen J-P (1991) Phys Rev Lett 66:2215–2218
8. Van Duijneveldt JS, Heinen AW, Lekerkerker HNW (1993) Europhys Lett 21(3):369–374
9. Bacri J-C, Perzynski R, Salin D, Cabuil V, Massart R (1987) J Chem Res (S) 130–131
10. Bacri J-C, Perzynski R, Salin D, Cabuil V, Massart R (1989) J Colloid Interface Sci 132(1):43–53
11. Massart R (1981) IEEE Trans Magn MAG-17:1247–1248
12. Jolivet J-P, Massart R, Fruchart J-M (1983) Nouv J Chimie 7:325–331
13. Bacri J-C, Perzynski R, Salin D, Cabuil V, Massart R, Pons J-N, Roger J (1986) In: Biophysical Effects of Steady Magnetic Fields. Springer Verlag, Berlin, pp59
14. Cabuil V, Perzynski R (in press) In: Berbovski (ed) Magnetic Fluids and Applications: Handbook. Begell House Inc. Publishers, NY
15. Bacri J-C, Perzynski R, Salin D, Cabuil V, Massart R (1986) J Magn Magn Mat 62:36–46
16. Bacri J-C, Perzynski R, Salin D, Cabuil V, Massart R (1987) J Magn Magn Mat 65:285–288
17. Bacri J-C, Perzynski R, Salin D, Servais J (1987) J Physique 48:1385–1391
18. Fruchart J-M, Bee A, private communication
19. Bacri J-C, Salin D (1982) J Phys Lett 43:L179–184
20. Charlot G (1966) In "Les Méthodes de la Chimie Analytique", Masson et Cie Ed. pp 737
21. Bibette J, Roux D, Nallet F (1990) Phys Rev Lett 65(19):2470–2473
22. Russier V, Douzi M (1994) J Colloid Interface Sci 162:356–371
23. Sano K, Doi M (1983) J Phys Soc Jap 52(8):2810–2815
24. Cebers AO (1982) Magn Gidrodin 2:42
25. Bushmanova SV, Ivanov AO, Buyevich Yu A (1994) Physica A 202:175–195

Progr Colloid Polym Sci (1995) 98:30–34
© Steinkopff Verlag 1995

J.-C. Bacri
A. Cebers
C. Flament
S. Lacis
R. Melliti
R. Perzynski

Fingering phenomena at bending instability of a magnetic fluid stripe

J.-C. Bacri · C. Flament · R. Melliti
Université Paris 7-Denis Diderot
(UFR de Physique)

Prof. J.-C. Bacri (✉) · C. Flament
S. Lacis · R. Melliti · R. Perzynski
Laboratoire d'Acoustique et Optique
de la Matière Condensée[1]
Université Pierre et Marie Curie
Tour 13, Case 78
4 Place Jussieu
75252 Paris Cedex 05, France

A. Cebers · S. Lacis
Institute of Physics
Latvian Academy of Sciences
Salaspils 1
2169 Riga, Latvia

[1] Associated with the Centre National de la
Recherche Scientifique

Abstract Patterns of magnetic fluid stripes are experimentally studied and compared to numerical simulations by boundary integral equation technique. Like in amphiphile monolayers and ferromagnetic films, we observe bending and alternated fingering of a single stripe. For bending deformations, a two-dimensional-smectic analogy is presented. This analogy is only valid for quasistatic variations of the external magnetic field strength. A fingering phenomenon occurs for non-quasistatic variations of the fields and for large stripes.

Key words Magnetic liquid – patterns – smectics – bending – fingering

Introduction

Due to the competition between repulsive and attractive interactions, complex patterns arise in different systems – ferromagnetic films, amphiphile monolayers, plane layers of magnetic fluids. It is well known that for definite conditions, those systems develop intricate labyrinthine patterns [1–5]. Generally, these patterns are similar whatever the physical system with different dynamics. In this work, we study the behavior of a stripe pattern of magnetic fluid (MF) in an Hele-Shaw configuration (cell thickness h) under a perpendicular magnetic field H: the control parameters are the ramp rate of magnetic field dH/dt and the width $2d$ of the stripe.

Theoretical analysis

Quasistatic description

At low ramp rates of the magnetic field, characteristic labyrinthine stripe patterns develop [2, 3]. A feature of the pattern consists in bending deformations of magnetic fluid stripes, which can be described [6] as deformations in two-dimensional (2D) smectics. Due to the long-range magnetic interaction, the effective surface tension of equilibrium MF stripe pattern is exactly equal to zero. The corresponding Hamiltonian of the system in that case includes the terms of curvature and compressional elasticity:

$$E = \frac{1}{2} B \int \left[\beta + \frac{\partial u}{\partial x} - \frac{1}{2}\left(\frac{\partial u}{\partial y}\right)^2 \right]^2 dS$$
$$+ \frac{1}{2} K_c \int \left(\frac{\partial^2 u}{\partial y^2}\right)^2 dS \,, \tag{1}$$

where $\beta = (l(H_1) - l(H_2)/l(H_1))$ is the relative deviation of the structure period $l(H)$ from the equilibrium when the magnetic field increases rapidly from H_1 to H_2. Elasticity constants B and K_c depend on the magnetization M of the MF, on its surface tension σ, and on the volume fraction occupied by the MF [6]. Hamiltonian (1) allows to describe the development of chevron patterns arising at $\beta > 0$ and observed for 2D ferromagnetic films [7]. The analogy with 2D smectics is valid if the stripes are only bending.

Here, for the first time, we are studying both experimentally and numerically the conditions at which the fingering of a MF stripe occurs. It is known [1] that the critical value of the magnetic Bond number $Bm = 2M^2h/\sigma$ at which bending deformations of wavenumber k develop can be found according to the relation ($\delta = 2d/h$-relative thickness of the stripe):

$$\frac{\sigma}{2M^2h} = \frac{F(kh)}{(kh)^2} = \frac{2(\gamma + \ln(kh/2) + \ln(\delta/\sqrt{1+\delta^2}) + K_0(kh) + K_0(\delta kh) - K_0(kh)\sqrt{1+\delta^2})}{(kh)^2} \tag{2}$$

where K_0 is McDonald function and γ the Euler constant. Large wavelength deformations occur when the effective surface tension of the stripe is equal to zero. The bending instability occurs at a critical magnetic Bond number Bm_c which decreases with δ. In an asymptotic development: $Bm_c = 2/\ln(\delta)$.

Dynamic study

Conditions for bending deformation development can be quite different at high ramp rates of magnetic field. The stripe is unstable with respect to different modes and one could expect then the development of the mode of largest growth increment. Due to the MF confinement in a thin layer the dynamics of instability may be described in an Hele-Shaw approximation [8, 9]. In that case the Darcy equation for the fluid motion in the flat layer, taking into account the long-range magnetic force, is:

$$-\alpha \vec{v} - \vec{\nabla} p + \frac{2M}{h} \vec{\nabla} \varphi = 0; \quad \text{div } \vec{v} = 0 \,, \tag{3}$$

where $\alpha = \eta/g$, $g = h^2/12$ being the permeability of the flat layer, $\vec{v}$ is the local velocity of the interface and the pressure p is given by the Laplace law ($p = ct + \sigma/R_c$, R_c the local

curvature of the stripe). The MF self-magnetic field potential φ on the boundary of the layer is given by the integral:

$$\varphi = - M \int dS' \left(\frac{1}{|\vec{\rho} - \vec{\rho}'|} - \frac{1}{\sqrt{(\vec{\rho} - \vec{\rho}')^2 + h^2}} \right). \tag{4}$$

The relation (4) is integrated along the interface between the MF and the cell wall.

Taking into account the kinematic boundary conditions for the free surface evolution, Eqs. (3) and (4) allow to calculate, in the framework of linear theory, the growth rate of the different modes. The most unstable wavenumber $k*$ for bending deformation is given by the following equation:

$$B_m = \frac{2M^2h}{\sigma}$$
$$= \frac{(k*h)^2 \left[3 - k*h\delta/sh(k*h\delta)\right]}{\left[1 - k*h\delta/sh(k*h\delta)\right] F(k*h) + k*hF'(k*h)} \,. \tag{5}$$

Figure 1 presents the dependence of the most unstable reduced wavenumber $k*h$ versus magnetic Bond number for two values of the relative thickness of the stripe. The wavenumber of the bending deformation increases with the magnetic field.

Numerical simulations

Numerical algorithm for the computer simulation of the MF stripe dynamics is based on conformal mapping methods [9] or boundary integral equation technique [10, 11]. In the case of boundary integral equation technique, the stream function ψ of the flow is found ($v_x = \partial\psi/\partial y$; $v_y = -\partial\psi/\partial x$) as a single layer potential

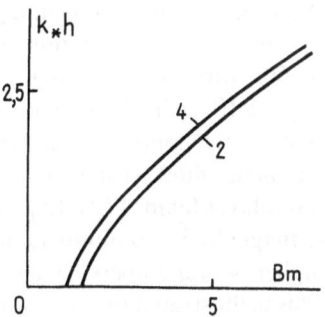

Fig. 1 Wavenumber of the fastest growing bending mode versus magnetic Bond number for two relative thicknesses of the magnetic stripe $\delta = 2$ and $\delta = 4$

$(L_+, L_-$ being the two lateral boundaries of the stripe):

$$\psi = \frac{1}{2\pi} \int_{L_+} f_+ \ln|\tilde{\rho} - \tilde{\rho}'|\, dl' + \frac{1}{2\pi} \int_{L_-} f_- \ln|\tilde{\rho} - \tilde{\rho}'|\, dl', \quad (6)$$

where f_+ and f_- are given by the system of coupled boundary integral equation:

$$\frac{1}{\alpha}\frac{d}{dl}\left\{\frac{\sigma}{R_c^+} + \frac{2M^2}{h}\int dS'\left(\frac{1}{|\tilde{\rho}_+ - \tilde{\rho}'|} - \frac{1}{\sqrt{(\tilde{\rho}_+ - \tilde{\rho}')^2 + h^2}}\right)\right\}$$

$$= -\frac{1}{2}f_+ + \frac{1}{2\pi}P\int_{L_+} f_+ \frac{\partial}{\partial n}\ln|\tilde{\rho}_+ - \tilde{\rho}'|\, dl'$$

$$+ \frac{1}{2\pi}\int_{L_-} f_- \frac{\partial}{\partial n}\ln|\tilde{\rho}_+ - \tilde{\rho}'|\, dl' \quad (7)$$

$$\frac{1}{\alpha}\frac{d}{dl}\left\{\frac{\sigma}{R_c^-} + \frac{2M^2}{h}\int dS'\left(\frac{1}{|\tilde{\rho}_- - \tilde{\rho}'|} - \frac{1}{\sqrt{(\tilde{\rho}_- - \tilde{\rho}')^2 + h^2}}\right)\right\}$$

$$= -\frac{1}{2}f_- + \frac{1}{2\pi}\int_{L_+} f_+ \frac{\partial}{\partial n}\ln|\tilde{\rho}_- - \tilde{\rho}'|\, dl'$$

$$+ \frac{1}{2\pi}P\int_{L_-} f_- \frac{\partial}{\partial n}\ln|\tilde{\rho}_- - \tilde{\rho}'|\, dl'.$$

Equations (7) are obtained applying the theorem for normal derivative of single-layer potential and accounting for the tangential component of the fluid velocity at the boundary, obtained from (3) (P being the Cauchy principal value). The solution of the set of boundary integral equations (7) allows to determine the velocities of the definite amount of marker points on the stripe boundary. Since it is impossible to consider an infinite stripe in the simulations, we assume that the stripe is periodical with period L. Long-range interactions are accounted for, only on the nearest neighbors.

Figure 2 shows the development of the stripe bending instability from an initial configuration of the two lateral boundaries ($x \in [-\pi/2, \pi/2]$, $\varepsilon = 0.1$):

$$y_+ = w + \varepsilon\cos(2x), \quad y_- = -w + \varepsilon\cos(2x) \quad (8)$$

at $\delta = 2$, $Bm = 2M^2h/\sigma = 3$, $\pi h/L = 0.2$, $w = \pi d/L$. According to Fig. 1 these parameter values slightly exceed the critical value for bending deformation development. Number of marker points employed for each boundary is 50, with a time step of 10^{-2} dimensionless unit ($\equiv hL\alpha 10^{-2}/2\pi M^2$ s). The configuration obtained resembles the configurations observed in the experiment for magnetic fluid stripes [2] and stripes in amphiphilic monolayer foams [12]. Experiments performed with higher magnetic Bond numbers show bending deformations of higher wavenumbers in agreement with expression (5). This is illustrated in Fig. 3 with $\delta = 2$, $Bm = 6$, $\pi h/L = 0.2$

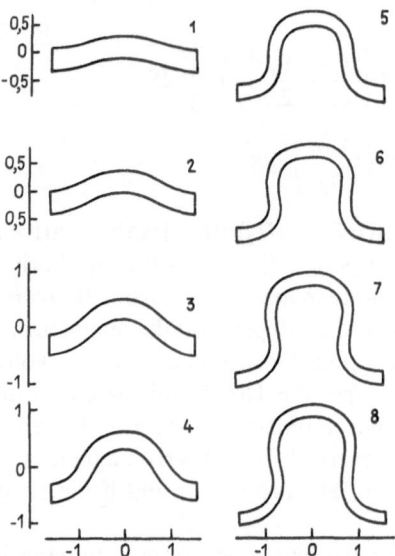

Fig. 2 Dynamics of the bending deformations at $\delta = 2$, $Bm = 2M^2h/\sigma = 3$, $\pi h/L = 0.2$ and initial conditions (8). Configurations shown are obtained at the following adimensional time moments 0.75 (1), 3.5 (2), 6.5 (3), 9 (4), 13 (5), 15.5 (6), 20 (7), 22.5 (8)

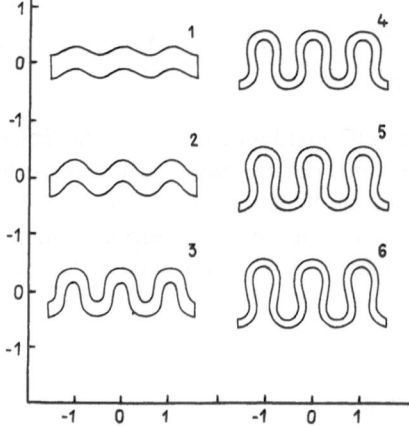

Fig. 3 Dynamics of the bending deformations at $\delta = 2$, $Bm = 6$, $\pi h/L = 0.2$ and other initial conditions (9). Configurations correspond to the following time moments 0.25 (1), 0.5 (2), 1 (3), 2 (4), 2.25 (5), 2.5 (6)

and initial configuration ($\varepsilon = 0.05$):

$$y_+ = w + \varepsilon\cos(6x), \quad y_- = -w + \varepsilon\cos(6x) \quad (9)$$

Time step in this case is 5×10^{-3} in dimensionless units and 80 marker points at each boundary are employed. An initial perturbation of wavelength about 60% of the one given by expression (5) is forced and the bending deformations which grow the fastest, roughly keep this same wavelength.

A further increase of the magnetic Bond number or of the stripe thickness, which according to Fig. 1 is equivalent, quite drastically changes the pattern deformation. As it is shown on Fig. 4 for $\delta = 4$, $Bm = 6$, $\pi h/L = 0.2$ with initial configurations (8), a periodical alternated array of fingers arises instead of the development of bending deformations (time step 0.01 dimensionless units, 50 marker points for each boundary). With initial configurations (9) ($\varepsilon = 0.05$) the patterns become rather intricate: tip splitting appears on Fig. 5 (time step 5×10^{-3}, 80 marker points for

each boundary). After the initial stage of the bending deformation mode development, a fingering occurs at the crests of bent stripe. For fingers formed, one can see the secondary tip instability leading to their bifurcation.

Comparison with experiments

We have also performed some experiments in order to compare with the numerical simulations. A MF is a colloidal suspension of magnetic particles. We use here an ionic MF (water based) with cobalt ferrite particles [13] of mean size of the order of 10 nm. The MF saturation magnetization is $M_s = 40$ kA/m with a particle volume fraction of 10%. A droplet of this MF is trapped between two perpendicular horizontal Plexiglass plates under a vertical magnetic field. The thickness of the cell is 0.75 mm and the MF droplet is face to face with an organic liquid in order to decrease the gravity effect, and make MF strongly non wetting with the solid surface. We have performed experiments with the same initial and final magnetic fields: $H_i = 8$ kA/m and $H_f = 25$ kA/m, the same ramp rate $dH/dt = 19$ kA/m/s but different thicknesses of the stripe. Figure 6 presents the evolution of a thin stripe $\delta = 2$ in function of time: bending instability is clearly shown. Figure 7 presents the same experiment with a thicker stripe $\delta = 3.5$; if an alternated fingering of the stripe is observed, there is no evidence of any tip splitting in the experiment. As in the numerical simulations, fingering occurs if the

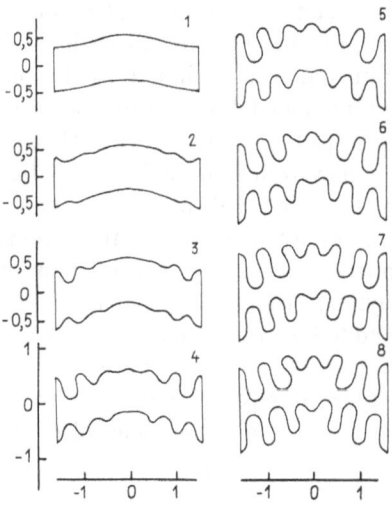

Fig. 4 Fingering of the MF stripe; dynamics at $\delta = 4$, $Bm = 6$, $\pi h/L = 0.2$ with initial conditions (8). Configurations correspond to the following time moments 0.25 (1), 1.5 (2), 1.75 (3), 2 (4) 2.25 (5), 2.5 (6), 2.75 (7), 3 (8)

Fig. 5 Fingering and tip-splitting of the MF stripe at $\delta = 4$, $Bm = 6$, $\pi h/L = 0.2$ with initial conditions (9). Configurations correspond to the following time moments 0.5 (1), 1 (2), 1.25 (3), 1.5 (4), 1.75 (5), 2 (6), 2.25 (7), 2.5 (8)

Fig. 6 Experimental picture of a MF thin stripe $\delta = 2$, showing the bending deformation. At the beginning $t = 0$, the magnetic field is $H_i = 8$ kA/m. After 0.9 s the field reaches its final value $H_f = 25$ kA/m. The equilibrium state is obtained at $t = 1.4$ s. The scale is given by the black bar (1 cm)

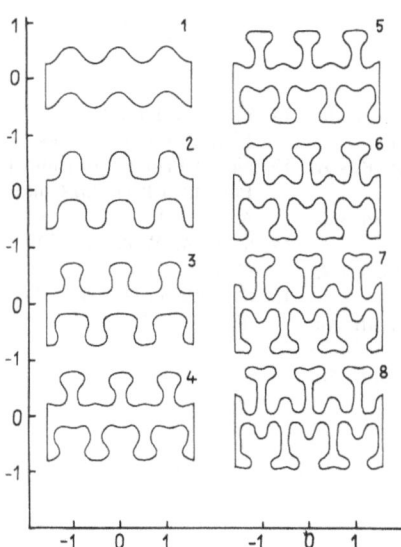

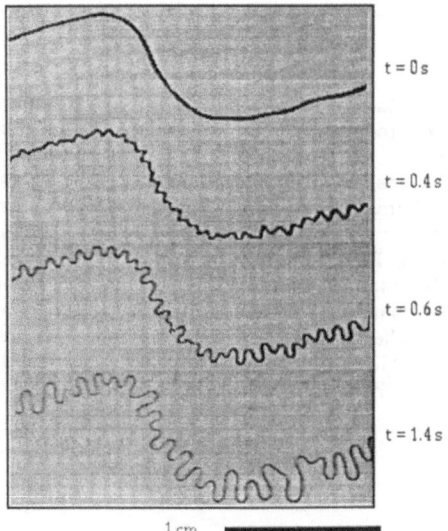

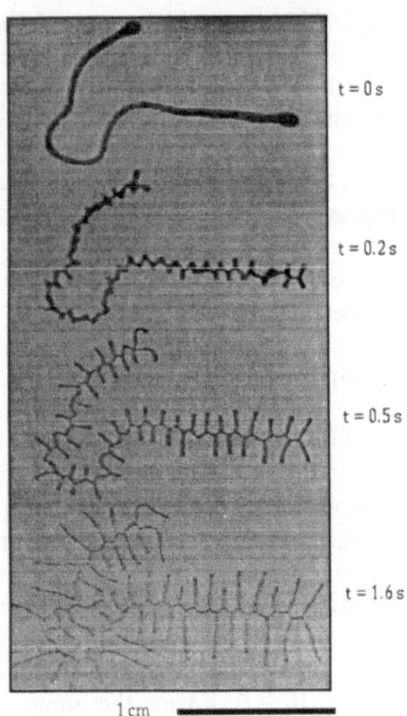

1 cm

Fig. 7 Same experiment as in Fig. 6 but with a thicker stripe $\delta = 3.5$ (fingering effect)

thickness of the stripe is increased, all the other parameters being constant.

Conclusion

Experimental and numerical simulation results reported here show clearly for the first time that 2D smectic analogy

for MF stripe patterns has limited validity. At high magnetic Bond numbers and with large stripe thicknesses, pattern fingering phenomena occur instead of the collective bending deformation mode of the stripe. This leads to the formation of patterns without any similarity with the 2D smectics. Experiments and numerical simulations are both performed with an isolated MF stripe. Alternated fingering patterns are observed with this single stripe: this means that it is not due to interactions between stripes as it is usually proposed [14]. It should be noted here that the possibility of the fingering phenomena at the deformation of the stripe domain patterns in thin ferromagnetic films is mentioned in [15], but not studied in detail. Some qualitative experimental results similar to the ones reported here are presented in [14]. During the fast demagnetization process (just opposite to the MF case !) of thin ferromagnetic film from saturated stage, patterns of stripe domains are observed, called there "set of disclination dipoles". This allows to conclude that complex pattern formation phenomena in the two different systems have the same physical origin. Self-magnetic field gradients of the stripes lead to fingering phenomena with a non-quasistatic variation of the control parameter. Due to the completely different physical nature of the systems the spatiotemporal scales of phenomena of course are quite different.

Acknowledgement We are greatly indebted to P. Lepert and J. Servais for their technical assistance. This work was supported by "Le Réseau Formation Recherche n° 90R0933 du Ministère de l'Enseignement Supérieur et de la Recherche" of France. Two of us (A. Cebers and S. Lacis) thanks "International Science Foundations" for financial support.

References

1. Cebers A, Maiorov MM (1980) Magnitnaya Gidrodinamika (in Russian) 1:27–35
2. Cebers A, Maiorov MM (1980) Magnitnaya Gidrodinamika (in Russian) 3:15–20
3. Rosensweig RE, Zahn M, Shumovich R (1983) J Mag Mag Mat. 39:127–132
4. Bacri J-C, Perzynski R, Salin D (1987) La Recherche 18:1150–1159
5. Langer S, Goldstein RE, Jackson DP (1992) Phys Rev A 46:4894–4904
6. Cebers A (1994) Magnitnaya Gidrodinamika (in Russian) (to appear)
7. Seul M, Wolfe R (1992) Phys Rev A 46:7519–7533
8. Cebers A (1981) Magnitnaya Gidrodinamika (in Russian) 2:3–15
9. Goldstein RE, Jackson DP, Cebers A (1994) Phys Rev E 50:298–307
10. Cebers A, Zemitis A (1983) Magnitnaya Gidrodinamika (in Russian) 4:15–26
11. Cebers A (1993) Magnitnaya Gidrodinamika (in Russian) 2:28–34
12. Stine KJ, Knobler CM, Desai CR (1990) Phys Rev Lett 65:1004–1007
13. Neveu-Prin S, Tourinho FA, Bacri J-C, Perzynski R (1993) Colloid and Surface A 80:1–10
14. Seul M, Wolfe R (1992) Phys Rev A 46:7534–7547
15. Sornette D (1987) J Physique (France) 48:151–153

Progr Colloid Polym Sci (1995) 98:35–38
© Steinkopff Verlag 1995

G.T. Dimitrova
Th.F. Tadros
P.F. Luckham
M.C. Taelman
P. Loll

Investigation of liquid crystalline phases using microscopy, differential scanning calorimetry and rheological techniques

G.T. Dimitrova · Dr. Th.F. Tadros (✉)
Zeneca Agrochemicals
Jealott's Hill Research Station
Bracknell
Berkshire RG12 6EY, United Kingdom

P.F Luckham
Imperial College of Science and Technology
Prince Consort Road
London SW7 2BY, United Kingdom

M.C. Taelman P. Loll
ICI Surfactants
Personal Care
Eveslaan 45
3078 Everberg, Belgium

Abstract Systematic studies on the phase diagrams of a series of nonionic surfactants namely Synperonic A7, A9, A11 and A20 are reported. The phase diagrams were first established using polarising microscopy, as a function of both concentration and temperature. For A7, a hexagonal phase was produced above 30% surfactant concentration which extended up to 55%. Above the latter concentration, a lamellar phase appeared. With A9, the appearance of the first liquid crystalline phase was shifted to higher concentrations but the phase diagram was qualitatively similar to that of A7. With A11, however, a cubic phase appeared between the hexagonal and lamellar phase. A20 showed a cubic phase as the first liquid crystalline phase

(above 25%) followed by a hexagonal phase, but no lamellar structures could be detected. These transitions could also be followed using differential scanning calorimetry (DSC) and rheology. Viscoelastic measurements showed these transitions very clearly. For example, above the transition temperature, the liquid crystalline system changed from predominantly viscous to predominantly elastic response. The phase diagrams were established using microscopy, DSC and rheology, and good agreement was obtained between the three methods.

Key words Nonionic surfactants
– liquid crystalline phases
– microscopy rheology differential scanning calorimetry

Introduction

Investigations of the structure of liquid crystalline phases that are produced in nonionic surfactant solutions are essential from a fundamental point of view. It is important to relate the structure of the units formed such as spherical or rod-shaped micelles and various liquid crystalline phases to the molecular architecture of the surfactant. One of the earliest ways of relating the structure of phases produced, to the molecular geometry is the packing ratio [1, 2]. This is simply related to the most probable shape that is produced in packing the molecules into aggregates.

The phase diagrams of many nonionic surfactants that are based on ethylene oxide (having the general formula C_nEO_m, where n is the number of carbon atoms in the alkyl chain and m is the number of ethylene oxide groups) have been investigated by several authors [3, 4], who used plane polarising light microscopy to show the various phases produced as a function of surfactant concentration and temperature. Light scattering, x-ray and small-angle neutron scattering are among the various other techniques which have been applied in these studies. The results described in this paper give a comprehensive study on the phase behaviour of four nonionic surfactants, namely Synperonic A7, A9, A11 and A20.

Experimental

Materials

Synperonic A surfactants were supplied by ICI Surfactants (Wilton, U.K.) and used as received. They are made by ethoxylation of Synperol alcohol that consists of 66% C_{13} and 34% C_{15} alkyl chains. The ethoxylation process gives rise to a wide distribution of ethoxylate chains and hence the numbers 7, 9, 11 and 20 represent an average of these ethoxylate chains. Water was doubly distilled from an all-pyrex apparatus. Samples were prepared by weight by adding water to the surfactant and then heating and stirring the mixture to give a homogeneous solution.

Microscopy investigations

A Leitz Ortholux microscope fitted with a controllable temperature stage was used. The temperature was measured by means of a small thermocouple placed between the microscope slide and the stage.

All photographs were taken on Ilford Pan F film which was normally exposed at +1 or +2 stops over the measured exposure.

Differential Scanning Calorimetry (DSC) Measurements

These were carried out using a heat flux Mettler TA4000 system. The "melting" of a liquid crystalline phase was accompanied by an endothermal peak which was automatically recorded.

Cloud point measurements

The cloud points of the surfactant solutions were determined as a function of concentration using turbidity measurements. A Pye Unicam spectrophotometer fitted with an SP 875 temperature programmer unit was used. The cloud point was taken as the temperature at which there was a sudden increase in absorbance with further increase of temperature.

Rheological measurements

Oscillatory (dynamic) measurements were carried out using a Bohlin VOR rheometer (Bohlin Reologie, Lund, Sweden). In these measurements, one initially fixes the frequency at 1 Hz and measures the moduli (G^*, G' and G'') as a function of the strain amplitude. This enables one to determine the linear viscoelastic region where the dynamic moduli are independent of the applied strain at any given frequency. Once the linear viscoelastic region is established, then measurements are made as a function of temperature at a fixed amplitude.

In steady-state measurements, the stress is measured as a function of the applied shear rate and the viscosity is calculated from the slope of the linear portion of the curve.

Results and discussion

Microscopy, cloud points and DSC measurements

The microscopy dilution experiments gave a crude picture and served as a very useful qualitative method for identification of the various liquid crystalline phases. These experiments involved diluting the dry surfactant with water on a microscope slide and then observing under

Fig. 1 Micrographs of the phase textures obtained at various A7 concentrations at 20 °C (magnification 150x) a) 30%; b) 35%; c) 40%; d) 45%; e) 70%; f) 80%

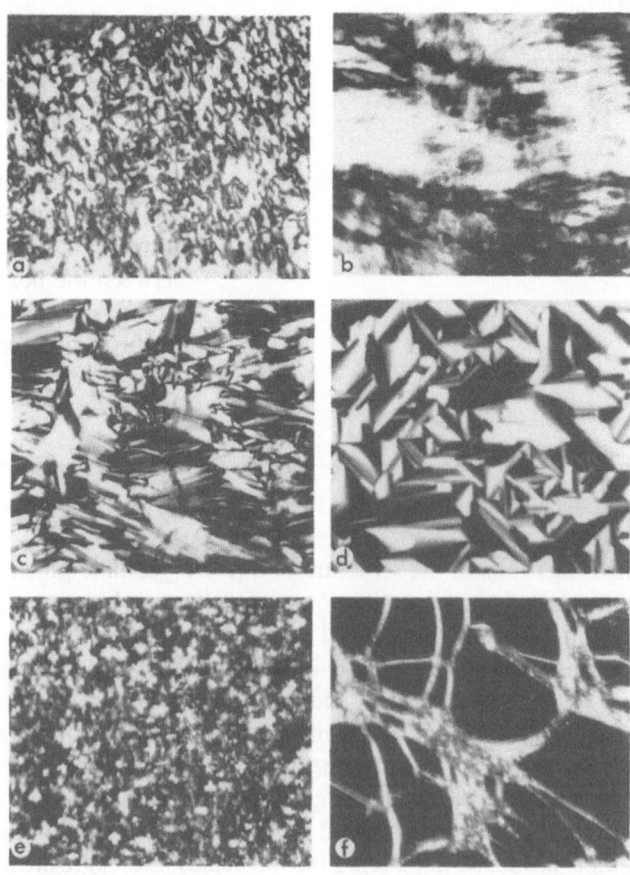

polarised light the textures of the phases formed [5, 6]. Synperonic A7 and A9 gave similar results i.e. appearance of lamellar followed by hexagonal phases as the surfactant was gradually diluted. However, with A11 a cubic isotropic phase (identified from the liquid isotropic phase by its noted high viscosity) appeared between the hexagonal and lamellar phases. The results using A20 showed a significant difference in phase behaviour i.e. appearance of hexagonal phase followed by a cubic phase but no lamellar structures were found. For more accurate determination of the structures, photographs were taken at various surfactant concentrations at constant temperature (20 °C). For example, the results for 30% A7 at 20 °C (Fig. 1a) show a mixture of isotropic solution and hexagonal structure. At 35% A7 (Fig. 1b), a non-geometric texture was produced. Such textures are commonly observed with the hexagonal phase. A typical angular texture of the hexagonal phase is shown in Fig. 1c for 40% A7. The typical mosaic texture of the lamellar phase is shown in Fig. 1e for 70% A7. In addition, photographs were taken at a fixed surfactant

concentration while increasing the temperature. This experiment allowed us to identify the structural changes which occur within the phase as the temperature was raised and to locate the "melting" temperature of the liquid crystalline phase. Using the above procedure one can establish the phase diagram of a surfactant-water system. These are illustrated in Figs. 2a–2d which also show the results obtained using DSC and viscoelastic measurements as well as the cloud point curves.

Rheological measurements

The influence of temperature on the rheological properties of the surfactant solutions is illustrated in Fig. 3 which shows the steady state viscosity results for dilute A7 solutions (L_1 region) in the range 10–25%. In all cases, the viscosity-temperature curves show a pronounced maximum which depends on the surfactant concentration. The maximum appears at a lower temperature as a surfactant

Fig. 2 Phase diagrams of Synperonic A/water systems a) A7; b) A9; c) A11; d) A20

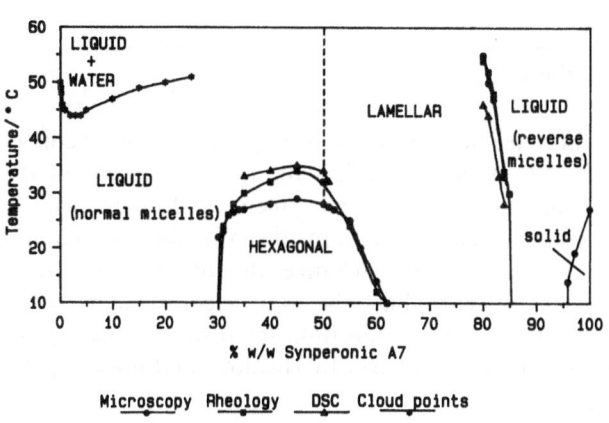

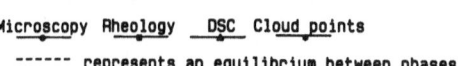

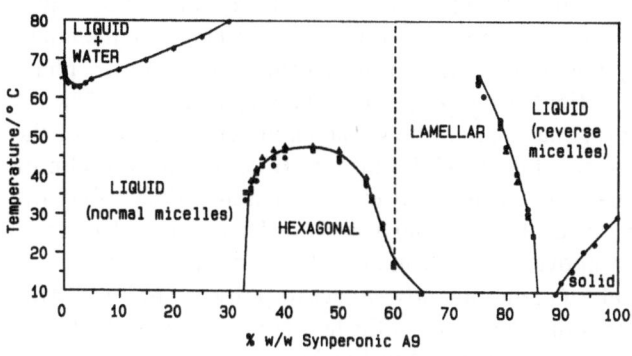

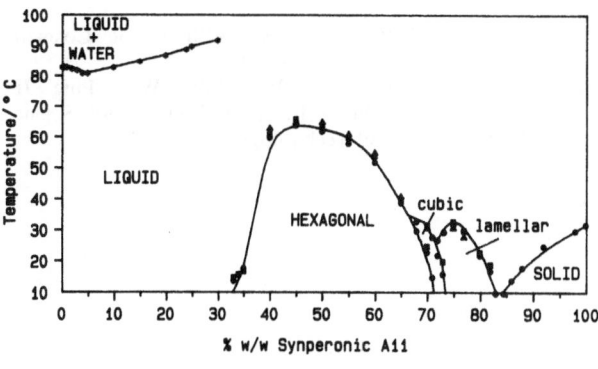

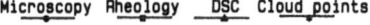

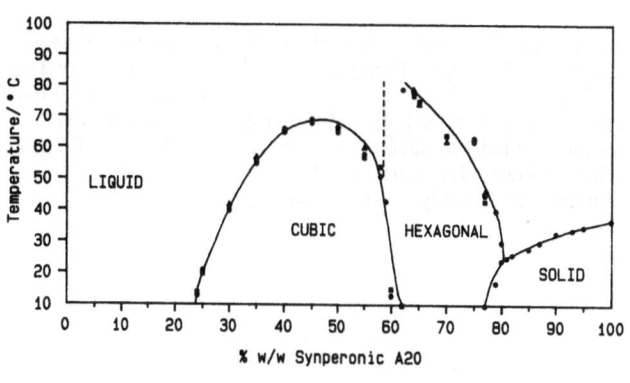

38

G.T. Dimitrova et al.
Investigation of liquid crystalline phases

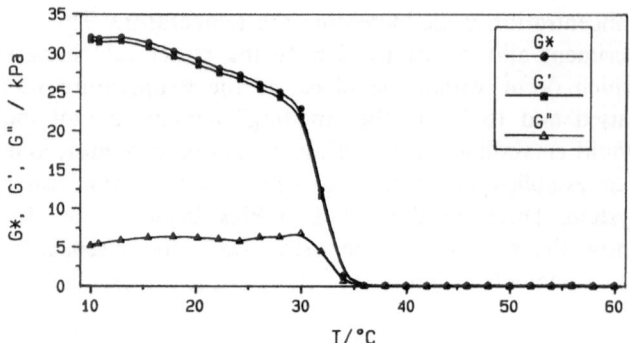

Fig. 3 Steady-state viscosity (at a shear rate of $92\,s^{-1}$) versus temperature for various A7 surfactant solutions

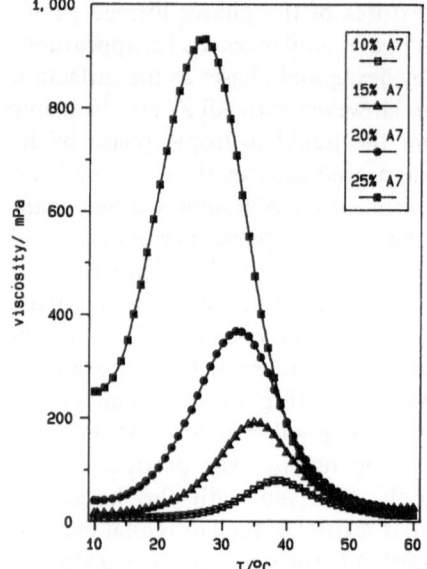

Fig. 4 Dynamic moduli (1 Hz) versus temperature at 35% A7

concentration is increased. There seems to be a correlation between the viscosity maxima and the cloud points of the surfactant solutions. The higher the cloud point, the lower the temperature maximum of the viscosity. These trends in viscosity suggest that some subtle structural changes may occur as the temperature of the surfactant soution is increased.

The viscoelastic-temperature measurements were carried out without imposing a steady-state shear since shearing causes some structural perturbation of the liquid crystals [7]. These measurements followed the phase transitions accurately. For example, Fig. 4 for 35% A7 (hexagonal phase), shows the highly elastic nature of the system below 32 °C. In this case $G' > G''$ and G' is almost equal to G^*. This highly elastic structure is the result of the presence of hexagonal rod-shaped structures. However, just above 32 °C, all moduli values decrease rapidly and finally a predominantly viscous response is produced. This marks the change from the hexagonal phase to the liquid phase as is illustrated in Fig. 2a.

Conclusions

The three techniques of microscopy, DSC and rheology which were used to establish the phase diagrams of the surfactants showed very good agreement. Both viscoelastic measurements and microscopy gave valuable information on the possible structural changes that occur in the surfactant-water system whilst the DSC measurements could only follow the phase transition temperatures, having already identified the phases by rheology and microscopy.

References

1. Israelachvili JN, Mitchell D, Ninham BW (1976) J Chem Soc Faraday Trans I, 72:1525
2. Israelachvili J (1984) In: Mittal KL, Botherel P (eds) Surfactants in Solution. Plenum Press, New York, Vol. 4, p 3
3. Mitchell DJ, Tiddy GJT, Waring L, Bostock T, McDonald MP (1983) J Chem Soc Faraday Trans I, 79:975
4. Nilsson P, Lindman B (1983) J Phys Chem 87:1377
5. Rosevear FB (1954) J Am Oil Chemists Soc 31:628
6. Rosevear FB (1968) J Soc Cosmetic Chemists 19:581
7. McKay KW, Miller WG, Puig JE, Frances EI (1991) J Dispersion Science and Technology 1:37

Progr Colloid Polym Sci (1995) 98:39–41
© Steinkopff Verlag 1995

A. Mourchid
A. Delville
P. Levitz

Phase diagram of dispersion of anisotropic particles

A. Mourchid · A. Delville · Dr. P. Levitz (✉)
Centre de Recherche sur la Matiére
Divisée, CNRS
1b Rue de la Férollerei
45071 Orléans Cedex 2, France

Abstract We discuss the phase diagram of aqueous dispersions of colloidal plate-like charged particles (300 Å × 10 Å). Particle concentration and ionic strength are the two parameters controlling the system. At fixed ionic strength, the suspensions undergo a sol/gel transition without macroscopic phase separation. Shear rheology is used to monitor this transition and to locate the appearance of the "mechanical gel" phase. Below a critical salt concentration at which flocculation appears, the transition line separating the liquid phase and the solid-like system shifts to lower volume fractions as the ionic strength increases. Direct inspection of this gel phase by cryofracture, MET and SAXS experiments shows correlated but well-separated particle populations. In order to check reversibility and equilibrium properties of this transition, the osmotic pressure was determined by an osmotic stress method. At fixed ionic strength, the osmotic pressure first increases at low particle concentration, then reaches a "pseudo plateau," and increases again for higher concentrations. The location of such a break in the equation of state of the suspension defines a thermodynamical transition line coinciding with the mechanical phase transition. In order to analyze the origin of this gel or glassy phase, the role of particle anisotropy, coupled with diffuse layer repulsion, will be discussed.

Key words Complex fluid – sol/gel transition – colloid stability – clay suspension

Introduction

Colloidal systems undergo phase transitions such as liquid-solid [1–3], order-disorder, sol-gel [4] or glass transitions [5], which are of important technical and scientific interest. The phase diagram of the colloidal suspensions is generally controlled by interparticle interactions. While temperature and ionic strength are two intensive variables able to modulate these interactions, the morphology of the particles and their anisotropy are more subtle parameters also playing a crucial role in colloidal stability and phase transitions [6, 7].

Disk-like particle such as (Na$^+$ or Li$^+$ homoionic swelling clays [8–9]) are well known to form a gel above a specific volume fraction. However, the addition of salt does not promote the liquid phase, as observed for spherical or rodlike particles, but lowers the particle volume fraction at which the gel or glassy phase appears [8, 9]. At first glance, this result seems to be contrary to what one

would expect from DLVO theory. Two opposite explanations are proposed in the literature:

i) a micro flocculation due to electrostatic or van der
Waals attractions between edges and faces of the plate-like
particles resulting in a linked structure, the so-called house
of cards configuration [10];
ii) a gelation mechanism, originally suggested by Norrish
[11] and supported by others [12–14], which stresses the
role of long-range electrostatic repulsion between overlapping double layers, as revealed by osmotic pressure
measurements, coupled with excluded volume interactions
induced by the specific anisotropy of the charged disklike
particles.

The choice between attractive and repulsive interparticle interactions during gelation of disk-like particles is
not easy, and the aim of this work is to provide experimental information on the phase diagram of plate-like
charged particles. More specifically, we will focus on suspensions of small (300 Å × 10 Å) synthetic clay particles
(laponite). This system has already been analyzed by different authors, and especially by Ramsay [13–15] and
Morvan [16]. In the present work, particle concentration
and ionic strength will be the two parameters controlling
the system. Shear rheology will be used to monitor the
sol/gel transition and to locate the appearance of the
"mechanical gel" phase. Self-organization of the gel phase
will be evidenced by cryofracture, MET and SAXS
measurements. Finally, the reversibility and the equation
of state of this colloidal system under gelation will be
investigated by an osmotic stress method.

Experimental results

Rheological behavior

As already shown by different authors, colloidal suspensions of laponite undergo a gel transition above some solid
fraction C_0. Below this value, the suspension is slightly
viscous and the elastic $G'(\omega)$ and loss modulus $G''(\omega)$ are
weak (around 10^{-1} Pa). Above C_0, the elastic modulus
increases markedly and the appearance of a yield stress is
observed. $G'(\omega)$ and $G''(\omega)$ do not show a strong variation
with frequency in the dynamical range between 10^{-2} and
10^{+2} rad/s. At fixed particle concentration, the elasticity
and, incidentally, the yield stress, increase with the ionic
strength. Experimental data fit a power law given by:

$$G'(0) = (C - C_0)^\alpha \qquad \text{(eq 1)}$$

Such a power law was already observed for colloidal
suspensions of clays [8, 13]. The exponent α (2.35) is not

sensitive to the ionic strength and the numerical determination of C_0 permits an estimation of the concentration
at which gelation appears, in good agreement with the
appearance of a yield stress. The resulting transition line
separates a viscous sol phase from a mechanical viscoelastic gel phase. As the ionic strength increases, the clay
concentration threshold is shifted towards lower values,
until flocculation occurs (at $I \approx 2 \ 10^{-2}$ M for NaCl).

Osmotic pressure and equation of state

The repulsive nature of the interparticle interaction should
be confirmed by the determination of the equation of state
of the suspension. The experimental results are obtained
by osmotic stress of diluted laponite suspensions against
Dextran solutions of fixed ionic strength at pH 10. Several
remarks can be made from these results:

- the existence of a net repulsion between the laponite
  particles, even in the concentrated regime above the
  sol/gel transition;
- the salt effect agrees with the predictions of the classical
  DLVO theory: adding salt reduces the screening length
  and thus reduces the swelling pressure [17–19];
- the appearance along the different curves of a break
  followed by a pseudo-plateau separating the liquid and
  the gel phase;
- a good reversibility of the stressed suspensions which
  can be reswollen to give back the original sol;
- no observation of a macroscopic phase separation along
  the pseudo-plateau.

While the two first observations are characteristic of colloidal suspensions of *charged* particles of any geometry
[17–19], the last observations are more interesting. The
location of the first break in the osmotic pressure coincides
reasonably well with the sol/gel transition line. The "mechanical" sol/gel transition appears to be directly linked
with an equilibrium phase transition.

Microstructure and particle organization

Cryofractures and TEM observations are performed on
laponite suspensions at clay concentrations of 3%, for an
ionic strength 10^{-4} M (pH stabilized at 10). Under these
conditions, the sample is in the gel phase. No aggregation
of the clay particles may be detected. Cryofractures show
a homogeneous dispersion of the particles without any
direct contact. The maximum size of the trace of the
laponite particles is around 400 Å, in good agreement with
former determinations. SAXS is another interesting way of

Progr Colloid Polym Sci (1995) 98:39–41
© Steinkopff Verlag 1995

quantifying the organization of the clay particles Scattering experiments from a 3.8% and 12.3% w/w gel phases at relatively high ionic strength (10^{-2} M NaCl) were performed. Three regions can be observed in these diffusion curves. Above $3\,10^{-2}\,\text{Å}^{-1}$, all curves superimpose (after absolute scale normalization), giving information on the shape of the individual particle (10 Å thick and 300 Å diameter). Between $2\,10^{-3}\,\text{Å}^{-1}$ and $3\,10^{-2}\,\text{Å}^{-1}$, the normalized scattering of the gel phases is lower than the scattering intensity of an isolated particle $[Ip(Q)]$. In the very low Q region, a positive divergence from $Ip(Q)$ is observed, running approximatively as Q^{-3}. These results, obtained at high ionic strength, are very similar to the observations of Morvan et al. [16]. However, the comparison between our results and that displayed in ref. [16] shows a marked difference for the 3.8% w/w sample, due to a variation of the ionic strength. This difference supposes the occurrence at constant particle number of large-scale density fluctuation which increases with the ionic strength.

In order to understand the intermediate regime, we have carried out Monte Carlo simulations of the organization of hard cylinders modeling the laponite particles (thickness 10 Å, diameter 300 Å). Clearly, simulations show how hard core interparticle interaction affects the angular correlation between the particles within correlation volume compatible with the particle size. In this case, the scattering exhibits a trend similar to the one experimentally observed in the intermediate Q range.

Discussion and conclusion

The gelation of clay suspensions was previously discussed in terms of mechanical and structural transitions. In this work, we have shown that this transition is reversible and identifiable thermodynamically. Particle concentration and ionic strength are the two variables controlling the sol/gel transition. Our experimental data do not provide any argument in favor of a local aggregation between particles. On the contrary, direct inspection by cryofracture and SAXS show a complete individualization of the clay particles in the gel phase. The osmotic pressure is always positive and increases for increasing clay concentration and decreasing ionic strength, while the gel phase can be swollen back to a weakly viscous sol. These results agree with repulsive interparticle interactions. The two main problems concern the origin of the sol-gel transition and the influence of the ionic strength on the continuous evolution of the transition line. A more extended discussion on this point will be published elsewhere.

Acknowledgments We thank Dr. Th. Zemb and Mr. J. Lambard (C.E.A. Saclay) for the SAXS measurements, and Drs. M. Dubois, B. Cabanne, C. Bonnet-Gonnet, M. Nabavi, R. Setton, and H. Van Damme, for helpful discussions The Monte Carlo calculations were performed locally at the CRMD on a workstation (HP 9000/720) for the simulations of hard disks, and on a Cray YMP supercomputer (IDRIS, Orsay) for the diffuse layer calculations.

References

1. Pieranski P (1983) Contemp Phys 24:25–73
2. Ottewill RH (1985) Ber Bunsenges Phys Chem 89:517–525
3. Okubo T (1994) Langmuir 10:1695–1702
4. Buscall R, Goodwin JW, Hawkins MW, Ottewill RH (1982) J Chem Soc Faraday Trans 78:2889–2899
5. van Megen W, Underwood SM (1994) Phys Rev E 49:4206–4220
6. Langmuir I (1938) J Chem Phys 6:873–896
7. Onsager L (1949) Ann NY Acad Sci 51:627–659
8. Sohm R, Tadros ThF (1989) J Coll Interface Sci 132:62–71
9. Rand B, Pekenc E, Goodwin JW, Smith RW (1980) J Chem Soc Faraday I 76:225–235
10. van Olphen H (1977) In: An Introduction to Clay Colloid Chemistry. Wiley, New York
11. Norrish K (1954) Discuss Faraday Soc 18:120–134
12. Forsyth PA, Marcelja JrS, Mitchell DJ, Ninham BW (1978) Adv Colloid Interface Sci 9:37–60
13. Ramsay JDF (1986) J Coll Interface Sci 109:441–447
14. Avery RG, Ramsay JDF (1986) J Coll Interface Sci 109:448–454
15. Ramsay JDF, Lindner P (1993) J Chem Soc Faraday Trans 89:4207–4214
16. Morvan M, Espinat D, Lambard J, Zemb Th (1994) Coll Surf A: Physiochem Engin Aspects 82:193–203
17. Dubois M, Zemb Th, Belloni L, Delville A, Levitz P, Setton R (1992) J Chem Phys 96:2278–2286
18. Delville A (1994) Langmuir 10:395–402
19. Israelachvili JN (1985) In: Intermolecular and Surface Forces. Acad Press, New York

Progr Colloid Polym Sci (1995) 98:42–46
© Steinkopff Verlag 1995

K. Loyen
I. Iliopoulos
U. Olsson
R. Audebert

Association between hydrophobic polyelectrolytes and nonionic surfactants. Phase behavior and rheology

Dr. K. Loyen (✉) · I. Iliopoulos
R. Audebert
Laboratoire de Physico-Chimie
Macromoléculaire
Université Pierre et Marie Curie
CNRS URA 278
E.S.P.C.I.
10 rue Vauquelin
75231 Paris Cedex 5, France

U. Olsson
Physical Chemistry 1
Chemical Center
Lund University
P.O. Box 124
22100 Lund, Sweden

Abstract We studied the association between hydrophobically modified poly(sodium acrylate) (HMPA) and nonionic surfactants of the oligoethylene glycol monododecyl ether type (mainly $C_{12}E_8$ and $C_{12}E_4$). The HMPA contains octadecyl chains randomly anchored on 1 or 3 mol% of the monomer units. The association has a large impact on the phase behavior and the rheological properties of the aqueous HMPA/nonionic surfactant mixtures.

Small amounts of HMPA dramatically increase the stability of micellar and lamellar phases of $C_{12}E_4$. In the presence of $C_{12}E_8$ small mixed micelles are formed, including the alkyl side chains of the polymer, acting as crosslinks between the polymer chains. On the other hand, $C_{12}E_4$ forms giant vesicle aggregates (above 23 °C) on the surface of which the polymer is adsorbed forming bridges between adjacent vesicles. By utilizing the intrinsic phase behavior of water/nonionic surfactant system, thermal gelation can be induced in the presence of polymer at the temperature corresponding to the formation of bilayers in the binary surfactant/water system.

Key words Nonionic surfactant – hydrophobic polymer – thermal gelation – phase behavior

Introduction

Complex aqueous solutions containing polymers and surfactants receive great attention because of their use in numerous practical applications. Sometimes, they are also considered as simple models of complex biological systems such as the cell membranes.

The behavior of polymer/surfactant mixtures is usually the result of a subtle balance between hydrophobic and hydrophilic interactions. As a rule, ionic surfactants associate cooperatively with oppositely charged polymers [1–3] and present no association with similarly charged polymers. Nonionic surfactants usually have a low affinity for ionic polymers. In many cases however, this general behavior can be altered by the presence of hydrophobic attractions between the polymer and the surfactant. The role of hydrophobic interactions becomes evident when the polymer contains a low fraction of very hydrophobic side groups [4]. Such polymers are referred to as Hydrophobically Modified (or Associating) Water Soluble Polymers (HMWSP), and they have been developed over recent years [5–10]. We have previously studied in detail the solution behavior of hydrophobically modified polyelectrolytes based on poly(sodium acrylate) (HMPA) [9, 11]. In aqueous solution the hydrophobic side groups (dodecyl or octadecyl chains) associate and form micellar type clusters. This association is reversible and endows intriguing rheological properties to these polyelectrolytes. For instance, the viscosity of their semidilute solutions increases upon addition of salt.

In this study, we focus on the interactions between HMPA and nonionic surfactants of the oligoethylene glycol monoalkyl ether type, C_mE_n, in particular $C_{12}E_8$

Progr Colloid Polym Sci (1995) 98:42–46
© Steinkopff Verlag 1995

and $C_{12}E_4$. The solution properties of these two surfactants are quite different. $C_{12}E_8$ forms small spherical micelles (a liquid phase) over a wide range of concentrations and temperatures [12]. The second, $C_{12}E_4$ forms giant micelles [13] at low temperature (below $\approx 7\,°C$) and bilayers (lamellar, vesicle and sponge (L_3) phases) at temperatures higher than 22 °C [14, 15].

$C_{12}E_8$ is a good candidate for studing the rheological consequences of the association of HMPA with small nonionic micelles. Essentially, no structural changes of the surfactant aggregates occur by varying the temperature or the surfactant concentration. On the other hand, $C_{12}E_4$ can be used for studing the association of HMPA with giant micelles and bilayers and to relate the rheological behavior of these mixed systems to the structural transitions of the surfactant aggregates.

Experimental

Poly(acrylic acid) (PA) was purchased from Polysciences. Its average molecular weight, given by the supplier, was 150000. The hydrophobically modified samples were prepared as described elsewhere [11]. They have the same polymerization degree as the precursor PA and a random distribution of the alkyl groups along their chain [9]. All the polymers were used in the fully neutralized sodium salt form. The typical structure of the modified samples is the following:

$$-(CH_2-CH)_{100-x}\quad(CH_2-CH)_x-$$
$$\begin{array}{cc} C{=}O & C{=}O \\ | & | \\ O^-Na^+ & HN{-}(CH_2)_{n-1}{-}CH_3, \end{array}$$

where x is the modification degree in mol% and n the number of carbon atoms on the alkyl side chain (here $n = 18$). For example, 1C18 denotes a modified poly(sodium acrylate) containing 1 mol% of octadecylacrylamide units.

$C_{12}E_3$, $C_{12}E_4$, $C_{12}E_5$, $C_{12}E_8$ were purchased from Fluka (>98% purity). SDS was obtained from Kodak Lab Chemicals as >99%. All surfactants were used without further purification.

Polymer surfactant mixtures were prepared as described elsewhere [16]. Most of the viscosity measurements (Fig. 1) were performed with a Contraves LS-30 viscometer. For very viscous mixtures ($\eta \geq 10$ Pa.s), the measurements were performed with a Carrimed Controlled Stress rheometer equipped with a cone and plate geometry (cone diameter 4 cm, angle 2°) and a solvent trap; the same pertains to the results depicted in Fig. 3. The oscillatory measurements were performed with

a Rheometrics controlled-strain rheometer equipped with a cone and plate geometry (cone diameter 5 cm, angle 2°). The shear storage, G', and the shear loss G'', moduli were recorded in the linear viscoelasticity range.

Results and discussion

Figure 1 displays the variation of the viscosity of a 1 wt% polymer solution (PA, 1C18 and 3C18) with the concentration in $C_{12}E_8$. This rheological behavior is similar to that observed when HMPAs are mixed with other surfactants (cationic or anionic) forming small spherical micelles [4, 17]. The spectacular viscosity enhancement is significant for a strong tendency of the polymer hydrophobic side chains to associate with the surfactants aggregates. In fact, the rheological behavior is strongly dependent on the molar ratio R of surfactant micelles to alkyl polymer side chains. The viscosity first increases with increasing R, passes a maximum corresponding to the concentration conditions under whose micelles solubilize alkyl side chains belonging to more than one single polymer chain. These mixed micelles act as crosslinkers between the polymer chains. When R increases more, the viscosity decreases rapidly as the number of alkyl side chains per aggregate becomes lower and the connectivity of the network is reduced. This mechanism is generally accepted to hold for mixtures of HMWSP with surfactants forming small micelles [4, 6, 9, 10, 16]. The non-modified PA exhibits no maximum in viscosity as a function of $C_{12}E_8$ concentration, which is an indication of non-association between PA and $C_{12}E_8$. We note also that the temperature

Fig. 1 Variation of the viscosity of 1% aqueous solution of precursor and modified polymers with $C_{12}E_8$ concentration ($T = 25\,°C$). All the measurements correspond to Newtonian viscosity except the cases $\eta > 10$ Pa.s for which the viscosity at shear rate 0.1 s^{-1} is given

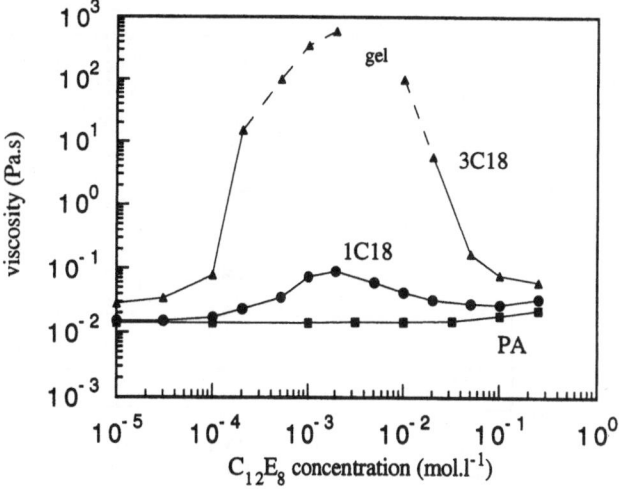

dependence of the rheological behavior of HMPA/$C_{12}E_8$ mixtures is trivial: the viscosity decreases with increasing temperature.

Since the HMPA associates with the nonionic surfactant aggregates, we expect polymer effects on the phase behavior of the surfactant. Those effects are quite clear for nonionic surfactants with a rich phase behavior like $C_{12}E_4$ [14].

The phase transitions for a 2 wt% solution of $C_{12}E_4$ as a function of temperature are shown in Fig. 2. The effect of several additives on these transitions is also depicted. In the absence of additive, the sequence of phases as a function of T is in very good agreement with that reported recently by Jonströmer and Strey [18]. At lower temperatures a liquid micellar phase (L_1) is formed whose phase separates in two liquid phases in equilibrium ($L_1' + L_1''$) above 7 °C. Above 21 °C bilayers are formed and stabilized, either as a lamellar phase (L_α, L_α'), or as a lamellar dispersion-presumably vesicles (L_α^+) [15]. At 53 °C formation of a liquid bilayer continuous phase (L_3) occurs whose phase separates into L_3 and a dilute aqueous phase ($L_3 + L_1$) above 58 °C.

Addition of 1C18 (0.6 wt%) strongly influences the phase behavior of the $C_{12}E_4$/water system. The miscibility gap ($L_1' + L_1''$) at lower temperatures disappears, the turbidity of the lamellar phase is strongly decreased, and its stability is increased at higher temperatures. The above results indicate that HMPA adsorbs on the surfactant film inducing electrostatic repulsions between the polymer coated micelles or bilayers [19]. Qualitatively similar

effects have been observed when SDS is added instead of 1C18 [19, 20]. The SDS concentration (0.015 wt%) is chosen in a way to have the [SDS]/[$C_{12}E_4$] ratio close to the [C_{18}]/[$C_{12}E_4$] ratio, where [C_{18}] is molar concentration of polymer side groups. On the other hand, addition of unmodified polymer (0.6 wt% PA) has no effect on the clouding temperature of the system. Furthermore, PA does not solubilize in the lamellar phase. Both these results are consistent with no association between PA and $C_{12}E_4$. Note that the repeat distance between $C_{12}E_4$ bilayers in the L_α phase (2 wt%) is about 1500 Å and the radius of gyration of PA in 0.14 M NaCl is about 800 Å and is expected to be much larger in water.

As the structure of $C_{12}E_4$ aggregates is strongly dependent on the temperature, one expects a noteworthy evolution of the rheological behavior of mixed systems containing $C_{12}E_4$ and 1C18 with temperature. The result is given in Fig. 3 curve b. For temperatures below the one corresponding to the formation of bilayers (T_L) the viscosity decreases with increasing temperature, as generally observed in aqueous systems containing HMWSP [21]. However, when the temperature increases above T_L the viscosity sharply increases and the system turns rapidly into a stiff gel.

These gelation phenomena can be described by the transitions of the shear storage G', and shear loss, G'', moduli upon heating. In Fig. 4, G' and G'' are plotted as a function of frequency, for the mixture corresponding to curve b of Fig. 3. At low temperature ($T = 17$ °C), G' and G'' are related to a typical viscous liquid behavior, G'' is higher than G' over most of the frequency range studied and the slopes of $\log G'$ and $\log G''$ versus log(frequency)

Fig. 2 Effects of various additives on the phase transitions of 2 wt% $C_{12}E_4$ solution in H_2O. L_1 is a liquid micellar phase. L_α and L_α' are lamellar liquid crystalline phases. L_3 is the "sponge phase" (liquid isotropic continuous bilayer). L_α^+ is an optically isotropic phase which exhibits streaming birefringence; it is referred to as a lamellar dispersion and is supposed to contain vesicles. L_1' and L_1'' are two coexisting micellar phases (one is dilute, the other concentrated)

Fig. 3 Temperature dependence of the viscosity of solutions containing 1 wt% 1C18 and 2 wt% nonionic surfactant. The different curves correspond to different surfactant mixing ratios as follows: curve a, $C_{12}E_4/C_{12}E_3 = 56.5/43.5$; curve b, pure $C_{12}E_4$; curve c, $C_{12}E_4/C_{12}E_5 = 65.8/34.2$; curve d, $C_{12}E_4/C_{12}E_5 = 25.6/74.4$. Open symbols represent the heating scan and full symbols the cooling scan (rate of heating/cooling = 0.5 °C per minute; shear rate = 10 s^{-1})

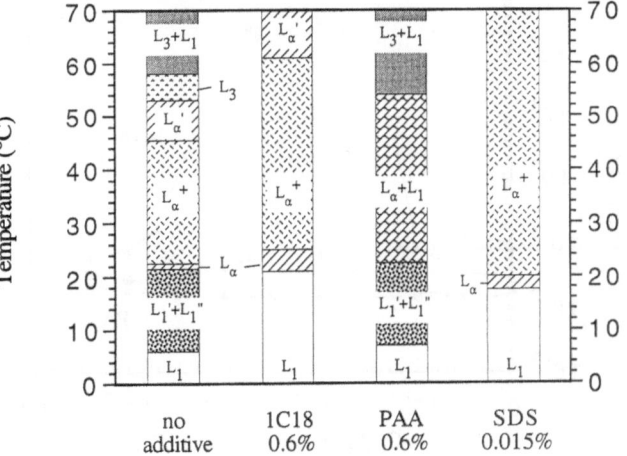

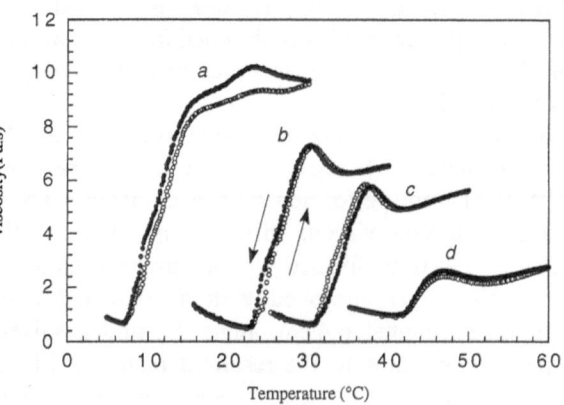

Progr Colloid Polym Sci (1995) 98:42–46
© Steinkopff Verlag 1995

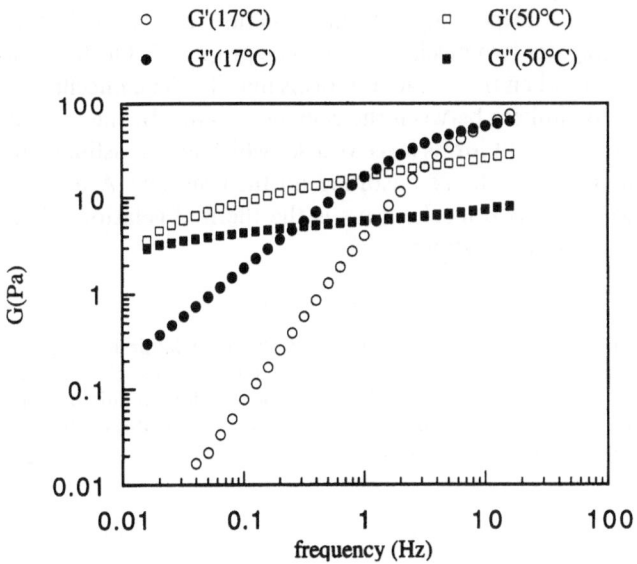

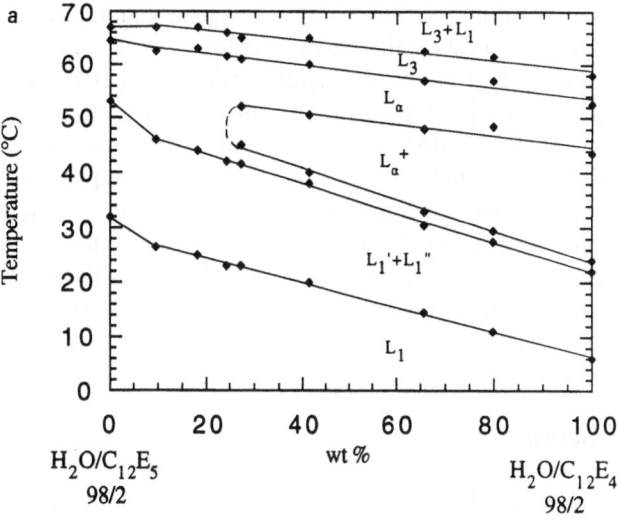

Fig. 4 Shear storage modulus (G') and shear loss modulus (G'') versus oscillation frequency at 17 and 50 °C for an aqueous solution containing 1 wt% 1C18 and 2 wt% $C_{12}E_4$

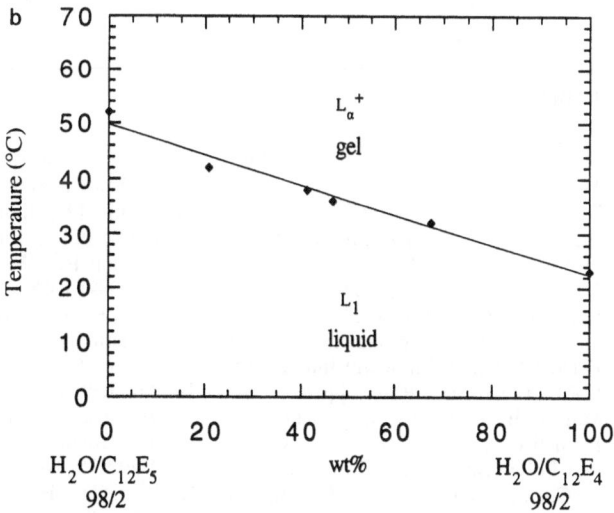

Fig. 5 a) Partial phase diagram of $C_{12}E_4/C_{12}E_5$/water systems at fixed total surfactant concentration = 2 wt%. b) Effect of addition of hydrophobically modified poly(sodium acrylate) 1C18 on phase diagram reported in (a). Polymer concentration is fixed at 1 wt%

approach 2 and 1, respectively [22]. At higher temperature ($T = 50$ °C), G' and G'' plots indicate an elastic behavior, G' is significantly higher than G'' over the entire measured frequency range and increases slightly with frequency.

It is noteworthy that the temperature of the thermal gelation coincides with the bilayer formation temperature of the $C_{12}E_4$/water binary system. We anticipate that this rheological behavior occurs also with other nonionic surfactant systems when they form a dilute lamellar phase. The different homologues of C_mE_n surfactants mix essentially ideally and, as a consequence, the various phase transitions can be shifted to higher or lower temperatures by using mixtures of high and low T_L surfactants and varying the mixing ratio as shown in Fig. 5a. Correspondingly, the gelation temperature of the C_mE_n/HMPA/water systems can be adjusted to the desired value by using the adequate surfactant mixture as shown in Figs. 5b and 3. We then conclude that the thermal gelation in these polymer/surfactant systems is clearly related to the formation of surfactant bilayers in the binary C_mE_n/water system. This process is reversible; essentially the same viscosity curves were found in the increasing and decreasing temperature scans (Fig. 3).

Although the details of the thermal gelation process and of the structure of the mixed aggregates above T_L are not fully understood, we can make some speculations. According to recent water self-diffusion results, vesicles are formed in the dilute aqueous $C_{12}E_4$ solutions above T_L. These vesicles are quite large. Their volume fraction is of about 30% at 0.05 mol l^{-1}, as estimated from the water

self-diffusion results [23]. Formation of vesicles is also reported for other C_mE_n surfactants [15, 24, 25]. Consequently, the schematic picture of the system above T_L in the presence of HMPA can be that of large vesicles cross-linked by the polymer chains adsorbed on the vesicle surface.

Below T_L, the C_mE_n/HMPA/water system contains micellar aggregates. Although they are certainly not spherical, they are still relatively small compared to the vesicles and they can act as cross-linkers between the polymer chains as described for the $C_{12}E_8$ system. However, because of the relatively high surfactant to polymer ratio (2 wt%/1 wt%), the number of alkyl polymer side-chains per micelle is low and the polymer chains are not crosslinked enough (the viscosity is low).

Thermal gelation in polymer/surfactant systems has been reported previously for mixtures of ethyl hydroxyethyl cellulose (EHEC) with ionic surfactant [26–30]. In that system, however, the temperature-dependent properties are related to the EHEC/water interactions (the EHEC/water system phase separates upon heating) while in our system the temperature-dependent properties are related to the nonionic surfactant/water interactions.

Conclusion

Hydrophobically modified polyelectrolytes can form viscous solutions or gels upon addition of nonionic surfactants. Depending on the structure of the surfactant aggregates two mechanisms can be proposed. The first one occurs when the surfactant forms micelles. The micelles act a crosslinkers between the polymer chains. In the second, the surfactant forms large vesicles which are crosslinked by the polymer chains adsorbed on their surface. A link between the two mechanisms is the thermal gelation of the $C_{12}E_4$/HMPA system.

Acknowledgment The Laboratoire de Physico-Chimie Macromoléculaire thanks LVMH-Recherche and CNRS for financial support. We thank A. Meybeck and J.F. Tranchant for stimulating discussions. U.O. acknowledges the Swedish Natural Science Council (NFR) for financial support.

References

1. Goddard ED (1986) Colloids Surf 19:301
2. Hayakawa K, Kwak JCT (1991) In: Rubingh D, Holland PM (eds) Cationic Surfactants Physical Chemistry. Marcel Dekker, New York, pp 189
3. Lindman B, Thalberg K (1992) In: Goddard ED, Ananthapadmanabham K (eds) Polymer-Surfactant Interaction. C.R.C. Press, Boca Raton, pp 203
4. Magny B, Iliopoulos I, Audebert R, Piculell L, Lindman B (1992) Prog Colloid Polym Sci 89:118–121
5. McCormick CL, Nonaka T, Johnson CB (1988) Polymer 29:731–739
6. Binana-Limbele W, Clouet F, François J (1993) Colloid Polym Sci: 748–758
7. Landoll LM (1982) J Polym Sci Polym Chem 20:443–455
8. Valint PL, Bock J (1988) Macromolecules 21:175–179
9. Wang TK, Iliopoulos I, Audebert R (1991) In: Shalaby SW, McCormick CL, Butler GB (eds) Water Soluble Polymers. Synthesis, Solution properties and Applications; ACS Symposium Series 467, American Chemical Society: Washington, DC, pp 218–231
10. Biggs S, Hill A, Selb J, Candau F (1992) J Phys Chem 96:1505–1511
11. Wang TK, Iliopoulos I, Audebert R (1988) Polym Bull 20:577–582
12. Nilsson PG, Wennerström H, Lindman B (1989) J Phys Chem 87:1377–1385
13. Henriksson U, Jonströmer M, Olsson U, Söderman O, Klose G (1991) J Phys Chem 95:3815–3819
14. Mitchell DJ, Tiddy GJT, Waring L, Bostock T, McDonald MP (1981) J Chem Soc, Faraday Trans 1 79:975–1000
15. Kunieda H, Nakamura K, Davis HT, Evans DF (1991) Langmuir 7:1915–1919
16. Sarrazin-Cartalas A, Iliopoulos I, Audebert R, Olsson U (1994) Langmuir 10:1421–1426
17. Iliopoulos I, Wang TK, Audebert R (1991) Langmuir 7:617–619
18. Jonströmer M, Strey R (1992) J Phys Chem 96:5993–6000
19. Iliopoulos I, Olsson U (1994) J Phys Chem 98:1500–1505
20. Douglas CB, Kaler EW (1994) J Chem Soc Faraday Trans 90(3):471–477
21. Biggs S, Selb J, Candau F (1993) Polymer 34(3):580–591
22. Ferry JD (1980) In: Viscoelastic Properties of Polymer. 3rd ed. John Wiley, New York
23. Olsson U, Strey R, unpublished results
24. Schomäcker R, Strey R (1994) J Phys Chem 98:3908–3912
25. Hofland HEJ, Bouwstra JA, Gooris GS, Spies F, Talsma H, Junginger HE (1993) J Colloid Interface Sci 161:366–376
26. Carlsson A, Karlström G, Lindman B (1990) Colloids Surf 47:147–167
27. Carlsson A, Karlström G, Lindman B, Stenberg O (1988) Colloid Polym Sci 266:1031–1036
28. Karlström G, Carlsson A, Lindman B (1990) J Phys Chem 94:5005–5015
29. Carlsson A, Lindman B, Watanabe T, Shirahama K (1989) Langmuir 5:1250
30. Carlsson A, Karlström G, Lindman B (1989) J Phys Chem 93:3673–3677

Progr Colloid Polym Sci (1995) 98:47–50
© Steinkopff Verlag 1995

F. Guillemet
L. Piculell
S. Nilsson
M. Djabourov
B. Lindman

Interactions between hydrophobically modified polyelectrolyte and oppositely charged surfactant. Mixed micelle formation

Dr. F. Guillemet (✉) · M. Djabourov
Laboratoire de Physique Thermique
CNRS URA 857
E.S.P.C.I.
10 rue Vauquelin
75231 Paris Cedex 5, France

F. Guillemet · L. Piculell · S. Nilsson
B. Lindman
Division of Physical Chemistry 1
Chemical Centre
University of Lund
P.O. Box 124
22100 Lund, Sweden

Abstract Aqueous mixtures of sodium dodecyl sulphate (SDS) with QUATRISOFT™ LM200, a cellulose derivative substituted with cationic hydrophobic side chains, have been investigated in the absence and in the presence of added salt. On adding SDS to a solution of polymer in the range 0.02–1%, liquid-liquid phase separation occurs near charge neutralization (for the same amount of polymer and surfactant charges in solution) for the salt-free mixture, and earlier in the presence of salt. In both cases, redissolution occurs upon further SDS addition. The total SDS concentration at redissolution increases linearly with polymer concentration, from a limiting value close to the CMC of the polymer-free solution, at vanishing polymer content. In a 1% solution, a very high viscosity is found on both sides of the two-phase area, showing the formation of mixed micelles between the polymer alkyl side chains and the surfactant molecules. The results are interpreted in terms of a binding isotherm of surfactant to polymer. The first stages of the isotherm involve binding of individual surfactants molecules to the mixed micelles, and the last stage, occurring when the free surfactant concentration approaches CMC, is a strong and cooperative binding related to the self-association of the surfactant. High SDS/polymer binding ratios seem required for redissolution. Such high binding ratios are only obtained close to the cooperative binding region, i.e., when the free surfactant concentration is close to the CMC.

Key words Mixed micelle – cationic hydrophobically modified polymer – anionic surfactant – associative phase separation

Introduction

Water Soluble Polymers containing a small number of hydrophobic groups are often referred to as Hydrophobically Modified Water Soluble Polymers (HMWSP). Since the contact between the hydrophobic side chains and water is unfavorable, these polymers have a strong tendency to associate together and with surfactants [1–4]. This association process can lead to a significant thickening of the solution, so that HMWSP can be used in various industrial applications where the control of the rheology of the solution is required: paints, pharmaceuticals, enhanced oil recovery, etc.

The phase behavior of HMWSP-surfactant mixture will depend on both the relative charge and hydrophobicity of the pair. In the case of oppositely charged

HMWSP and surfactant aqueous mixtures, the association is promoted by both hydrophobic and electrostatic interactions. Thus, gel formation and phase separation phenomena have been observed in such systems [1–6].

In the present study, we consider the interaction of a cationic hydrophobically modified cellulose polymer with an anionic surfactant, sodium dodecyl sulphate (SDS). We discuss how the formation of mixed micelles between the polymer alkyl side chains and the surfactant can be related to the viscosity and the solubility of these solutions.

Experimental

The polymer, QUATRISOFT™ LM200, supplied from Union Carbide, is the chloride salt of a N, N-dimethyl-N-dodecyl derivative of hydroxyethyl cellulose. The structure scheme of the polymer is shown in Fig. 1. The average molecular weight is 100 000 g/mol and the degree of side-chain substitution, τ, is 5.4 per 100 sugar residues [7]. SDS was obtained from BDH. Sodium chloride (NaCl) was obtained from Fluka.

The preparation of polymer-surfactant mixtures and the determination of the phase diagram are described elsewhere [7]. The Newtonian viscosity at 25 °C of the 1% polymer solution was measured with a controlled stress rheometer equipped with a cone-and-plate geometry. SDS concentrations were measured by sulphur analysis, using filament-pulse-pyrolysis-gas-chromatography with a sulphur photometric detector [8].

Results and discussion

Figure 2 displays the viscosity of a 1% polymer solution as a function of surfactant or salt concentration. The polymer

Fig. 1 Chemical structure of Quatrisoft LM200. $\tau = 5.4$ mol.%

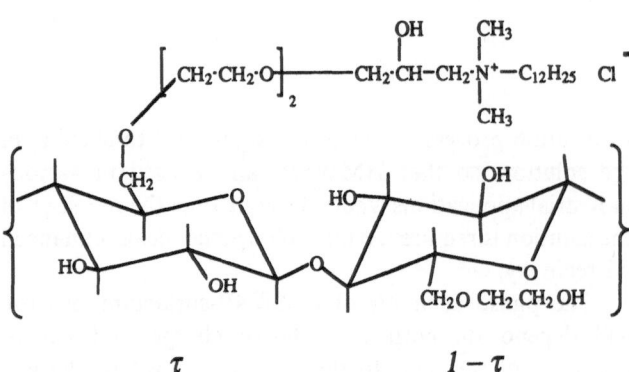

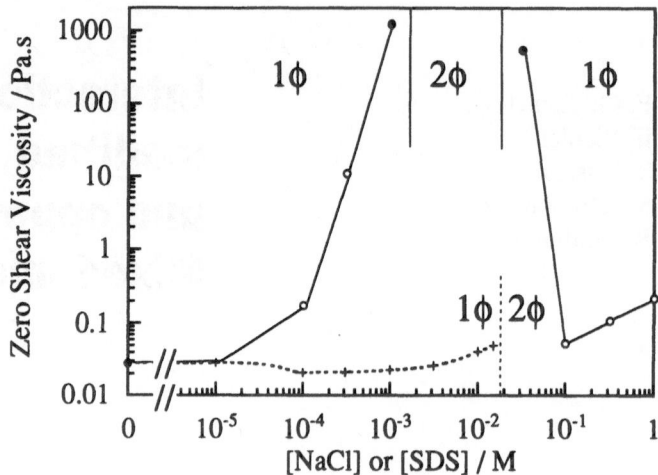

Fig. 2 Zero shear viscosity of a 1% LM200 solution as a function of added SDS (open circles) or NaCl (crosses) at 25 °C. The filled circles correspond to the samples where the Newtonian plateau could not be reached, so that the viscosity at 0.01 s^{-1} has been reported

solution exhibits a rather constant viscosity as NaCl is added. The slight increase of viscosity at 10 mM NaCl is related to the enhanced ionic strength of the solution: the repulsion of the polymer charges is screened so that the association of the alkyl chains is promoted [9]. At 15 mM NaCl, the system phase separates. This phase separation with salt addition evidences that the polymer derives its solubility from the presence of charges.

A very different behavior is found with SDS addition. The viscosity first increases dramatically and then the solution phase separates. At high SDS concentration, the solution is again monophasic and the viscosity drops down to values close to that found in the absence of surfactant. This peak in viscosity is generally attributed to the formation of micellar type mixed aggregates between the polymer alkyl side chains and the surfactant molecules [2, 3, 5]. This type of structure occurs over a wide range of concentrations and the stoichiometry of these mixed aggregates should vary greatly. At 0.1 mM SDS, there is one surfactant molecule for 20 polymer alkyl side chains, while at 10 mM SDS there are 50 surfactant molecules for 1 polymer alkyl side chain.

The formation of mixed micelles between the polymer and the surfactant is promoted by both hydrophobic and electrostatic interactions. Pyrene fluorescence measurement on this polymer has shown the existence of very hydrophobic microdomains in solution already at low polymer concentration (0.1%) [7]. At 1% polymer, the hydrophobic probe senses, in its solubilization site, a medium of a polarity close to the one of the hydrophobic core of a cationic surfactant micelle. This shows that, even in

Progr Colloid Polym Sci (1995) 98:47–50
© Steinkopff Verlag 1995

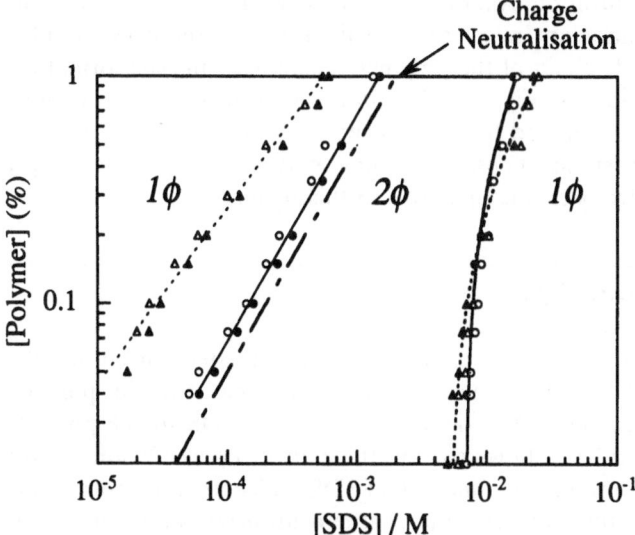

Fig. 3 Phase diagram of the LM200-SDS-Water system without added salt (circles and solid lines) or in the presence of 10 mM NaCl (triangles and dotted lines). The charge neutralization line corresponds to an equal amount of polymer and surfactant charges. The lines on the redissolution side are computed with Eq. (1)

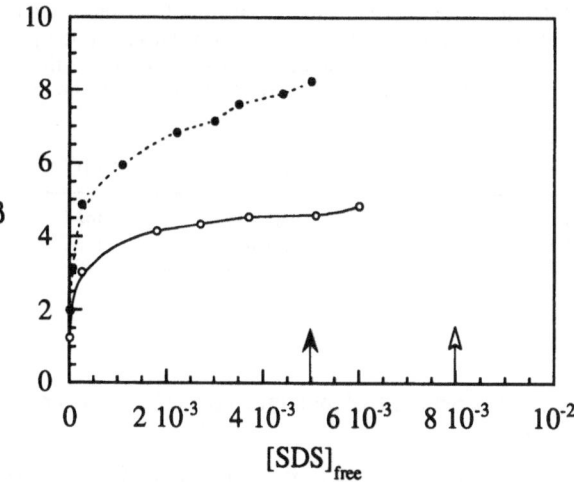

Fig. 4 Binding isotherm of SDS to LM200. Open and filled symbols correspond to samples without added salt and in the presence of 10 mM NaCl, respectively. The arrows correspond to the CMC of the surfactant

the absence of surfactant, the polymer has "micellized," forming in this way strongly hydrophobic microdomains.

Figure 3 illustrates the phase diagram of the polymer-surfactant system, without added salt and in the presence of 10 mM NaCl. In a wide range of compositions, an associative phase separation occurs [10], with two phases in equilibrium: the top phase is water-like whereas the bottom phase is translucent, gel-like, and rich in both polymer and surfactant. The two-phase region is delineated by two boundaries, one at low surfactant concentrations, referred to as the coacervation line, and the other at high surfactant concentrations, referred as to the redissolution line. With salt addition, the coacervation line is shifted to lower surfactant concentrations, while redissolution requires more surfactant at high polymer concentration, but less surfactant at low polymer concentration. At the charge neutralization, for systems containing the same amount of polymer and surfactant charges, the SDS concentration determination shows that there is less than 10^{-6} M SDS in the top phase (for 1% polymer and 2 mM SDS). Also, up to charge neutralization, all the added surfactant binds to the polymer, in the hydrophobic microdomains. Therefore, the parallelism between the coacervation and charge neutralization lines indicates that phase separation occurs at a fixed charged density (0.27) which is higher when salt is added (0.7 with 10 mM NaCl).

To know if redissolution occurs also at a fixed charge density of the polymer-surfactant complex, it is necessary to know the fraction of bound SDS to the polymer in the

two-phase region. We have built the binding isotherm of SDS to the polymer as follows: The binding ratio, β, is defined as the ratio between the concentration of bound surfactant and the concentration of polymer alkyl side chains. The free surfactant concentration is approximated to the SDS concentration in the top phase, assuming the concentration of polymer in the top phase to be negligible, and thereby the amount of bound surfactant in the top phase. Figure 4 displays the binding isotherm of SDS to a 1% polymer solution, in the two-phase region, without added salt and in the presence of 10 mM NaCl. At charge neutralization, the binding is complete, but with further surfactant addition, the binding becomes less favorable. Indeed, the insoluble polymer-surfactant complex becomes negatively charged, so that the association of the anionic surfactant is no longer promoted by electrostatic interaction. In the presence of 10 mM NaCl due to the screening of the electrostatics, the binding is less unfavorable. In all this anti-cooperative binding region, the free surfactant concentration builds up in solution up to concentrations close to the CMC of the surfactant. Similar binding isotherms have been observed in mixtures of oppositely charged protein or polyelectrolyte and surfactant [11, 12].

The binding isotherm depends on salt concentration in solution and, therefore, we can assume that, in a first approximation, it does not depends on polymer concentration [7] (the counterions of the polyelectrolyte contribute to the ionic strength of the solution, but their concentration is much lower than the ones of the others ions on solution). Thus, redissolution occurs at fixed binding ratio, β_{redis}, and free surfactant concentration,

50

F. Guillemet et al.
Oppositely charged hydrophobic polyelectrolyte and surfactant

Table 1 Redissolution coordinates as extracted from the fit of the redissolution lines with Eq. (1), without added salt and in the presence of 10 mM NaCl. The free surfactant concentration at redissolution is compared to the CMC of the surfactant found in the literature [13].

Salt concentration	β_{redis}	$[surf]_{free,redis}$	CMC
0 mM NaCl	5.0	6.8 mM	8 mM
10 mM NaCl	9.5	5.1 mM	5 mM

$[surf]_{free}$, independent of polymer concentration, [polymer]. Hence,

$$[Surf]_{redis} = [surf]_{free,redis} + \beta_{redis} * [polymer] . \qquad (1)$$

The redissolution lines have been fitted with Eq. (1) in Fig. 3. The values of the fit are reported in Table 1. Equation (1) gives a natural explanation for the crossing of the redissolution lines in the presence and absence of added NaCl. At enhanced ionic strength, the charge density required for the redissolution is larger, while the free surfactant concentration is lower. The free surfactant concentration at redissolution is very close to the CMC of the surfactant, suggesting that redissolution occurs close to the micellization of the surfactant (the free surfactant concentration cannot exceed the CMC). Therefore, when the surfactant micellizes, it will solubilize the cationic alkyl side chain of the polymer. This mixed micelle formation, shown in Fig. 2 (high viscosity after the two-phase region), is promoted by both the entropy of mixing and the electrostatics. Thus, the binding of SDS to the polymer is very strong and cooperative in this region.

Conclusions

In the range of compositions investigated, mixed micelles of SDS molecules and alkyl side chains of the polymer dominate. The binding isotherm of SDS to the polymer has been shown to have three distinct parts: Before charge neutralization, individual SDS molecules bind completely to the polymer. After charge neutralization, the binding is weak and anti-cooperative. The concentration of free SDS builds up in solution up to the micellization of the surfactant aided by the polymer. The binding is then very strong and cooperative. Finally, in having considered the existence of mixed micelles and the different stages of the binding process, it has been possible to understand that the solubility of the polymer-surfactant complexes is essentially controlled by the charged density of the mixed micelles, at a given ionic strength.

References

1. Magny B, Iliopoulos I, Audebert R, Piculell L, Lindman B (1992) Prog Colloid Polym Sci 89:188–121
2. Biggs S, Selb J, Candau F (1992) Langmuir 8:838–847
3. Tanaka R, Meadows J, Williams PA, Phillips GO (1992) Macromolecules 25:1304–1310
4. Goddard ED, Leung PS (1992) Colloids Surf. 65:211–219
5. Dualeh AJ, Steiner CA (1990) Macromolecules 23:251–255
6. Goddard ED (1992) J Colloid Interface Sci. 152:578–581
7. Guillemet F, Piculell L, Submitted to J Phys Chem
8. Almen P, Ericsson I, Submitted to Langmuir
9. Magny B, Thesis, Paris (1992)
10. Piculell L, Lindman B (1992) J Colloid Interface Sci. 41:149–178
11. Jones MN, Manley P (1980) J.C.S. Faraday I 76:654–664
12. Ohbu K, Hiraishi O, Kashiwa I (1982) J Am Oil Chem Soc 59:108–112
13. van Os NM, Haack JR, Ruppert LAM (1993) Physico-Chemical Properties of Selected Anionic, Cationic and Non-ionic Surfactants. Elsevier, Amsterdam

Progr Colloid Polym Sci (1995) 98: 51–56
© Steinkopff Verlag 1995

Surfactant-polymer interaction in thermo-induced gelling systems of ethyl (hydroxyethyl) cellulose (EHEC) studied by NMR self-diffusion, dynamic light scattering and rheology

H. Walderhaug
B. Nyström
F.K. Hansen
B. Lindman

Dr. H. Walderhaug (✉) · B. Nyström
F.K. Hansen
Department of Chemistry
University of Oslo
P. O. Box 1033 Blindern
0315 Oslo, Norway

B. Lindman
Physical Chemistry 1
Chemical Center
University of Lund
P. O. Box 124
221 00 Lund, Sweden

Abstract Two EHEC (i.e. ethyl (hydroxyethyl) cellulose)/surfactant/water systems have been investigated by NMR self-diffusion, dynamic light scattering, and rheology. An anionic surfactant, sodium dodecyl sulfate (SDS), and a cationic surfactant, hexadecyl trimethylammonium bromide (CTAB), have been used. The surfactant concentration has been kept at 4.0 mmolal, and the EHEC concentration at two different levels; at 0.5 wt% which is a dilute system that does not gellify on heating, and at 1.0 wt% semidilute concentration. Both semidilute systems go through a reversible sol-gel transition at 37 °C. Dynamic light scattering and rheological stress relaxation measurements show that the strength of the chain association is higher for the EHEC/SDS system than for the corresponding EHEC/CTAB system. From rheological oscillation measurements on the incipient gels it is inferred that the polymer network structure is rather inhomogeneous, and that the gel strength is higher for

the former as compared to that for the latter system. The amount of surfactant adsorbed to the polymers is determined by NMR self-diffusion measurements. The amount of surfactant bound to the polymer is practically independent of temperature for both the dilute and semidilute systems over a wide temperature range. However, the level of surfactant binding is higher for the EHEC/CTAB system than that for the corresponding EHEC/SDS system. In the analysis of the interaction situation, the interplay between surfactant induced associations and enhanced polymer-polymer interactions must be considered. Strong polymer-polymer interactions most probably evolve at higher temperatures due to gradually deteriorated thermodynamic conditions.

Key words Thermoreversible gelling system – EHEC – surfactant – dynamic light scattering – NMR self-diffusion – stress relaxation – oscillation

Introduction

The system ethyl (hydroxyethyl) cellulose (EHEC)/ionic surfactant/water belongs to a group of systems that in the semidilute concentration regime forms thermo-reversible, transparent, and stiff gels on heating to a certain temperature [1–5]. The sol-gel cycle can be repeated a number of times without destruction of the system, or gel capacity. Aqueous solutions of EHEC have found wide-

52

H. Walderhaug et al.
Surfactant/polymer interaction in thermo-induced gelling systems

spread use in industrial formulations such as latex paints and wallpaper paste where it is used as a rheology modifier or so-called thickener.

From a fundamental point of view, systems of this kind are complex and have attracted a lot of interest during recent years. Various approaches have been tried to unravel the molecular mechanisms behind the phase behavior of these polymer/surfactant systems (see refs. [1–5] and literature cited therein). These studies include thermodynamic phase behavior by means of cloud-point determinations and related phase separation investigations, ion activities as measured by ion selective electrodes, rheology, molecular diffusivity as measured by radioactive tracer techniques [6], NMR self-diffusion, and fluorescence methods. In the present study, we have combined results from NMR self-diffusion, dynamic light scattering and rheology (stress-relaxation and oscillation) on the *same* systems. For this purpose, we have chosen a well-characterized sample of EHEC (see below), dissolved it in water at two concentrations, i.e. 0.5 wt% and at 1.0 wt% in the presence of either an anionic surfactant, i.e. sodium dodecyl sulfate (SDS), or a cationic surfactant, i.e. hexadecyl trimethylammonium bromide (CTAB). The 0.5 wt% EHEC solutions are considered to be dilute, i.e. below the overlap concentration, c^*, whereas at 1.0 wt% EHEC, we are in the semidilute concentration range, i.e. above c^* for this EHEC sample. In the presence of a surfactant the dilute solutions do not gellify on heating, whereas the corresponding semidilute solutions form a gel at a temperature of 37 °C [7]. The surfactant concentration has been kept constant at 4 mmolal in all systems.

The surfactant binds to the polymer, giving the formed complex polyelectrolyte properties. On the basis of the results from the above-mentioned investigations, it has been argued that the surfactant binds to the polymer as micelles. In this process, the hydrophobic junctions necessary for the evolution of the gel network are generated. This view stresses the importance of the surfactant in the gel-forming process and an increased amount of surfactant is expected [3–6] to bind to the polymer upon raising the temperature. Our systems have been investigated over a wide temperature range from 22° to 40 °C. For the semidilute samples, we start well below the gelation temperature and follow them well inside the gel zone. The corresponding dilute systems are investigated for comparison. With the NMR self-diffusion technique the amount of surfactant bound to the polymer has been determined. With dynamic light scattering the polymer chain dynamics is determined and compared with the findings from the rheological measurements.

Experimental

The NMR Fourier Transform Pulsed Field Gradient Spin Echo (FT-PGSE) method for determining self-diffusion is described in detail elsewhere [8]. The experiments were conducted on a Bruker CXP 200 spectrometer equipped with a proton diffusion probe head from Cryomagnetics Inc., and a home-built gradient driver unit. The dynamic light scattering apparatus consists of an argon ion laser (Spectra Physics Model 2020) operating at 488 nm with vertically polarized light, a thermostated sample cell and an ALV-5000 multiple tau digital correlator. The stress relaxation and oscillation measurements were performed on a Bohlin VOR Rheometer System with a cone-and-plate geometry with a cone angle of 5° and a diameter of 30 mm.

The EHEC sample was manufactured by Berol Nobel AB, Stenungsund, Sweden. The degree of substitution was $DS_{ethyl} = 1.9$ per anhydroglucose unit, and the molar substitution of ethylene oxide groups was $MS_{EO} = 1.3$ per anhydroglucose unit. The number average molecular weight M_n for this polydisperse sample ($M_w/M_n \simeq 2$) is ca. 100 000. The surfactants CTAB and SDS were obtained from Fluka and were used as received. The dilute EHEC solution was dialyzed against pure water to remove salt and was then freeze-dried. After freeze-drying the polymer was re-dissolved in water. Samples were prepared by weighing the components and the solutions were then stirred at room temperature until the solutions became homogeneous. Distilled, de-ionized water was generally used as solvent, except in the NMR experiments, where heavy water (Fluka, 99.8 atom % D) was used to prevent a disturbing solvent signal.

Results and discussion

NMR self-diffusion

The echo intensity of an NMR signal, I, in a pulsed field gradient experiment is given by the following relation [9]:

$$I(\delta)/I(0) = \exp(-2\Delta/T_2)\exp(-(\gamma g \delta)^2(\Delta - \delta/3)D), \quad (1)$$

where $I(0)$ is the intensity of the signal immediately after a conventional 90° radio-frequency (rf) pulse, T_2 is the spin-spin relaxation time, Δ is the time interval between the onsets of two pulsed magnetic field gradients, corresponding closely to the physical observation time of the diffusion process, with amplitude g and duration δ. The parameter γ is the magnetogyric ratio of protons and D is the self-diffusion coefficient. Fourier transformation of the second half of the spin echo formed after the 90°–180° rf sequence (time interval Δ) yields the possibility to determine individual self-diffusion coefficients of several components in one experiment.

Surfactant self-diffusion

For a surfactant molecule that resides either in solution or is adsorbed to a polymer molecule the observed self-diffusion is interpreted as a time average of two processes, i.e., free diffusion of monomeric surfactant molecules in the aqueous solution and, when adsorbed, diffusion of the polymer molecules. This view constitutes a two-sites model, and in its most crude form the observed diffusion coefficient, D_{obs}, is given by:

$$D_{obs} = (1 - p)D_f + pD_b , \qquad (2)$$

where p is the fraction of the total amount of surfactant adsorbed to the polymer, D_f is the diffusion coefficient of the free surfactant monomer, and D_b is the diffusion coefficient of the polymer/surfactant complex. For the present systems $D_b \ll D_f$ (see below), and p is given by:

$$p = 1 - (D_{obs}/D_f) . \qquad (3)$$

In this study the degree of surfactant binding, p, has been determined from the measured diffusion data with the aid of Eq. (3).

Figure 1 displays the degree of surfactant binding of the systems EHEC/SDS ($c_{EHEC} = 0.5$ and 1.0 wt%; $c_{SDS} = 4.0$ mmolal) and EHEC/CTAB ($c_{EHEC} = 0.5$ and 1.0 wt%; $c_{CTAB} = 4.0$ mmolal) at several temperatures up to 40 °C. In addition, the corresponding data of a more polar EHEC sample ($M_n = 150\,000$, cloud point = 68 °C, $DS_{ethyl} = 1.1$ and $MS_{EO} = 2.1$), at an EHEC concentration of 1.0 wt%, is included. For the hydrophobic EHEC it is found that the value of p is weakly or insignificantly temperature dependent in the dilute and semidilute regimes for both systems. No change in the level of surfactant binding to the polymer is observed as the sol-gel transition (at 37 °C for both systems) is passed. We generally find a higher degree of surfactant binding for the EHEC/CTAB systems, where more than 95% of the surfactant is adsorbed to the polymer than for the corresponding EHEC/SDS systems. In spite of this, the EHEC/SDS system forms a stiffer gel and also at lower temperatures exhibits stronger chain interactions than the corresponding EHEC/CTAB system (see below). It should be noted from the results in Fig. 1 that the rather hydrophilic EHEC in the presence of SDS (this system does not form a gel in the considered temperature range) shows a different pattern of behavior as regards the surfactant binding

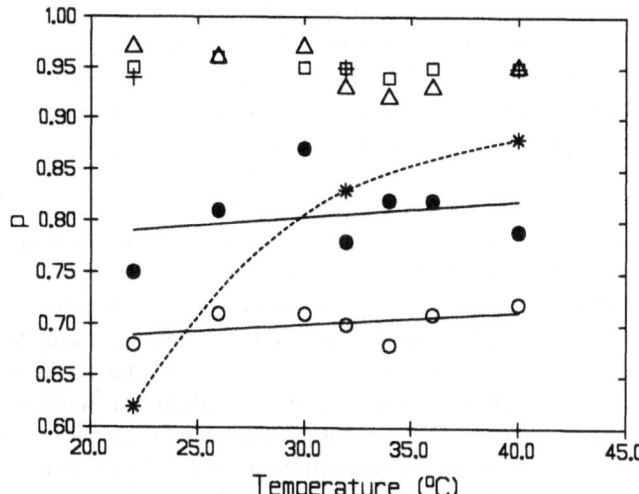

Fig. 1 The degree of surfactant binding (see Eq. (3)) to the polymer as a function of temperature at a fixed surfactant concentration of 4.0 mmolal for EHEC/SDS/D$_2$O at EHEC concentrations of 0.5 wt% (○) and 1.0 wt% (●), for the more polar EHEC/CTAB/D$_2$O sample at a concentration of EHEC at 1.0 wt% (*), for EHEC/CTAB/D$_2$O at EHEC concentrations of 0.5 wt% (□) and 1.0 wt% (△), and for the more polar EHEC/CTAB/D$_2$O sample at a concentration of EHEC at 1.0 wt% (+). The dashed curve and the solid lines are drawn to show the trends

versus temperature. In this case, there is a pronounced increase in surfactant binding with temperature.

The conclusion that can be drawn from these findings is that *there is no direct correlation between the amount of surfactant adsorption to the EHEC polymers and the systems' ability to form gels*. The gelation process seems to be governed by an intricate interplay between polymer-surfactant and polymer-polymer interactions. In this context, it is important to take into account the change of the *thermodynamic* properties of the system with temperature. It has been shown from cloud point (cp) determinations [10] that the cp of the EHEC/surfactant system increases as an ionic surfactant is introduced (the cp of the pure EHEC in water is ca. 34 °C for our particular sample). The change of the cloud point curve toward higher temperature is more pronounced in the presence of CTAB than for the corresponding EHEC/SDS system. When pure EHEC/water solutions reach the cp, a progressively more cloudy solution is obtained as the temperature is further increased. At this stage, strong attractive chain-chain interactions evolve and eventually phase separation occurs. The adsorbed surfactant modifies the temperature-induced interaction situation. In fact, when heating the EHEC (1%)/SDS (4 mmolal) gel to about 40 °C a turbidity gradually appears, whereas no incipient phase separation occurs for the corresponding EHEC(1%)/CTAB

(4 mmolal) gel, at least for temperatures below 50 °C. These observations suggest that the phase separation phenomenon may have an important influence on the gelation process. The delicate interplay between gelation and macroscopic phase separation has recently been discussed [11] for thermoreversible gelation systems. It should be noted that Zana et al. in their work on the EHEC/CTAB [5] and EHEC/SDS [6] systems, using fluorescence quenching, studied the size of the bound surfactant micelles by determining surfactant aggregation numbers as a function of temperature. They argued that the aggregation number became smaller on heating. It is possible that a reorganization of the surfactant on the polymer with increasing temperature occurs. We reserve a more detailed discussion on NMR self-diffusion results from these systems to a separate work [12]. Below, the EHEC/surfactant interactions will be discussed further in connection with the analysis of dynamic and rheological properties of the systems.

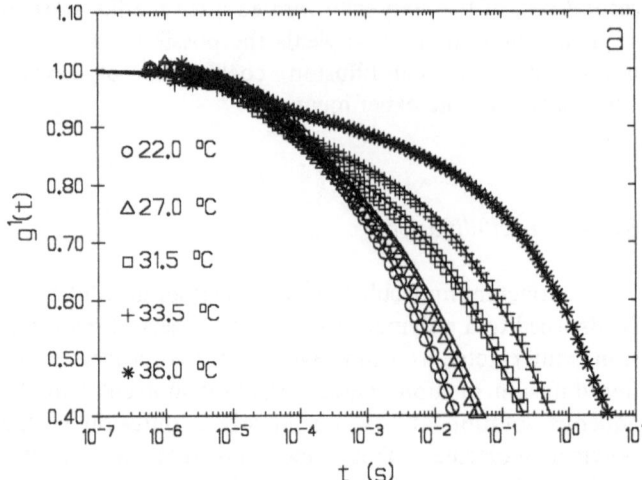

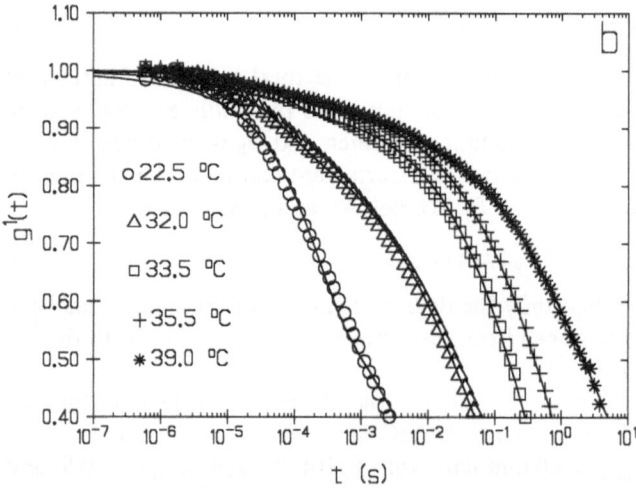

Fig. 2 Plot of the first-order electric field correlation function versus time (every second data point is shown) for the systems EHEC/SDS (a) and EHEC/CTAB (b). The EHEC concentration is 1.0 wt% and the surfactant concentration is 4.0 mmolal. The curves have been fitted by using Eq. (4). The scattering angle is 90°

Dynamic light scattering

In Fig. 2 is shown the first-order electric field correlation function, $g^1(t)$, as a function of time and temperature for the two gelling systems EHEC/SDS (Fig. 2a) and EHEC/CTAB (Fig. 2b). The decay of $g^1(t)$ displays two processes occurring on different time scales. The relaxation function can be described by the following relationship:

$$g^1(t) = (1 - A)\exp(-t/\tau_f) + A\exp(-(t/\tau_{se})^\beta), \qquad (4)$$

where A is the amplitude and τ_f is the relaxation time for the fast process. The slow mode is described as a distribution of relaxation processes in form of a stretched exponential, where τ_{se} denotes an effective relaxation time and β ($0 < \beta \leq 1$) is a measure of the width of the distribution of relaxation times. A mean relaxation time, τ_s, can be calculated by:

$$\tau_s = (\tau_{se}/\beta)\Gamma(1/\beta), \qquad (5)$$

where Γ denotes the gamma function. Typical values for β are close to 0.45 for both systems, except at the highest temperatures where a slight decrease to ca. 0.35 is observed.

By comparing the results presented in Figs. 2 a and b, it can be seen that the slow relaxation time is shifted more toward longer times for the EHEC/SDS system than for the corresponding EHEC/CTAB system at all the investigated temperatures.

In Fig. 3 is shown the extracted fast and mean slow relaxation times for the two systems as a function of temperature. It can be seen that the fast relaxation time shows a decreasing trend with increasing temperature and

is practically the same for the two systems at a given temperature. The fast mode has been observed [13] to be diffusive under all conditions, yielding a cooperative diffusion coefficient D_c ($\tau_f^{-1} = D_c q^2$, where q is the scattering vector), which reflects a concerted motion of the polymer chains relative to the solvent. The slow process (ascribed to chain disengagement relaxation) is generally not diffusive in nature, but exhibits a complex pattern of behavior, which seems to be related to the viscoelastic response of the system (see below). The slow mean relaxation time increases strongly with temperature and, moreover, the EHEC/SDS system relaxes slower than the corresponding EHEC/CTAB system for all temperatures. We may also

Progr Colloid Polym Sci (1995) 98:51–56
© Steinkopff Verlag 1995

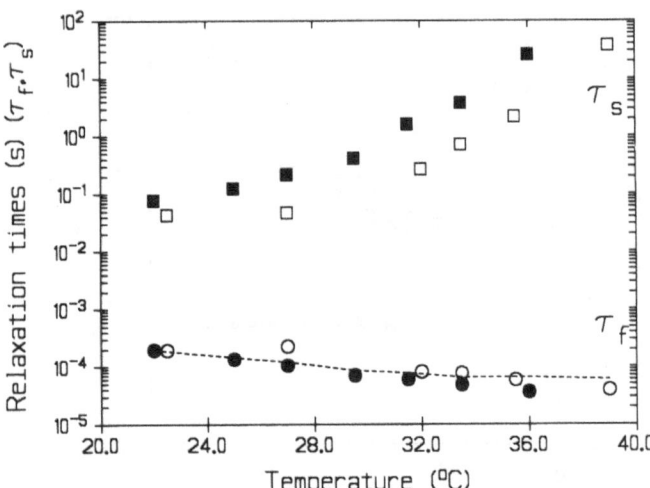

Fig. 3 Temperature dependences of the fast (τ_f) and the slow (τ_s) relaxation times for the systems EHEC/SDS (●, ■) and EHEC/CTAB (○, □). The EHEC and surfactant concentrations are the same as in Fig. 2. The values of τ_f and τ_s have been obtained from a fit of the data in Fig. 2 with the aid of Eq. (4), together with Eq. (5). The dashed line indicates the behavior when the change of solvent viscosity with temperature is accounted for

note that the difference in time scale between the fast and slow processes spans 3–5 orders of magnitude depending on temperature.

Rheology

Stress relaxation

In Fig. 4 is displayed the relative shear relaxation modulus, $G(t)/G(0)$, where $G(t) = \sigma(t)/\gamma_0$, i.e. the ratio of the stress to the (constant) strain, as a function of time and temperature for the two semidilute (i.e. gelling) systems EHEC/SDS (Fig. 4a) and EHEC/CTAB (Fig. 4b). As can be seen, there is a pronounced decrease in the viscoelastic relaxation rate when entering the gel zone for both systems. This finding parallels the results from dynamic light scattering (see above) in that the slow relaxation process gets slower with increasing temperature. A strict analysis of the data in Fig. 4 has not been performed, but a power-law $G(t) \sim t^{-n}$ is implied for the gel-zone data [13]. It has also been found [13] that the values of G at elevated temperatures are significantly higher for EHEC/SDS system than those for the corresponding EHEC/CTAB system. This finding is a quantification of the statement that the EHEC/SDS gel is stiffer than the corresponding EHEC/CTAB gel under otherwise identical conditions [7], a statement alluded to in several places in this paper.

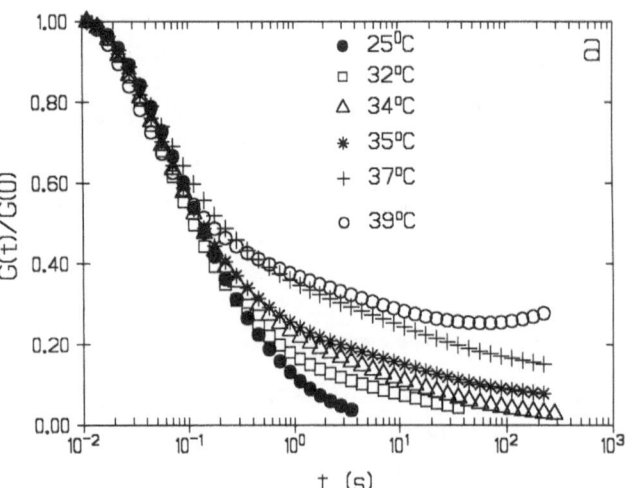

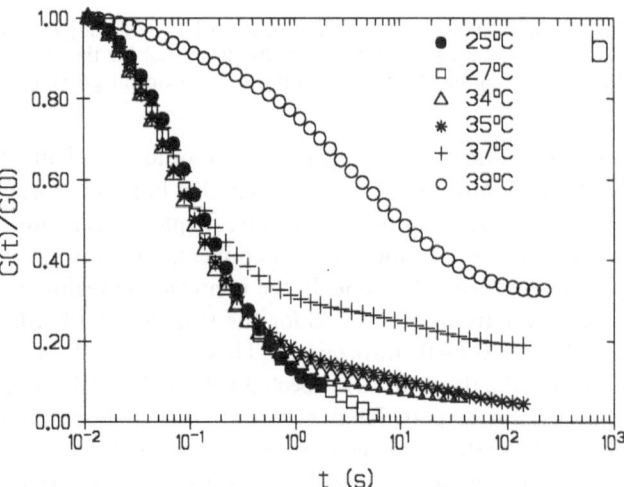

Fig. 4 Plot of the relative shear relaxation modulus as a function of time for the systems EHEC/SDS (a) and EHEC/CTAB (b) at the temperatures indicated. The EHEC concentration is 1.0 wt% and the surfactant concentration is 4.0 mmolal

Oscillation measurements

An implication of the statement above that the shear relaxation modulus decays as a power-law $G(t) \sim t^{-n}$ is that in the linear response domain the shear storage modulus, G', and the shear loss modulus, G'', will follow the same power-law in frequency, viz. $G'(\omega) \sim G''(\omega) \sim \omega^n$ [14].

In Fig. 5 are the results, in the incipient gel zone, from oscillation experiments on the two systems EHEC/SDS (Fig. 5a) and EHEC/CTAB (Fig. 5b) depicted in the form of log-log representations. In addition, the frequency dependence of the loss tangent, tan δ, is illustrated in Fig. 5. For both systems, tan δ is found to be practically independent of frequency over the studied region. This finding is

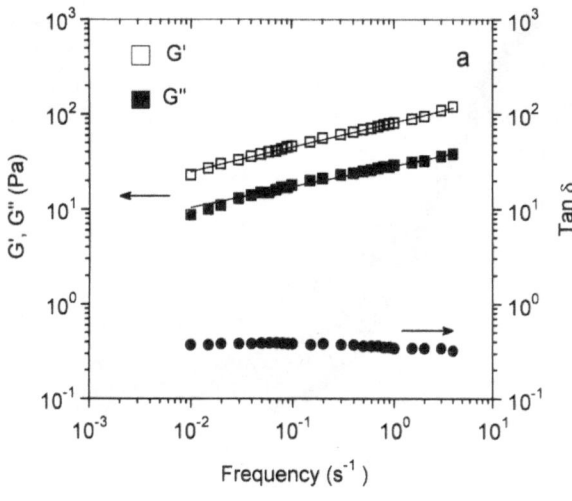
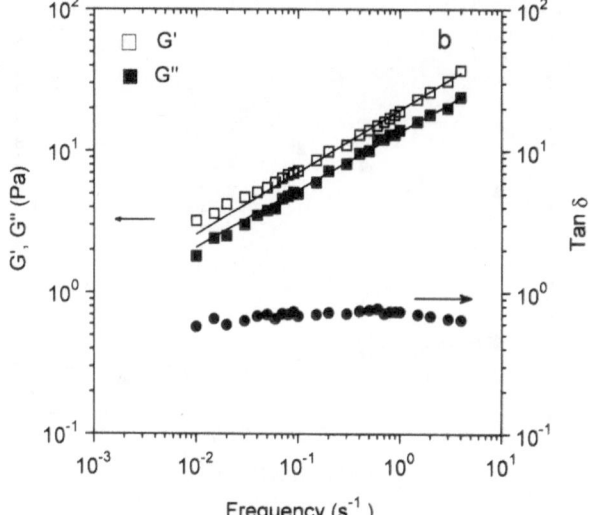

Fig. 5 Log-log plots of storage modulus (G'), loss modulus (G''), and loss tangent (tan δ) as a function of frequency (ω) for the systems EHEC/SDS (a) and EHEC/CTAB (b) in the incipient gel state (at 37 °C). The straight lines correspond to least-square fits in the frequency domain 0.01 to 4 Hz (see text)

an expected and characteristic feature of an incipient gel [15]. The results for G' and G'' reveal that both the storage and loss moduli can be described by frequency-dependent power-laws, where the scaling exponents are, within experimental error, the same. In the incipient gel regime the values of n are 0.24 and 0.43 for the systems EHEC/SDS and EHEC/CTAB, respectively. These values values are far below the theoretically expected value of 0.72 (electrical analogy) for a percolated network [16], and therefore demonstrate that these gels are not of a percolated nature. In a recent theoretical model [17], where screening of excluded volume effects is taken into account all values of the relaxation exponent in the range $0 < n < 1$ are possible for a fractal dimension d_f in the physically acceptable range

$1 \leq d_f \leq 3$. In this model, low values of n observed for these systems suggest that screening of excluded volume interactions plays an important role. This tendency in n leads to lower fractal dimensions for the present systems than the percolation prediction ($d_f = 2.5$). The difference in the value of n for these systems may indicate that the network structure of the two gelling systems is different [7]. Finally, we may note that the values of G' and G'' are consistently higher for the EHEC/SDS than for the EHEC/CTAB system. This trend suggests that the gel strength is higher for the former system. A more comprehensive rheological study of these systems is in progress [7].

References

1. Goddard ED (1989) Colloids Surf 19:255–300
2. Nagarajan R (1986) Adv Colloid Interface Sci 26:205–264
3. Carlsson A (1989) PhD Thesis, University of Lund
4. Lindman B, Carlsson A, Gerdes S, Karlström G, Piculell L, Thalberg K, Zhang K (1993) In: Walstra P, Dickinson E (Eds) Food Colloids and Polymers: Stability and Mechanical Properties. The Royal Society of Chemistry, London, pp 113–125
5. Zana R, Binana-Limbelé W, Kamenka N, Lindman B (1992) J Phys Chem 96:5461–5465
6. Kamenka N, Burgaud I, Zana R, Lindman B (1994) J Phys Chem 98:6785–6789
7. Nyström B, Walderhaug H, Hansen FK, Lindman B, Langmuir, in press
8. Stilbs P (1987) Prog Nucl Reson Spectrosc 19:1–45
9. Stejskal EO, Tanner JE (1965) J Chem Phys 42:288–292
10. Karlström G, Carlsson A, Lindman B (1990) J Phys Chem 94:5005–5015
11. Tanaka F, Stockmayer WH (1994) Macromolecules 27:3943–3954
12. Walderhaug H, Nyström B, Hansen FK, Lindman B, J Phys Chem, in press
13. Nyström B, Lindman B, Macromolecules, in press
14. Chambon F, Winter HH (1985) Polym Bull (Berlin) 13:499–503
15. Te Nijenhuis K, Winter HH (1989) Macromolecules 22:411–414
16. de Gennes PG (1979) In: Scaling Concepts in Polymer Physics. Cornell University Press: Ithaca, NY
17. Muthukumar M (1989) Macromolecules 22:4656–4658

Progr Colloid Polym Sci (1995) 98:57–62
© Steinkopff Verlag 1995

MIXED SYSTEMS

U. Kästner
H. Hoffmann
R. Dönges
R. Ehrler

A comparison of several samples of modified hydroxyethyl cellulose and their interactions with surfactants

Dr. U. Kästner (✉) · H. Hoffmann
Universität Bayreuth
Physikalische Chemie I
95440 Bayreuth, FRG

R. Dönges · R. Ehrler
Hoechst AG
Werke Kalle-Albert
Rheingaustraße 190
65174 Wiesbaden, FRG

Abstract We compared the solution behavior and the macroscopic properties of samples of hydroxyethyl cellulose (HEC) which were modified by perfluoroalkyl chains (F-HMHEC), by cationic groups (cat-HEC), and by hydrophobically modified cationic side chains (cat-HMHEC). It is shown that the viscosity increases by hydrophobic and or ionic modification of the HEC, and the entanglement concentration, c^*, decreases by hydrophobic substitution of the HEC.

On addition of ionic surfactants the viscosity of the modified HEC solutions increases even more. Solutions with viscoelastic properties and sometimes yield stress values are found for the F-HMHEC samples with anionic and cationic surfactants and the cat-HEC and cat-HMHEC samples with oppositely charged (anionic) surfactants. Interactions between the cat-HMHEC samples and surfactants of the same charge are shown in surface tension measurements. The results of the observed polymer-surfactant-interactions are interpreted in terms of association between the oppositely charged and/or the hydrophobic parts of both components.

Key words Modified Hydroxyethyl Cellulose (F-HMHEC, cat-HEC, cat-HMHEC): – phase behavior – polymer-surfactant-interactions – rheological results – surface tension measurements

Introduction

Water-soluble cellulose derivatives like hydroxyethyl cellulose (HEC) are used in many industrial applications. Because of their biological compatibility, they are used as thickeners in food and cosmetic products, protective colloids in polymerization processes, and very recently as thickeners in water-based paints. To optimize the properties for the different applications the HEC samples are often modified by ionic or hydrophobic groups.

Aqueous solutions of water soluble HEC have high viscosities. The chemical modification of these polymers with hydrophobic or ionic side chains causes a further increase in viscosity. Such systems have been investigated systematically during recent years. Hydrophobically modified HEC samples interact by hydrophobic association of their side chains. Added surfactant molecules bind on these hydrophobic parts and strengthen the cross-links. Highly viscous polymer gels with yield stress values are formed [1–3]. Systems with similar properties are obtained with ionically modified HEC by the addition of oppositely charged surfactants [3–5]. Recently, some new results have been reported on samples of HEC which were hydrophobically and ionically modified [6, 7]. These polymers are synthesized in order to combine the properties of the single modified polymers.

The main emphasis of this paper is a comparison of the macroscopic properties of three differently modified samples of HEC in aqueous media and their interactions with added ionic surfactants. The samples are hydrophobically (F-HMHEC), cationically (cat-HEC) or both hydrophobically and cationically (cat-HMHEC) modified HEC derivatives.

Methods and materials

Static light-scattering experiments were carried out on a photometer KMX 6 (Chromatix) which allows measurements at 6°–7° and 174°. To determine the dimensions of the molecules, we used an electric birefringence apparatus (for details see ref. [8]). Surface tension measurements were made with a Lauda tensiometer. All rheological measurements were recorded using an oscillating capillary rheometer Paar OCR-D and a Bohlin CS rheometer.

Four samples of a neutral perfluoralkyl hydroxyethyl cellulose (F-HMHEC) with the substituent

$$-CH_2-CHOH-CH_2-O-CH(CH_2Cl)-CH_2-$$

$$O-(CH_2)_2-(CF_2)_{7-11}-CF_3$$

and the basic hydroxyethyl cellulose (HEC), four samples of cationically modified HEC (cat-HEC) with the substituent

$$-CH_2-CHOH-CH_2-N(CH_3)_3Cl$$

and three samples of cationically and hydrophobically modified hydroxyethyl cellulose (cat-HMHEC) with the substituent

$$-CH_2-CHOH-CH_2-N(CH_3)_2(C_{12}H_{25})Cl$$

were used for our measurements without further purification. The molecular weights and the degrees of molar substitution of these samples are listed in Table 1.

The used surfactants are lithium-perfluorononanoat ($C_8F_{17}CO_2Li$), sodium-dodecyl-sulfate ($C_{12}H_{25}SO_4Na$: SDS) and dodecyl-threemethylammonium-chloride ($C_{12}H_{25}N(CH_3)_3Cl$: $C_{12}TAB$).

Experimental results and discussion

The aqueous solutions of the polymers

All samples except the F-HMHEC derivatives are water-soluble at 25 °C in the investigated concentration range. The F-HMHEC samples 1) and 2) are water-soluble up to 1 wt% while the higher modified samples 3) and 4) are soluble only below 0.3% and 0.1%, respectively.

Aqueous polymer solutions are characterized by a critical concentration, c^*, at which the polymer chains start to entangle and to form networks. This concentraton can be determined by measurements of the zero-shear-viscosity of these solutions with increasing polymer concentration. Figure 1 shows the double-log-plot of these measurements for different samples of HEC.

The viscosity of the unmodified sample HEC starts to rise at about 0.4 wt%. The viscosity of the F-HMHEC sample with the same molecular weight increases at polymer concentrations above 0.1%. This lower concentration c^* is the result of an intermolecular association of the hydrophobic side chains of the polymer coils.

The cat-HEC samples behave like a polyelectrolyte and the viscosity increases below c^* with an exponent of 0.5 following the power low equation (1):

$$\eta^0 \sim (c/c^*)^{1/2} . \tag{1}$$

Fig. 1 The zero-shear viscosity of the unmodified and modified HEC samples with increasing polymer concentration: × HEC; ▲ F-HMHEC 2); □ cat-HEC 4); and ⊠ cat-HMHEC 3). For the neutral polymers HEC and F-HMHEC x_1 is about 0.2 and for the polyelectrolytes cat-HEC and cat-HMHEC x_1 is 0.5. x_2 is for all polymer solutions about 4.0

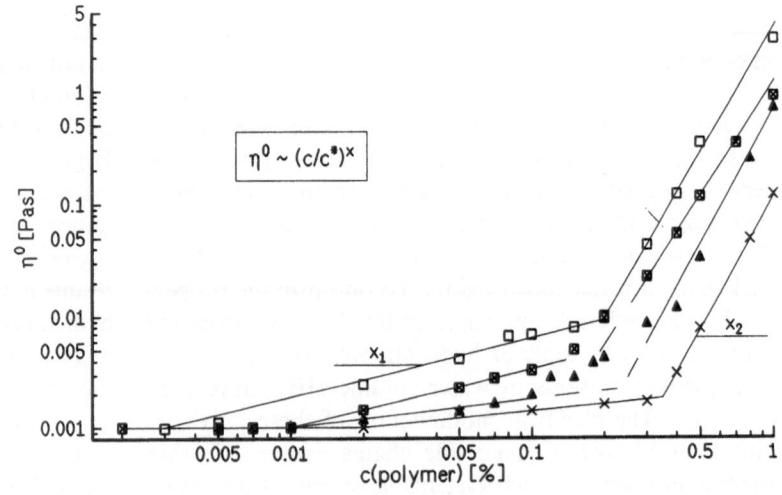

Above c^* the polyelectrolyte forms a network structure through simple entanglements of their chains and behaves now in the same way like unmodified HEC.

The cat-HMHEC sample also shows polyelectrolyte behavior at low polymer concentrations. The critical concentration c^* is lower than c^* for the sample which is only cationically modified. The decrease of c^* is likely due to intermolecular associations of the hydrophobic parts of this sample.

The slope of the viscosity as a function of polymer concentration above c^* is about 4.0 for all polymer samples. This exponent nearly corresponds to the predicted exponent of 5.5 for the reptation model of polymers [9, 10]:

$$\eta^0 \sim (c/c^*)^{5.5} . \tag{2}$$

Therefore, the dynamics of stress relaxation is characterized by a curvilinear diffusion of the polymer along its own contour length.

Electric birefringence measurements give some information about conformational changes in the polymer solutions with increasing concentration. The reduced Kerr constants for all polymer samples remain constant until c^* is reached. From the electric birefringence decay, we determined a relaxation time τ which is relate to a rotational diffusion coefficient D_r and a particle length L [8]:

$$D_r = 1/(6\tau) = 3kT[\ln p - 0.76 + 7.5(1/\ln(2p)$$
$$- 0.27)]^2/(\pi\eta L^3) , \tag{3}$$

where kT is the thermal energy, η is the solvent viscosity, and p is the axial ratio.

A comparison between this length L at a polymer concentration below c^*, the contour length l_k (is determined from the molecular weight by:

$$l_k = z \cdot 0.63 \text{ nm} , \tag{4}$$

where z is the number of monomer units per chain and 0.63 nm is the approximate length of a monomer unit), and the radius of gyration R_G (determined from light scattering) gives some information about the conformation of the polymer in aqueous solutions. The parameters are listed in Table 1.

For the HEC and the F-HMHEC samples L corresponds to R_G and the contour length is more than 10 times higher than these parameters. These polymers have a coiled conformation.

For the cat-HEC samples 1) and 2), L corresponds to their contour length l_k. For the samples 3) and 4) with higher molecular weights the difference of both parameters is less than a factor of 2. These polyelectrolytes have an almost stretched conformation.

The cat-HMHEC samples show values of L lower than l_k, but still higher than R_G. Therefore, they show the

Table 1 Characteristic parameters of the unmodified and modified samples of hydroxyethyl cellulose

Sample	$M_w^{a)}$ [g/mol]	$MS^{b)}$	l_k [nm]	L [nm]	R_G [nm]
HEC	500 000	–	1 200	82	53
F-HMHEC 1)	500 000	0.0007	1 200	82	53
F-HMHEC 2)	500 000	0.0027	1 200	82	53
F-HMHEC 3)	500 000	0.0031	1 200	–	–
F-HMHEC 4)	500 000	0.0043	1 200	–	–
cat-HEC 1)	33 000	0.27	71	80	–
cat-HEC 2)	43 000	0.21	102	115	–
cat-HEC 3)	120 000	0.10	260	160	–
cat-HEC 4)	150 000	0.11	300	185	–
cat-HMHEC 1)	950 000	0.011	2 300	155	60
cat-HMHEC 2)	950 000	0.021	2 300	160	60
cat-HMHEC 3)	950 000	0.027	2 300	180	60

$^{a)}$ M_w is the molecular weight determined from light-scattering experiments.
$^{b)}$ MS means the degree of molar substitution per cellulose monomer for the different side chains of the modified HEC samples. The degree of molar substitution of hydroxyethyl per cellulose monomer is about 1.9 to 2.6 for all samples.

properties of both polymer types mentioned above: the single chain is coiled with stretched parts of the length of a few monomer units.

Interactions with added ionic surfactants

Interactions between polymers and ionic surfactants are based on associations of the hydrophobic or ionic parts of the polymer with the hydrophobic tails or oppositely charged head groups of the surfactant molecules. Figures 2, 3 and 4 show the influence on the viscosity by the addition of ionic surfactants to the different polymer solutions (1 wt%).

The unmodified sample HEC does not show an influence on its macroscopic properties with increasing ionic surfactant concentration. The surfactant molecules do not seem to interact with the highly flexible and hydrophilic HEC compound.

The 1% solutions of the F-HMHEC samples 3) and 4) become water-soluble as soon as a small amount of surfactant is added. The viscosity of the F-HMHEC samples passes over a maximum with increasing concentration of anionic or cationic surfactants (Fig. 2). The surfactant molecules interact with the hydrophobic cross-links of the polymer network. Therefore, the number of effective cross-links increases (an increase of the storage modulus G' is observed [3]) and the viscosity rises. The value of maximum viscosity increases thereby with an increasing degree

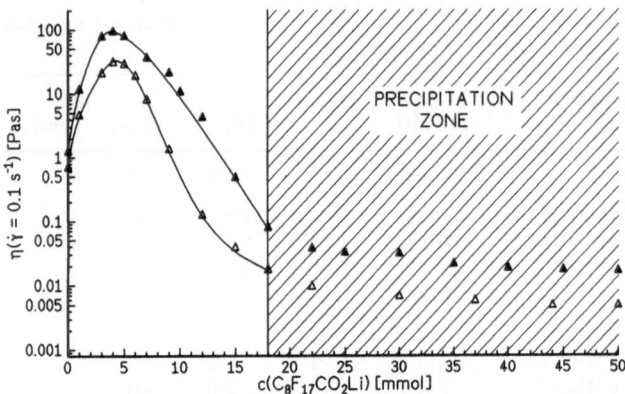

Fig. 2 The viscosity at a constant shear rate of 0.1 s^{-1} of the 1 wt% solutions of ▲ F-HMHEC 4) and △ F-HMHEC 2) as a function of the lithium-perfluorononanoat concentration

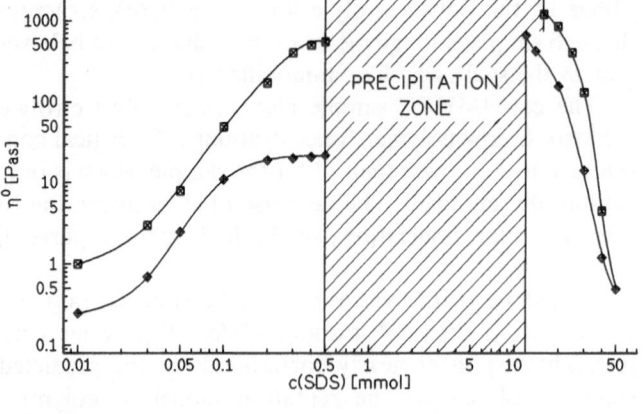

Fig. 4 The zero-shear viscosity of the 1 wt% solutions of ⊠ cat-HMHEC 3) and ⊕ cat-HMHEC 1) as a function of the SDS concentration

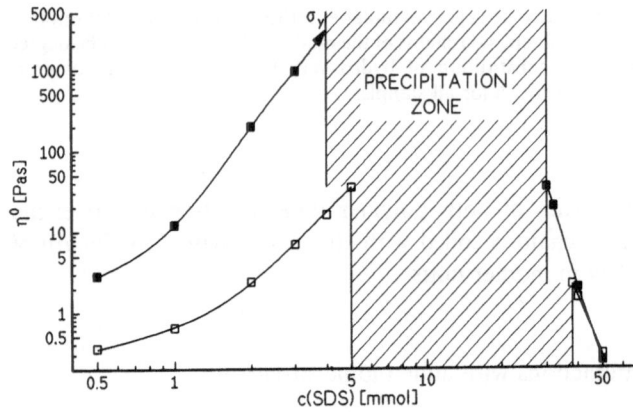

Fig. 3 The zero-shear viscosity of the 1 wt% solutions of ■ cat-HEC 4) and □ cat-HEC 2) as a function of the SDS concentration. σ_y marks the yield stress behavior of the cat-HEC 4) sample with 4 mmol SDS

of hydrophobic substitution. After passing the maximum the cross-links open again with increasing surfactant concentration because of an increasing saturation of the hydrophobic side chains with surfactant molecules and the viscosity decreases. At higher surfactant concentrations phase separation occurs. The phase separation seems to be a special case of depletion flocculation [11, 12]. Thus, the observed flocculation is the result of the large osmotic pressure of the surfactant micelles. They compress the polymer into a dense phase.

The cat-HEC samples show an increase in viscosity by addition of oppositely charged surfactants (Fig. 3). The anionic head groups of the surfactant molecules interact with the cationic substitutents of the polymer while the hydrophobic interactions between the tails of the surfactant molecules connect different polymer backbones. The increase in viscosity corresponds thereby with the increas-

ing molecular weight of the samples. For sample 4) with the highest molecular weight (and a lower substitution degree than sample 2)!) a yield stress value (σ_y) is observed. At about 100% of charge neutralization phase separation occurs. At higher surfactant concentrations the solution becomes single phase again. Surfactant micelles now adsorb on the charged polymer chains. Some backbones are still connected and the viscosity is higher than the viscosity of the pure polymer solution (without added surfactant).

The addition of oppositely charged surfactants to the cat-HMHEC samples results in a similar behavior of the viscosity as observed for the cat-HEC samples (Fig. 4). There are, however, two remarkable differences: the plateau value of the viscosity at a charge neutralization of about 125% before phase separation and the higher value of the viscosity after phase separation. The plateau value corresponds to the situation in which the charges are neutralized but the hydrophobic parts are not saturated with surfactant molecules. More surfactant molecules associate and the viscosity shows a maximum like the F-HMHEC samples with added ionic surfactant. In the resolubilization region micelles adsorb on the charges and additional small surfactant aggregates connect the hydrophobic side chains of the polymer. The number of effective cross-links is now optimized and a maximum value of the viscosity is observed.

To complete the results of our investigations on the cat-HMHEC samples, we emphasize that these samples also interact with surfactants of the same charge. We found, however, no increase in viscosity, but rather a change of the surface tension which is sensitive to polymer-surfactant-interactions.

Figure 5 shows the results of these measurements on a sample of cat-HMHEC with added oppositely charged surfactant (Fig. 5a) and equally charged surfactant

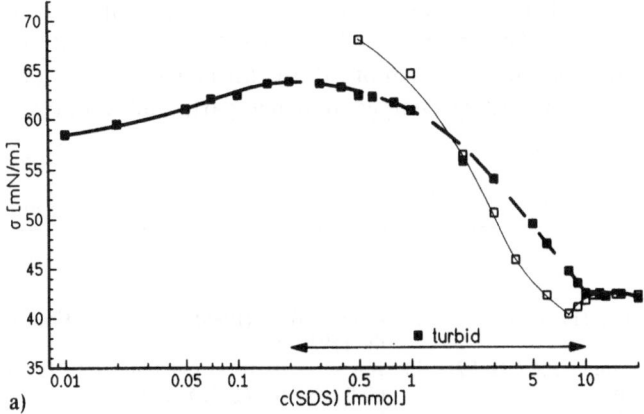

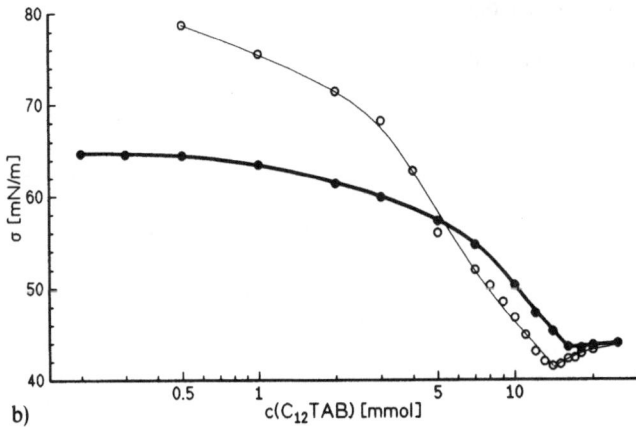

Fig. 5 The surface tension as a function of: a) the SDS-concentration of □ water and ■ cat-HMHEC 3) 0.1 wt% b) the C_{12}TAB-concentration of ○ water and ● cat-HMHEC 3) 0.1 wt%

actions between the polymer and the surfactant molecules of the same charge [7].

Conclusions

The aqueous unmodified and modified HEC solutions are characterized by a critical concentration, c^*, at which the polymer backbones start to entangle. Hydrophobically modified HEC solutions (F-HMHEC, cat-HMHEC) show lower values of c^* according to intermolecular associations of their hydrophobic side chains.

The neutral polymers (HEC, F-HMHEC) are coiled in solution while the charged polymer (cat-HEC) is in an almost stretched conformation. For the both hydrophobically and cationically modified sample (cat-HMHEC), with comparable low substitution degrees, we found a coiled conformation with some stretched parts over a few monomer units.

Modified HEC solutions interact with ionic surfactant molecules:

- Solutions of hydrophobically modified HEC (F-HMHEC) interact with anionic and cationic surfactants. The viscosity passes over a maximum and phase separation occurs with excess surfactant concentration. The strength of the gels increases with increasing hydrophobic substitution.

- Cationically modified HEC solutions (cat-HEC) interact with oppositely charged surfactants. The viscosity rises to a first maximum. At about 100% of charge neutralization phase separation occurs. With excess surfactant concentration resolubilization takes place and the viscosity shows a second maximum with a lower value than the first one. The values of the viscosity maxima increase with increasing molecular weight and they are nearly independent of the ionic substitution degree.

- Both hydrophobically and cationically modified HEC solutions (cat-HMHEC) interact with anionic and cationic surfactants. By addition of oppositely charged surfactants the viscosity shows a first plateau maximum. Phase separation occurs at about 125% of charge neutralization. The resolubilization zone shows a second maximum of the viscosity which is remarkably higher than the first one. The values of the first maximum viscosity increase with increasing substitution degree while the values of the second maximum viscosity are mainly influenced by the molecular weight. Interactions between this polymer and ionic surfactants of the same charge are found by surface tension measurements. For our investigated systems, we could not find an increase in viscosity.

(Fig. 5b). The polymer solution has a concentration of 0.1% in order to avoid the complications by gel and precipitate formation. The surface tension of the cat-HMHEC-SDS-system increases because the surfactant molecules associate to the polymer (see Fig. 5a). The solution becomes turbid at the maximum value of the surface tension. The surface tension then decreases due to the increasing amount of free surfactant molecules (saturation of the polymer). The surface tension of the polymer-surfactant-system becomes equal to the surface tension of the water-surfactant-system above the critical micellar concentration (CMC) of the surfactant, and the polymer-surfactant-solution becomes clear again.

The surface tension of the same cat-HMHEC sample with increasing concentration of cationic surfactant does not show such a maximum (Fig. 5b). But the decrease of the surface tension for low surfactant concentrations is lower than in the case without polymer and a crossover between the water-surfactant- and the polymer-surfactant-curve is observed. This crossover reflects direct inter-

62

U. Kästner et al.
Interactions of modified hydroxyethyl cellulose with surfactants

Interactions between polymers and ionic surfactant molecules are formed by association of the oppositely charged or/and the hydrophobic parts of both components. The increase of the viscosity and the formation of strong hydrogels are due to hydrophobic associations of the hydrophobic tails of the surfactant molecules which connect different polymer backbones. These associations are more or less reinforced by attractive interactions of the oppositely charged polymer parts and surfactant headgroups.

References

1. Tanaka R, Meadows J, Phillips GO, William PA (1992) Macromolecules 25:1304–1310
2. Goddard ED (1986) Colloids Surf 19:255–300
3. Kästner U, Hoffmann H, Dönges R, Ehrler R (1994) Colloids Surf A 82:279–297
4. Goddard ED, Leung PS, Padmanabhan KPA (1991) J Soc Cosmet Chem 42:194–34
5. Dualeh AJ, Steiner CA (1991) Macromolecules 24:112–116
6. Goddard ED, Leung PS (1992) Langmuir 8:1499–1500
7. Goddard ED, Leung PS (1992) Colloids Surf 65:211–219
8. Schorr W, Hoffmann H (1981) J Phys. Chem 85:3160–3167
9. Cates ME (1987) Macromolecules 20:2289–2296
10. Cates ME (1988) J Phys France 49:1593–1600
11. van de Pas JC, Buytenhek CJ (1992) Colloids Surf 68:127–139
12. Vincent B, Edwards J, Emmett S, Jones A (1986) Colloids Surf 18:261–281

Progr Colloid Polym Sci (1995) 98:63–68
© Steinkopff Verlag 1995

MIXED SYSTEMS

A. Tahani
H. van Damme
P. Levitz

Study of mixtures of hydrolyzed polyacrylamide-non-ionic surfactant in aqueous phase and in the kaolin-water interfacial surface

Dr. A. Tahani (✉) · H. van Damme
P. Levitz
Centre de Recherche sur la Matière Divisée
CNRS, 1b de la Férollerie
45071 Orléans Cedex 2, France

Abstract This study addresses the problem of the co-adsorption of a non-ionic surfactant (TX100) and a polymer (hydrolysed polyacrylamide) on a mineral surface (kaolin). This type of system is involved in the formulation of an inverse emulsion in oil recovery.

The mixture of reactants in the aqueous solution is first analyzed, essentially by spectroscopic methods. This is followed by the study of the interaction of each individual molecule on kaolin. The physico-chemical approach is centered on obtaining the mixed isotherms in terms of salinity. We did not find evidence for any interaction between the two molecules in the aqueous solution, in the pH range 3 to 7. The mixed adsorption results show that both molecules coexist in a same interfacial volume, with no perturbation. More elaborate analyses of sequential adsorption nevertheless disclose slight differences.

Key words Polyelectrolyte – non ionic surfactant – kaolin – interaction – adsorption

Introduction

Hydrolyzed polyacrylamide, generally conditioned inside emulsion, is used in tertiary oil recovery. Degradation of the injected plug is usually responsible for surfactant-polymer loss process as this can affect the permanence of low interfacial tension between oil and water, and viscosity or charge of the plug. The flow through porous media of this mixture depends on the interactions, both in solution and at the interface of the solid. Thus, the adsorption of mixed polymer-surfactant systems depends on the type of the interaction between different species in the solution and at the interface.

The nature of the solid plays an important role in the adsorption of mixed polymer-surfactant systems.

Theoretical studies [1–3] concerning polyelectrolyte adsorption on a solid surface show that the electrostatic energy of polyion-colloid interactions depends funda-mentally on three variables: i) the charge density of the polymer, ii) the charge density of solid surface, and iii) the ionic strength of the medium.

The presence of surfactant strongly affects the polymer adsorption on the solid surface. Literature shows that mixed adsorption depend on the nature of each constituent [4–7].

Kronberg et al. [4], studied the adsorption of polyacrylate in the presence of nonyl octyl phenol (NP-EO$_{10}$) on Na-kaolin, and reported that surfactant adsorption changes according to ionic strength; these variations were interpreted by conformational changes of the polymer and by the formation of a complex PAA-NP-EO$_{10}$ at pH 7. Bocquenet and Siffert [5] studied the adsorption of polyacrylamide (PAM) in the presence of Na-dodecylbenzene sulfonate (NaDBS) on kaolin and illite, and their results show a decrease of polyacrylamide adsorption in the presence of surfactant. In that case, the ionic surfactant seems to compete with the polyelectrolyte for the exchangeable

64

A. Tahani et al.
Adsorption of mixed system TX100-HPAM on the Kaolin

cations. Tadros [6] found an increase of polyvinyl acid (PVA) on silica in the presence of cetyl trimethyl ammonium bromide CTAB and NaDBS. At high pH values, the increase in adsorption of PVA is attributed to the CTA$^+$ ions which favor interactions with PVA chains to the SiO$^-$ surface. In the presence of NaDBS, as well as for high and low pH value, the effect is interpreted as due to the formation of a PVA-NaDBS complex. Somasundaran and Moudgil [7] found weak interactions in a mixture of sulfonated polyacrylamide (PAMs) and dodecylsulfonate in solution, but at the interface of hematite, a depressing effect on PAMs adsorption was found when surfactant was added, depending on the effect of order of reagent addition [7].

The primary objectives of this paper are to present the results of experiments on the study of the interactions between the non ionic surfactant TX100 and polyelectrolyte HPAM in solution and at the interface of non swelling clay minerals.

Materials and methods

The polymer is a hydrolyzed polyacrylamide (HPAM) with a hydrolysis degree τ_h of 27%. Its mean molecular weight is $7.5 \cdot 10^6$. The detailed characteristics are given in ref. [8]. The non ionic surfactant is the alkyl benzene polyoxyethylene surfactant TX100 from Rhom and Hass. Pyrene used is from Aldrich and of analytical grade.

The substrate is a kaolin Supreme (English China Clay). Its surface area is $16 \ m^2 \cdot g^{-1}$, and zero point charge using potentiometric titration [9] is pH 4.

The amount of polymer in solution was determined using amide group titration [10]. For dilute supernatant, we must check that surfactant does not perturb the polymer counting. Surfactant and pyrene concentrations were determined by UV absorption using the spectrophotometer ACTA-MIV. The aggregation number of TX100 was determined using time-resolved fluorescence decays, using a single photon counting method, and the numerical analysis has been described elsewhere [11]. The steady-state fluorescence data are recorded on a Perkin–Elmer LS-5 spectrofluorimeter.

Adsorption of the polymer and surfactant species at the solid-liquid interfacial region was measured using the solution depletion method. Details of the adsorption procedure are given below. 0.25 g of kaolin was dispersed in 5 cm^3 of water with required salinity. The required amount of polymer and/or surfactant solution was then added and the suspension was agitated for 24 h. In sequential adsorption of mixture system, the first reagent added to the kaolinite suspension was equilibrated for 24 h, and then the second reagent was added and the mixture was equilibrated for an additional 24 h.

Experimental results

Interactions in aqueous phase

In Fig. 1, we report the results of extrinsic pyrene fluorescence solubilized in TX100 micelles in the presence or absence of HPAM at pH 7. The aggregation number of surfactant molecules in the presence of the polymer remains the same as for the surfactant alone, and the value measured for surfactant concentration varying between $10^{-3} \ Mol \cdot L^{-1}$ and $10^{-2} \ Mol \cdot L^{-1}$ is about 100 ± 20.

At pH 7, the results obtained do not point to specific interactions between polyelectrolyte and non-ionic surfactant. Perturbations observed by adding NaCl confirm the results concerning salts. These results are not surprising since, at pH 7, the acrylate groups are completely dissociated and there is no protonated sites able to associate to the ether groups of TX100. At pH 3, where HPAM is completely protonated (pKa = 4.8), the results obtained concerning an aggregation number of TX100 are given in Fig. 2. There is no change of the mean aggregation number after polymer addition, and polarity inside the micelle as probed by pyrene does not change.

It is clear that analysis of the mixture of TX100-HPAM in aqueous solution is a necessary step in the study of mutual interactions of the solid/liquid interface. Absence of any specific interaction in solution will be an important information in order to understand the specific nature of

Fig. 1 Evolution of III/I ratio (polarity detected by pyrene) as function of TX100 concentration. Salt effect compared to HPAM effect (at 400 ppm) pH = 7

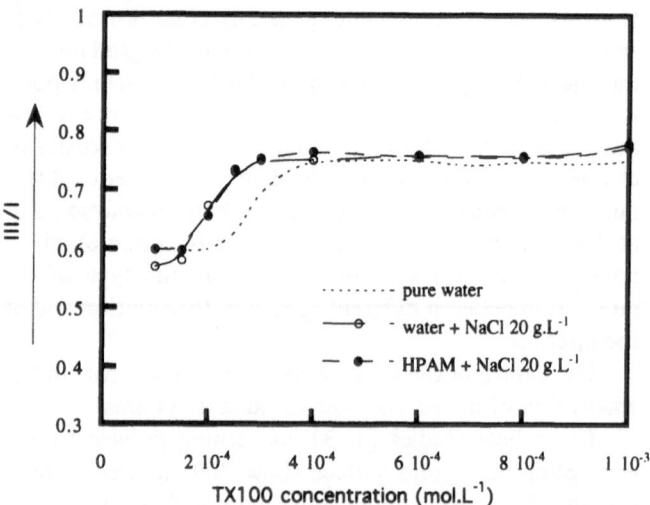

Progr Colloid Polym Sci (1995) 98:63–68
© Steinkopff Verlag 1995

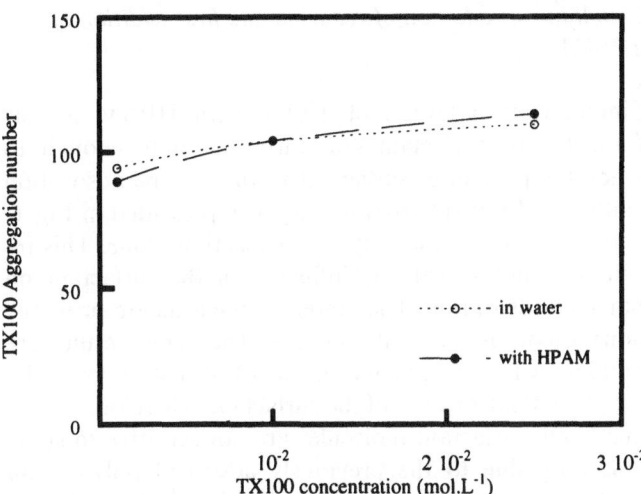

Fig. 2 Evolution of the mean aggregation number of TX100 as function of its concentration in presence and absence of HPAM (at 400 ppm). pH = 3

an eventual interaction during mixed adsorption at the solid/water interface.

Interactions at the solid liquid interface

TX100 adsorption alone

In Fig. 3, we show the adsorption isotherms of TX100 on kaolin for different ionic strengths. In the absence of salt, the TX100 adsorption isotherm on Supreme kaolin is quite different to the adsorption isotherm on silica [11, 12], but

maximum adsorption is about the same, as compared to adsorption on Charentes kaolin [13]. The two steps observed are possibly due to adsorption on lateral and basal surfaces [13]. However, a directly related shape between the two steps of isotherm and the two different surfaces of kaolin is a problem not clearly settled in the literature [13, 14]. The influence of salinity on adsorption is marked by an increase of amount adsorbed in the whole concentration range, and the displacement of the plateau toward the lowest concentrations agrees with the decrease of critical micelle concentration (in the figure the arrows indicate the position of the CMC).

HPAM adsorption alone

The adsorption of HPAM increases continuously with salinity as shown in Fig. 4, and this shows that the adsorption level of HPAM on Supreme kaolin is essentially governed by electrostatic repulsions between the negatively charged clay and the polymer. Due to the high affinity of amide group for kaolinite [15], and to the subsistence of positive charges on kaolin surface at pH 7 [16], a non-adsorption regime for HPAM is not found, even at zero salinity.

The partial desorption occurring upon reducing the ionic strength of the dispersion medium can be attributed to the reduction in charge screening causing an increase in the intermolecular lateral repulsions within the adsorbed layer [17]. The flocculation of kaolinite suspensions by bridging occurs instantaneously upon addition of the polymer and its adsorption onto a large number of particles. The flocculation effect is vigorous when ionic

Fig. 3 Adsorption isotherm of TX100 on kaolin for different ionic strength. $S/L = 0.01$. pH = 6–7

Fig. 4 Adsorption isotherm of HPAM on kaolin for different ionic strength. $S/L = 0.01$. pH = 6–7

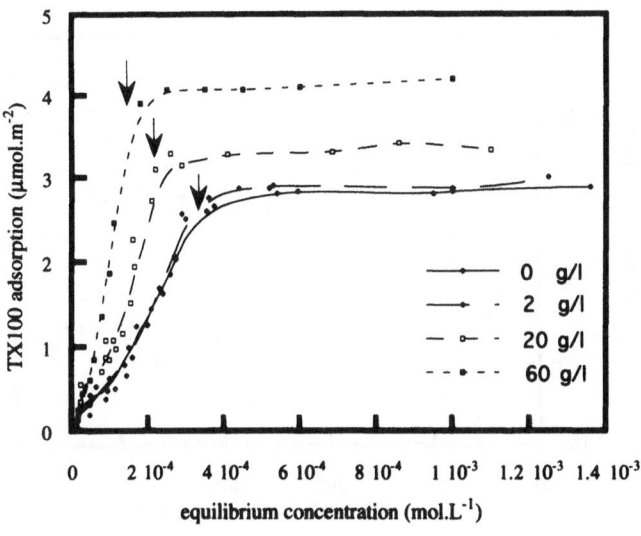

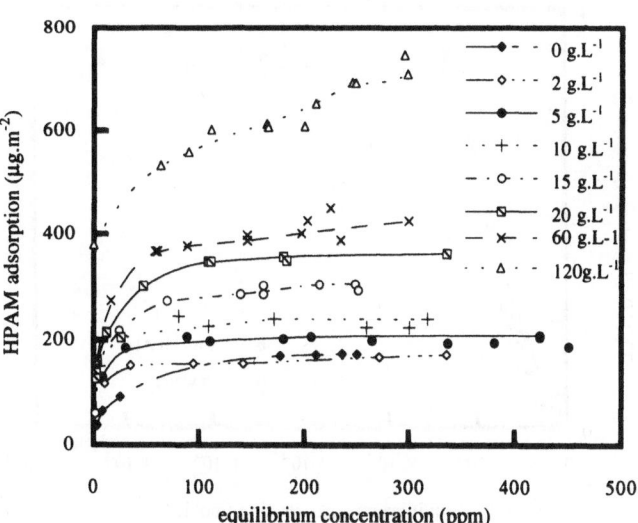

strength and polymer concentration increase or, for a given concentration of polymer, when the solid/liquid ratio increases [17].

TX100 adsorption from mixture

The adsorption isotherm of TX100 on Supreme kaolin from a mixture of reagents compared to the adsorption isotherm of TX100 alone are presented in Fig. 5. Polymer concentration was fixed in the plateau region of the adsorption isotherm ($Ci = 400$ ppm). We note that the adsorption isotherm of TX100 is superimposed on the one obtained without polyelectrolyte. This means that the adsorption of TX100 is practically not modified by the presence of polymer on the surface. The shape of the isotherm is not modified by TX100 adsorption from the mixture. Thus, the associative adsorption mechanism of surfactant on the kaolin surface is unchanged in the presence of polyelectrolyte, namely, polymer is adsorbed at a few anchoring points, and does not compete for the same adsorption sites on kaolin with the non ionic surfactant.

HPAM adsorption from mixtures

The effect of order of reagent addition on HPAM adsorption on kaolin is examined in this part.

Simultaneous addition of reagents and first addition of HPAM

Simultaneous addition of TX100 with HPAM is performed with surfactant concentration quite enough to reach the plateau of surfactant isotherm. The adsorption isotherm of HPAM from mixture is represented in Fig. 6, and compared to the polymer adsorption alone. This result does not reveal any influence of the surfactant on polymer adsorption. The segregation effects are probably nonexistent on kaolinite surface. The same results are obtained when the polymer is adsorbed first, followed by subsequent adsorption of the surfactant. Therefore, on the one hand, surfactant molecules are not sensitive to steric hindrance due to the previously adsorbed polymer on kaolinite surface and, on the other hand, the polyelectrolyte is not displaced by subsequent TX100 adsorption.

Sequential addition (I surfactant/II polymer)

When the surfactant is adsorbed first, followed by polyelectrolyte adsorption, the adsorption level of HPAM from the mixture changes with ionic strength, but surfactant adsorption remains the same compared to the results obtained for separate surfactant adsorption. The results of HPAM adsorption from mixtures compared to its individual adsorption occurring in the plateau region are drawn versus ionic strength in Fig. 7.

Fig. 5 Mixed adsorption isotherm of TX100 on kaolin. $S/L = 0.01$, $20\,\mathrm{g \cdot L^{-1}}$ NaCl, pH = 6–7

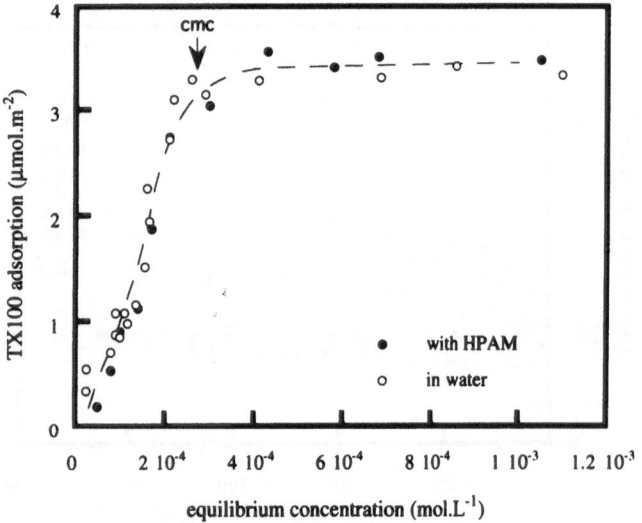

Fig. 6 Mixed adsorption of HPAM on kaolin, simultaneous addition of reactants. $S/L = 0.01$, $20\,\mathrm{g \cdot L^{-1}}$ NaCl, pH = 6–7

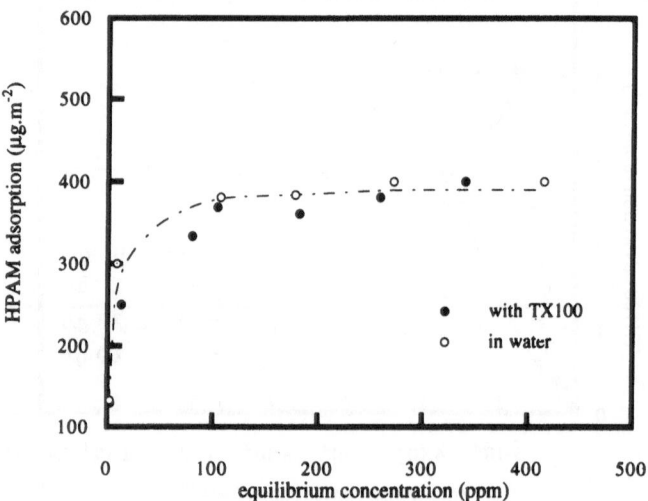

Progr Colloid Polym Sci (1995) 98: 63–68
© Steinkopff Verlag 1995

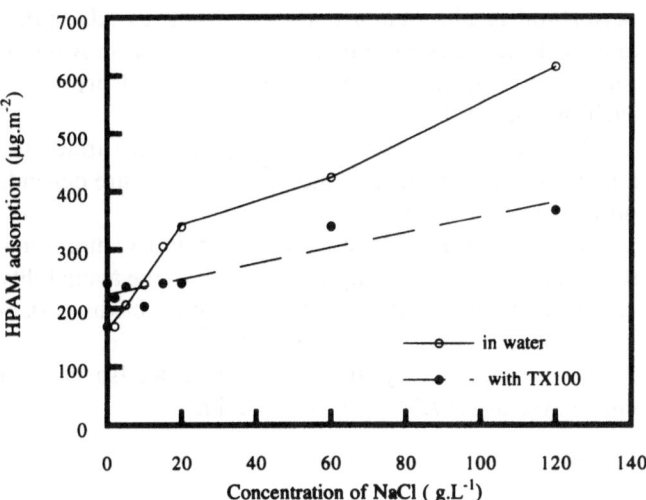

Fig. 7 Evolution of the maximal adsorption level of HPAM on kaolin as function of salinity. Effect of preadsorbed TX100. $S/L = 0.01$, pH = 6–7

Discussion

These results show that there is no interaction between a non-ionic surfactant and HPAM 27%, and confirm the theoretical results of Ruckenstein [18] and Nagarajan [19]. They agree with experimental results of Saito on polyacrylate-non ionic surfactant [20, 21]. In order to understand this lack of interactivity, the analysis of the interaction between polyacrylic acid (PAA) and non-ionic surfactant is desirable. In this case, Saito showed that a specific interaction exists between non-dissociated acrylic groups of the polymer and the polar chain of non-ionic surfactant by hydrogen bonds reinforced by hydrophobic interactions [21]. This specific interaction disappears when the dissociation of the acrylic groups exceeds 20%, i.e., when the pH value is larger than the dissociation constant pk_a (≈ 5). The absence of any interaction in the HPAM-TX100 system can be attributed to deficiency of acrylic sites (27% at maximum) on HPAM able to be protonated and likely to form hydrogen bonds with the ether function of the non-ionic surfactant TX100. Furthermore, the free energy gain due to hydrogen bond formation between the acidic function of HPAM and polar head groups of TX100 does not compensate the free energy lost through steric repulsion between polar head groups of surfactant molecules due to the approach of polymer chains and to the relaxation of polymer chains by Brownian movement.

At the interface, experimental results show that simultaneous adsorption of surfactant and polymer from a mixture is comparable to that of each constituent adsorbed alone on the surface. This fact is also observed in the case of simultaneous addition and in the case of previous adsorption of HPAM followed by TX100. This is due to the adsorption mechanism of each constituent, although the polyelectrolyte is adsorbed at some anchoring points, leaving room for surfactant molecules to be adsorbed onto the surface and to form aggregates in equilibrium with the chemical potential of the monomers in solution. If there were competition for the same adsorption sites, sequestration effects would be observed, inducing mutual depression in the adsorption levels of both molecules.

When the surfactant is previously adsorbed on the surface, the subsequent polymer adsorption depends on the ionic strength. For a given ionic strength, surfactant adsorption does not induce a significant modification of adsorption with different orders of addition of reagents. In this study, we deal with two molecular objects with very different sizes: surfactant molecules, with a medium size and a tendency to aggregate both in solution and at the interface [11], and large polyelectrolyte able to connect different solid particles. Therefore, both can coexist without mutual perturbation in the same interfacial volume.

Surfactants characterized by small sizes adsorb, whatever the interface state. The most important change of behavior is in the adsorption kinetics and, in turn, the diffusion of surfactant molecules toward the surface is modified by the presence of the polymer. The adsorption kinetics of the surfactant is, in this, case delayed. In Fig. 8, we report the results of kinetic adsorption of TX100 by simultaneous addition of both reagents HPAM-TX100 on kaolinite. Obviously, the adsorption kinetic of TX100 is slowed down by the presence of HPAM.

Fig. 8 Adsorption kinetic of surfactant alone and in mixed system on kaolin. Direct addition of the mixture on kaolin. $S/L = 0.01$, $20 \, g \cdot L^{-1}$ NaCl, pH = 6–7

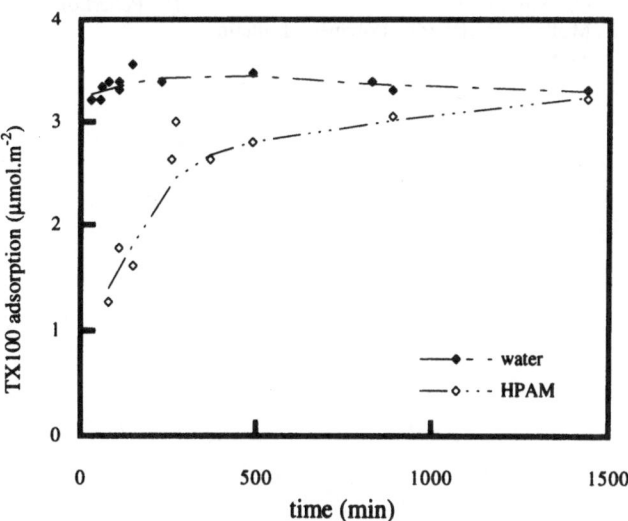

In order to explain the results obtained, we have admitted that kinetic adsorption of HPAM is always faster than that of surfactant (instantaneous flocculation by addition of polymer).

At week ionic strength ($\leq 5 \, g \cdot L^{-1}$) when the polymer is adsorbed by itself, its configuration is probably flat, thus amplifying negative electrostatic repulsions between polymer and the negatively charged surface, resulting in weaker adsorption level of HPAM. Previous adsorption of surfactant screens the negative charges on kaolinite surface by steric effects, and facilitates the approach of HPAM, thus leading to an increase of the adsorption level. At medium or high ionic strength ($\geq 5 \, g \cdot L^{-1}$), previous adsorption of TX100 decreases the adsorption of HPAM, and in such cases, surfactants act partly as steric sequester for subsequent polyelectrolyte adsorption.

Conclusion

No interactivity was detected in aqueous phase between non-ionic surfactant and hydrolyzed polyacrylamide. Experimental results obtained on simultaneous adsorption and by the first adsorption of the polymer had shown that the reagents adsorb in the mixture like the adsorption of each one on kaolin.

When the surfactant is preadsorbed on the surface, the adsorption level of the polymer from the mixture depends on the ionic strength.

No effect of HPAM on the adsorption of non-ionic surfactant was obtained, apparently due to the availability of sufficient surface sites for the adsorption of the surfactant molecules, even on polymer-coated surface.

Kinetic factors play an important role in adsorption of mixed system HPAM-TX100 on kaolin.

Acknowledgments This research program was supported by Project GS N° 91N80/0083 PIRSEM of the CNRS-ARTEP. We thank C. Noik and D. Defives of IFP for their help, and R. Setton for correcting the manuscript.

References

1. Hesselink FTh (1977) J Col Int Sci 60:448
2. Scheutjens JMHM, Fleer GJ (1979) J Phys Chem 83:12
3. Scheutjens JMHM, Fleer GJ (1980) J Phys Chem 179:84
4. Kronberg B, Kuortti J, Stenius P (1986) Col Sur 18:411
5. Boquenet Y, Siffert B (1984) Col Sur 9:147
6. Tadros ThF (1974) J Col Int Sci 46:528
7. Moudgil B, Sommasundaran P (1985) Col Sur 13:87
8. Muller G (1984) Polymèr Bulletin 11:391
9. Lamarche JM (1989) Thèse d'état ès-Sciences Physiques, Besançon, French
10. Scoggins MW, Miller JW (1975) Analytical chemistry 4:152
11. Levitz P, Van Damme H (1984) J Phys Chem 88:2228
12. Partyka S, Lindheimer M, Zaini S, Keh E, Brun B (1986) Langmuir 2:10
13. Denoyel R, Rouquerol J (1991) J Col Int Sci 143:555
14. Kronberg B, Stenius P, Thorssel Y (1984) Col Surf 12:113
15. Pefferkorn E, Jean Chonberg AC, Chauveteau G, Varoqui R (1990) Col Polymer Sci 137:66
16. Lecourtier J, Chauveteau G (1985) 3^{èm} Col Eur RAP, Rome, n° 32
17. Lecourtier J, Lee LT, Chauveteau G (1990) Col Sur 47:212
18. Ruckenstein E, Huber G, Hoffman H (1987) Langmuir 3:382
19. Nagarajan R (1985) Col Surf 13:1
20. Saito S, Tanigushi J (1973) J Col Int Sci 44:144
21. Saito S (1979) Colloid Polym Sci 257:266

Progr Colloid Polym Sci (1995) 98:69–74
© Steinkopff Verlag 1995

MIXED SYSTEMS

Effect of the addition of block copolymers on the formation and stability of vesicles (liposomes) prepared using soybean lecithin

K. Kostarelos
P.F. Luckham
Th.F. Tadros

K. Kostarelos · P.F. Luckham
Th.F. Tadros
Imperial College of Science and Technology
University of London
London SW7 2BY, United Kingdom

Dr. K. Kostarelos (✉) · Th.F. Tadros
Zeneca Agrochemicals
Jealott's Hill Research Station
Bracknell
Berkshire RG12 6EY, United Kingdom

Abstract Soybean lecithin disperses into water, forming multilamellar liposomes which, on sonication, produce vesicles of the order of 40–50 nm (diameter), as determined by Photon Correlation Spectroscopy (PCS). The effect of concentration of lecithin and sonication time was systematically investigated. Vesicles were then prepared by incorporation of A–B–A block copolymers of polyethylene oxide (PEO) and polypropylene oxide (PPO), i.e. (PEO–PPO–PEO). Experiments using electrophoresis showed a reduction in the ζ-potential of the vesicles on incorporation of the block copolymer which can be attributed to the shift of the shear plane. Differential Scanning Calorimetry experiments showed changes in the transition temperature of the lipids.

Such results provide some evidence of incorporation. The effect of the molecular weight of the PEO and PPO chains on the vesicle size was systematically studied by using various molecules to prepare the vesicles. Initial addition of these block copolymers causes an increase in the size of the vesicles. This increase continues until a certain concentration of block copolymer is reached, after which a decrease in size is observed. The initial increase was thought to be due to the incorporation of the block copolymer onto the vesicle bilayer. Various models are presented to describe this incorporation.

Key words Liposomes – synperonics – incorporation – PCS – ζ-potential

Introduction

When phospholipids are dispersed in water, they form vesicular bilayer structures best known as liposomes. Since their discovery [1] they have been the subject of considerable research because of the special features they acquire such as: a) the ability to carry both hydrophilic and lipophilic molecules, b) the variety of sizes and structures of liposomes depending on the method of preparation selected, c) their biocompatibility and biodegradability. Their applications include pharmaceuticals (drug carriers), cosmetics, agrochemicals, research tools (e.g., as model cell membranes). The major drawback of these vesicular colloids is their limited physical and biological stability. Indeed serious attempts have been made in order to increase their physical stability, for example, by freeze-thawing techniques [2], and mainly their biological stability, i.e., increasing blood circulation and serum stability, inhibiting protein and macrophage binding, etc. The strategies proposed include: **a)** stiffening of the lipid bilayer by either selection of phospholipids with high transition temperatures (T_c), so that the gel-packing of the lipid chains induces rigidity at low temperatures [3], or inclusion of cholesterol to reduce the mobility of the lipid chains [4], or by polymerization of specific polymerizable groups included in the phospholipid molecules [5, 6]; **b)** modification of the liposome surface to avoid attraction or recognition, by using phospholipids containing charged or uncharged head groups; adding molecules with hydrophillic chains able to extend in the aqueous phase (gangliosides [7], various surfactants [8, 9]) and grafting polyethylene glycol on the phospholipid head group [10].

The main objective of this work is to study the role of added A–B–A block copolymers of the Polyethylene (PEO)–Polypropylene(PPO)–Polyethylene(PEO) type on the formation and stability of the vesicles prepared using soybean lecithin. These block copolymers are well known to sterically stabilize (mainly dispersions [11–15]. It is likely that the hydrophobic component, namely, the PPO, will be strongly adsorbed and/or incorporated inside the vesicle, leaving the PEO chains dangling in solution, thus providing an effective steric barrier. Incorporation of the anchoring group in the bilayer will cause geometric constraints that are thought to govern the aggregation process of surfactants into vesicles as well as micelles, microemulsions, etc. [16].

Materials and methods

Liposomes were prepared with the sonication method, using a Kerry ultrasonic bath. L-α-Dimiristroylphosphatidyl choline (Sigma, ∼ 48% DMPC purity) was mixed with surfactants (ICI Surfactants, Belgium, Synperonics) and both were dispersed in water. The Synperonics are block (triblock) copolymers of the following family structure:

(Polyethylene oxide)$_X$ (Polypropylene oxide)$_Y$ (Polyethylene oxide)$_X$

$$(PEO)_X \quad - \quad (PPO)_Y \quad - \quad (PEO)_X$$
$$A \quad - \quad B \quad - \quad A$$

The triblock copolymers used are shown in Table 1 with some of their characteristic values and the numbers of PEO and PPO units that constitute their lipophilic and hydrophilic chains. All materials were used without any further purification.

Preparation of the vesicles

Vesicles can be formed using a wide variety of techniques and methods already established [17]. In this study the sonication method was used as it was found to produce after approximately 240 min of sonication time, liposomes of 40–50 nm mean diameter. Each dispersion after sonication was filtered through 0.2 μm pore size filters

Table 1 Characteristics of the Synperonics

Synperonic PE grade	Molecular weight	Cloud point, °C (10% aqueous)	Structure
L35	1900	78–82	$(EO)_{12}(PO)_{16}(EO)_{12}$
L61	2090	15–19	$(EO)_{3}(PO)_{30}(EO)_{3}$
L75	4150	84–90	$(EO)_{26}(PO)_{35}(EO)_{26}$
P105	6500	–	$(EO)_{38}(PO)_{54}(EO)_{38}$
PF68	8350	–	$(EO)_{75}(PO)_{30}(EO)_{75}$
PF127	12000	–	$(EO)_{98}(PO)_{67}(EO)_{98}$

(Millipore). The optimum concentration of phospholipids for producing the liposomes was found to be 2% (w/w). The pH of the dispersions was monitored during sonication without any considerable changes, fluctuating between 5 and 6. Therefore, after systematic study the standard vesicle preparation procedures adopted were 240 min sonication time, and 2% (w/w) of lecithin. The samples containing copolymers were prepared exactly as above. The synperonics were mixed initially with the lecithin, letting the two surfactants (lecithin and block copolymer) interact from the start while dispersing in water; so each sample was separately prepared (mixed, sonicated, filtered).

Characterization of vesicles

The characterization of the particles consisted of dynamic light scattering using the PCS 4700 (Malvern) with an Argon laser beam at 48 nm wavelength, interfaced with the

7032CE Correlator for the size distribution analysis. All measurements were performed at 90° angle and 25 °C. Electrophoresis experiments were carried out with the Zetasizer III (Malvern), the voltage applied was 150 mV, and the Hückel approximation was used to derive the zeta potential from the electrophoretic mobility. The samples were diluted approximately 100 times with a 10^{-5} KCl electrolyte solution before each measurement. Differential scanning calorimetry experiments were undertaken using a Heat Flux 4000 DSC (Mettler) system. All samples weighed 300 mg (± 5 mg). Changes in the mean particle diameter, ζ-potential, and the critical transition temperature (T_c) of the liposomal systems were monitored by each technique, respectively.

Progr Colloid Polym Sci (1995) 98 : 69–74
© Steinkopff Verlag 1995

Results and discussion

The photon correlation spectroscopy (PCS) results indicated a gradual increase in the mean diameter of the vesicles with increasing synperonic concentration, until a plateau value was reached (Fig. 1). Further increase of the surfactant concentration above a certain value ($\sim$ 5 wt% for the F127 depicted) led to reduction of the liposome diameter. The increase in vesicle size may be due to adsorption onto the liposome bilayer and/or incorporation of the block copolymer molecules into the bilayer. The decrease above a certain concentration of block copolymer may be attributed to solubilization of the bilayer with the consequent formation of mixed micelles [18, 19, 20]. However, recent cryo-TEM experiments did not show its solubilization effect and it is likely that the reduction could be due to formation of separate synperonic micelles [21].

Additional evidence of adsorption was provided by electrophoresis experiments with the observed decrease in the vesicle ζ-potential (Fig. 2), as the F127 concentration was increased. The extension of the polyoxyethylene oxide chains from the bilayer to the aqueous medium, causing

Zeta potential (mV)

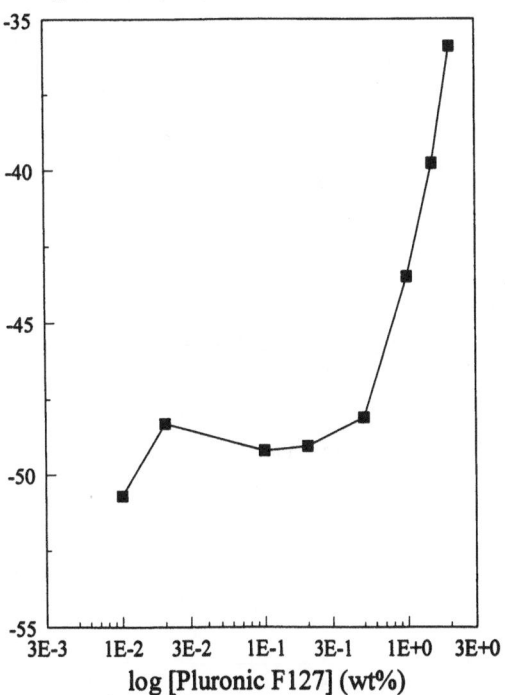

Fig. 2 Changes in vesicle ζ-potential with increasing synperonic F127 concentration (wt%). The smooth increase indicates the shift of the shear plane

a shift of the Stern layer, was considered responsible for this effect [15].

Differential Scanning Calorimetry (DSC) using F127 Synperonic could provide us only with a crude picture that the phase transition pattern of the lipid bilayer was changed with addition of the pluronic, but the exact shifts and changes in the transition temperatures of the lipid mixture used (low level of DMPC purity) could not be detected because the endothermic peaks produced were not clear enough. Incorporation of the polypropylene oxide chain inside the liposomal bilayer and, consequently, disruption of the structure and/or the mobility of the lipid chains, may cause their transition temperature patterns to deviate.

Experiments using series of triblock copolymers (Synperonics)

The different number of units in each homopolymer chain (namely the PEO and PPO) constitutes the series of synperonics, where each member follows certain physicochemical attributes (e.g., solvency, micellisation, etc.), relative to which homopolymer chain exerts more influence on the molecule. The pattern of incorporation

Fig. 1 Mean particle diameter of the vesicle (nm) when adding synperonic PF127, against concentration (wt%). All samples were sonicated for $\sim$ 240 min and the lecithin concentration kept constant at 2 (wt%)

Particle Diameter(nm)

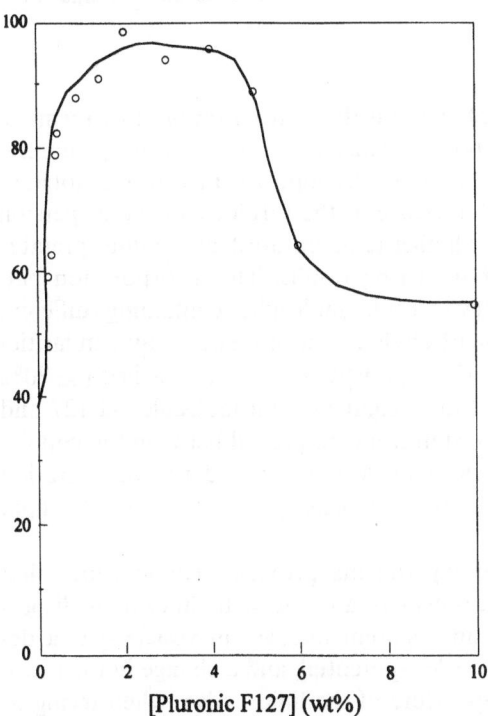

Particle Diameter (nm)

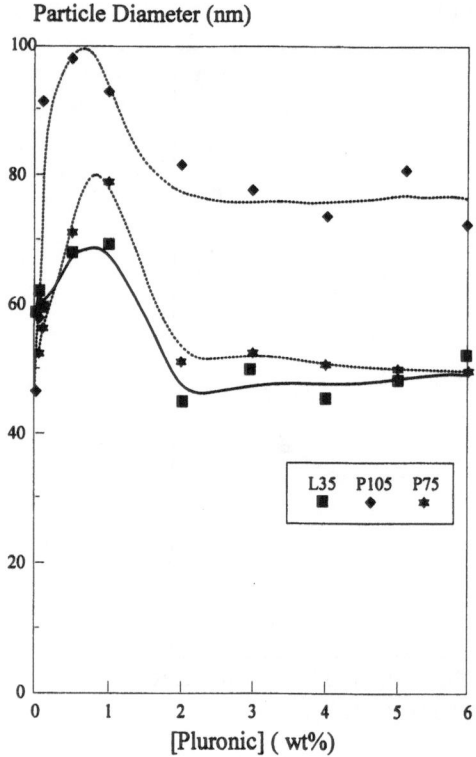

Fig. 3 Vesicle size (nm) as measured by the PCS when the L35, P75, and P105 synperonic molecules are included in the dispersion mixture against their concentration. The synperonics used contain the same percentage of PEO in the total M.wt (% of PEO = ~ 50%), varying in the PPO units/chain: 16, 35 and 54, respectively, for L35, P75 and P105

Particle Diameter(nm)

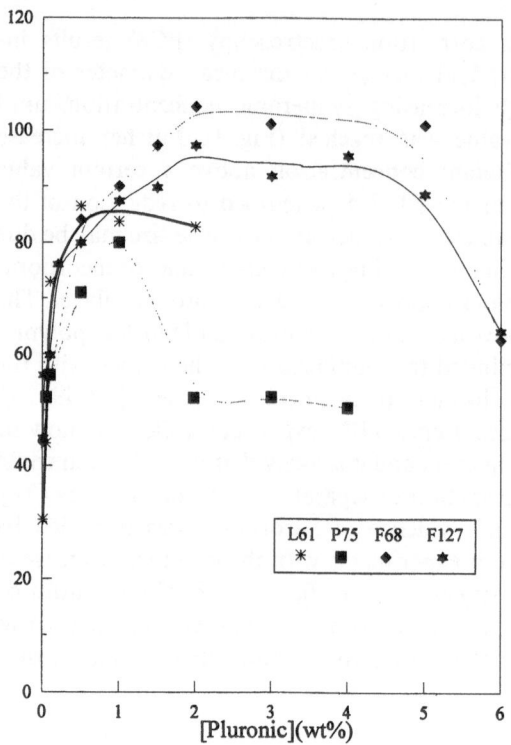

Fig. 4 Same as in Fig. 3, the vesicle diameter (nm) depicted against synperonic concentration (wt%). In this figure, L61, P75, and F68 have approximately the same PPO units/chain = 34, while differing in the PEO units/chain content: 3, 26, and 75, respectively. The turbidity of the dispersions containing > 0.5 − 1 (wt%) L61 increases dramatically (extremely high scattering), indicating the immiscibility of the molecule. Note that F68 and F127 show similar incorporation patterns (F127: PEO units/chain = 98 and PPO units/chain = 65)

was studied as a function of the copolymer molecule size and concentration, by using various synperonics. The possibility of existing dependencies between these patterns and the PEO or PPO polymer chains was examined. In the first case (Fig. 3) the PEO chains of the synperonics used (L35, P75, P105) were maintained at around the same percentage of the total molecular weight. The peak in mean particle diameter at about 1% (w/w) and the identical reduction with further increases of synperonic concentration for all three molecules, indicate that the polyethylene oxide chains play a crucial role as far as the interaction of these molecules with the liposomes is concerned. The "peak" particle diameter differences seem to be in accordance with the different size of the polypropylene oxide chains of the synperonics.

In the next case (Fig. 4), the selected synperonic molecules contained the same PPO homopolymer chain of 30 units (M.wt. ~ 1740). L61 was too hydrophobic even to solubilize in water, and only diminutive amounts could be incorporated in the dispersion, presumably by solubilization in the lipid bilayer of the vesicles. The increase in particle size up to 1 wt% of pluronic concentration is

evident (Fig. 5). Around this concentration, L61 seems to saturate the liposome bilayer. Because of containing minimal hydrophilic chains the copolymer separates, noted as a considerable increase in the turbidity of the dispersion (particularly at higher temperatures), and uninterpretable dynamic light scattering results. The incorporation patterns of the synperonic molecules containing different percentages of PEO chains do not show any similarities unless sufficiently large PEO chains are reached (≥ 70% polyethylene oxide weight in total molecule − F127 and F68). A relationship seems to prevail between the copolymer's polyethylene oxide content and its concentration required to solubilize (at least partly) the vesicular lipid bilayer.

The above experiments provide clear evidence that when block copolymers are added to liposomal dispersions there is an apparent increase in vesicle size, a decrease in the vesicle ζ-potential, and a change in the phase transition temperature of the lipids. Also, when trying to

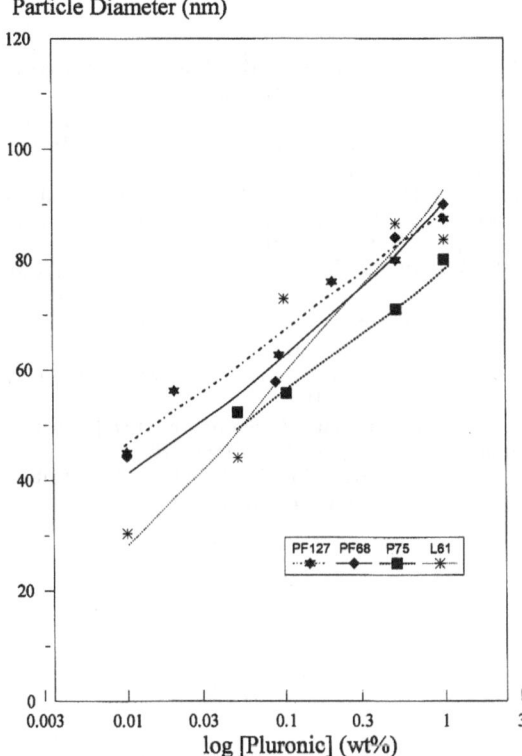

Particle Diameter (nm)

PF127 PF68 P75 L61

log [Pluronic] (wt%)

Fig. 5 The evident increase of the vesicle size when using low concentrations of synperonics is portrayed. Note that the L61 with hardly any hydrophilic chains prompts diameter increase

assess the most influential homopolymer part of the synperonics as to the incorporation patterns followed, the PEO hydrophilic chains seems to be the major factor governing this process, probably by attracting phospholipid molecules in the formation of mixed micelles rather than vesicles leading to the disruption of the bilayer.

Three different models of incorporation may be introduced and they are schematically shown in Fig. 6. In model A, the PPO is assumed to become sandwiched between the lipophilic layers, leaving the PEO chains at the outside and inside, respectively, of the vesicle. In model B, the PPO chain is assumed to become incorporated inside the bilayer in a flat configuration, leaving the PEO chain dangling in solution. In model C, the molecule is simply adsorbing on the surface of the vesicle. The results presented so far cannot give any concrete evidence as to which of the models, or any of their combinations, is the correct one. Experiments are being carried out to elucidate which model is the most likely to be correct, and the results will be presented in the near future.

Fig. 6 Models of triblock copolymer (synperonic) incorporation onto liposomal surfaces

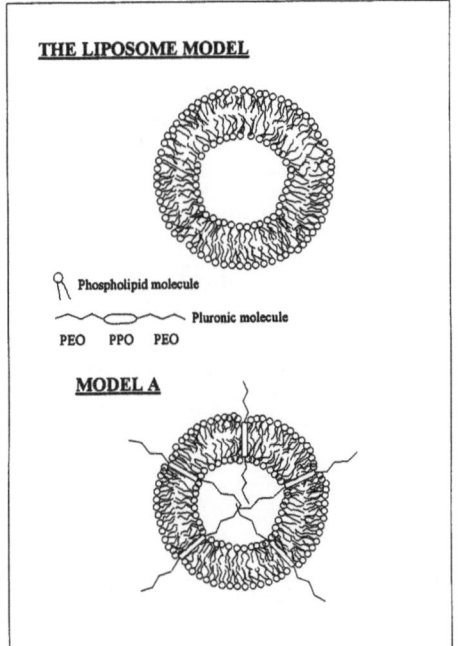

THE LIPOSOME MODEL

Phospholipid molecule

Pluronic molecule

PEO   PPO   PEO

MODEL A

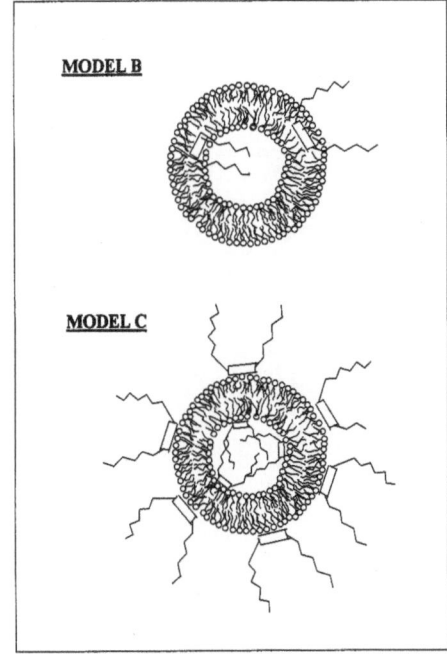

MODEL B

MODEL C

References

1. Bangham AD, Standish MM, Weissman G (1965b) J Mol Biol 13:238–252
2. Crommelin DJA, van Bommel EMG (1984) Pharm Res 4:159–165
3. Chapman D, Williams RM, Ladbrook BD (1967) Chem Phys Lipids 1:445
4. Papahadjopoulos D (1973) In: Prince LM, Sears DF (eds) Biological Horizons in Surface Science, Chapt 5, pp 184, Academic press
5. Albrecht O, Johnston DS, Villaverde C, Chapman D (1982) Biochim Biophys Acta 687:165
6. Bader H, Dorn F, Hupfer B, Ringsdorf H (1985) Adv Polym Sci 64:1–62
7. Allen T, Chonn A (1987) FEBS Lett 223:42–46
8. Kronberg B, Dahlman A, Carlfors J, Karlsson J, Artursson P (1990) J Pharm Sci 79:667–671
9. Virden JW, Berg JC (1992) J Coll Int Sci 153(2):411–419
10. Papahadjopoulos D, Allen TM, Gabizon A, Mayhew E, Matthay K, Huang SK, Lee K-D, Woodle MC, Lasic DD, Redemann C, Martin FJ (1991) Proc Natl Acad Sci USA 88:11460–11464
11. Stolnik S, Davies MC, Illum L, Davis SS, Boustta M, Vert M (1994) J Contr Release, 30:57–67
12. Kuo PL, Okamoto M, Turro NJ (1987) J Phys Chem 91:2934–2938
13. Claesson P (1993) Colloids and Surfaces 77:109–118
14. Malmsten M, Linse P, Cosgrove T (1992) Macromolecules 5:2474
15. Muller RH (1991) Colloidal Carriers for Controlled Drug Delivery and Targeting, CRC Press
16. Israelachvili JN, Mitchel DJ, Ninham BW (1977) Biochim Biophys Acta 470:185–201
17. Szoka F, Papahadjopoulos D (1980) Ann Rev Biophys Bioeng 9:467–508
18. Edwards K, Almgren M (1991) J Coll Int Sci 147(1):1–21
19. Ueno M, Akechi Y (1991) Chem Lett 1801–1804
20. Helenius A, Simons K (1975) Biochim Biophys Acta 415:29–79
21. Kostarelos K, Luckham PF, Tadros ThF (to be published)

Progr Colloid Polym Sci (1995) 98:75–78
© Steinkopff Verlag 1995

J.A. McDonald
A.R. Rennie

Scattering studies of mixed micelles formed from $C_{16}TAB$ and $C_{12}E_6$ surfactants

Dr. J.A. McDonald (✉) · A.R. Rennie
Department of Physics
Cavendish Laboratory
Cambridge University
Madingley Road
Cambridge CB3 0HE, United Kingdom

Abstract Scattering studies have been performed on a mixed micellar system containing a cationic surfactant $C_{16}TAB$ (hexadecyl trimethylammonium bromide) and a non-ionic surfactant $C_{12}E_6$ (hexaethylene glycol monododecyl ether). Light-scattering measurements indicate that the addition of salt (KBr) initiates unidirectional growth of the mixed micelles into elongated aggregates. These elongated micelles exhibit some flexibility and can be characterised by scaling laws used to model polymer solutions. Small-angle neutron scattering (SANS) measurements indicate that the micelles usually prefer to exist as a single population of homogeneous aggregates, although evidence is obtained to suggest that some degree of de-mixing may occur in more dilute surfactant solutions containing salt.

Key words Mixed micelles – light scattering – small-angle neutron scattering – growth effects – micelle composition

Introduction

Micelles of cationic surfactants often undergo a transition from spheres to rods in the presence of salt and form flexible worm-like chains [1–4]. This process of unidirectional growth has been attributed to screening of the charged head-groups by the salt ions which favours packing of the surfactant molecules into rod-like micelles [5]. In this paper we examine salt-induced micelle growth from an aqueous mixed surfactant system containing the cationic surfactant $C_{16}TAB$ (hexadecyl trimethyl ammonium bromide) and the non-ionic surfactant $C_{12}E_6$ (hexaethylene glycol monododecyl ether) at $C_{16}TAB$ mole fraction of 0.55. A combination of light and small-angle neutron scattering has been used to obtain information on overall dimensions and shape of the micelles. In addition, the contrast variation method has been used to determine the composition of the micelles.

Light scattering measurements

Static light scattering

The excess scattered intensity from a dilute dispersion of weakly interacting micelles may be approximated by:

$$K(c - c_0)/\Delta R_\theta = [1 + (\langle R_g^2 \rangle_z \, q^2/3)]/M + 2B_2(c - c_0), \quad (1)$$

where c is the total surfactant concentration, c_0 is the critical micelle concentration, K is a constant, M is the weight-average micelle mass, B_2 is the second virial coefficient and q is the scattering wave vector; $q = 4\pi n_0/\lambda_0 \sin(\theta/2)$, where θ is the scattering angle and λ_0/n_0 is the wavelength of incident light in the sample. The parameter ΔR_θ is the Rayleigh ratio of the micelles which is obtained from the difference between the Rayleigh ratios of the surfactant solution and the solvent medium. R_g is the radius of gyration.

Dynamic light scattering

In dynamic light scattering experiments the apparent diffusion coefficient (D_q) is related to the decay rate of the time correlation function (Γ) by:

$$D_q = \Gamma/q^2 . \tag{2}$$

The apparent diffusion coefficient may contain contributions from both rotational and translational motion. In order to obtain the pure translational diffusion coefficient (D_0) it is necessary to extrapolate the apparent diffusion coefficient to zero scattering wave vector. The hydrodynamic radius (R_h) of the micelles may then be calculated from the translational diffusion coefficient according to the Stokes–Einstein equation.

$$R_h = kT/6\pi\eta D_0 , \tag{3}$$

where η is the solvent viscosity, k is the Boltzmann constant and T is the absolute temperature.

Results

For solutions containing KBr the reciprocal scattering intensity (expressed as $K(c - c_0)/\Delta R_\theta$) is first seen to decrease as a function of surfactant concentration before passing through a minimum and then slowly increasing again. This behaviour is reflected by the increase and then decrease in R_g indicated in Fig. 1. The initial decrease in reciprocal scattering intensity can be attributed to a process of micelle growth provided that interactions between the micelles are negligible (i.e. $B_2 \sim 0$). This assumption is considered to be reasonable given that KBr is expected to screen out electrostatic repulsions from dissociated $C_{16}TAB$ molecules. The mass of the micelles (M) can then

be estimated from the minimum in reciprocal scattering intensity. Using this approach micelle masses are found to be 2.3×10^5 in 0.1 M KBr and 1.3×10^6 in 0.5 M KBr. These values compare with a lower micelle mass of $\sim 7 \times 10^4$ measured from SANS for the mixed micelles in pure water.

The concentration where the minimum in reciprocal scattering intensity occurs may be identified as a "crossover" concentration, c^*, between a dilute and semi-dilute regime. This treatment, originally developed for semiflexible polymers, has also been used to describe cationic micelles which undergo similar growth processes in the presence of salt to form worm-like chains. Below c^*, in the dilute region, the micelles are viewed as diffusing almost independently of each other, whereas on passing into the semi-dilute region micelle growth has proceeded to an extent that the micelles start to overlap and form an entangled network. In the semi-dilute region the scattering is now dominated by correlation lengths (ξ) which characterise the entangled network and not by the size of individual micelles. This idea of micelle growth is supported by dynamic light scattering results. The translational diffusion coefficient (D_0) is found to pass through a minimum at roughly the same concentration as the maximum in R_g (Fig. 1).

Scaling behaviour can provide useful information on the conformation and flexibility of micelles. The scaling relationship between micelle mass and radius of gyration is sensitive to micelle shape [6]:

$$R_g \sim M^n . \tag{4}$$

Figure 2 shows a double logarithmic plot of R_g versus micelle mass for micellar solutions in 0.5 M KBr. The exponent of 0.53 derived from this plot is characteristic of a random coil structure. Rigid, thin rods are expected to give an exponent close to unity.

Fig. 1 Variation of radius of gyration and translational diffusion coefficient with surfactant volume fraction in 0.5 M KBr: R_g (●); D_0 (◆)

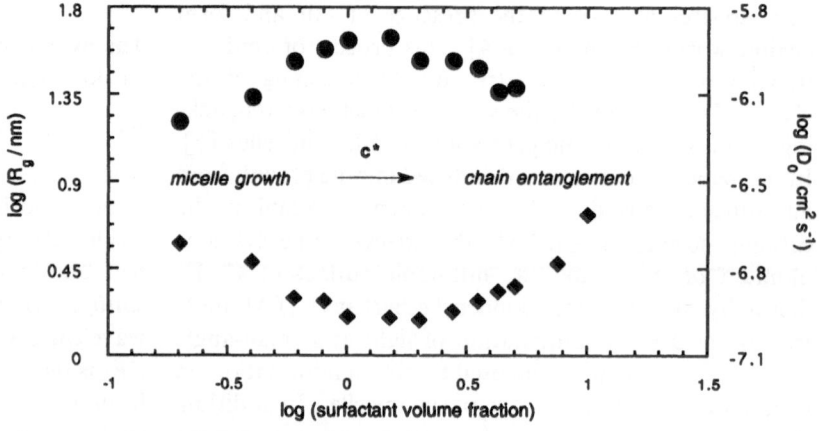

c^* is the cross-over concentration

Progr Colloid Polym Sci (1995) 98:75–78
© Steinkopff Verlag 1995

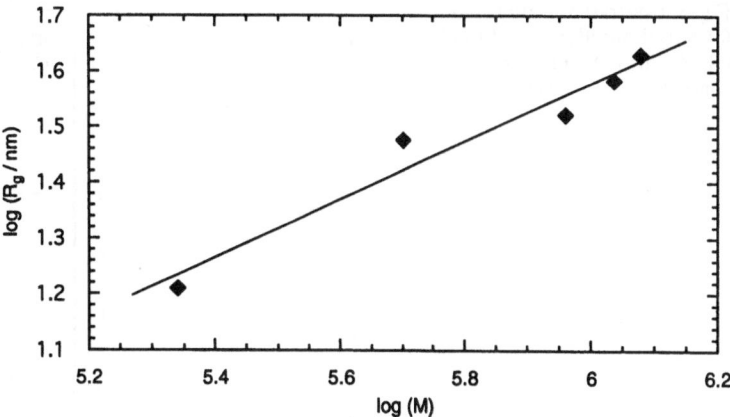

Fig. 2 Scaling relationship between radius of gyration (R_g) and micelle mass (M) in 0.5 M KBr

Small-angle neutron scattering

Analysis of the scattering profile

For a monodisperse system of interacting micelles the scattered intensity $I(q)$ as a function of the scattering wave vector q is given by:

$$I(q) = K (\rho_M - \rho_S)^2 \, v^2 \, c_M \, M \, P(q) \, S(q) \,, \qquad (5)$$

where K is a calibration constant, M is the micelle mass as before, c_M is the micelle concentration, and v is the specific volume of the micelle; ρ_M and ρ_S are the scattering length densities of the micelle and solvent, respectively. $P(q)$ is the particle form factor and describes the scattering contribution from individual micelles. $S(q)$ is the structure factor and describes the scattering contribution from inter-micellar interactions. For a dilute solution where interactions are negligible $S(q) = 1$. In these experiments mixtures of fully hydrogenated $C_{12}E_6$ and partially deuterated d-C_{16}h-TAB were used as the surfactant components. By conducting the experiments with solutions containing this surfactant combination in both D_2O and H_2O solvents it was possible to focus on the scattering contributions from each surfactant.

Determination of micelle composition

Following from Eq. (5), it can be shown that:

$$I(q)^{1/2} \sim V(\rho_m - \rho_s) \,, \qquad (6)$$

where V is the micelle volume. In the so-called contrast variation method, the variation of $I(0)$ with $\rho_m - \rho_s$ is determined by changing ρ_s with H_2O/D_2O mixtures. A plot of $I(0)^{1/2}$ versus ρ_s should be linear and pass through zero when the solvent scattering length density

matches that of the micelle. The value of ρ_m determined can then be compared with the value calculated from the scattering length increments assuming a truly mixed micelle. This comparison provides an indication of whether the micelles are homogeneous or if de-mixing into separate micelle populations is occurring.

Results

For mixed micelles in pure water the scattering data were analysed assuming a form factor for spherical monodisperse particles and a structure factor based on the mean spherical approximation (MSA). Applying the MSA it is assumed that particle interactions are dominated by Coulombic repulsion from dissociated C_{16}TAB molecules. Micellar radii of ~ 2.8 nm with inter-micellar potentials in the region of 70 mV were extracted from this analysis. Values were independent of surfactant concentration and solvent deuteration. Radii were usually slightly lower in H_2O than in D_2O (by $\sim 10\%$) suggesting that the $C_{12}E_6$ molecules extend further from the centre of the micelle than the C_{16}TAB molecules. Applying the contrast variation method to 0.05 M surfactant solutions in pure water a linear plot was obtained with a matchpoint (i.e. $\rho_m = \rho_s$) close to that calculated for a homogeneous micelle (Fig. 3). It would seem that the micelles are truly mixed in these solutions.

Adding KBr to the mixed micellar solutions has a marked effect on the scattering intensity profile. The interaction peak, previously observed for micelles in pure water, is now absent, indicating that the KBr has effectively screened out any inter-micellar Coulombic repulsion. Good fits to the data are obtained by assuming a form factor for simple rods of length ~ 9 nm. The radius of the rods extracted from this analysis (~ 2 nm) is a little lower than for the spherical micelles in pure water,

Fig. 3 Contrast variation plots for mixed micelles: no KBr (♦); 0.1 M KBr (◇). Surfactant concentration = 0.05 mol dm^{-3}

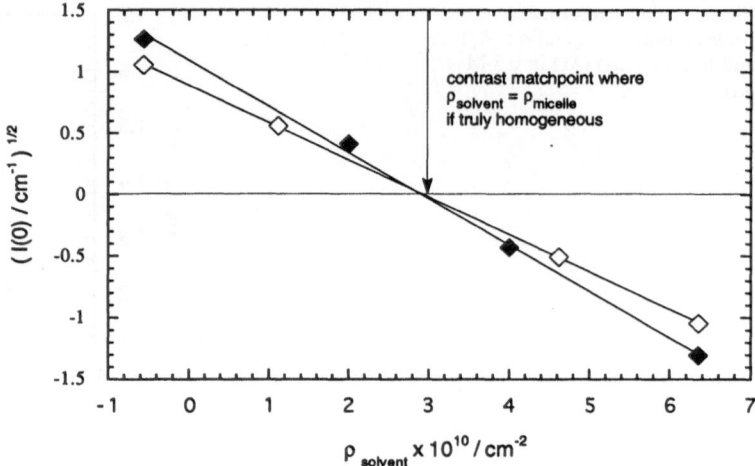

suggesting that the surfactant molecules occupy a smaller cross-sectional area when packed into a cylindrical aggregate. Applying the contrast variation method to solutions containing 0.05 M of surfactant a linear plot was also obtained, indicating that the micelles are homogeneous (Fig. 3).

At lower surfactant concentrations (0.01 M) the apparent rod length increases to 15 nm in D_2O solutions but is only 6 nm in H_2O solutions. This disparity in micelle rod-length, depending on which surfactant is being emphasised in the scattering profile, suggests that some form of de-mixing may be occurring. We offer two possibilities to account for this behaviour. One explanation is that the screening effect of the KBr favours closer packing of the $C_{16}TAB$ molecules with the net result that the $C_{12}E_6$ molecules become excluded and begin to form a separate micelle population. Alternatively, some form of surfactant segregation may be taking placing *within* the micelles. For example, non-ionic surfactant molecules may concentrate towards the ends of the elongated micelle with $C_{16}TAB$ molecules becoming localised in the middle section of the micelle. Such a process of internal segregation would be consistent with the longer rod lengths found when analysing the scattering contribution from the hydrogenated non-ionic surfactant in D_2O. At present, we cannot distinguish which of these segregation processes is actually occurring.

Conclusions

By employing a combination of scattering methods, we are able to show that $C_{12}E_6$ and $C_{16}TAB$ form mixed micelles which undergo a process of unidirectional growth in the presence of salt (KBr). Similar behaviour has been observed in some pure cationic micellar systems. Growth occurs as screening of charges on the $C_{16}TAB$ molecules allows closer packing of the molecules into a cylindrical structure. Micelle growth is also a function of surfactant concentration if salt is also present. These elongated micelles are not completely rigid but exhibit some flexibility. The scaling relationship between R_g and micelle mass indicates that a random coil structure is preferred.

Small-angle neutron scattering (SANS) through the technique of contrast variation is an appropriate method for focusing on the scattering contributions of different surfactants in a mixed solution and hence for determining the composition of mixed micelles. Using SANS, we find that the mixed micelles examined in this study are usually homogeneous, although there is evidence to suggest that more dilute solutions undergo some form of de-mixing in the presence of salt.

Acknowledgements We are grateful to the Department of Trade and Industry in conjunction with Unilever plc, ICI and Schlumberger Cambridge Research for funding this work.

References

1. Candau SJ, Hirsch E, Zana R (1984) J Phys (Paris) 45:1263
2. Candau SJ, Hirsch E, Zana R (1985) J Colloid Interface Sci 105:521
3. Imae T, Kamiya R, Ikeda, S (1985) J Colloid Interface Sci 108:215
4. Imae T, Ikeda S (1986) J Phys Chem 90:5216
5. Clint JH (1992) Surfactant Aggregation; Blackie: London
6. DeGennes PG (1979) Scaling Concepts in Polymer Physics; Cornell University Press: Ithaca NY

Progr Colloid Polym Sci (1995) 98:79–84
© Steinkopff Verlag 1995

MIXED SYSTEMS

W. Richtering
K. Berend

Dynamics of monodisperse and bidisperse polymer latices

Dr. W. Richtering (✉) · K. Berend
Institut für Makromolekulare Chemie
Albert-Ludwigs-Universität Freiburg
Stefan-Meier-Straße 31
79104 Freiburg, FRG

Abstract Dynamic properties of monodisperse and bimodal mixtures of electrostatically stabilized polymer latices have been studied at high concentration. Macroscopic properties were measured by rheometry and viscous as well as elastic behavior was investigated. Information on particle diffusion was obtained by means of fiber optical quasi elastic light scattering (FOQELS). In monodisperse systems, short time self-diffusion of particles at high ionic strength could be described by theoretical calculations for hard spheres although shear viscosity showed deviations from hard sphere behavior. Elastic properties were observed by rheology at high volume fractions and the existence of a yield stress could also be optically detected by FOQELS. In bimodal mixtures, a "probe diffusion", i.e., diffusion of a small number of large spheres in a matrix of small particles could be measured by FOQELS. Probe diffusion was strongly slowed down with increasing volume fraction and the concentration dependence was similar to that of viscosity. Variation of particle size ratio and composition allowed for the possibility to measure diffusion of small and/or large particles in the mixture by FOQELS.

Key words Latex – colloidal dispersion – diffusion – bimodal dispersion – viscosity

Introduction

Concentrated polymer dispersions have been intensely studied with respect to their rheological behavior. In electrostatically stabilized polymer latices the interaction potential depends on the ionic strength and colloidal crystals can be observed at low particle volume fraction and low ionic strength. Sterically stabilized hard sphere systems can be used to investigate the glass transition [1]. Information on microscopic dynamic properties, i.e., the diffusive motion of the particles, is of great interest in order to correlate local dynamics with macroscopic properties of the sample. Diffusion coefficients are usually determined by means of quasi elastic light scattering (QELS), but this technique requires the solution to be optically transparent, which is not the case for concentrated dispersions. Therefore, only dispersions for which index matching of colloidal particles and suspension medium is feasible, can be studied by QELS at high concentrations. A new technique to determine diffusion coefficients in turbid solutions has recently been developed by Horn and coworkers and uses a fiber optical QELS (FOQELS) to detect the back scattered light. Wiese and Horn demonstrated the usefulness of this technique [2].

The objective of this contribution is to compare rheological data and diffusion coefficients of concentrated polymer dispersions. There are only few studies that compare results obtained from the same system. To this end, we investigated i) suspensions of monodisperse electrostatically stabilized polymer latices, and ii) bimodal mixtures of these monodisperse dispersions at different composition and size ratio of particles.

Theoretical considerations for both rheology and light scattering have been frequently presented in the literature [3–5] and the reader is referred to the indicated references for further information. QELS probes particle movements on a length scale given by $2\pi/q_{exp}$ (q being the scattering vector at the experimental conditions), which has to be compared to the position of the first maximum of the structure factor q_m [5]. q_m is related to the distance between particles and their surrounding nearest neighbors in the system and, to a first approximation, is proportional to $\rho_N^{-1/3}$ (ρ_N being the particle number per volume). If $q_{exp} \geq q_m$, FOQELS measures the short time self-diffusion coefficient D_{st}^s. The particle displacements are small in comparison to the interparticle spacing. If $q_{exp} < q_m$ collective diffusion (combined diffusion of many different particles) is measured. In conventional QELS q_{exp} is variable due to the different scattering angles, whereas it is fixed in FOQELS experiments.

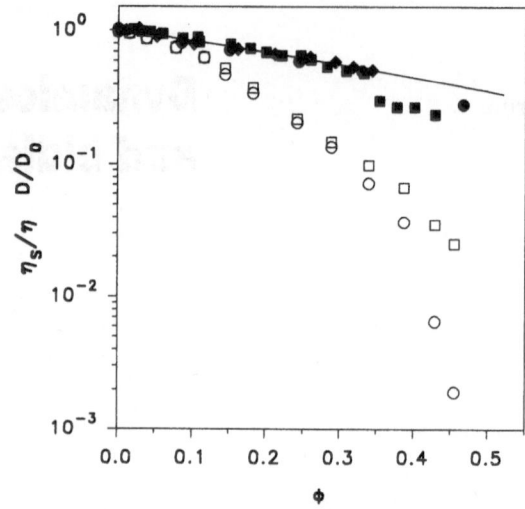

Fig. 1 Comparison of concentration dependence of normalized diffusion coefficient (filled symbols) and relative viscosities (open symbols, $\bigcirc$: η_0, $\square$: η_∞) for 164 nm latex

Experimental

Synthesis and characterization of monodisperse latices have been described elsewhere [6]. Particle diameters were 304, 164, and 64 nm, respectively. Particle size distributions were narrow. Rheological tests were performed with a Bohlin CS 10 rheometer. A self-constructed FOQELS set-up was used to measure diffusion coefficients at high concentration. The optical set-up is equivalent to that described by Wiese and Horn [2]. Time-correlation functions were measured with an ALV-5000 correlator. A HeNe laser was used as light source, thus defining the experimental scattering vector at $q_{exp} = 2.647 \cdot 10^5$ cm^{-1}.

Results and discussion

Monodisperse latices

Figure 1 shows a comparison of normalized diffusion coefficient and viscosity of concentrated monodisperse polymer latices versus particle volume fraction. The viscosity data are given as inverse relative viscosity at low $(\eta_0/\eta_S)^{-1}$ and high shear rate $(\eta_\infty/\eta_S)^{-1}$, respectively, η_S being the viscosity of water. The diffusion data in Fig. 1 agree nicely with theoretical predictions by Beenakker and Mazur [7] for short time self-diffusion of hard spheres. The concentration dependence of the relative viscosities (at high and low shear rate) cannot be satisfactorily be described by the Dougherty-Krieger or Quemada equations [8, 9], which are known to describe hard sphere suspen-

sions. Obviously, deviations from hard sphere interaction influence viscosity, but not short time self-diffusion.

When the particle interaction potential was increased further, with increasing concentration and/or decreasing ionic strength, dispersions became viscoelastic. A transition from liquid like to solid like behavior was found in a narrow concentration range. Elastic properties could be detected by various rheometrical tests:

i) A yield stress was observed when a ramp of increasing shear stresses was applied to the samples.
ii) In constant stress, i.e., creep experiments, a spontaneous recoverable compliance was found.
iii) A plateau modulus was observed in dynamic-mechanical experiments.
iv) Low shear rate viscosity diverged.

The solidification transition could also be detected by FOQELS. Close to the transition, the decay of the time correlation function became much broader. For elastic samples the time correlation function no longer decayed completely and strong count rate fluctuations were observed during the measurements. Figure 2 shows two intensity autocorrelation functions corresponding to a liquid (top) and solid (bottom) sample, respectively, as well as G' and G'' data of the same samples. The predominantly viscous sample showed elastic properties only at the highest frequencies accessible (G' and G'' have about the same magnitude there) and was clearly viscous at lower frequencies. The higher concentrated sample showed elastic properties even at the smallest frequency (G' stayed constant and was always higher than G'') and the time

Progr Colloid Polym Sci (1995) 98:79–84
© Steinkopff Verlag 1995

Fig. 2 Time-correlation function and storage and loss modulus of a liquid-like sample (top) and a solid-like sample (bottom)

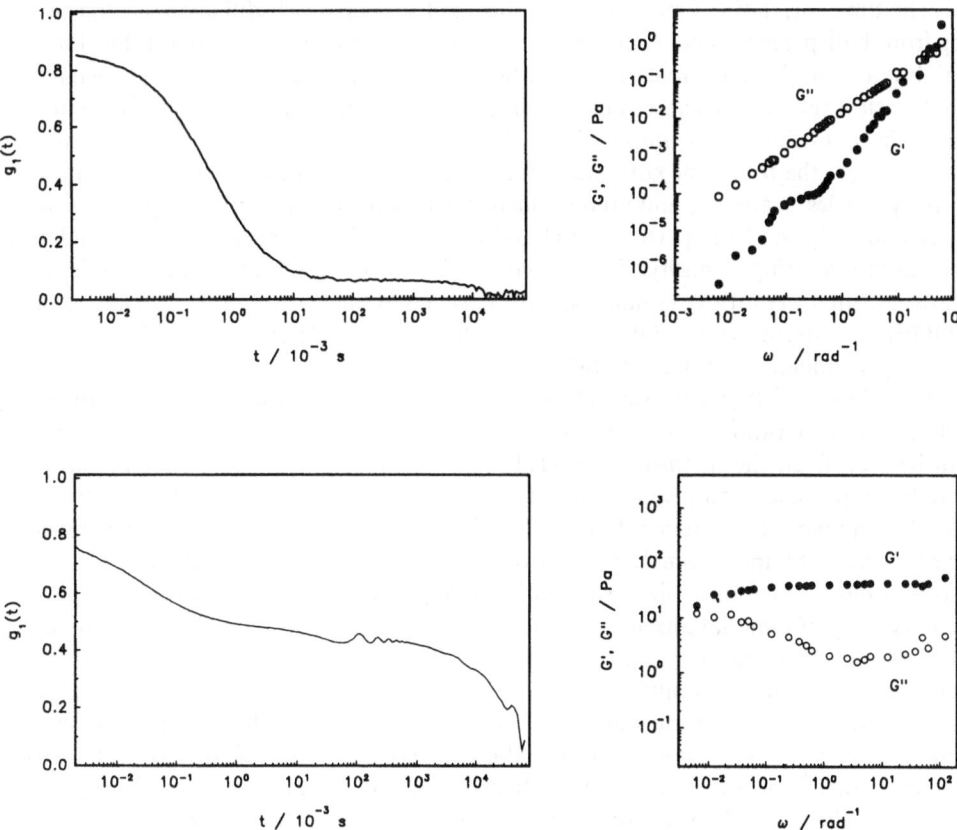

correlation function did not decay completely. Such a behavior is also observed for gel-like samples (e.g., cross-linked swollen networks) and is typical of nonergodic systems [10]. Thus, FOQELS provides a convenient optical tool to determine the existence of elastic properties in unperturbed samples.

Bimodal latices

The viscosity of colloidal suspensions can be influenced by mixing particles of different sizes [11, 12]. If a small number of large particles is replaced by small particles, the viscosity is decreased in comparison to the monodisperse latex at the same total volume fraction of suspended particles. The decrease of viscosity is mainly caused by an increase of the possible maximum volume fraction. The maximum volume fraction depends on the size ratio of the spheres and on the composition of the latex, i.e., on the ratio of volume occupied by small and large particles, respectively. If the total volume fraction is kept constant, a minimum in viscosity is observed when the small particles contribute approximately 25% to the total volume fraction [12]. On the other hand, viscosity is hardly affec-

ted if a few large particles are added to a suspension of small spheres.

We investigated bimodal mixtures at three different size ratios $\lambda = a_L/a_S$ (a_L, a_S radius of large and small particles, respectively) in the range of 1.8 to 4.6 and, additionally, the composition of the samples with regard to the content of small and large particles was varied. Table 1 gives a list of mixtures investigated in this study.

Table 1 Bimodal mixtures of monodisperse latices. m_S and m_L denote the partial mass fraction (in %) of small and large spheres in the mixture. $N_S:N_L$ denotes the number ratio and $I_S:I_L$ the intensity ratio at back scattering

Mixture	$a_S:a_L$	$m_S:m_L$	$N_S:N_L$	$I_S:I_L$
64 nm/164 nm	2.56	79.25 : 20.75	64.3 : 1	1 : 1.8
		50 : 50	16.8 : 1	1 : 7
		24.1 : 75.9	5.3 : 1	1 : 15
64 nm/304 nm	4.75	81.1 : 18.9	460 : 1	5.1 : 1
		50.3 : 49.7	108 : 1	1.2 : 1
		24.5 : 75.1	35 : 1	1 : 2.5
164 nm/304 nm	1.85	79.6 : 20.4	24.8 : 1	32.8 : 1
		50.3 : 49.7	6.4 : 1	8.5 : 1
		17.4 : 82.3	1.4 : 1	1.77 : 1

The time correlation function of dynamic light scattering from bidisperse samples is no longer a single exponential, but a sum of exponentials, each of which decays with a relaxation time that corresponds to the specific particle size. The contribution in amplitude of the different exponentials to the total time correlation function is related to the particles' scattering amplitude, which is determined by size and shape of the particles [13] and is reflected in the static scattering intensity. Table 1 gives a list of the scattering intensity ratios of small and large particles at the FOQELS scattering vector. Obviously, the intensity ratio is strongly influenced by the size ratio.

In a 79.3:20.7 (w/w) mixture of 64 and 164 nm particles, a number ratio of $N_S : N_L = 64:1$ is present, but as can be seen from the intensity ratio (1:1.76), both small and large particles contribute significantly to the time correlation function and two well separated diffusion coefficients were obtained from Laplace inversion. Figure 3 shows a plot of the normalized slow diffusion coefficient and viscosity of bimodal mixtures of this size ratio versus total particle content. For the sake of clarity the fast diffusion coefficient was omitted in the plot. Comparison to Fig. 1 reveals a significant difference between bimodal and monodisperse dispersions. The slow diffusion coefficient did not follow the theoretical predictions for short time self-diffusion at high concentrations, but decreased much stronger. The shape of the curve is similar to that of viscosity.

Recently, Imhof and coworkers [14] reported on a comparison between zero shear viscosity and long time self-diffusion coefficient in concentrated suspensions of silica particles. The long time self-diffusion coefficient, D_{lt}^s, describes the Brownian motion of particles in the long time limit. That means that particles move over large distances compared to the interparticle separation, and thus have enough time to leave the "cage" formed by surrounding particles. Imhof et al. observed a very similar concentration dependence of D_{lt}^s and η_0. Since we had obtained comparable results for the slow diffusion coefficient in bimodal latices, we wondered whether this could be associated with a long time self-diffusion.

As discussed previously, the ratio of interparticle separation and the length scale probed by FOQELS ($2\pi/q_{exp}$) decides whether collective or single particle motions are observed. In bimodal latices, different particle distances have to be discussed: the mean interparticle distance for all particles and the distance between particles of same size. In the case of the 79.3:20.7 mixture of 64 and 164 nm particles, the mean interparticle distance and the separation of small particles is smaller than $2\pi/q_{exp}$, whereas the distance between 164 nm particles is much larger. Two diffusion processes were observed in the FOQELS experiment on the mixtures which can be explained as follows: Since the number of large particles is very small, and thus the mean interparticle spacing almost identical with the separation of small particles, the fast diffusion is attributed to a collective motion of the small spheres, which is also observed in the pure 64 nm latex. The concentration dependence in both cases, pure latex and mixture, is fairly well in agreement with theoretical predictions [7] for collective diffusion. The slow diffusion coefficient is attributed to self-diffusion of the large particles, which is hindered due to the presence of the small spheres and thus decreases rapidly with increasing concentration.

Similar effects were observed for mixtures with the same size ratio but different composition (50:50 w:w), i.e., containing more large particles. The interparticle distance increases with the fraction of large spheres, but particle separation for the small particles is still smaller than $2\pi/q_{exp}$. So, again, two diffusion processes were observed, but since the number of small particles is lower, the slow diffusion of the large particles is not as strongly hindered as in the 79.3:20.7 mixture. A comparison also reveals that the dispersion viscosity is decreasing with increasing fraction of large particles.

Results obtained from a 50:50 mixture at a different size ratio (64 nm and 304 nm particles) are displayed in Fig. 4. Again, two diffusion coefficients were found that could be associated with mutual diffusion of small and self-diffusion of large spheres, respectively. Only the latter was included in Fig. 4. Again, the self-diffusion of the large particles deviated from the short time behavior and was close to the viscosity data.

Fig. 3 Comparison of normalized diffusion coefficient and inverse relative viscosities of bimodal mixture of 64 and 164 nm latices at a composition of 80:20 (w:w)

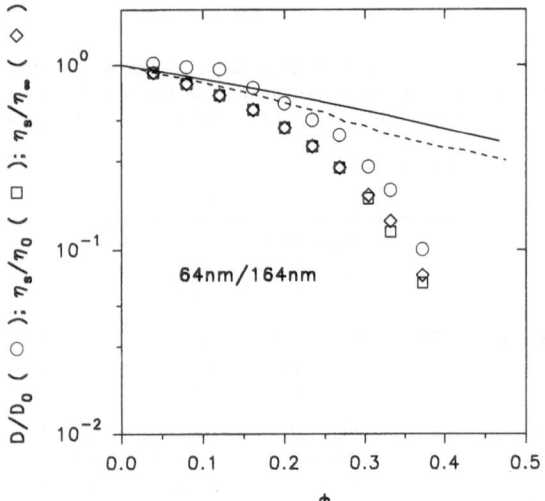

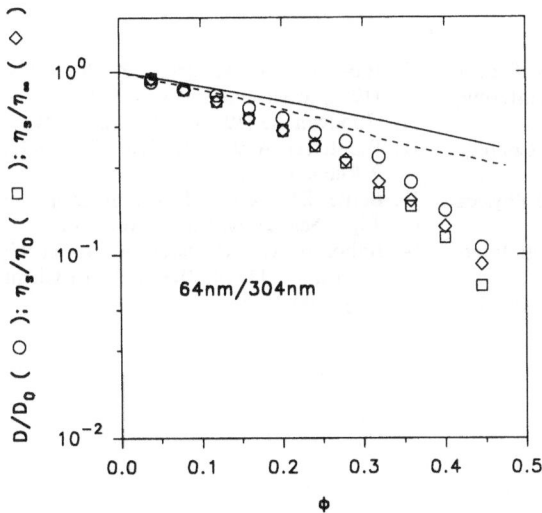

Fig. 4 Comparison of normalized diffusion coefficient and inverse relative viscosities of bimodal mixture of 64 and 304 nm latices at a composition of 50:50 (w:w)

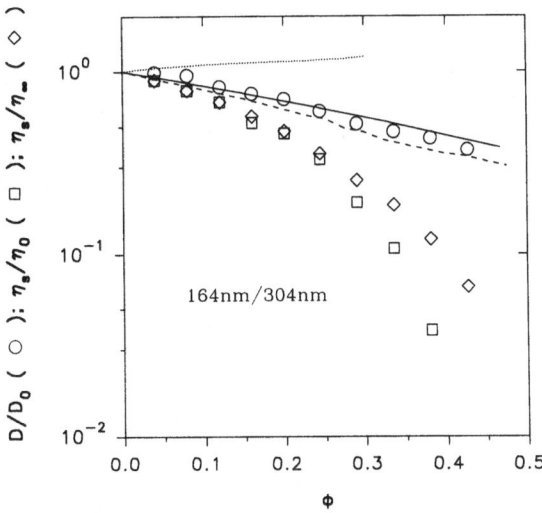

Fig. 5 Comparison of normalized diffusion coefficient and inverse relative viscosities of bimodal mixture of 164 and 304 nm latices at a composition of 80:20 (w:w)

In all mixtures discussed so far, only a small number of large particles was present in the dispersion, nevertheless contributing significantly to the scattering intensity. FOQELS could detect two diffusion processes, the slower one being the diffusion of large spheres hindered by the surrounding small particles. The concentration depend-

ence of this "probe diffusion" was similar to that of viscosity and agreed well with the findings by Imhof et al. [14] for long time self-diffusion. It appears that "probe diffusion" of large particles in a bimodal dispersion with a much larger number of small particles resembles long time self-diffusion in monodisperse suspensions.

More complicated behavior, however, was found with mixtures of 164 and 304 nm particles (see Fig. 5, where results for a 80:20 mixture are presented). In this case scattering of large particles is extremely weak in comparison to that of the small ones (see Table 1) and thus only one diffusion coefficient could be measured, which was basically identical to that found in pure 164 nm particle dispersions and which corresponds to short time self-diffusion as shown in Fig. 1. Still, only one diffusion process could be observed when the fraction of large particles was increased. However, interpretation became more difficult, since at higher fractions of large particles they do contribute to the measured correlation function. Mean interparticle spacing and particle separations for the two types of particles individually were of the order of $2\pi/q_{\mathrm{exp}}$, so that for both diffusion processes self-diffusion was expected. Inverse Laplace transformation did not yield two separate diffusion coefficients, but only one, which appeared to be a weighted average of the two expected. Diffusion of 164 nm particles dominated in both the 80:20 and 50:50 mixtures (Intensity ratios of 32.8:1 and 8.5:1, respectively, number ratios 24.8:1 and 6.4:1). At a composition of 20:80, the measured diffusion coefficient was predominately influenced by the large particles, but the contribution from the small ones was far from negligible (intensity ratio small to large 1.78:1, number ratio 1.35:1). So, the results from FOQELS measurements at these conditions cannot be as clearly interpreted as for the mixtures discussed previously.

All the examples discussed in this paper demonstrate that FOQELS provides an interesting tool to study dynamic properties of bimodal latices. Variation of particle sizes and size ratios offers the possibility to simultaneously observe diffusion of small and/or large particles. Comparison between rheology and diffusion in bimodal suspensions provides information on possible relations between macroscopic and microscopic dynamic properties. Shear viscosity mainly depends on maximum volume fraction which is a function of the size ratio. Diffusion of particles in the long time limit seems to depend on the shell of nearest neighbors and further variation of size ratio and composition might give more information on the processes that control particle mobility at different concentrations.

Acknowledgement We thank the Deutsche Forschungsgemeinschaft for financial support. K.B. thanks the Fonds der chemischen Industrie for a fellowship.

References

1. Russel WB, Saville DA, Schowalter WR (1989) Colloidal Dispersions, Cambridge University Press, Cambridge
2. Wiese H, Horn D (1991) J Chem Phys 94:6429
3. Buscall R, Goodwin JW, Hawkins MW, Ottewill RH (1982) J Chem Soc Faraday Trans I 78:2873
4. Tadros ThF (1989) Progr Colloid Polym Sci 79:120
5. Pusey PN, Tough RJA (1985) In: Pecora R (ed) Dynamic Light Scattering, Plenum Press New York
6. Berend K, Richtering W (1995) Colloids Surfaces A, accepted
7. Beenakker CWJ, Mazur (1984) Physica 126A:349
8. Krieger IM (1972) Adv. Colloid Interface Sci 3:111
9. Quemada D (1977) Rheol Acta 16:82
10. Joosten JGH, McCarthy JL, Pusey PN (1991) Macromolecules 24:6690
11. Goodwin JG (1975) Colloid Sci 2:245
12. Woutersen ATJM, de Kruif CG (1992) J Rheol 37:681
13. Berne BJ, Pecora R (1976) Dynamic Light Scattering, Wiley, New York
14. Imhof A, van Bladeren A, Maret G, Mellema J, Dhont JKG (1994) J Chem Phys 100:2170

Progr Colloid Polym Sci (1995) 98:85–88
© Steinkopff Verlag 1995

R. Antón
F. Mosquera
M. Oduber

Anionic-nonionic surfactant mixture to attain emulsion insensitivity to temperature

Prof. R. Antón (✉) · F. Mosquera
M. Oduber
Lab. FIRP
Ingenieria Quimica
Universidad de Los Andes
Mérida 5101, Venezuela

Abstract It is well known that anionic surfactants become more hydrophilic as temperature increases, whereas nonionic surfactants show the opposite trend. By a proper mixing of both types of surfactants it is possible to produce intermediate situations and, eventually, insensitivity to temperature. This mixing principle is applied to both the phase behavior and emulsion properties.

The temperature/water-to-oil ratio phase diagram is mapped for different mixtures of an alkyl aryl sulfonate and an ethoxylated alkyl phenol. The experimental evidence indicates that there is a continuous variation from the case 100% of anionic to 100% nonionic. The three-phase zone behaves as a rotating band which finally flips upside down.

The emulsion inversion line undergoes a transition, with an anionic-nonionic intermediate mixture exhibiting an inversion line independent of the temperature. For this anionic-nonionic mixture, the three-phase behavior region is an extremely extended band, and it is associated with an "abnormal" emulsion type all over the temperature and composition range. These emulsions are extremely unstable.

Key words Surfactant – formulation – emulsion – inversion

Introduction

The effect of the temperature on the behavior of surfactant-oil-water systems has been known for more than 40 years [1]; it has been used as a standard way to produce formulation changes and associated transitions, particularly with nonionic surfactants [2, 3]. It is known that anionic surfactants become more hydrophilic when the temperature is increased, while nonionic ones behave the opposite way.

The temperature and formulation effects have been translated into algebraic expressions, so-called correlations for optimum formulation, which render the physico-chemical conditions necessary for a three-phase microemulsion-oil-water system to occur. These correlations have been proposed for both anionic and nonionic surfactant systems [4, 5].

As mentioned recently [6], a way to quantify the effect of temperature is to measure the shift in optimum value of a formulation variable versus temperature. The experimental data provided the following average effects for anionic and nonionic systems:

$$\left.\frac{\partial \text{ACN}^*}{\partial T}\right|_{\text{AI}} = -0.06 \pm 0.01 \text{ and } \left.\frac{\partial \text{ACN}^*}{\partial T}\right|_{\text{NI}}$$
$$= +0.38 \pm 0.1 , \tag{1}$$

where ACN* is the Alkane Carbon Number of the oil phase that results in a three-phase behavior in a given physico-chemical environment, and T the temperature in °C. These relations indicate that the effects are opposite, since the derivatives bear different signs; moreover, the effect on nonionic systems is several times stronger than with anionic ones.

However, the use of ACN as the yardstick variable is inconvenient because the series of liquid alkane ACN is restricted from 5 to 16, a situation that considerably reduces the range of experimentally attainable data. The salinity of the aqueous phase is preferred as a formulation variable which exhibits a much larger range of variation; additionally, it is cheaper and easier to handle [7, 8].

Phase behavior with anionic-nonionic mixtures

Early work to compare anionic and nonionic systems [6, 9] indicated some variations on the value of the temperature effect with nonionic surfactants. It was not known whether these slight discrepancies were due to experimental errors or actual differences. This can be resolved by keeping track of the degree of ethoxylation as shown in the following. Figure 1 indicates the variation of the optimum salinity (expressed as the Neperian logarithm of the weight percentage of NaCl in brine) versus temperature for systems containing both alkyl aryl sulfonates and ethoxylated nonyl phenol surfactants with variable degree of ethoxylation symbolized by EON (Ethylene Oxide

Number, i.e., average number of ethylene oxide groups per molecule). These data exhibit essentially three features:

A) First, the optimum formulation line, taken as the temperature located at the center of the three-phase behavior region, follows a linear Ln S vs T variation for all systems (anionic or nonionic). This is a slight change with respect to the correlation previously published [5] that mentioned a linear S vs T relationship from very few data.

B) Second, the slope is found to depend upon the average number of ethylene oxide groups (EON) of the nonionic surfactant. A curve fitting indicates the following empirical relationship for this system, valid from EON = 6 to 11.

$$\left.\frac{\partial \text{Ln} S}{\partial T}\right|_{\text{NI}} = -0.008 \,(\text{EON} - 2.9)\,. \qquad (2)$$

This means that the temperature effect is slightly larger on the longer ethylene oxide chains than on the shorter ones. This is no surprising, since the hydration of the ethylene oxide groups is probably lower in long folded chains than in short chains.

C) The slope is the same for the two sulfonated surfactants; comparing to the case of the nonionic systems, the sign is opposite, and the absolute value is much lower, in accordance to previous results [4]:

$$\left.\frac{\partial \text{Ln} S}{\partial T}\right|_{\text{AI}} = +0.009 \pm 0.003\,. \qquad (3)$$

If the two types of surfactants are present in a mixed system, the two antagonist effects of temperature should compensate each other in some way. A simple linear model has been tested recently with some success [6]. However, both the possibility of fractionation of the nonionic surfactant oligomers [10], and the possibility of anionic-nonionic interaction [11], introduce additional complexities. As a consequence, it can be said that it is possible to find a mixed system independent from temperature, but that its composition cannot be exactly deduced from a linear mixing rule based on mole fraction contributions.

Figure 2 shows the effect of the temperature on the optimum formulation of different anionic-nonionic mixtures. According to the previously mentioned values of the temperature effects, the mixture insensitive to temperature should contains 26 (± 5) mole % of nonionic component, while the data indicate that the mixture whose optimum salinity does not change with temperature actually contains 38 mole % of it. However, it is obvious that the linear mixing rule applies quite well to this system, in spite of this numerical discrepancy between the actual and effective composition.

Fig. 1 Optimum formulation for systems containing anionic and nonionic surfactants as Neperian Logarithm of the optimum salinity versus temperature

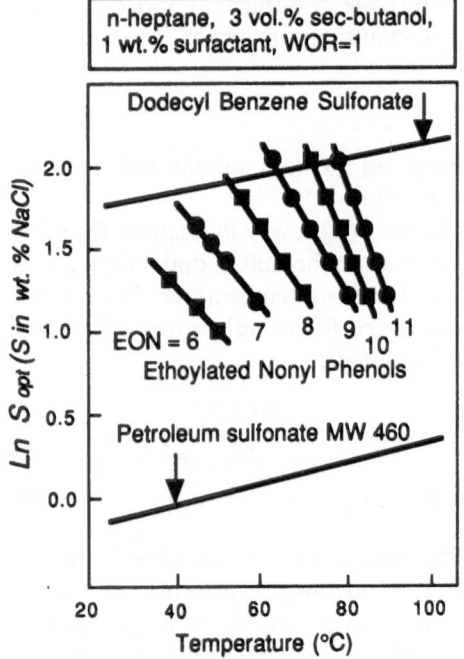

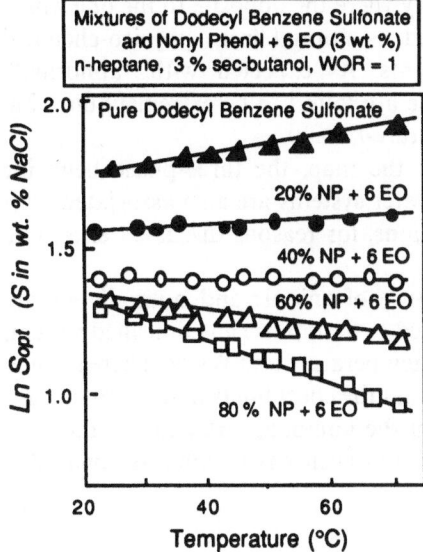

Fig. 2 Optimum formulation for systems containing anionic and nonionic surfactant mixtures as Neperian Logarithm of the optimum salinity versus temperature

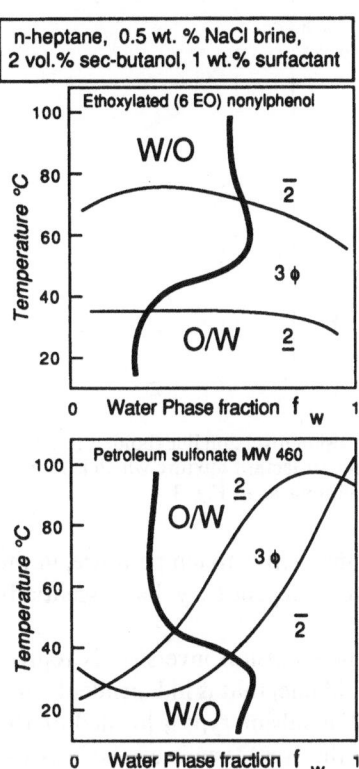

Fig. 3 Phase behavior and emulsion inversion line (bold) for systems containing nonionic (top) and anionic (bottom) surfactants on a temperature – composition diagram

Emulsion inversion on a formulation-WOR map

The phase behavior and emulsion type are studied on a formulation-water-to-oil ratio (WOR) bidimensional scan [12]. It has been shown that there exists a general phenomenology, which describes the emulsion properties on such a diagram, whenever the formulation variable is the brine salinity, the surfactant parameter or the temperature [12–14]. Figure 3 shows such type of diagram for both nonionic and anionic systems.

The two-phase behavior is indicated according to notation $\underline{2}$ when the affinity of the surfactant for the aqueous phase dominates (Winsor I system), while it is symbolized as $\bar{2}$ when the opposite occurs (Winsor II system). Winsor III or 3 systems exhibit three-phase behavior, i.e., a microemulsion in equilibrium with both water and oil excess phases. Winsor III or 3ϕ systems are found in the neighborhood of optimum formulation, which is here an optimum temperature, i.e., the temperature at which the surfactant dominant affinity switches from one phase to the other.

In nonionic systems such as in Fig. 3 (top), Winsor I or $\underline{2}$ type (respectively II or $\bar{2}$) is found below (respectively above) the optimum temperature with the associated O/W (respectively W/O) emulsion. This is true for this system in the central region, when $30\% < f_w < 70\%$. Note that the inversion line (bold line) is in staircase form, going up from left to right.

Figure 3 (bottom) shows the bidimensional temperature-f_w map for an anionic system. The general aspect is

similar, but the temperature effect is inverted with respect to the previous case i.e., Winsor I or $\underline{2}$ (respectively II or $\bar{2}$) and O/W (respectively W/O) emulsions are located at a temperature higher (respectively lower) than the optimum temperature. As a consequence, the inversion line (bold line) is staircase like, going down from left to right.

It is worth noting that in the central region of the two previous maps ($30\% < f_w < 70\%$), the phase behavior and the emulsion type are opposite. A combination or superposition of the two antagonist cases might thus lead to some insensitivity of the phase behavior to temperature.

In order to analyze this combination, the two surfactants are mixed in different amounts, and a bidimensional temperature-f_w map is determined for each nonionic-anionic mixture. The anionic and nonionic surfactants are selected for exhibiting the same optimum temperature (45 °C) at $f_w = 50\%$, so that the temperature effects are strictly opposite. This case corresponds to the double insensitivity of the phase behavior versus temperature and composition which was reported elsewhere (6).

Figure 4 indicates the phase behavior map and the emulsion inversion line for a mixture containing 38% of nonionic surfactant. The three-phase zone is a vertical band, with Winsor I or ($\underline{2}$) case on the left (low f_w), and Winsor II ($\bar{2}$) phase behavior on the right (high f_w). This

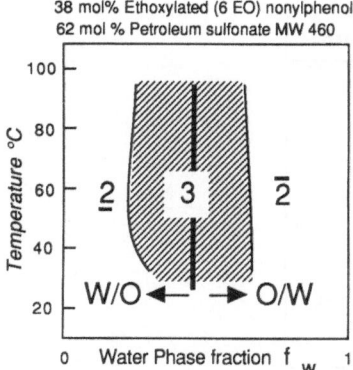

Fig. 4 Phase behavior and emulsion inversion line (bold) for systems containing the nonionic-anionic surfactant mixture which is insensitive to temperature; same conditions as in Fig. 3

system phase behavior is insensitive to temperature, in the sense that the temperature dependency has essentially vanished.

On the other hand, the emulsion inversion is represented by a vertical line (bold line) that is independent from the temperature. The O/W emulsion type is located on the right side (high f_w), while the W/O type is on the left (low f_w). The emulsion type corresponds to the general trend which is valid at high oil or water content, i.e., the external phase is the phase which is present in the higher volume amount. However, it should be pointed out that the phase behavior, and thus the surfactant affinity balance, tends to favor the other type of emulsion, since Winsor I ($\underline{2}$) is normally associated to O/W emulsions and Winsor II ($\overline{2}$) to W/O ones [12–13].

This observation indicates that the intermediate mixture case corresponds to the so-called "abnormal" [12] situation all over the map, but at the center. In effect, the emulsion type is everywhere the opposite to the one which would be considered as normal from physico-chemical formulation arguments. As expected with "abnormal" emulsions [14], these are found to be extremely unstable all over the temperature-f_w map.

At the center of the map, the three-phase behavior microemulsion-oil-water systems are also associated with very unstable emulsions, for reasons discussed elsewhere [15].

The proper mixture of anionic and nonionic surfactants may thus be used to destabilize an emulsified system, over a wide range of temperature and WOR. There is some additional evidence that the effect holds in the present case from 25% to 75% of the nonionic surfactant in the mixture, an indication that insensitivity to composition is also reached [6].

Conclusions

Nonionic-anionic surfactant mixing allows the attainment of insensitivity to temperature for both the phase behavior and the emulsion inversion.

When the proper nonionic-anionic mixture is used, the produced emulsion is always of the "abnormal" type, i.e., the opposite of the one which would be expected from physico-chemical formulation arguments.

As a consequence these emulsions are found to be extremely unstable, over a wide range of variations of temperature, WOR and surfactant mixture composition.

Acknowledgments The Lab. FIRP research program at Universidad de Los Andes is funded by the University Research Council CDCHT and the Industrial Sponsor Group: CORIMON, HOECHST de Venezuela, INTEVEP, and PROCTER & GAMBLE de Venezuela.

References

1. Winsor P (1954) Solvent Properties of Amphiphilic Compounds, Butterworth, London
2. Shinoda K (1967) Solvent Properties of Surfactants Solutions, M Dekker, New York
3. Shinoda K (1968) Proceedings Int Congress Surface Activity 5th, 2:275, Barcelona, Spain
4. Salager JL, Morgan J, Schechter RS, Wade WH, Vasquez E (1979) Soc Petrol Eng J 19:107
5. Bourrel M, Salager JL, Schechter RS, Wade WH (1980) J Colloid Interface Sci 75:451
6. Antón RE, Salager JL, Graciaa A, Lachaise J (1992) J Dispersion Sci Technology 13:565
7. Salager JL, Bourrel M, Schechter RS, Wade WH (1979) Soc Petrol Eng J 19:271
8. Antón RE, Andérez JM, Graciaa A, Lachaise J, Salager JL, 8th Int Symposium Surfactants in Solution, Gainsville FLA USA, June 1990
9. Bourrel M, Salager JL, Schechter RS, Wade WH (1978) Colloques Nat CNRS "Physicochimie des composés amphiphiles", 938:337
10. Graciaa A, Lachaise J, Sayous JG, Grenier P, Yiv S, Schechter RS, Wade WH (1983) J Colloid Interface Sci 93:474
11. Rubingh DN, Holland N, Eds (1991) Cationic Surfactants – Physical Chemistry, Chap 4, M Dekker
12. Salager JL, Miñana-Pérez M, Pérez-Sanchez M, Ramirez-Gouveia M, Rojas C (1983) J Dispersion Sci Technology 4:313
13. Antón RE, Castillo P, Salager JL (1986) J Dispersion Sci Technology 7:319
14. Miñana-Pérez M, Jarry P, Pérez-Sanchez M, Ramirez-Gouveia M, Salager JL (1986) J Dispersion Sci Technology 7:331
15. Antón RE, Salager JL (1986) J Colloid Interface Sci 111:54

Progr Colloid Polym Sci (1995) 98:89–93
© Steinkopff. Verlag 1995

MIXED SYSTEMS

J.A. Maroto
F.J. de las Nieves

Colloidal stability in homo- and hetero-coagulation processes. Comparison between theoretical and experimental data

J.A. Maroto · Prof. F.J. de las Nieves (✉)
Biocolloids and Fluid Physics Group
Department of Applied Physics
University of Granada
18071 Granada, Spain

Abstract In this work the colloidal stability of aqueous dispersions of uniform spherical cationic and anionic particles has been studied by turbidity measurements in homocoagulation and heterocoagulation processes. Two polymer colloids prepared by surfactant free emulsion polymerization were used throughout this work. Anionic and cationic latexes of similar surface charge densities, but with different particle sizes were selected. Once the adequate wavelength and the particle concentration into the cell were selected, the coagulation rate constants were estimated for the heterocoagulation and homocoagulation (cationic and anionic latexes) processes. By using the kinetic rate constants provided by the Muller and Smoluchowski equations, the Stern–Grahame model of the electric double layer, the total potential energy from the DLVO theory and some expressions from Reerink and Overbeek, we have compared the experimental and theoretical values of the homocoagulation and heterocoagulation rate constants, and also the diffuse potentials at the critical coagulation concentration.

Key words Heterocoagulation – homocoagulation – polymer colloids

Introduction

The heterocoagulation experiments permit to interpret the stability of colloidal dispersions consisting of more than one kind of particle. In some applications appear particles or aggregates with different sign of the surface charge and particle sizes, which make unsuitable the homocoagulation experiments, but which increases the interest of the heterocoagulation studies. In this work, we have studied the colloidal stability of aqueous dispersions of uniform spherical cationic and anionic particles by turbidity measurements in homocoagulation and heterocoagulation processes. Anionic and cationic latexes of similar surface charge densities but with different particle sizes were se-

lected, although the particle size is of the same order of magnitude in order to talk about heterocoagulation more so than deposition. By simple theoretical considerations it is possible to conclude that the usual stability factor, W, is not useful in heterocoagulation experiments and the hetero-coagulation rate constant, K_{D12}, has to be calculated from only one direct experiment. This implies to determine the adequate experimental conditions (wavelength and particle concentration into the cell) which provide satisfactory values of the rate constants. For that reason, in a first step, we have determined the stability factor and the rate constants of the homocoagulation processes (K_{D11} and K_{D22}). These rate constants were compared with those obtained by the Smoluchowski [1] and Muller [2] equations. For the case of the anionic latex, we can compare the diffuse potential obtained by the Stern–Grahame theory [3] of the electric double layer and those experimentally obtained by two different methods: a) from the experimental value of the critical coagulation concentration, CCC, by using the relationships derived by Reerink and Overbeek [4]; b) from the CCC values, but now by numerical calculations of the total interaction energy, V_T, imposing the mathematical conditions for V_T at the CCC (where V_T is zero), or the condition of maximum for V_T at a distance, H_0, between the surfaces. Finally, we have also compared the theoretical values of the log W/log C slopes, by using again the Reerink and Overbeek relationships [4].

Theory

During the homocoagulation process of the species i, the stability factor is defined as [5]:

$$W_{ii} = \frac{K_{Di,F}}{K_{Di,S}} . \tag{1}$$

If we have a heterocoagulation process of the species 1 and 2, with different particle sizes, a_1 y a_2, besides the factors W_{11} and W_{22} (Eq. (1)), the factor W_{12} has to be taken into account [6]:

$$W_{12} = \frac{K_{D12,F}}{K_{D12,S}} , \tag{2}$$

where S and F mean slow and fast coagulation rates. Thus, the total stability factor W_T is a certain average which corresponds to the following equation [6]:

$$\frac{1}{W_T} = \frac{n^2}{W_{11}} + \frac{(1-n^2)}{W_{22}} + \frac{2n(1-n)}{W_{12}} , \tag{3}$$

where

$$n = \frac{N_1}{N_1 + N_2}, \tag{4}$$

N_1 and N_2 being the volume concentrations of each colloidal species.

By using the relationship between the turbidity and particle concentration,

$$\tau = \sum_i C_i N_i \tag{5}$$

together with the kinetic equation for the coagulation derived by Von Smoluchowsk [1],

$$\frac{dN_n}{dt} = \frac{1}{2} \sum_{j+i=n} K_{ij} N_i N_j - N_n \sum_{i=1}^{\infty} K_{ni} N_i \tag{6}$$

and also the equations derived by Smoluchowski [1] and Muller [2] for the fast rate constants during the formation of dimers in a homo- and hetero-coagulation process,

$$K_{Di,F} = \frac{4}{3} \frac{kT}{\eta} \tag{7}$$

$$K_{Dij,F} = \frac{1}{3} \frac{kT}{\eta} \frac{(a_i + a_j)^2}{a_i a_j} , \tag{8}$$

we can find for the homocoagulation process that:

$$\frac{(d\tau/dt)_{0,F}}{(d\tau/dt)_{0,s}} = \frac{K_{Di,F}}{K_{Di,S}} = W_{ii} . \tag{9}$$

Thus, the stability factor W_{ii}, is observable by optical procedure. However, in the case of a heterocoagulation process, following a similar procedure, it is possible to show [7] that:

$$W_{HET} \neq \frac{(d\tau/dt)_{0,HOMO,F}}{(d\tau/dt)_{0,HET}} \tag{10}$$

Thus, the only equation to relate the heterocoagulation rate constant to experimental data is:

$$\left(\frac{d\tau}{dt}\right)_{0,HET} = K_{D12} N_1 N_2 (C_{D12} - C_{S1} - C_{S2}) , \tag{11}$$

where C_{S1}, C_{S2} and C_{D12} are the scattering sections of monomers and dimers, respectively, the latter ones formed by particles of the species 1 and 2.

In order to relate the surface charge density of the polymer particles to the diffuse potential at the solid-liquid interface, we have employed the Stern and Grahame [4] model of the electric double layer, which provides the following equation:

$$\sigma_0 = [8\varepsilon n(\infty)kT]^{1/2} \operatorname{senh} \left[\frac{e_0 \psi_d}{2kT}\right], \tag{12}$$

where σ_0 is the surface charge density, Ψ_d is the diffuse potential, $n(\infty)$ the ion concentration in the bulk solution, e_0 the electron charge, and k and T the Boltzmann constant and the absolute temperature, respectively.

To estimate the total interaction energy between colloidal particles, the DLVO theory [5, 8] can be used. This theory provides the form of the electric V_{EL} and attractive V_A potentials. For two particles of different size, a_i and a_j, the attractive potential is:

$$V_A(H_0) = -A a_i a_j / 6(a_i + a_j) H_0 . \tag{13}$$

For the electric potential, and in the case of homocoagulation, we can use the equation:

$$V_{EL}^{(I)}(H_0) = 2\pi\varepsilon a_i \left(\frac{4kT}{ve_0} \gamma\right)^2 \ln(1 + \exp(-\kappa H_0)) , \tag{14}$$

where

$$\gamma = \tanh\left(\frac{ve_0 \psi_d}{4kT}\right) . \tag{15}$$

Progr Colloid Polym Sci (1995) 98:89–93
© Steinkopff Verlag 1995

Using the low potential approximation of Debye–Huckel, Eq. (14) can be written as:

$$V_{EL}^{(II)}(H_0) = 2\pi\varepsilon\dot{a}_i\psi_d^2 \ln\left[1 + \exp(-\kappa H_0)\right]. \qquad (16)$$

By assuming the distances between the particles to be much larger than the Debye length, Eq. (14) can be approximated to the following:

$$V_{EL}^{(III)}(H_0) = \frac{32\pi\varepsilon k^2 T^2 a_i \gamma^2}{e_0^2} \exp(-\kappa H_0). \qquad (17)$$

The previous equations have to fulfill that at the CCC conditions the total interaction energy, V_T, is zero and also has to verify that V_T displays a maximum at the position H_0.

Finally, we have employed some equations derived by Reerink and Overbeek [4], which relate the experimental results from colloid stability measurements to the Hamaker constant or to the electric potential γ:

$$A = -(24kT/\kappa_c a)(d\log W/d\log n(\infty)) \qquad (18)$$

$$-\frac{d\log W}{d\log n(\infty)} = \left(\frac{16\pi\varepsilon kT}{ee_0^2}\right)\frac{a}{v^2}\gamma^2, \qquad (19)$$

where κ_c is the reverse of the Debye length at the CCC conditions.

Materials and methods

Two polymer colloids of sulfate and amidine surface groups with similar surface charge densities ($-6.9\,\mu C/cm^2$ y $6.5\,\mu C/cm^2$, respectively), but different particle sizes (297 ± 3 y $177 \pm 7\,nm$, respectively) were used in the homo- and hetero-coagulation measurements. The anionic commercial latex (RP) was from Rhone Poulenc, while the cationic latex (JA3) was synthesized following the recipe described in [9].

The methods and procedure to obtain a complete characterization of the latexes were described in [10]. From the previous electrokinetic characterization, we have translated the electrophoretic mobility results to zeta potential data by several theories. The zeta-potential values are shown in Fig. 1. In this figure we also show the diffuse potential obtained by (12).

The homo- and hetero-coagulation rates were evaluated from the changes in turbidity versus time with a spectrometer (Milton Roy Spectronic 601). The methods and procedures for the heterocoagulation experiments were previously described [7].

Results and discussion

Figure 2 shows the stability factor W versus the electrolyte concentration (KBr) for the anionic latex RP. From the linear relationship between $\log W$ and $\log C$, we can obtain the intersection point with the abcissa (which is defined as the critical coagulation concentration, CCC = 190 mM) and the slope of the linear plot ($[d\log W/d\log C] = -1.55$). From the surface charge density value of the R–P latex and by Eq. (12), we can obtain the diffuse potential of the solid-liquid interface at that specific concentration, $-57.0\,mV$. By Eqs. (14), (16), and (17) we have plotted the total interaction energy, $V_T(H_0)$, versus the distance using the results of R − P latex and taking into account the mathematical conditions that

Fig. 1 Diffuse (+), Smoluchowski (◇), and Baran (*) potentials versus κa, for RP latex and a 1:1 electrolyte (KBr)

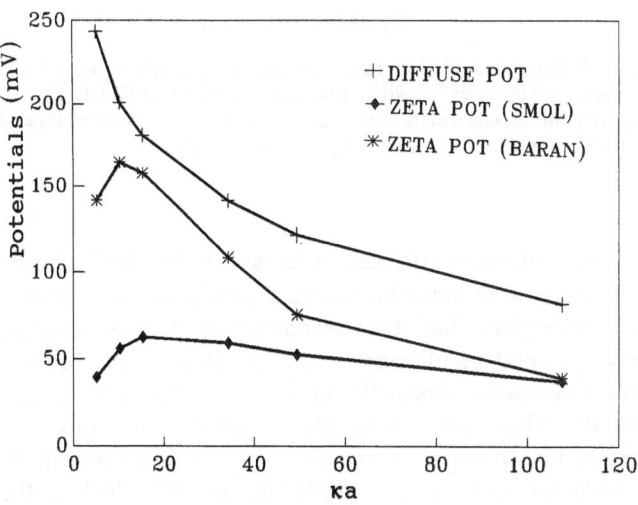

Fig. 2 Stability factor (W) versus KBr concentration for latex RP at pH = 6

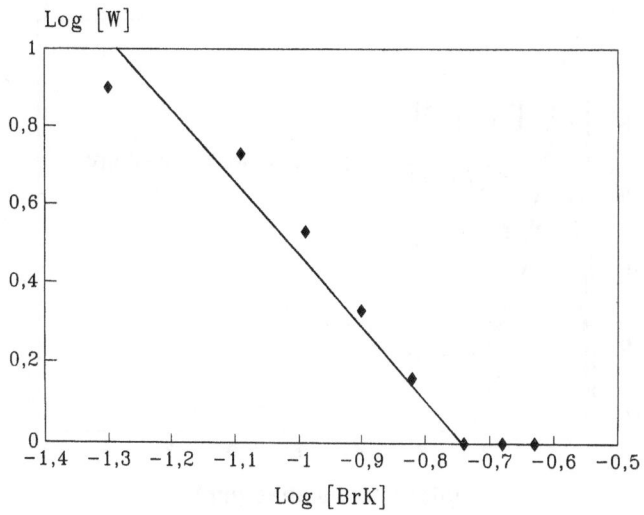

V_T has to fulfill (V_T is zero at the CCC and displays a maximum at the distance H_0). Figure 3 shows the curves obtained for V_T from Eqs. (14), (16), and (17). The diffuse potentials for the latex R–P at the CCC, when κa is 211, were: $-36.0\,mV$ from Eq. (14), $-34.1\,mV$ from Eq. (16) and $-33.0\,mV$ from Eq. (17). These potentials are similar but very different from that found by using Eq. (12). However, if we compare the diffuse potential calculated by the equations of V_T with the zeta-potential values shown in Fig. 1, the results seem to be more similar. These results are in agreement with those shown by several authors [10–12]. These authors use the zeta-potential values (obtained from mobility measurements) as the diffuse potential, in the place of those calculated by Eq. (12), when the DLVO theory is applied.

Using Eq. (18) and including the theoretical Hamaker constant for the polystyrene water system ($1.37\,10^{-20}\,J$) [13], it is possible to obtain a theoretical value for the slope of the curves $\log W/\log C$ of -29.25, more than one order of magnitude higher than that experimentally found. However, if we include this theoretical value of the slope in Eq. (19) the diffuse potential is $-32.8\,mV$, in good agreement with those diffuse potentials theoretically calculated by using the energy curves from Fig. 3. Thus, the DLVO theory seems to predict the diffuse potential values at the CCC conditions. However, the slope of the stability curves are overestimated, even by as much as one order of magnitude, as Matijevic et al. reported [11].

Figures 4 and 5 show the heterocoagulation and homocoagulation rate constants (K_{D12} and K_{D11}, respectively) estimated from Eqs. (11) and (9), versus the particle

Fig. 3 Total interaction energy as a function of interparticle distance (H) calculated by several equations: Pot III (13 and 17), Pot II (13 and 16) and Pot I (13 and 14)

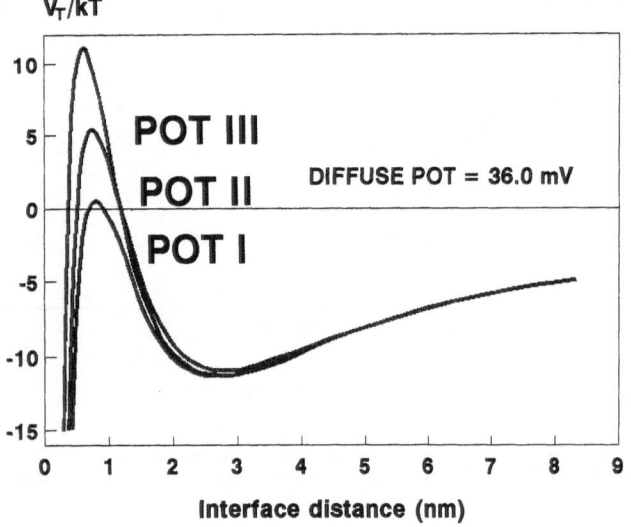

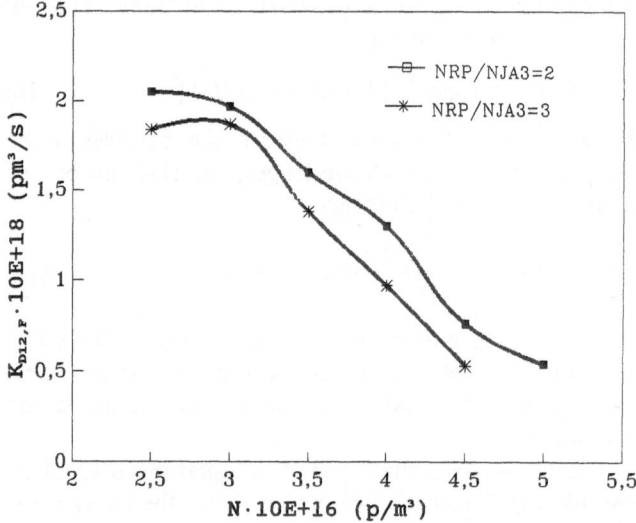

Fig. 4 Heterocoagulation rate constant (K_{D12}) values versus the total particle number (N) for the ratios: (∗), [NRP]/{NJA3] = 3; (□), [NRP]/[NJA3] = 2

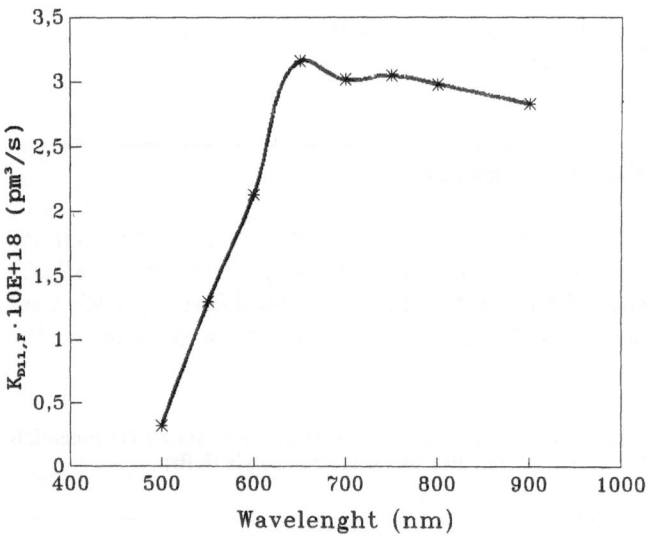

Fig. 5 Homocoagulation rate constant (K_{D11}) values versus the wavelength for latex RP with a total particle number of $2.1\cdot10\,E+16$ (p/m³) and an electrolyte concentration of $2.1\cdot10\,E+16$ (p/m³) and an electrolyte concentration of [KBr] = 200 mM

concentration into the cell. Acceptable values for both rate constants were only obtained at a certain particle concentration region. Thus, it is very important to select adequate experimental conditions (resp. wavelength and particle concentration), especially in heterocoagulation experiments, where the results are obtained from only one direct measurement, while for homocoagulation experiments, we can use the stability factor, W, which is the

Progr Colloid Polym Sci (1995) 98:89–93
© Steinkopff Verlag 1995

Table 1 Heterocoagulation rate constants (K_{D12}) values against different ratios [NJA3]/[NRP], for two runs with different N and λ values.

$\lambda = 770$ nm NJA3/NRP	$N = 6'5 \cdot 10^{16}$ p/m^3 $K_{D12} \cdot 10^{18}$ pm^3/s
10	2'08
6	2'24
3	2'14
2	2'04
1	1'97
1/2	1'10
1/3	0'98
1/6	0'58

$\lambda = 600$ nm NJA3/NRP	$N = 4 \cdot 10^{16}$ p/m^3 $K_{D12} \cdot 10^{18}$ pm^3/s
6	2'11
2	2'08
1	2'04
1/2	1'45
1/3	0'97
1/6	0'34

ratio between the initial slope of different measurements. The K_{D12} values obtained from Eq. (11) were $(2-2.1) \cdot 10^{-18}$ pm^3/s, while for homocoagulation the rate constant values were $2.25 \cdot 10^{18}$ pm^3/s for latex JA3, and $3 \cdot 10^{-18}$ pm^3/s for latex R–P. Table 1 shows the K_{D12} values obtained in heterocoagulation experiments for several ratios between cationic and anionic particles. For the cases where the number of anionic particles versus the number of cationic particles is 2 or 3 ([NRP]/[NJA3] = 2, or 3), the rate constant values were discordant in comparison with the other values. Figure 4 shows that by varying the measurement conditions it is possible to obtain K_{D12} values in agreement with those found in the better experience when the concentration of both types of particles is the same. By comparing the heterocoagulation and homocoagulation rate constants, we can see that they are relatively similar and in agreement with the results of James et al. [14]. The theoretical rate constant values obtained by Eqs. (7) and (8) were: $K_{\text{HOMO, SMOL}} = 1.1 \cdot 10^{-17}$ pm^3/s and $K_{\text{HETE, MULL}} = 1.18 \cdot 10^{-17}$ pm^3/s. As can be seen, the theoretical values are one order of magnitude larger than those experimentally obtained. This type of discrepancy has usually been found in the literature [15, 16], and several explanations have been offered related to the hydrodynamic effects or the discreteness of the surface charge.

Acknowledgments This work is supported by the Comisión Interministerial de Ciencia y Tecnología (CICYT), project MAT93-0530-C02-01, and by Junta de Andalucía (G.I. 1218).

References

1. Von Smoluchowski M (1906) Ann Phys 21:756
2. Muller H (1926) Kolloid Z 38:1
3. Grahame DC (1950) J Chem Phys 18:403
4. Reerink H, Overbeek JThG (1954) Disc Faraday Soc 18:74
5. Verwey EJ, Overbeek JThG (1948) In: Theory of the Stability of Lyophobic Colloids, Elsevier Publ. Co.
6. Healy R, Hogg TW, Fuersteneau DW (1966) Trans Faraday Soc 62:1638
7. Maroto JA, de las Nieves FJ (1994) Colloids Surfaces A (in press)
8. Derjaguin BV, Landau LD (1941) Acta Physicochim 14:633
9. Hidalgo-Alvarez R, de las Nieves FJ, van der Linde, Bijsterbosch BH (1986) Colloids Surfaces 21:259
10. Carrique F, Salcedo J, Cabrerizo MA, Gonzalez-Caballero F, Delgado AV (1991) Acta Polimerica 42:6
11. Kihira H, Ryde N, Matijevic E (1992) Colloids Surfaces 64:317
12. Bastos D, de las Nieves FJ (1994) Colloid Polym Sci 272:592
13. Prieve DC, Russel WB (1988) J Colloid Interface Sci 125:1
14. James RO, Homola A, Healy TW (1977) J Chem Soc Faraday Trans 173:1436
15. Kihira H, Ryde N, Matijevic E (1992) J Chem Soc Faraday Trans 88:2379
16. Fernández A, Martín A, Callejas J, Hidalgo-Alvarez R (1994) J Collod Interface Sci 162:257

Progr Colloid Polym Sci (1995) 98:94–98
© Steinkopff Verlag 1995

P. Fischer
H. Rehage

Quantitative description of the non-linear flow properties of viscoelastic surfactant solutions

P. Fischer (✉) · H. Rehage
Institut für Umweltanalytik
Universität Essen, FB 8
Universitätsstraße 3
45141 Essen, FRG

Abstract We have systematically studied the non-linear flow properties of certain viscoelastic surfactant solutions. While the dynamic features of these systems can be described by a modified reptation theory, one obtains in the regime of large velocity gradient stress overshoot in start-up flow experiments and shear thinning properties in the steady-state regime. These properties of viscoelastic surfactant solutions can be described with a non-linear Maxwell material, first proposed by Giesekus. From these models it is easy to calculate steady-state values of the shear viscosity $\eta(\infty, \dot{\gamma})$ and the first normal stress coefficient $\Psi_1(\infty, \dot{\gamma})$. It is often observed that these properties obey viscometric functions such as the Yamamoto relation and the Gließle mirror relationships. It turns out that

the experimental results are in fairly good agreement with theoretical predictions of the one-mode Giesekus model, especially by adjusting the only free parameter with $\alpha = 0.5$. For these conditions, the Cox–Merz rule, the Yamamoto relation, and both Gließle mirror relations are automatically derived from the theoretical model. It is very interesting that the non-linear Maxwell material can be used for the qualitative description of non-linear flow properties of viscoelastic surfactant solutions, such as the shear viscosity of the first normal stress difference.

Key words Rheology – viscoelastic surfactant solutions – non-linear Maxwell model – one-mode Giesekus model

Introduction

Viscoelastic surfactant solutions are used as simple model systems for rheological studies because of their interesting flow properties. Due to the limited lifetime of the micellar aggregates, often monoexponential relaxation functions are observed. The dynamic features of these solutions can be described using the theoretical equations of an ideal Maxwell material [1]. At experimental conditions, where the reptation time is of the same order of magnitude as the average lifetime, often stretched exponential relaxation functions are obtained [2].

The relations hold especially in the regime of small shear stresses. The dynamic features of these systems can be described by a modified reptation theory, which was first proposed by Cates [3].

In the regime of large velocity gradients, however, large deviations occur in transient and static experiments. One generally obtains stress overshoot in start-up flow experiments and shear thinning properties in the steady-state regime. It is often observed that the shear viscosity coincides pretty well with the magnitude of the complex viscosity. In viscoelastic surfactant solutions, these correlations are very similar to the one observed in polymer solutions, where it is usually designed as the Cox–Merz rule [4]. In the regime of high shear rates non-linear effects such as first normal stress differences occur as well. These features and the transient properties of viscoelastic surfactant solutions can be described with a non-linear Maxwell material, which was first proposed by Giesekus [5]. In this model, the concept of configuration-dependent tensional mobility leads to equations which can be easily compared with the experimental results.

From the one-mode Giesekus model it is easy to calculate steady-state values of the shear viscosity $\eta(\infty, \dot{\gamma})$ and the first normal stress coefficient $\Psi_1(\infty, \dot{\gamma})$. It is often observed that these functions obey the Yamamoto relation [6] and the Gleißle mirror relationships [7, 8]. These equation, although of empirical nature, allow to calculate steady-state values from transient experiments or inverse. Rheological properties may be determined over many orders of magnitude as a function of the shear rate $\dot{\gamma}$ using these relations.

One-mode Giesekus model

The one-mode Giesekus model gives a detailed relation between rheological parameters and the mobility tensor β. The concept of configuration-dependent tensional mobility leads to molecular models predicting the characteristic flow behavior of viscoelastic fluids. Giesekus used a simple constitutive equation (1) in which two assumptions are introduced. The first one is concerned with the stress in the deformed network structure (2) and the second one describes a linear dependence of the mobility tensor (3)

$$\beta * S + \eta(\vartheta C/\vartheta t) = 0 \tag{1}$$

$$S = \mu(C - 1) \tag{2}$$

$$\beta = 1 + \alpha(C - 1) = (1 - \alpha)1 + \alpha C \tag{3}$$

with $\mu \approx nkT$, $\eta = $ viscosity, C as configuration tensor, and S as stress tensor. Introducing Eqs. (2) and (3) into Eq. (1), one obtains (4) as new constitutive equation

$$[1 + \alpha(C - 1)] * (C - 1) + \eta(\vartheta C/\vartheta t) = 0 . \tag{4}$$

For simple shear flow, four equations can be derived:

$$\alpha(C_{11}^2 + C_{12}^2) + (1 - 2\alpha)C_{11} - (1 - \alpha) - 2\tau\dot{\gamma}C_{12} = 0$$

$$\alpha(C_{22}^2 + C_{12}^2) + (1 - 2\alpha)C_{22} - (1 - \alpha) = 0$$

$$\alpha C_{33}^2 + (1 - 2\alpha)C_{33} - (1 - \alpha) = 0$$

$$\alpha(C_{11} + C_{22})C_{12} + (1 - 2\alpha)C_{12} - 2\tau\dot{\gamma}C_{22} = 0 , \tag{5}$$

with $C_{11} - C_{22} = N_1$ as the first normal stress difference, $C_{33} - C_{22} = N_2$ as the second normal stress difference, and $C_{12} = \sigma$ as the shear stress. For start-up flow three

coupled differential equations are easily obtained,

$$\dot{\sigma} + [\alpha(N_1 - 2N_2) + 1]\sigma = \tau\dot{\gamma}(1 - N_2)$$

$$\dot{N}_1 + [\alpha(N_1 - 2N_2) + 1]N_1 = 2\tau\dot{\gamma}\sigma$$

$$\dot{N}_2 + [(1 - \alpha N_2) + 1]N_2 = \alpha\sigma^2 . \tag{6}$$

These theoretical results from the general one-mode Giesekus model for steady and transient shear flows are compared to experimental results using a Runge–Kutta program to calculate the theoretical data.

Experimental section

The rheological experiments are performed using a Rheometrics RFS II rheometer. The 0.02 rad cone and plate geometry was used, with a diameter of 50 mm. To avoid evaporation, a spherical solvent trap was used during the experiments. The investigated frequency range was 0.01 rad/s to 100 rad/s. The cone and plate geometry allows to apply shear rates from 0.05 1/s up to 5000 1/s. Temperatures are maintained at 20 °C. For all investigations a solution of 60 mmol/L cetyltrimethylammonium-bromide and 350 mmol/L sodiumsalicylate (CTAB–NaSal 60–350) was used.

Results and discussion

We have systematically investigated the non-linear rheological behavior of viscoelastic surfactant solutions in the regime of very large shear rates and in the regime of start-up flow. Typical non-linear properties such as shear-thinning and normal stresses are observed. In transient shear flow a shear stress overshoot depending on the amount of shear rate was clearly observed. Typical frequency sweep and rate sweep experiments are obtained for CTAB–NaSal 60–350. In the linear viscoelastic regime, a monoexponential stress decay function is obtained. The experimental data can be described qualitatively by the Maxwell material. At these conditions, one obtains the relaxation time τ from the intersection point of $G'(\omega, \gamma)$ and $G''(\omega, \gamma)$. The zero shear viscosity $\eta(0, 0)$ is calculated from the magnitude of the complex viscosity $\eta^*(\omega, \gamma)$ at infinite small shear deformations. From rate sweep experiments, one obtains the transient values of the shear viscosity $\eta(\infty, \dot{\gamma})$ and the first normal stress function $N_1(\infty, \dot{\gamma})$.

Data analysis

For start-up flow sweeps we have used the following procedure to normalize the data:

Normalized times $s = t/\tau$
Normalized shear stress $\eta_n^+ = \sigma/\tau\dot{\gamma}$
Normalized first normal force coefficient $\Psi_{1n}^+ = \Psi_1/\tau\eta_0$,

with t as real time, τ as relaxation time, η_0 as zero shear viscosity, determined from $\eta(\infty, \dot{\gamma})$ for $\dot{\gamma} \to 0$, and from $\eta^*(\omega, \gamma)$ for $\omega \to 0$. Ψ_1 denotes the first normal stress coefficient. This coefficient is determined using the relationship $\Psi_1 = N_1/2\dot{\gamma}^2$.

For steady-state shear flow such as rate sweeps and frequency sweeps, we have used another procedure to treat the experimental data:

Normalized rate $\chi = \tau\dot{\gamma}$
Normalized frequency $\zeta = \tau\omega$
Normalized shear viscosity $\eta_n = \eta/\eta_0$
Normalized dynamic viscosity $\eta_n^* = \eta^*/\eta_0$
Normalized elasticity coefficient $G'/\omega_n^2 = (G'/\omega^2)/(G'/\omega_0^2)$

Here, ω denotes the angular frequency, $\dot{\gamma}$ the shear rate, η_0 the zero viscosity, and G' the storage modulus.

Steady shear flow

The normalized results of frequency sweeps and steady shear flow experiments and the theoretical predictions of the Giesekus model are compared in Fig. 1. A very good agreement between experimental results and the theoretical predictions is obtained using parameter $\alpha = 0.5$. This holds for all investigated rheological properties, such as $\eta^*(\omega, \gamma)$, $\eta(\infty, \dot{\gamma})$, G'/ω^2, $\Psi_1(\dot{\gamma})_{YM}$, and $\Psi_1(\infty, \dot{\gamma})$.

The semiempirical Cox–Merz rule combines linear and non-linear flow properties by comparing dynamic viscosity $\eta^*(\omega, \gamma)$ and steady-state values of the shear viscosity

$\eta(\infty, \dot{\gamma})$ at equal values of shear rate γ and frequency ω. It is interesting to denote that for parameter $\alpha = 0.5$ the Cox–Merz rule is informally obtained from the non-linear Giesekus model.

By using the Yamamoto-relation [7] it is possible to obtain the normal stress coefficient $\Psi_1(\dot{\gamma})_{YM}$ from transient experiments after cessation of steady-state shear flow (7)

$$\Psi_1(\dot{\gamma})_{YM} = 2 \int \eta_0^-(t, \dot{\gamma})\,dt . \tag{7}$$

The good agreement of the first normal stress coefficient $\Psi_1(\infty, \dot{\gamma})$ obtained from rate sweep experiments and the first normal stress coefficient $\Psi_1(\dot{\gamma})_{YM}$ obtained from the Yamamoto relation together with the theoretical predictions of the Giesekus model is shown in Fig. 1. For this purpose the Yamamoto data are normalized according to $\Psi_1(\dot{\gamma})_{YMn} = \Psi_1(\dot{\gamma})_{YM}/\tau\eta_0$. The normalized elasticity coefficient G'/ω_n^2 also corresponds with the Giesekus model as shown in Fig. 1 for a solution of CTAB–NaSal 60–350.

Start-up flow

The normalized results of transient start-up shear flow experiments and the predictions of the one-mode Giesekus model are compared in Figs. 2–5. In these experiments the applied steady shear rate is increased systematically from 0.1 1/s to 10.0 1/s. A very good comparison to the theoretical predictions is again obtained by using parameter $\alpha = 0.5$. This holds for shear viscosity $\eta(\infty, \dot{\gamma})$ and first normal stress coefficient $\Psi_1(\infty, \dot{\gamma})$.

Both mirror relations introduced by Gleißle [5] are derived by experimental observations. The first relation (8) combines the shear stress growth coefficient $\eta_0^+(t, \dot{\gamma})$ with

Fig. 1 Normalized steady shear properties, normalized first normal stress coefficient $\Psi_{1n}(\dot{\gamma})_{YM}$ derived from the Yamamoto relation, and normalized elasticity coefficient $G'/\omega_n^2(\omega, \gamma)$ (dots) as function of angular frequency ω and shear rate γ for CTAB–NaSal 60–350 and $T = 20$ °C compared to theoretical predictions by one-mode Giesekus model (lines)

Fig. 2 Normalized data of stress growth coefficient $\eta_n^+(t, \dot{\gamma})$, reflected shear viscosity $\eta(\infty, \dot{\gamma})$, first normal stress growth coefficient $\Psi_{1n}^+(t, \dot{\gamma})$, and reflected first normal stress coefficient $\Psi_{1n}(\infty, \dot{\gamma})$ (dots) as function of normalized time s at shear rate $\dot{\gamma} = 0.1$ 1/s for CTAB–NaSal 60–350 and $T = 20$ °C compared to theoretical predictions by one-mode Giesekus model (lines)

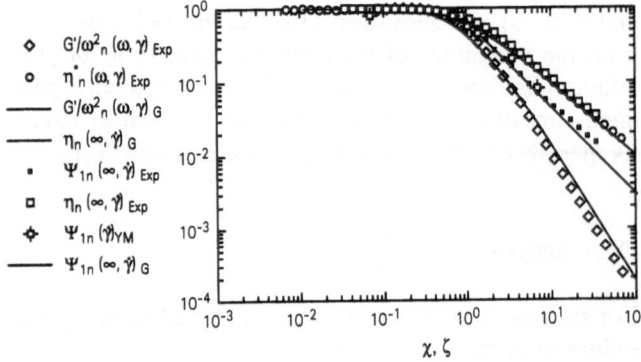

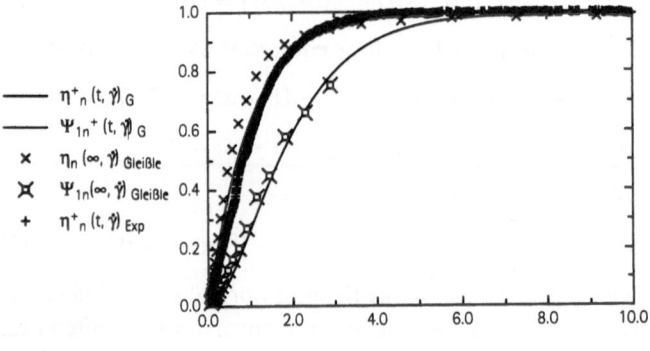

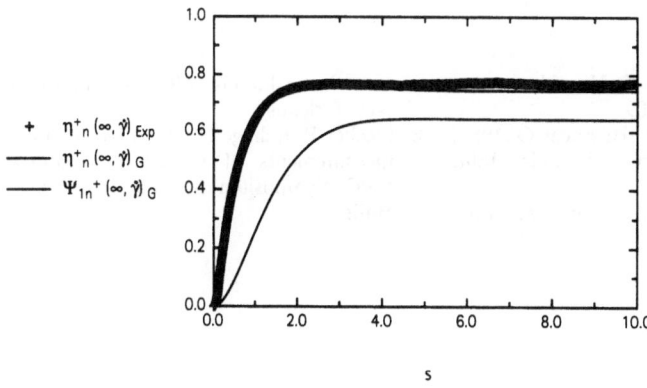

Fig. 3 Normalized stress growth coefficient $\eta_n^+(t, \dot{\gamma})$ and normalized first normal stress growth coefficient $\Psi_{1n}^+(t, \dot{\gamma})$ (dots) as function of normalized time s at shear rate $\dot{\gamma} = 1.0$ 1/s for CTAB–NaSal 60–350 and $T = 20\,°C$ compared to theoretical predictions by one-mode Giesekus model (lines)

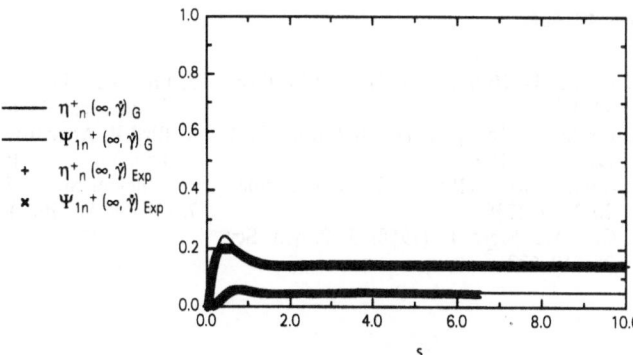

Fig. 5 Normalized stress growth coefficient $\eta_n^+(t, \dot{\gamma})$ and normalized first normal stress growth coefficient $\Psi_{1n}^+(t, \dot{\gamma})$ (dots) as function of normalized time s at shear rate $\dot{\gamma} = 10.0$ 1/s for CTAB–NaSal 60–350 and $T = 20\,°C$ compared to theoretical predictions by one-mode Giesekus model (lines)

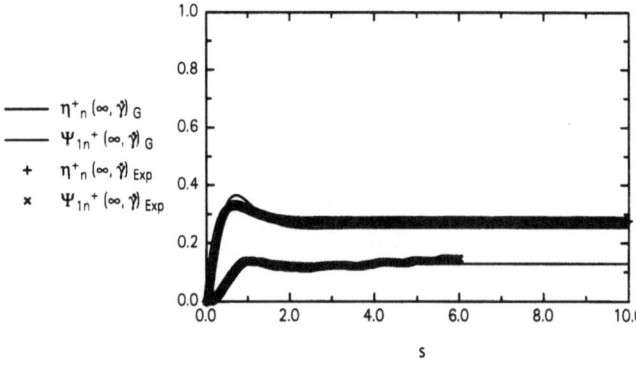

Fig. 4 Normalized stress growth coefficient $\eta_n^+(t, \dot{\gamma})$ and normalized first normal stress growth coefficient $\Psi_{1n}^+(t, \dot{\gamma})$ (dots) as function of normalized time s at shear rate $\dot{\gamma} = 5.0$ 1/s for CTAB–NaSal 60–350 and $T = 20\,°C$ compared to theoretical predictions by one-mode Giesekus model (lines)

ical predictions of the non-linear Maxwell material with parameter $\alpha = 0.5$.

Errors

A systematic error of 5% due to the inaccuracy of measuring time, normal forces and torque is assumed to hold for all rate sweep and frequency sweep data. Due to the electronic lag time of the torque transducer and the inertia of all mechanical parts a calibration for each investigated viscosity is performed with Newtonian fluids of the same viscosity as the surfactant solution.

the shear viscosity $\eta(\infty, \dot{\gamma})$. To obtain a correlation, one has to plot $\eta(\infty, \dot{\gamma})$ as a function of $1/\dot{\gamma}$

$$\eta_0^+(t) = \eta(\infty, \dot{\gamma}) \quad \text{with } t = 1/\dot{\gamma} . \tag{8}$$

The second Gleißle relation (9) combines the normal stress growth coefficient $\Psi_{10}^+(t, \dot{\gamma})$ with the normal stress coefficient $\Psi_1(\infty, \dot{\gamma})$. To obtain a correlation one has to plot $\Psi_1(\infty, \dot{\gamma})$ as a function of $1/\dot{\gamma}$:

$$\Psi_{10}^+(t, \dot{\gamma}) = \Psi_1(\infty, \dot{\gamma}) \quad \text{with } t = k/\dot{\gamma}; \quad k = \text{shift factor} . \tag{9}$$

For the investigated model system CTAB–NaSal 60–350 both mirror relations are plotted in Fig. 2. It is worthwhile to mention that both relations coincide with the theoret-

Summary

In this paper, we have systematically investigated the non-linear flow properties of viscoelastic surfactant solutions. It turns out that the experimental results are in a fairly good agreement with the theoretical predictions of the one-mode Giesekus model, especially by adjusting the only free parameter with $\alpha = 0.5$.

For these conditions, the Cox–Merz rule, the Yamamoto relation, and both Gleißle mirror relations are automatically derived from the theoretical mode. It is very interesting to state that the non-linear Maxwell material, first proposed by Giesekus, can be used for the qualitative description of non-linear flow properties of viscoelastic surfactant solutions, such as the shear viscosity or the first normal stress difference.

References

1. Rehage H, Hoffmann H (1992) Mol Phys 74:933–973
2. Fischer P, Rehage H (1994) Tenside Surf Det 31:99–109
3. Cates M (1987) Macromolecules 20:2289–2296
4. Cox W, Merz E (1958) J Polym Sci 28:619–622
5. Giesekus H (1982) J Non-Newtonian Fluid Mech 11:69–109
6. Bird R, Armstrong R, Hassager O (1987) Dynamics of Polymer Liquids. John Wiley & Sons, NY
7. Gleißle W (1981) The mirror relation for viscoelastic liquids. AlChE Symposium, New Orleans
8. Fischer P, Rehage H (1994) Normalforce measurements of surfactant solutions. DPG Symposium in Polymer Science, Halle

Progr Colloid Polym Sci (1995) 98:99–102
© Steinkopff Verlag 1995

A.B.D. Brown
R.C. Ball
S.M. Clarke
J. Melrose
A.R. Rennie

Model hard sphere systems under flow, a novel experimental approach

A.B.D. Brown · R.C. Ball · S.M. Clarke
J. Melrose · Dr. A.R. Rennie (✉)
Polymers and Colloids Group
Cavendish Laboratory
Madingley Road
Cambridge CB3 OHE, United Kingdom

Abstract Experimental determination of the flow-induced structures of particles in suspension have been employed previously to provide a stringent test of models of non-Newtonian behaviour of flowing systems. To date, most work has employed light and neutron scattering on micron and sub-micron sized particles. Here we report on a novel experimental technique that provides structural data for 0.32 mm diameter polystyrene balls suspended in gelatine under static and shear flow conditions. The results indicate significant structural differences between the static and sheared sample. One of the features observed is a pronounced layering at the walls of the plates for the sheared samples. Oscillatory and steady shear are both found to give ordering but with different particle arrangements. Some proposed developments of the technique are outlined.

Key words Rheology – hard spheres – structures – dispersions

Introduction

Colloidal dispersions exhibit a wide range of rheological behaviour which should be related to the microscopic properties of the suspensions, particularly the interparticle structure. Recently, well characterised, monodisperse samples of sub-micron sized particles have been employed in both rheological and structural studies [1–3]. Determination of the (real-space) structure has principally been by the interpretation of (reciprocal space) scattering patterns, a method which has inherent ambiguities. Other approaches to determining the inter-particle structures have included computer simulation based on a range of models of greater or lesser complexity. Experimentally, by using larger spheres, we can determine the structure of a suspension directly in real space. Here we report structural data for 0.32 mm diameter polystyrene balls suspended in a neutrally buoyant fluid both at rest and under conditions of controlled shear stress. A novel feature of this work is that the suspending fluid gels during the shear so

that a solid sample is obtained for sectioning and subsequent analysis.

Experimental

The experimental set up is illustrated in Fig. 1. In this simple plate-plate rheometer the bottom plate is held stationary while the top plate is rotated by the imposition of a controlled, constant stress. In this manner the experiment could be performed at constant Peclet number, Pe, where Pe is defined as:-

$$6\pi\eta\gamma a^3/8KT = Pe, \tag{1}$$

where η is the viscosity, γ is the strain rate, a is the particle diameter and K is the Boltzmann constant. For a Newtonian fluid, the strain rate is equal to the stress, τ divided by the viscosity, η:

$$\tau/\eta = \gamma \tag{2}$$

Fig. 1 Schematic diagram of the plate-plate rheometer employed in this study. The disk radius was 50 mm, and interplate separation was varied between 3–30 mm

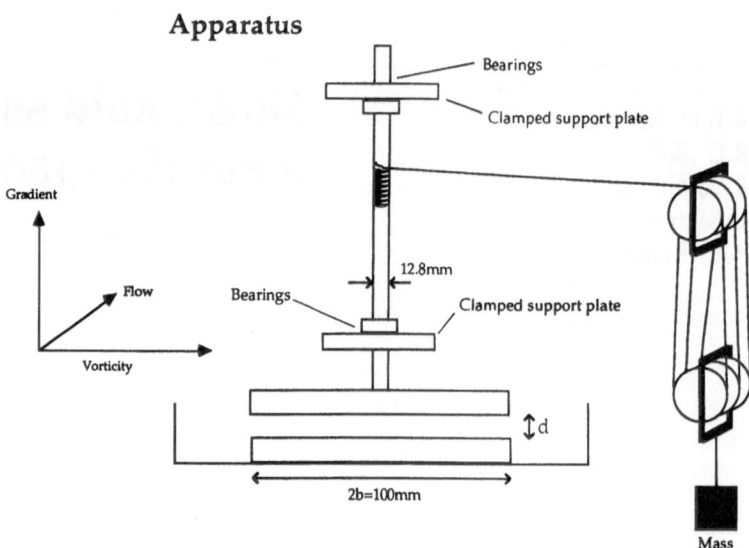

Apparatus

Combining (1) and (2) it is clear that for constant applied stress and varying viscosity, *Pe* will be constant. The bottom plate sits in a Petri dish to prevent the loss of sample fluid. The upper plate is attached to a shaft held by two self-centring bearings which can be adjusted to align the plates. The plate separation can be adjusted and dissembled completely to remove the sample for detailed examination.

Polystyrene spheres with a nominal diameter of 0.32 mm were obtained from Fisons Ltd. and passed through a series of fine wire meshes to reduce the polydispersity to about 3%. The suspending medium consisted of 7.3% gelatine, 6.9% glycol and 85.8% water by weight. The particles were neutrally buoyant in this medium with little evidence of creaming over the time scale of the experiments. The volume fractions studied were 1%, 50% and 55%, although only the results at 50% are presented here. The suspension was prepared at a temperature of approximately 80 °C at which the suspending medium is completely fluid. After transferring to the rheometer at 28 °C the dispersion was continuously sheared until it sets, thus fixing the particles in the shear-induced structure. The final temperature was 24 °C. The size of the particles is so large that the imposed shear of these experiments is always more significant than the random Brownian motion of the spheres ($Pe = 10^{14}$).

Details of the particle arrangements in the gel were obtained via sectioning in the planes of interest (flow-vorticity, vorticity-gradient and gradient-flow). The particle positions were collected by scanning photographs of these sections and then locating the particle centres by inspection with the program "NIHImage". It is clear that sectioning does perturb some of the surface layer of particles. Therefore any loose particles or others that appear

to have been disturbed are excluded from further analysis. The particle positions were used to calculate radial and linear distribution functions. Average particle distributions were calculated by taking one plate as the origin and determining the number of particles at a given distance from that plate. This parameter is expected to reveal any ordering normal to the plate surfaces.

Results

Sections of samples at a plate spacing of 10 mm in the shear gradient-vorticity plane are shown in Fig. 2(a–c). Figure 2(a) illustrates an unsheared suspension showing an amorphous structure. In Fig. 2(b) a sample that has undergone steady shear is illustrated. This sample shows layering at the surface of the plates which is not evident in the static sample. There is also an amorphous region in the middle of the sheared sample. This layering is reflected more clearly in the linear average given in Fig. 3. The number of layers has been studied as a function of plate separation. At the smallest separation there is layering all across the interplate gap while at a separation of 12 mm, four to five layers are observed. Detailed analysis of the amorphous region did not reveal significant anisotropy.

The position of particles in adjacent layers also shows some differences between samples undergoing steady shear or oscillatory motion. With steady shear (Fig. 2(b)) the particles in one layer sit in the troughs of adjacent layers forming a dense structure. In small strain oscillatory runs (Fig. 2(c)) the particles sit directly on top of their neighbours in the layering region. Figure 2(c) also shows a dislocation in the regular array.

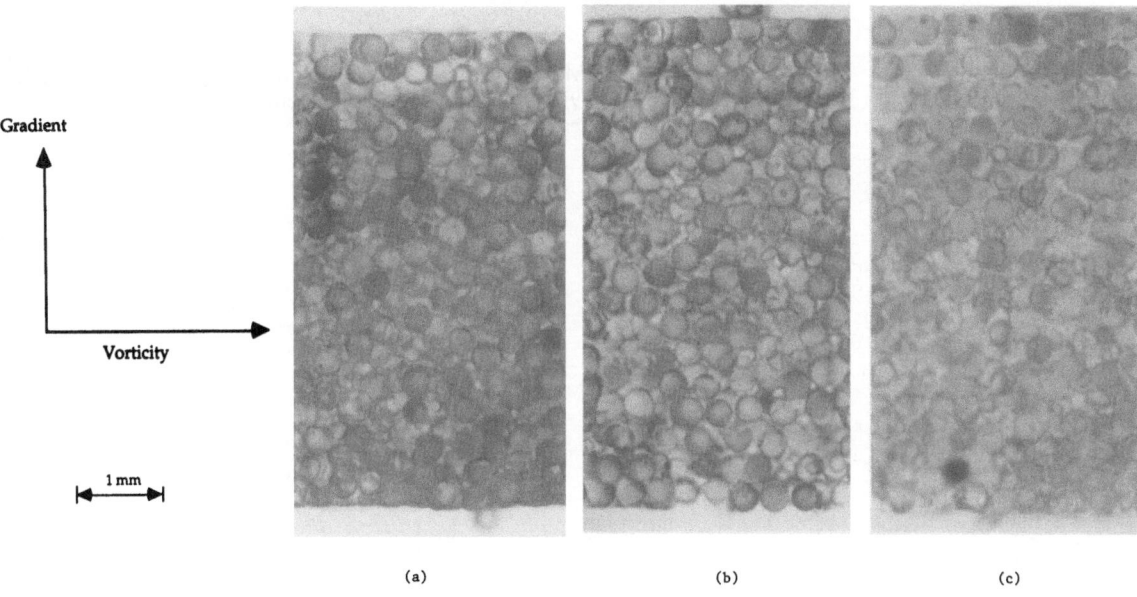

Gradient

Vorticity

1 mm

(a)   (b)   (c)

Fig. 2 Sections in the vorticity-gradient plane at 10 mm inter-plate separation. The volume fraction of spheres was 50% for all these samples. a) static, b) after steady shear and c) after oscillatory shear

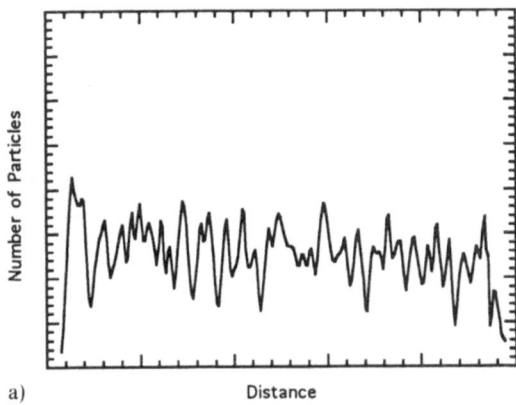

Number of Particles

a)   Distance

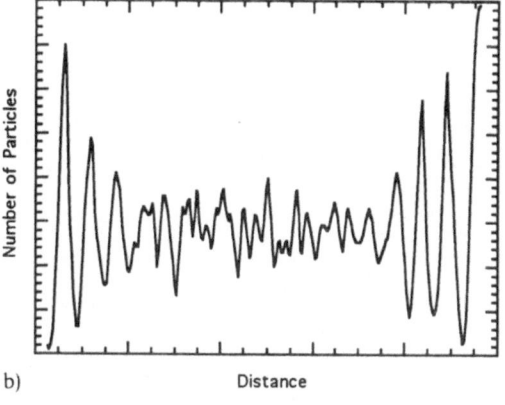

Number of Particles

b)   Distance

Fig. 3 Positional average for a sample in the vorticity-gradient plane a) at rest and b) after steady shear. The interplate separation was 11 mm and the volume fraction of particles was 50%. The peaks and troughs arise from layering at the plate surface

Conclusions

This paper reports a new and novel experimental method to directly study the structure of particulate dispersions under flow. The results above indicate significant differences in structure between the sheared and unsheared suspensions. The most significant features being a layered region close to the plates of the rheometer as well as differences in the influence of steady and oscillatory shear. Future work is planned employing NMR imaging tech- niques. This method should enable the three-dimensional arrangements of particles to be determined without any sectioning. Other developments may employ a more rigid polymeric media than gelatine to fix the particles. It is desirable to understand the effect of the non-Newtonian fluid before making detailed comparison with simulations and experiments on colloidal dispersions.

Acknowledgement This work was supported by ECCI & EPSRC.

102
A.B.D. Brown et al.
Model hard spheres under flow

References

1. Heyes DM, Melrose JR (1993) Journal of Non-Newtonian Fluid Mechanics 46:1
2. Clarke SM, Melrose J, Rennie AR, Ottewill RH, Heyes D, Mitchell PJ, Hanley HJM, Straty GC (1994) Journal of Physics: Condensed Matter 6:A333
3. Ackerson BJ, Pusey PN (1988) Physical Review Letters 61:1033

Progr Colloid Polym Sci (1995) 98:103–105
© Steinkopff Verlag 1995

A. Ponton
D. Quemada

Rheology of concentrated colloidal suspension: indirect prediction of interaction potential

Dr. A. Ponton (✉) · D. Quemada
Laboratoire de Biorhéologie et
d'Hydrodynamique
Physico-chimique (L.B.H.P.)
Université Paris VII
CNRS URA 343
2, Place Jussieu
75251 Paris Cedex 05, France

Abstract The non-Newtonian viscosity of colloidal dispersions at low electrolyte concentration is analyzed with an effective interaction potential proposed by Buscall. Experimental data at several volume fractions are compared to the model in order to first obtain interaction potentials versus surface-to-surface distance between particles, and then viscosity curves as function of shear-rate.

Key words Concentrated colloidal suspensions – interparticle interactions – rheology

Introduction

In dispersions of colloidal particles in an aqueous medium at low electrolyte concentration, electrostatic forces are dominant. They are responsible for an increase of the viscosity of such suspensions, which may be several times the viscosity of a hard sphere system at the same concentration. The physical origin of this phenomenon is generally believed to be due to the presence of an ionic double layer around each particle. This surface excess charge is linked to a surface potential associated with the adsorbed ions.

The present paper reports on an analysis of viscosity measurements in electrostatically colloidal silica suspensions in terms of an effective interaction potential, first introduced by Buscall. The model allows us to obtain a direct relationship between macroscopic flow behavior and microscopic features concerning the particles and their interactions.

Model

The key idea which supports this model is that colloidal particles (true radius a) with their double ionic layer have an effective radius a_{eff} depending on shear-stress σ (or shear-rate) and interact via a repulsive potential V_{eff}. In first approximation, such a suspension behaves as a hard sphere-like dispersion, i.e., spherical particles of size a_{eff} and volume fraction Φ_{eff}, immersed in a fluid solvent.

The system viscosity η may be then evaluated by:

$$\eta = \eta_f \left(1 - \frac{\Phi_{eff}}{\Phi_m}\right)^{-2}, \tag{1}$$

where η_f is the fluid solvent viscosity, Φ_m the maximum packing fraction and

$$\Phi_{eff} = \Phi \left(\frac{a_{eff}}{a}\right)^3, \tag{2}$$

The fictitious radius a_{eff} is related to the effective interaction potential V_{eff}.

The shear-stress dependence of V_{eff} has been introduced by Buscall. By considering competition between energies associated to shear forces and Brownian motion, he proposed the following expression [1] for $V_{eff}(a_{eff})$:

$$V_{eff}(a_{eff}) \approx kT + \frac{\sigma a_{eff}^3}{C}, \tag{3}$$

where k is the Boltzmann constant, T the temperature, and C the phenomenological constant.

Thus, an indirect estimation of the effective interaction potential is possible from rheological measurements with the set of equations (1), (2), and (3) by calculating $a_{\mathrm{eff}}(\sigma)$ from the data $\eta(\Phi_{\mathrm{eff}})$.

Moreover, if theoretical potential is known, the viscosity as function of shear-rate may be calculated.

Results

The model has been applied to viscosity data of colloidal suspension of silica particles ($a = 50$ nm) bearing negative charges (charge density $q = 25 \pm 3$ mC·m^{-2}) and dispersed in an aqueous phase (permittivity ε) low electrolyte concentration (symmetrical, valence z, ionic strength $I = 4 \cdot 10^{-4}$ M at pH = 8). Flow measurements have been performed [2] by varying the volume fraction of particles at constant ionic strength. At these experimental conditions, the colloidal interactions are dominated by electrostatic repulsion. The interaction potential can be then reduced to the electrostatic one. In the case of thin double layer, it is given by:

$$V_{\mathrm{el}}(r) = 32\,\pi\varepsilon\left(\frac{kT}{ze}\right)^2 a \tanh^2\left(\frac{\Psi_s ze}{4kT}\right)$$
$$\times \exp\left\{-\kappa_{\mathrm{eff}}(r - 2a)\right\}, \qquad (4)$$

where $\kappa_{\mathrm{eff}}^{-1}(q)$ is the Debye length modified in order to account for concentration effects, e the electronic charge, and Ψ_s the surface potential of particles [3]. The latter has been previously determined [4] from the interpretation of the low-shear viscosity with a structural model: ($\Psi_s \approx -100$ mV).

The identification between the two potentials $[V_{\mathrm{eff}}(2a_{\mathrm{eff}}) = V_{\mathrm{el}}(2a_{\mathrm{eff}})]$, normalized by thermal energy, has been performed for different values of Φ between 25% and 32% by fixing the charge density to its measured value and by searching the C-parameter with a least square method. As an example, this identification is illustrated in Fig. 1 for two volume fractions. It should be noted that the agreement is rather good throughout the whole range of surface-to-surface distance between particles $h = 2(a_{\mathrm{eff}} - a)$. The phenomenological constant C has a power law dependence on volume fraction. When this variation is established, the effective potential (Eq. (3)) is fully determined. It is then possible to use the model in a reverse way to predict viscosity as function of shear-rate (Eqs. (3), (2), and (1)). Figure 2 shows the results for three volume fractions. We note that flow curves are correctly reproduced and a good fit is obtained between the cal-

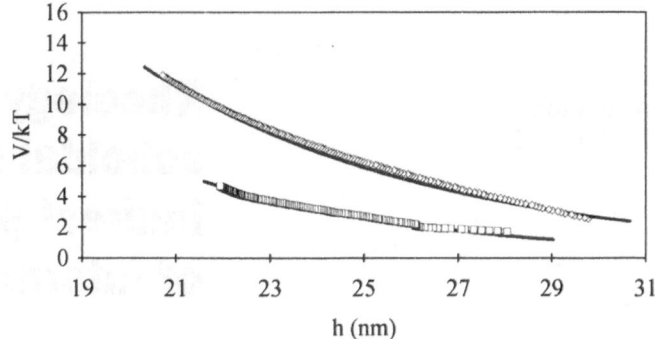

Fig. 1 Variations of effective Buscall potential V_{eff} ($\square$ $\Phi = 25\%$; $\diamond$ $\Phi = 29.6\%$) and electrostatic potential V_{el} (solid lines) as function of surface-to-surface distance between particles

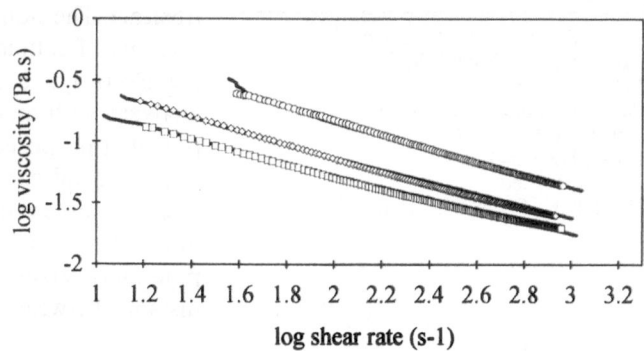

Fig. 2 Flow curves from experimental results ($\square$ $\Phi = 27.6\%$; $\diamond$ $\Phi = 28.6\%$; $\circ$ $\Phi = 31.2\%$) and calculated values (solid lines)

culated values and the experimental data. Thus, the model is a useful way to predict viscosity in the case where theoretical interaction potential between particles is known.

Conclusion

Viscosity measurements of electrostatically stabilized silica suspensions have been analyzed with a model relating the non-Newtonian viscosity to an effective interaction potential. In the case studied here (low electrolyte concentration), colloidal interactions are mainly electrostatic in nature and the theoretical potential is known. Thus, after a complete determination of the effective potential for various fractions of particles, we have successfully reproduced experimental flow curves.

These results suggest to use or to modify this model for sterically stabilized suspensions, which we are now working on.

Progr Colloid Polym Sci (1995) 98:103–105
© Steinkopff Verlag 1995

References

1. Buscall R (1992) Proc XIth Int Congr on Rheology, Brussels, Belgium: 591–594
2. Rheological measurements were carried out with Neel O, Lafuma F (Laboratoire de Physico-Chimie Macromoléculaire, CNRS URA 278, Ecole Supérieure de Physique et Chimie Industrielles de Paris)
3. Russel WB, Saville DA, Schowalter WR (1989) in Colloidal Dispersions, Cambridge Univ. Press, Cambridge pp 118–120
4. Ponton A, Quemada D, Neel O (1994) Cahiers de Rhéologie, XII 2:71–79

Progr Colloid Polym Sci (1995) 98:106–110
© Steinkopff Verlag 1995

R. Hofmann
Th. Förster
W. von Rybinski
A. Wadle

Rheological properties of fine disperse o/w-emulsions

Dr. R. Hofmann (✉) · Th. Förster
W. von Rybinski · A. Wadle
Henkel KGaA
TTR-Physical Chemistry
Building Z1
Henkelstraße 67
40191 Düsseldorf, FRG

Abstract Rheological parameters are important for the characterization of dispersions and emulsions regarding storage stability and application properties. Fine disperse o/w-emulsion concentrates prepared by means of the phase inversion temperature (PIT) method were characterized by dynamic and steady-state rheology, photo correlation spectrocopy for oil droplet size distribution and their phase behaviour. Stable emulsion concentrates show a remarkable yield value and marked shear-thinning flow behaviour. For cosmetic applications as lotions or creams these o/w-emulsion concentrates were diluted up to 80 weight percent water and the desired viscosity was adjusted with a polymer or a self-bodying agent. It could be shown that fatty alcohol thickeners build up a similar gel structure as polymers, when used at higher concentrations. Closer inspection of the rheological profiles reveals remarkable differences in gel properties which can be correlated with advantages in application properties.

Key words o/w-emulsions – rheology – self-bodying agents – thickening mechanism – PIT process

Introduction

Emulsions are dispersions of at least two liquid phases immiscible with each other, whose physicochemical properties and composition can vary considerably, depending on the field of application. Usually, cosmetic emulsions contain polar and nonpolar emollients which are stabilized by a mixture of hydrophilic emulsifiers and hydrophobic coemulsifiers. The selection of the emulsifier/coemulsifier system and the production process are essential for the quality of the finished emulsion. For cosmetic emulsions it is optimal if a fine disperse emulsion is obtained with a minimal quantity of emulsifier. The minimal quantity of emulsifier contributes to the formulation of skin compatible products, and the fine dispersity supports the physical stability. Usually, for economic reasons, emulsions are produced from an emulsion concentrate in a two-stage process and then diluted [1–3]. Here the phase inversion temperature (PIT)-process can be advantageous [2–4]. The usual consistencies – well flowable lotion or paste-like cream – are adjusted in a thickening step, thus reaching the rheological property profile and the necessary physical long-term stability. As a thickener, oleochemical self-bodying agents such as longer chain fatty alcohols or mono-glyceride esters [5–13] or polymers [14–17] are commonly used in cosmetic emulsions.

Concerning the general rheological behaviour of emulsions, e.g., as a function of the content of internal phase or the dispersity, broad experience is available [18, 19]. However, it is still not known how different thickeners influence the rheological properties of emulsions, in particular of PIT emulsions.

This work intends to elucidate the influence of a polyacrylate thickener and a fatty alcohol, as fundamentally different thickening systems, on the rheological properties of fine disperse o/w emulsions. In order to ensure good comparability, we used a single emulsion concentrate, which was diluted with water to obtain the desired final concentration; at the same time the viscosity was adjusted with polymer or fatty alcohol.

Experimental part

Test system

An emulsion concentrate of the following composition served as a basis for the preparation of cosmetic model emulsions:

3.1%   Glycerol monosterate (Cutina GMS)
6.9%   Polyoxyethylene-12-Cetylstearyl alcohol (Eumulgin B1)
22.5% Di-n octyl ether (Cetiol OE)
22.5% 1, 3-Diisooctylcyclohexane (Cetiol S)
5%   Glycerol
40%   Water (deionized)

The specified emulsifiers and oils are commercial cosmetic raw materials from Henkel KGaA. The emulsion concentrate was produced according to the PIT process (2–4); the phase inversion temperature was 80 °C. Fine disperse, blue emulsions were obtained, since during the preparation a bicontinuous microemulsion phase with extremely low interfacial tension occurs between the oil and the aqueous phase (3).

The phase diagram in Fig. 1 shows the phase behaviour for this model system in dependence upon the temperature and water content of the emulsion. In addition, Fig. 1 contains information about the consistency of the emulsions, measured as the yield value at 25 °C. In the case

of a low water content (< 40%), high-viscous gels with high yield values result.

In order to avoid difficulties with the incorporation of the thickeners, a water content of 40% was selected for sample preparation. Finally, the samples were diluted to 80% and at the time thickened. At 70 °C, an oleochemical self-bodying agent, i.e., cetylstearyl alcohol (Lanette O, Henkel KGaA), was incorporated, while a polymer (sodium polyacrylate, product PNC 400 from 3V Sigma, Bergamo, Italy) was added at 30 °C.

Particle size determination

The size of the oil droplets were determined with Zeta-Sizer-3 instrument (Malvern Instruments, U.K.) at 25 °C by means of dynamic light scattering with a dilution 1:1000 with dist. water. The following mean particle sizes were found:

Emulsion concentrate, diluted (40% water content)	213 nm
Emulsion, diluted to 80% water content	159 nm
Emulsion diluted to 80% water content and thickened with 2.6% cetylstearyl alcohol	153 nm

The particle size remains constant during the storage time of 5 months, therefore the emulsions are stable against coalescence.

Sedimentation profiles by means of ultrasonic measurements

When an emulsion creams, oil accumulates in the top part of the emulsion. This leads to a density gradient, which can

Fig. 1 Phase diagram of the test emulsion (Emulsifier/Oil ratio given in figure)

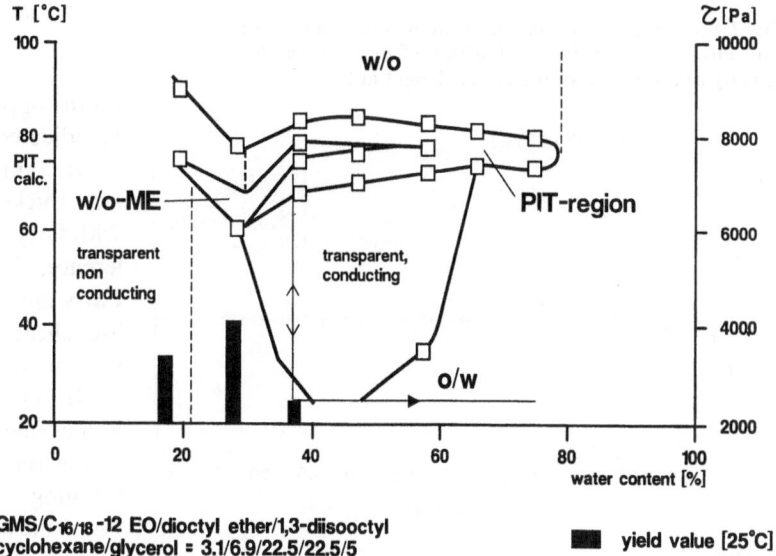

GMS/C$_{16/18}$-12 EO/dioctyl ether/1,3-diisooctyl cyclohexane/glycerol = 3.1/6.9/22.5/22.5/5

■ yield value [25°C]

108

R. Hofmann et al.
Rheology of o/w-emulsions

be detected through locally resolved measurement of the ultrasonic velocity. In this way sedimentation profiles were recorded as a function of storage time.

Figure 2 shows sedimentation profiles for the low-viscous emulsion, diluted to 80% water, for which creaming was most probable. Even after 5 months of storage, no creaming can be found. The same applies to the thickened emulsions: They are also stable against creaming.

Rheological measurements

The rheological properties were determined by means of a shear-rate controlled rotational rheometer, Rheometrics Fluids Spectrometer RFS II (Rheometrics, Piscataway, New Jersey, USA), using a cone-plate measuring system. The shear-rate dependent viscosity data were determined through recording of flow of viscosity curves. Dynamic oscillating measurements (Strain-Sweep and Frequency-Sweep experiments) provided information on the visco-elastic behaviour [20].

Results and discussion

The unthickened emulsion, diluted to 80% water content is low-viscous and shows Newtonian flow behaviour – as is to be expected in the case of a content of internal phase of only 15%. Viscosity is only achieved through addition of a thickener. Here, the two types of thickener differ clearly in their activity (see Fig. 3).

Cetylstearyl alcohol only forms an appreciable viscosity at a threshold value of over 2% (5) while low quantities of polyacrylate are sufficient for thickening. Depending

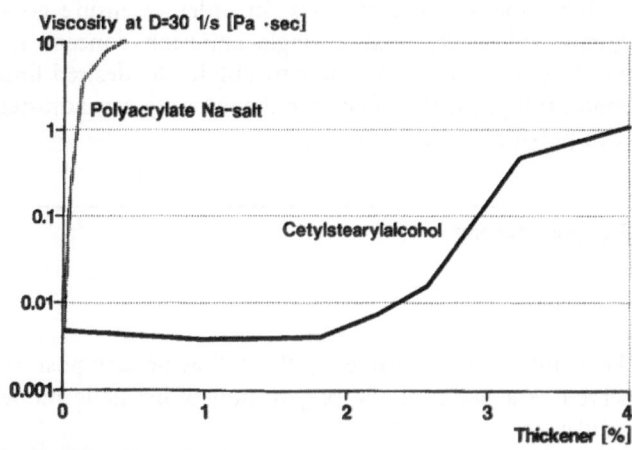

Emulsion composition after dilution
Cutina GMS/Eumulgin B1/Cetiol OE/Cetiol S/Glycerol/Water = 1/2,3/7,5/7,5/1,7/80

Fig. 3 Viscosity at $D = 30$ 1/s of the diluted emulsion as a function of the thickener concentration at 25 °C

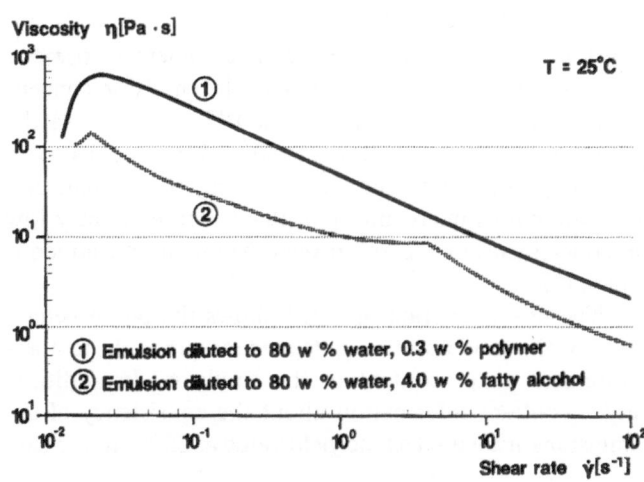

Fig. 4 Viscosity curves of the thickened emulsions at 25 °C

on the application, with both substances the viscosity can be adjusted as desired for lotion or creams.

A comparison of the complete flow curves of an emulsion thickened with polymer and an emulsion thickened with fatty alcohol (see Fig. 4) reveals clear qualitative differences in the thickening behaviour with yield value. Fatty alcohol causes a more complex flow behaviour: At low shear rates shear-thinning flow behaviour is found with an yield value which changes at a shear rate of 30/s.

In general, it can be said that through incorporation of a thickener a gel structure is built which leads to a change of the rheological behaviour from Newtonian to shear-thinning and to viscoelasticity. The strain-sweep curves in Fig. 5 show the viscoelastic properties of two diluted emulsions in dependence on the concentration of the thickener

Fig. 2 Sedimentation profile of the diluted test emulsion measured with ultrasonic velocity as a function of measured with ultrasonic velocity as a function of the sample height at 25 °C

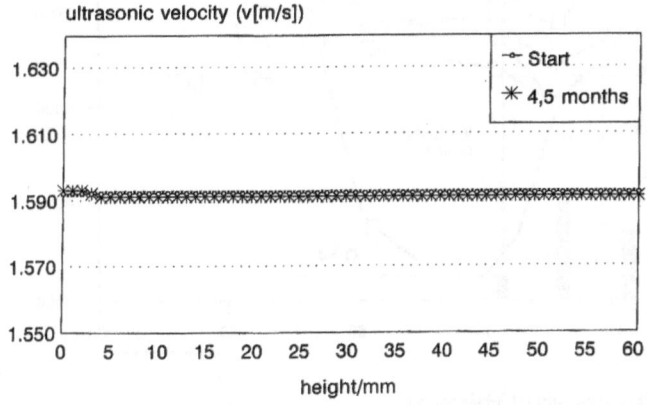

T=25°C

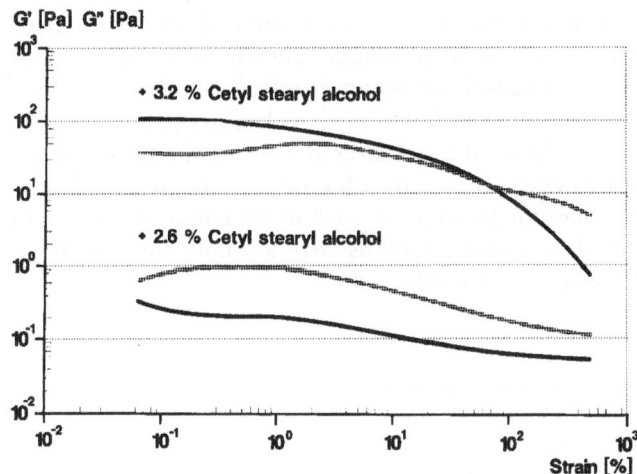

Fig. 5 Strain-sweep-curves of two diluted emulsions with 2, 6 and 3, 2 w% cetyl stearyl alcohol at 25 °C

7.5   parts Cetiol OE
7.5   parts Cetiol S
1.7   parts Glycerol
X%   thickener

Table 1 compares the rheological parameters from dynamic experiments for the two thickener systems. The polymer thickener already builds a gel structure at much lower concentrations than cetylstearyl alcohol, as can be seen from the high G'- and the low tan Delta-values.

However, for the characterization of the gel structure, not only the values G'- and the tan delta are important, but also their dependence upon the mechanical stress. In the case of comparable absolute values of G' and tan delta the gel networks formed with the two thickener types can be clearly distinguished by means of their strain-sweep-curves (see Fig. 6). The polyacrylate thickener forms an elastic gel

cetylstearyl alcohol. In the emulsion thickened with 2.6% fatty alcohol the behaviour is still mainly viscous (Young's modulus $G' <$ viscosity modulus G''), while at higher thickener concentrations a clear gel structure develops: Young's modulus G' in the linear viscoelastic range increases by a factor 400 and is now clearly higher than the viscosity modulus G''. This tendency continues with a further increases of the cetylstearyl alcohol concentration.

Frequency-dependent experiments also show this drastic change of the rheological properties. While in the case of 2.6% cetylstearyl alcohol G'' is higher than G' over the entire experimentally accessible range (0.001 rad/s to 100 rad/s), this ratio is inverted with a 3.2% solution.

Emulsion formulation 80% water (20-X)% consisting of:

1   part Cutina GMS
2.3   parts Eumulgin B1

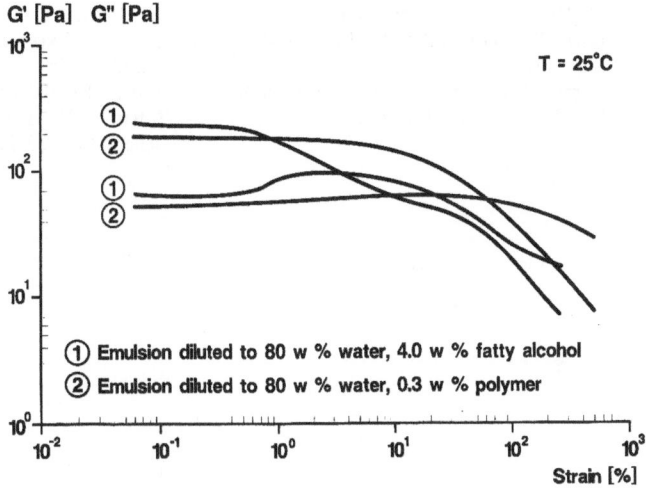

Fig. 6 Strain-sweep-curves of polymer thickened and a cetyl stearyl alcohol thickened emulsions at 25 °C

Table 1 Viscoelastic properties of thickened emulsions

Thickener	G'/Pa	G''/Pa	tan Delta	crit. strain
0%	0.1	1	10	< 0.1%
0.05% Polyacrylate	0.1	0.1	1	< 0.1%
0.1% Polyacrylate	200	60	0.3	12%
0.3% Polyacrylate	400	80	0.2	18%
0.5% Polyacrylate	800	120	0.15	20%
0.9% Cetylstearyl alcohol	0.1	1	10	< 0.1%
1.8% Cetylstearyl alcohol	0.1	0.9	9	< 0.1%
2.2% Cetylstearyl alcohol	0.1	0.9	9	< 0.1%
2.6% Cetylstearyl alcohol	0.3	2.4	8	< 0.1%
3.2% Cetylstearyl alcohol	120	60	0.5	1%
4.0% Cetylstearyl alcohol	272	95	0.35	1%

110

R. Hofmann et al.
Rheology of o/w-emulsions

with a large linear viscoelastic range, i.e., with high critical-strain-values; consequently, this gel shows completely elastic behaviour, even in the case of strong deformation. On the other hand, the fatty-alcohol-containing gel already collapses in the case of small deformations (low critical-strain-values in Table 1). This basic difference can also be perceived in practice: The large linear viscoelastic range of polymer-thickened emulsions expresses itself in the stickiness of such emulsions.

The test series shows how dynamic rheological experiments can be used to obtain valuable information about the structure formation and therefore also about cosmetic aspects in an emulsion from viscoelasticity parameters. On the one hand, this gel structure forms the basis for the desired viscosity structure, but on the other hand it can be the cause for phenomena such as stickiness. By means of suitable thickeners such as fatty alcohols, this unwanted effect can be avoided.

References

1. Lin TJ (1978) J Soc Cosmet Chem 29:117
2. Lin TJ, Akabori T, Tanaka S, Shimura K (1981) Cosmet Toil 96:31
3. Förster Th, Tesmann H (1991) Cosmet Toil 106:49
4. Förster Th, Schambil F, Rybinski WV, Dispersion (1992) J Sci Technol, 13:183
5. Förster Th, Schambil F, Tesmann H (1990) Int J Cosmet Sci 12:33
6. Barry BW, (1975) Adv Colloid Interface Sci, 5:37
7. Barry BW, Saunders GM (1972) J Colloid Interface Sci 5 38:616
8. Talman FAJ, Rowan EM (1970) J Pharm Pharmae, 22:338
9. Davies PJ, Talman PAJ, Expo Congres (1983) Int Techn Pharm 1:337
10. Davis SS (1969) J Pharm Sci 58:418
11. Eccleston GM (1986) Pharmacy International 63–70
12. Fukushima S, Takahashi M, Yamaguchi M (1976) J Colloid Interface Sci 57:201
13. Krog N, Larsson K (1968) Chem Phys Lipids 2:129
14. Pal R (1992) J Rheol 36:1245
15. Otsubo Y, Prud'homme RK (1994) Rheol Acta 33:29
16. Chen C-R, Zatz JL (1992) J Soc Cosmet Chem 43:1
17. Gladwell N, Grimson MJ, Rahalkar RR, Richmond P (1984) J Chem Soc, Faraday Trans 2, 81:643
18. Sherman P (1968) in: Emulsion Science, Sherman P (Ed), chap 4, 217–347
19. Tadros ThF, (1992) in: Emulsions – A Fundamental and Practical Approach, J Sjöblom (Ed), NATO ASI Series C, Kluwer 363:173
20. Barnes H, Hutton JF, Walters K (1989) An Introduction to Rheology, Elsevier, New York

Progr Colloid Polym Sci (1995) 98:111–116
© Steinkopff Verlag 1995

J. Castle
A. Merrington
L.V. Woodcock

Phase behaviour in the uniform shear of idealised suspensions

Dr. J. Castle (✉) · A. Merrington
L.V. Woodcock
Department of Chemical Engineering
University of Bradford
Bradford BD7 1DP, United Kingdom

Abstract The fluid mechanics of ideal dense suspensions of monodisperse latex particles has been measured in simple rheometric geometries under controlled stress conditions, thereby providing a means of testing computer simulation predictions for some constitutive rheological properties.

A new observation of rheopectic behaviour is described whereupon an initially amorphous dense hard-sphere suspension at low stresses at first begins to flow, but gradually slows down as the structure becomes ordered, and eventually stops. There is a critical stress for each sample, characterised by its volume fraction, above which, this phenomenon does not occur. That is the effective yield stress of a static colloidal crystal in the Couette rheometer.

This experimental observation is shown to be consistent with predictions of a phase diagram underlying the Couette flow of hard-sphere suspensions, and can be interpreted as the effect of a shear perturbation on the otherwise equilibrium, or static, phase diagram of the classical hard-sphere fluid.

Other indications of the phase behaviour in shear flow are also found, such as i) evidence for the formation of two-phase regions analogous to thermodynamic coexistence, ii) a shear thickening transition from order to disorder with increasing shear, and iii) the non-reproducibility of very dense suspension of the Bingham yield, plastic viscosity, and critical shear thickening, which, it is suggested, is due to random faceting.

Key words Phase behaviour – shear – idealised suspensions

Introduction

In order to predict flow processes of a colloidal suspension in a given geometry, two ingredients are required: i) the constitutive transport properties of the material, and ii) solutions to the conservation equations of fluid mechanics [1].

A two-component non-Newtonian fluid such as a colloidal suspension, however, presents hitherto insuperable difficulties. The basic problem stems from the fact that, during flow, frictional resistances arise from the repulsive or osmotic forces acting between the solid particles themselves. This leads to deformation rate dependent osmotic pressure and granular "temperature" (particle kinetic energy) gradients. Without any knowledge of the deformation rate dependent osmotic properties of these materials, which form part of the constitutive rheology, equations of fluid mechanics cannot be solved. Without a solution of the equations of fluid mechanics, the appropriate constitutive rheology (e.g. stress versus shear-rate flow curves for homogeneous laminar shear flow) cannot be measured by conventional rheometry because the unknown concentration and velocity profiles cannot be determined. These difficulties are now well-known and have been reviewed recently [2].

At present, different rheometer geometries, such as coaxial cylinder, cone and plate and capillary, give widely differing flow curves for dense suspensions. The discrepancies diverge towards maximum solids content, and/or at high shear rate. This is because sheared colloidal suspensions, even in simple laminar flows, are generally inhomogeneous in both particle concentration and velocity gradient. Osmotic pressures arising from forces between particles, which increase with both rate of shear and with solids fraction, lead to particle migrations, and inhomogeneities. The constitutive rheological relations for suspension flow are therefore inaccessible by present experimental rheometry.

One way forward is via computer simulations. By using tricks such as the Lees–Edwards [3] boundary conditions, driving surfaces can be eliminated and a uniform shear gradient imposed. Once the equations of motion of the particles are specified, it is then straightforward to calculate the shear stress as a function of shear rate. At present, almost all the computational effort, using a variety of methods involving Brownian, Stokesian and Newtonian equations of motion, is confined to monodisperse hard spheres in Newtonian media [4].

Idealised suspensions

The foremost objective is to calculate the constitutive rheology, to begin with the stress as a function of shear rate for homogeneous laminar shear flow, for the simple monodisperse frictionless hard-sphere suspension in an otherwise Newtonian fluid of given viscosity. The object of the experimental effort is to produce the nearest it is possible to get to the ideal hard-sphere suspension, and to measure its fluid mechanics (not in the first instance its rheology!) in simple rheometer geometries, for comparisons with the computer simulation predictions.

An electron micrograph of the nearest thing to the idealised "hard-sphere" monodisperse suspensions presently attainable is illustrated in Fig. 1. The experimental method used to produce these particles is similar to that described by Almog et al. [5] This is a suspension of highly monodisperse polystyrene latex, with a mean particle diameter of approximately 2 μm and a standard deviation of about 2%. The particles are sterically stabilised so that the surface charge is as close to zero as possible, and the particles are readily dispersed in water. With the addition of a tiny amount of sucrose the aqueous suspension can be made neutrally buoyant. This is the simplest imaginable real suspension, but an inspection of the behaviour of the dense suspension under a microscope indicates that there can be both ordered and disordered phases.

The equilibrium phase coexistence of hard-sphere colloidal suspensions was first demonstrated by Pusey and van Megen [6]. It seems reasonable to assume that the

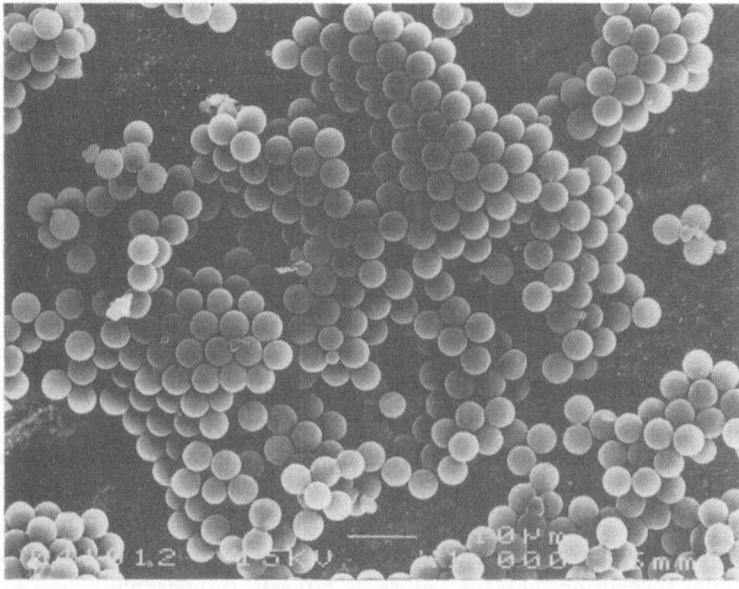

Fig. 1 Scanning electron micrograph of a sample of the sterically-stabilised well-characterised polystyrene colloidal particles; when suspended in aqueous media this is presently the nearest experimental system to the idealised hard-sphere model suspension. The mean particle diameter is approximately 2 μm

Progr Colloid Polym Sci (1995) 98:111–116
© Steinkopff Verlag 1995

effect of shear flow on the rheological properties may also involve a complex phase behaviour. Of particular importance for experimental effects involving order-disorder transitions are the characteristic relaxation times for equilibration. For hard-sphere suspensions these characteristic times scale roughly as diameter cubed. The ordering experiments of Pusey and van Megen [6] were for 0.5 μm PMMA and took several days. The corresponding time for the present 2-μm particles would therefore be 64 times longer, i.e. several months. The present polystyrene lattices can therefore be assumed to be in the disordered state at the outset of all these rheometric experiments, even for packing fractions exceeding the equilibrium freezing density of the hard-sphere fluid (48%). The effect of shear forces is to impart kinetic energies to the particles over and above the Brownian motion at equilibrium, it reduces the relaxation times for nucleation, for example, and leads to complex rheological effects.

Rheopexy

Two different results can be seen in constant stress experiments on the same sample as shown in Fig. 2. A homogenised sample of monodisperse polystyrene latex, at a solids packing fraction of 55%, was examined in a coaxial cylinder Deer rheometer at differing stresses. For higher stresses, there is an initial instrumental response as the rotating bob (inner cylinder) accelerates and then it reaches a maximum

rate of shear strain. For stresses below a certain critical stress, however, after the initial acceleration of the strain rate, it reaches a maximum value, and then the rate of strain begins to slow down until eventually the rotating cylinder stops altogether. At this point the suspension is effectively a solid. This experiment shows that the effective viscosity of the suspension in the rheometer is increasing, or shear thickening, with time. This type of complex rheological phenomena is called "rheopexy" and is quite rare, compared to the much more common time-dependent shear thinning effect of thixotropy.

This novel rheopectic effect was first reported by Merrington and Woodcock [7], and was explained as a transition from an initially metastable disordered fluid to the stable ordered gel state that the fluid would prefer to be in, given the time and means to equilibrate. The results were obtained from a series of cone-and-plate constant stress rheometric measurements on a monodisperse polystyrene latex suspension similar to that shown in Fig. 1 [8]. Further evidence for this interpretation was subsequently provided by Rodriguez et al. [9], who apparently observed the effect independently in a parallel plate flow geometry. They were able to confirm the transition to an ordered structure by observing Bragg diffraction patterns from index matched suspensions.

This phenomenon can be interpreted in terms of the equilibrium hard-sphere phase diagram (Fig. 3). At 55%, the system would be ordered at equilibrium. In the present experiments, however, for solid fractions exceeding the equilibrium fluid freezing concentration (about 45%) the

Fig. 2 Rheometer printouts for the shear rate as a function of time for controlled (constant) stress experiments using coaxial cylinder geometry in a Deer rheometer; note that the increasing time scale goes from right to left. The upper result, higher stress, approaches a steady shear rate with time, whereas the lower result, lower stress, shows rheopexy (time-dependent shear thickening) and the flow eventually stops

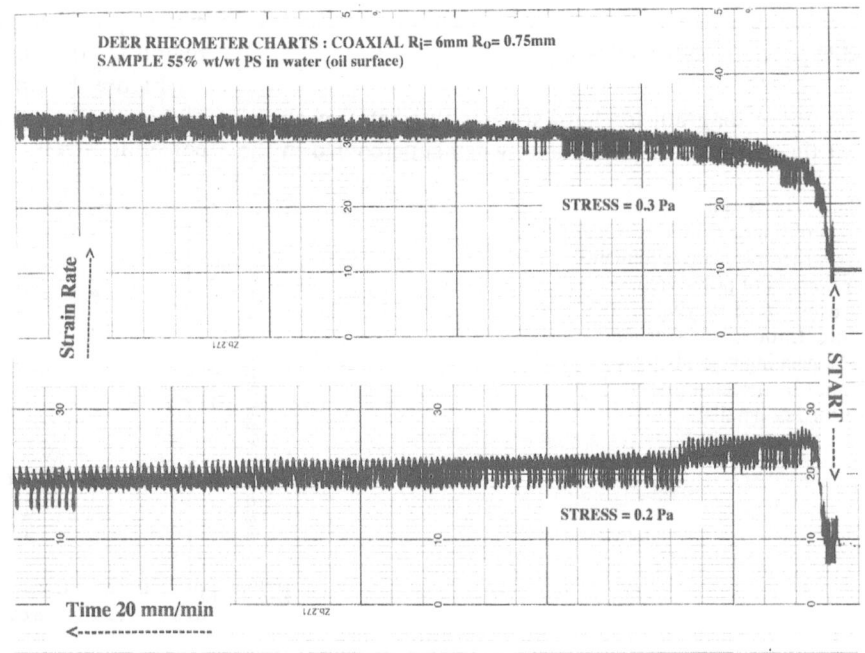

Fig. 3 Static equilibrium state behaviour of the idealised hard-sphere suspension; at constant volume fraction and temperature (T) the osmotic pressure (p) varies inversely as the thermal energy per particle (kT, where k is Boltzmann's constant), as indicated on the right-hand scale. σ is the particle diameter

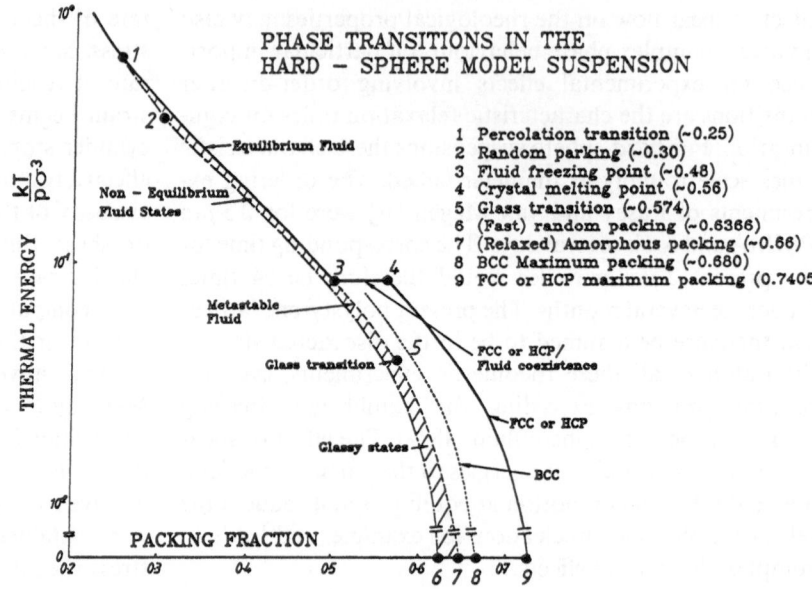

suspension is initially in the "supercooled" metastable amorphous fluid region, because for particles greater than 1 μm in diameter have crystallisation nucleation times greatly exceeding the time scale of the experiments. When the system is sheared at low stress, the suspension equilibrates to the ordered gel after several minutes. When the system is subjected to a higher stress, however, constant flow is observed. At stresses which are higher than the Bingham yield stress of the colloidal crystal the suspension continues to flow after it has ordered. The Bingham yield stress for the present 55% polystyrene colloidal crystal is evidently around 0.25 Pascal, i.e. between the two stresses of the results shown in Fig. 2.

Other phase transition effects

The phase diagram for hard-spheres (Fig. 3) also shows that the static hard-sphere fluid, when supercooled in the

metastable region, undergoes a glass transition. This is the point where the viscosity and self diffusion coefficients diverge, to infinity and zero respectively. At this point the fluid becomes effectively an amorphous solid which itself has a finite shear modulus and would support a stress and behave like a Bingham solid. By performing a range of controlled stress experiments on "fresh" samples of suspensions in the concentration range from 40 to 60% it is possible to measure a characteristic Bingham yield stress for the time scale of the experiment. This is taken to be simply the smallest perceptible stress for which any strain is observed over a fixed period of 1 min.

Some results for the yield stresses over a range of concentration are shown in Fig. 4; it is clear that the yield stress approaches zero around the known glass transition region for the hard-sphere fluid, i.e. around 58%.

Figure 5 shows an interesting flow curve which is obtained when a sample at 60% is subjected to higher constant stress. The flow curve shows three distinct

Fig. 4 The Bingham yield stress obtained from controlled stress rheometer measurements for a series of polystyrene suspensions of varying concentration; below around 58% packing a yield stress cannot be detected (see ref. [9] for further details)

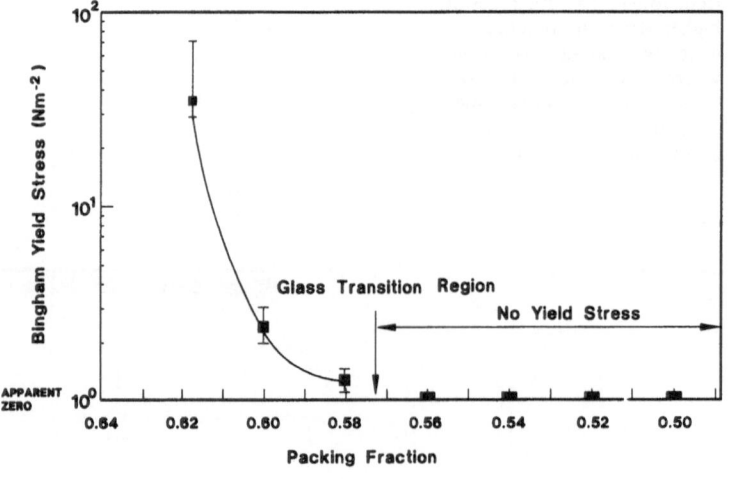

Progr Colloid Polym Sci (1995) 98:111–116
© Steinkopff Verlag 1995

Fig. 5 Rheograph of
a controlled stress experiment
at 60% packing obtained using
cone and plate geometry on
a Rheometrics controlled stress
instrument; this result is
indicative of a two-phase
coexistence in the sheared
sample where the instrument
detector fluctuations are high

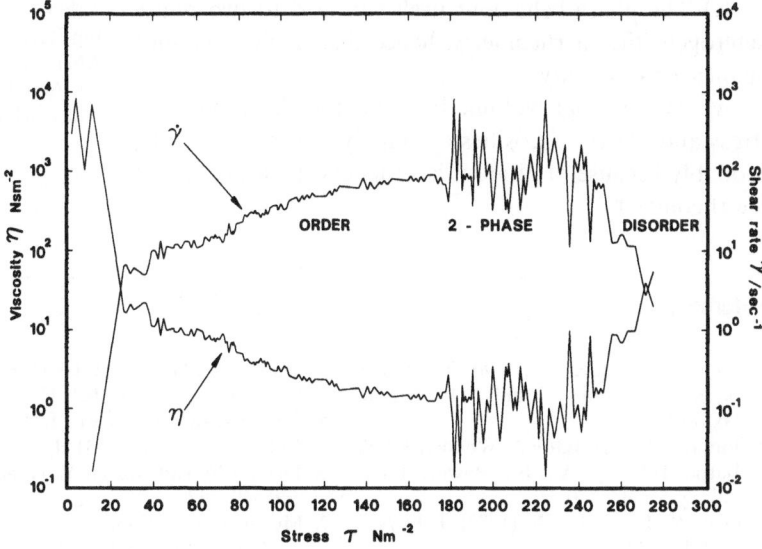

regions, each characterised by differences in the fluctuations in the response detector. Initially, the suspension starts off as a disordered fluid, it then shear thins and becomes an "ordered fluid"; at a higher stress there is a sudden sharp increase in fluctuations, which may be indicative of a two-phase region. This occurs at constant shear rate and constant effective viscosity, and finally the sample suddenly shear thickens; it is suggested to be a "disordered solid".

Figure 6 shows a remarkable set of experimental results on the same plot. Flow curves from dense suspension

rheometry are notoriously non-reproducible, often for reasons which we now understand, as mentioned earlier and reviewed in reference [2]. In these experiments, however, the flow curve of a very dense (62%) sample has been repeatedly measured, but starting anew each time, by the same person, using the same instrument, same settings, same geometry, and the same starting sample. The results show that the Bingham yield stress, the plastic viscosity, and the shear thickening transition can vary over many orders of magnitude, in a random manner, from one experiment to another.

This remarkable non reproducibility may now also be explained be the hard sphere phase transition. Initially, at 62%, the sample is in the metastable amorphous solid state, but when the shear cell is loaded it experiences various shear forces and begins to form random faceted crystallites that span the gap of the rheometer in various orientations. The rheology of crystallites can be extremely sensitive to the plane of shear. For example, for the BCC crystal at close packing, the yield stress on the 111 slip plane is zero, but it is infinity in all other slip directions!

Fig. 6 Rheographs of a range of "identical" experimental measurements of the flow as a function of stress of the same sample of polystyrene latex suspension at 62% packing; the very wide non reproducibility of the results can be explained by random faceting of the nucleation during loading of the sample and during flow

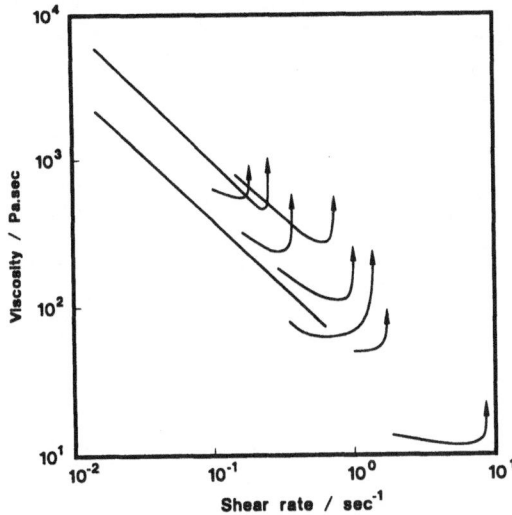

Conclusions

i) Simple hard-sphere suspensions in laminar Coutte flow exhibit rich phase behaviour, and have a complex phase diagram yet to be determined.

ii) This derives logically from the effect of shear upon the static behaviour of hard-sphere suspensions.

iii) The order-disorder behaviour gives rise to novel rheological effects like rheopexy, which are complex and which complicate the determination of constitutive rheological relations.

iv) The phase behaviour itself can lead to two-phase heterogeneities in rheometry; hence the rheology cannot be measured directly!

v) At very high volume fractions the Bingham yield stress and plastic viscosities are highly non-reproducible, probably because of random faceting effects in the gap of the rheometer.

Acknowledgements We wish to thank the UK Science and Engineering Research Council for the award of a Research Studentship (to AM) and a postdoctoral Research Grant (to JC and LVW) and Dr. Alex Lips of Unilever Research for sponsorship of this research, many helpful discussion, and the use of laboratory facilities.

References

1. See e.g. Schowalter W (1978) "Mechanics of Non-Newtonian Fluids" (Pergamon: Oxford)
2. Jomha AI, Merrington A, Woodcock LV, Barnes HA, Lips A (1991) Powder Technology 65:343–370
3. Lees A, Edwards SF (1972) J Phys C 5:1921–1929
4. Hopkins A, Woodcock LV (1990) J Chem Soc Faraday Trans 86:2121–2132
5. Almog Y, Reich S, Moshe L (1982) British Polymer Journal 14:131–136
6. Pusey PN, van Megen W (1986) Nature 320:340–342
7. Merrington A, Woodcock LV (1991) Proceedings of the A.S.M.E.-J.S.M.E. Fourth International Symposium on Liquid-Solid Flows, ed. M.C. Roco, pp 109–115 (Portland, Oregon, U.S.A. June 1991)
8. Merrington A (1992) Thesis: University of Bradford
9. Rodriguez BE, Wolfe MS, Kaler EW (1993) Langmuir 9:12–13

Progr Colloid Polym Sci (1995) 98:117–123
© Steinkopff Verlag 1995

Short-time dynamics and sedimentation of charge-stabilized suspensions

G. Nägele
B. Mandl
R. Klein

Dr. Nägele (✉) · B. Mandl · R. Klein
Fakultät für Physik
Universität Konstanz
P.O. Box 5560
78434 Konstanz, FRG

Abstract In this work, we discuss the combined effects of the electrostatic and hydrodynamic interaction (HI) on the short-time dynamics and the sedimentation velocity of charge-stabilized suspensions. For this purpose the measurable hydrodynamic function $H(q)$ is calculated by using two methods based, respectively, on a renormalized density fluctuation expansion, and on a pairwise-additivity approximation of the hydrodynamic mobility tensors. It is shown that $H(q)$ can deviate considerably from the free diffusion coefficient, D^0, for systems of volume fractions Φ as low as 10^{-3}, and that these effects are more pronounced for collective diffusion than for self-diffusion. For deionized suspensions, the sedimentation coefficient, $H(0)/D^0$, and the short-time self-diffusion coefficient, $D^{s,\,short}$, are found at low Φ to scale, respectively, as $\Phi^{1/3}$ and $\Phi^{4/3}$. Furthermore, we analyze the dependence of $H(0)$ and $D^{s,\,short}$ on the amount of added electrolyte.

Key words Short-time dynamics – charge-stabilized suspensions – hydrodynamic interaction – sedimentation

Introduction

In this contribution, we investigate the combined effects of the electrostatic and hydrodynamic interactions on the short-time dynamics of charge-stabilized suspensions. Information on the short-time suspension dynamics is contained in the first cumulant $\Gamma^{(1)}(q)$ of the dynamic scattering function, which can be probed by dynamic light scattering (DLS). The first cumulant is related to the hydrodynamic function $H(q)$ by $\Gamma^{(1)}(q) = q^2 H(q)/S(q)$, where $S(q)$ denotes the static structure factor and q is the scattering wave number. As a consequence, it is possible to measure $H(q)$ by combining DLS measurements of $\Gamma^{(1)}(q)$ with static light scattering measurements of $S(q)$ [1].

It is important to realize that $H(q)$ depends both on the solvent-mediated hydrodynamic interaction (HI), and on the equilibrium microstructure, which is quantitatively described by the radial distribution function $g(r)$. The observed qualitative differences in the radial distribution functions of suspensions of charged and uncharged colloidal spheres lead to quite different behaviour of the corresponding $H(q)$. In fact, because of the more developed positional correlations in charge-stabilized suspensions, the effects of HI are substantially more pronounced in these systems than in hard-sphere suspensions at the same volume fraction.

We use two distinct methods of calculating the $H(q)$ of monodisperse suspensions. The first method consists of a pairwise-additivity approximation for the many-body HI combined with an expansion of the two-particle diffusivity tensors in powers of (a/r), i.e., the ratio of the radius a and the center-of-mass distance r between two colloidal particles. This method is referred to as PA-theory [2–4]. The second method has been developed by Beenakker and Mazur [5] and amounts to an expansion of $H(q)$ in powers of the renormalized density function $\delta\gamma$. Following earlier work of Genz and Klein [6], we will evaluate the $\delta\gamma$-expansion to lowest order.

118
G. Nägele et al.
Short-time dynamics and sedimentation in suspensions

Short-time dynamics and DLS

Polarized DLS experiments from monodisperse suspensions of N identical and spherically shaped particles probe the dynamic structure factor $S(q,t)$. The dynamical behavior of $S(q,t)$ depends crucially on the time range of the experiment. Most of the DLS experiments are restricted to correlation times $t > 10^{-6}$ s $\gg \tau_B$, where $\tau_B = M/\zeta^0$ is the momentum relaxation time of a colloidal sphere of mass M and friction coefficient $\zeta^0 = 3\pi\eta\sigma$, immersed in a fluid of shear viscosity η. Here, $\sigma = 2a$ denotes the particle diameter, and ζ^0 is related to the Stokesian diffusion coefficient by $D^0 = k_B T/\zeta^0$. For typical suspensions, one finds that $\tau_B \approx 10^{-9} - 10^{-8}$ s. As a consequence, inertial effects arising from the momentum relaxation of the spheres and from the unsteady solvent flow decay too fast to be resolved by the photon correlator, so that only the relaxations of the colloidal particles positions are probed. This fact allows for a coarse-grained configuration-space description of Brownian motion on the basis of the generalized Smoluchowski equation (GSE). On this level of description, hydrodynamic interaction can be considered as infinitely fast [7, 8].

At short times, $S(q,t)$ is known to decay exponentially according to

$$S(q,t) = S(q)\exp\left[-q^2 D_{\text{eff}}(q)t\right], \quad \tau_B \ll t \ll \tau_I \qquad (1)$$

with a wavenumber-dependent effective diffusion coefficient, $D_{\text{eff}}(q)$, related to the first cumulant, $\Gamma^{(1)}(q)$ of $S(q,t)$ by

$$\Gamma^{(1)}(q) = q^2 D_{\text{eff}}(q) = q^2 \frac{H(q)}{S(q)}. \qquad (2)$$

The structural relaxation time $\tau_I = a^2/D^0$ is the time needed for a colloidal sphere to diffuse a distance roughly to its radius. Typically, one finds $\tau_I \approx 10^{-3}$ s such that the short-time scale $\tau_B \ll t \ll \tau_I$ is well separated from the long-time scale $t \gg \tau_I$. Application of the GSE leads to the well-known expression [7]

$$H(q) = \left\langle \frac{1}{N} \sum_{i,j=1}^{N} \hat{\mathbf{q}} \cdot \mathbf{D}_{ij}(\mathbf{R}^N) \cdot \hat{\mathbf{q}} \, e^{i\mathbf{q}\cdot[\mathbf{R}_i - \mathbf{R}_j]} \right\rangle \qquad (3)$$

for the hydrodynamic function $H(q)$. Here, $\mathbf{R}_i$ is the position of the center of the ith colloidal sphere, $\mathbf{q}$ is the scattering wave vector and $\hat{\mathbf{q}} = \mathbf{q}/q$. The bracket indicates an equilibrium ensemble average. The positively valued function $H(q)$ depends on the diffusivity tensors $\mathbf{D}_{ij}(\mathbf{R}^N)$ which describe the HI. Notice that $\mathbf{D}_{ij}(\mathbf{R}^N)$ depends in general on the configuration $\mathbf{R}^N$ of all N particles, and this adds to the complexity of HI. Moreover, $H(q)$ also depends on the equilibrium distribution, which is in general quite different for different types of interparticle potentials.

It is instructive to split $H(q)$ into a self-part ($i = j$) and a distinct part ($i \neq j$)

$$H(q) = D^{s,\text{short}} + H^d(q), \qquad (4)$$

where the self-part

$$D^{s,\text{short}} = \langle \hat{\mathbf{q}} \cdot \mathbf{D}_{11}(\mathbf{R}^N) \cdot \hat{\mathbf{q}} \rangle \qquad (5)$$

is the short-time self-diffusion coefficient, $D^{s,\text{short}}$, and the distinct part is given by

$$H^d(q) = (N-1)\langle \hat{\mathbf{q}} \cdot \mathbf{D}_{12}(\mathbf{R}^N) \cdot \hat{\mathbf{q}} \, e^{i\mathbf{q}\cdot[\mathbf{R}_1 - \mathbf{R}_2]} \rangle. \qquad (6)$$

The distinct part $H^d(q)$ becomes vanishingly small for $q \gg q_m$, where q_m is the position of the main peak of $S(q)$ [2]. For deionized suspensions, this peak is situated roughly at $q_m \approx 2\pi/\bar{r}$, where $\bar{r} = n^{-1/3}$ is the mean particle distance, and n is the concentration of colloidal particles. Therefore, for $q \gg q_m$ it is found that $H(q) \to D^{s,\text{short}}$ and $D_{\text{eff}}(q) \to D^{s,\text{short}}$, i.e., the self-diffusion coefficient is probed by DLS experiments performed at large wavenumbers. In the opposite limit $q \ll q_m$, $D_{\text{eff}}(q)$ reduces to the short-time collective diffusion coefficient $D^{c,\text{short}} = D_{\text{eff}}(q \to 0) = H(0)/D^0$, which describes the initial decay of long-wavelength density fluctuations.

The hydrodynamic function $H(q)$ is exactly known only for vanishing HI, where $H(q) = D^{s,\text{short}} = D^0$. In this case, the effective diffusion coefficient reduces to $D_{\text{eff}}(q) = D^0/S(q)$ and its q-dependence is entirely due to $S(q)$. When HI becomes important, $H(q)$ is different from D^0. According to Eq. (2), this can be experimentally verified by separately measuring $S(q)$ with static light scattering, and the first cumulant $\Gamma^{(1)}(q)$ with DLS. From the PA-theory and the lowest order $\delta\gamma$-expansion, it is found for deionized suspensions that $H(q)$ can deviate considerably from D^0 at volume fractions Φ as low as 10^{-3}. These effects are more pronounced for collective diffusion, since $H(0)/D^0$ can become substantially smaller than one [2–4]. However, when electrolyte is added to the suspension, these effects become less pronounced. The study of $H(0)/D^0$ is particularly interesting since, under suitable conditions, this quantity is equal to the sedimentation coefficient [9, 10], which is the ratio of the ensemble averaged sedimentation velocity of a colloidal sphere in the suspension to its value at infinite dilution. Thus it is possible to measure $H(0)/D^0$ alternatively by sedimentation experiments.

Calculation of the hydrodynamic function

For calculating $H(q)$, it is necessary at first to specify the diffusivity tensors $D_{ij}(\mathbf{R}^N)$. In principle, the diffusivity tensors can be obtained by solving the stationary Stokes equation for N hard spheres with appropriately specified boundary conditions. This is, however, an extremely diffi-cult task and important advances towards practical nu-merical results have been achieved only very recently [11].

However, for small volume fractions and sufficiently low content of excess electrolyte, it is reasonable to assume pairwise additivity of the HI. In this case

$$D_{ij}(\mathbf{R}^N) = \delta_{ij}\left[D^0\mathbf{1} + \sum_{l=1}^{N}{}' \mathbf{A}(\mathbf{R}_i - \mathbf{R}_l)\right]$$
$$+ (1 - \delta_{ij})\mathbf{B}(\mathbf{R}_i - \mathbf{R}_j) , \qquad (7)$$

where the term $l = i$ is excluded. The two-particle hydro-dynamic tensors $\mathbf{A}(\mathbf{r})$ and $\mathbf{B}(\mathbf{r})$ are known as expansions in powers of (a/r), where r is the distance between a pair of spheres [12]. We only quote the leading terms

$$\mathbf{A}(\mathbf{r}) = -\frac{15}{4}D^0\left(\frac{a}{r}\right)^4 \hat{\mathbf{r}}\hat{\mathbf{r}} + O(r^{-6}) \qquad (8)$$

$$\mathbf{B}(\mathbf{r}) = \frac{3}{4}D^0\left(\frac{a}{r}\right)[\mathbf{1} + \hat{\mathbf{r}}\hat{\mathbf{r}}] + O(r^{-3}) , \qquad (9)$$

where $\hat{\mathbf{r}} = \mathbf{r}/r$. The leading contribution to $\mathbf{B}(\mathbf{r})$ is the well-known Oseen tensor. The short-time tracer-diffusion coefficient $D^{s,\,\text{short}}$ and the distinct part $H^d(q)$ are expressed in terms of the tensors $\mathbf{A}$ and $\mathbf{B}$ as

$$D^{s,\,\text{short}} = D^0 + n\int d^3r\,[\hat{\mathbf{q}}\cdot\mathbf{A}(\mathbf{r})\cdot\hat{\mathbf{q}}]\,g(r) \qquad (10)$$

$$H^d(q) = n\int d^3r\,[\hat{\mathbf{q}}\cdot\mathbf{B}(\mathbf{r})\cdot\hat{\mathbf{q}}]\cos(\mathbf{q}\cdot\mathbf{r})g(r) , \qquad (11)$$

where the number density n of colloidal particles is related to the volume fraction Φ by $\Phi = \pi n\sigma^3/6$. The substitution of the expansions of $\mathbf{A}(\mathbf{r})$ and $\mathbf{B}(\mathbf{r})$ into Eqs. (10–11) leads to rather lengthy expressions, which are available upto terms proportional to $(a/r)^{11}$ in $\mathbf{A}$ and $\mathbf{B}$ [2].

The only further input needed to calculate $D^{s,\,\text{short}}$ and $H(q)$ is the radial distribution function $g(r)$. Contrary to the case of suspensions of hard spheres, where the radial distribution function has its maximum at contact, $r = \sigma^+$, $g(r)$ is practically zero for charge-stabilized spheres for all r up to the position r_m of the next neighbor shell, where it has a rather pronounced peak. Therefore, rapid conver-gence is expected for the series of integrals in Eqs. (10–11), contrary to hard-sphere suspensions where it is necessary to consider many terms in the power series for $\mathbf{A}(\mathbf{r})$ and $\mathbf{B}(\mathbf{r})$.

It is possible to give a qualitative discussion of the form of $H(q)$ (for $\Phi \leq 0.1$) on the basis of the expressions (10–11), together with the leading terms for $\mathbf{A}(\mathbf{r})$ and $\mathbf{B}(\mathbf{r})$ given in Eqs. (8–9). For example, the effect of HI on $D^{s,\,\text{short}}$ can be expected to be smaller for charged colloidal particles than for uncharged ones. This is due to the $(a/r)^4$ behavior of the leading term of $\mathbf{A}(\mathbf{r})$, which is small at those values of r where $g(r)$ is different from zero. However, the behavior

of $H(q)$ at small q, which is mainly determined by $H^d(q)$, is more strongly influenced by HI, because the leading (Oseen) contribution to $\mathbf{B}(\mathbf{r})$ decays only as (a/r). In con-clusion, one expects for charge-stabilized suspensions only weak effects of HI on $D^{s,\,\text{short}}$, but its influence at $q \leq q_m$ will be at least comparable to the case of hard-sphere suspensions at the same volume fraction. In fact, explicit calculations of $H(q)$ show that the effects of HI on $H(q)$ for $q \leq q_m$ are substantially stronger for charge-stabilized col-loids, particularly when all excess ions have been removed from the suspension [2–4].

So far, our discussion was based on the assumption of pairwise additive HI (PA-theory). It has been shown [13] that higher than two-body terms, which contribute to $H(q)$ to order Φ^2, first appear in $\mathbf{D}_{ij}$ and $\mathbf{D}_{ii}$ to order r^{-4} and r^{-7}, respectively. Hence, the above conclusion concerning the weak influence on HI on $D^{s,\,\text{short}}$ will not be changed.

The results of $H(q)$ obtained from the PA-theory have been compared with the lowest order form of the $\delta\gamma$-expansion [5, 6]. In this more elaborate method, many-body contributions of the HI are accounted for in an approximate way by a partial resummation of higher or-der correlations. The PA-theory and $\delta\gamma$-expansion results of $H(q)$ are found to be in good agreement for volume fractions $\Phi \leq 0.05$ [3, 4, 14].

Results and discussion

Our results on the short-time dynamics of charge-stabi-lized suspensions are based on the effective macrofluid model. In this model, the effective pair potential, $u(r)$, between two identical colloidal spheres is described as the sum of a hard-sphere potential and a screened Coulomb potential of the form

$$\beta u(r) = \begin{cases} \infty , & r < \sigma \\ L_B(Z^{\text{eff}})^2\,\dfrac{\exp[\kappa\sigma]}{(1 + \kappa\sigma/2)^2}\,\dfrac{\exp[-\kappa r]}{r}, & r > \sigma \end{cases} \qquad (12)$$

Here, $\beta = 1/(k_B T)$, $L_B = \beta e^2/\varepsilon$ is the Bjerrum length, e the elementary charge, ε the dielectric constant of the suspending fluid, and the effective valency of a colloidal particle is denoted by Z^{eff}. The Debye-Hückel screening parameter κ, is defined by

$$\kappa^2 = 4\pi L_B|Z^{\text{eff}}| + 8\pi L_B n_s = \kappa_c^2 + \kappa_s^2 , \qquad (13)$$

where it is assumed that the counterions are monovalent. Furthermore, n_s is the number density of an added 1–1 electrolyte. The second equality in Eq. (13) defines the two constituents of κ, viz. κ_c and κ_s, which originate from the counterions and added electrolyte (e.g., salt) ions, respectively.

120
G. Nägele et al.
Short-time dynamics and sedimentation in suspensions

The structure functions $g(r)$ and $S(q)$, which are needed to calculate $H(q)$ in the PA-theory and $\delta\gamma$-expansion, are obtained for the pair potential (12) by solving the Ornstein–Zernike integral equation in combination with the hypernetted chain (HNC) and rescaled mean spherical (RMSA) closure schemes [15].

The discussion will be restricted to the small-q and large-q limits of $H(q)$, since these are relevant, respectively, to measurements of the sedimentation coefficient and $D^{s,short}$. For a discussion of the q-dependence of $H(q)$ consult Refs. [1–4]. Moreover, we are primarily interested in dilute to moderately concentrated suspensions of charge-stabilized particles, where $\Phi \leq 0.15$.

We first analyze the behavior of $H(0)/D^0$ as a function of the volume fraction. Figure 1 displays the PA-theory and lowest order $\delta\gamma$ results for $H(0)/D^0$ for the case of no added electrolyte. Two distinct values $Z^{eff} = 250, 350$ for the effective valency have been selected to illustrate that $H(0)$ is almost independent of Z^{eff}. The figure also includes the result $H(0)/D^0 = 1 - 6.55 \times \Phi + O(\Phi^2)$, which holds for a suspension of uncharged hard spheres [9]. The system parameters are characteristic of suspensions of charge-stabilized silica spheres, which have been investigated very recently by sedimentation experiments [16]. It should be noted that the results of the PA-theory and $\delta\gamma$-expansion are very similar for $\Phi < 0.05$. The PA-theory result for $H(0)$ starts to fail for volume fractions $\Phi > 0.05$, which illustrates the restricted validity of describing HI as pairwise additive. The comparison with the hard sphere result (dashed-double dotted line) clearly shows that the effect of HI on $H(0)$ is strongly enhanced by the presence of long ranged electrostatic repulsion, even for volume fractions as low as 10^{-3}. For example, at $\Phi = 0.04$ one observes a value of $H(0)/D^0$ smaller than one by as much as 60%. When $\Phi \leq 10^{-3}$, it is sufficient to use the simple Oseen approximation for calculating $H(q)$, provided that the particle hard-core is masked by the electrostatic repulsion. This requires Z^{eff} to be sufficiently large. Within the Oseen approximation, it follows that $D^{s,short} = D^0$.

At sufficiently small volume fractions, the Φ-dependence of $H(0)$ can be parametrized as

$$\frac{H(0)}{D^0} = 1 - a^c \Phi^{b^c} \, . \tag{14}$$

For hard spheres, $a^c = 6.55$ and $b^c = 1$, whereas for deionized suspensions the values $a^c = 1.82$ ($a^c = 1.72$) and $b^c = 0.33$ ($b^c = 0.33$) are obtained from the PA-theory ($\delta\gamma$-expansion). This difference in the behavior of charged and uncharged suspensions is quite remarkable. It is further interesting to note that the same exponent $b^c = 1/3$ has been found for the sedimentation velocity of rigid lattices of spherical particles, with values of a^c in the range $[1.76 - 1.79]$ [17]. Recent measurements of the sedimentation velocity in deionized suspensions of silica spheres are in agreement with the theoretical finding $b^c = 1/3$ [16].

The non-analytic behavior of $H(0)$ at small volume fractions ($\Phi \leq 0.05$) can be understood qualitatively from simple arguments based on the PA-theory. If the true radial distribution function is crudely approximated by a unit step function

$$g(r) = \Theta(r - \sigma^{EHS}) \, , \tag{15}$$

where $\sigma^{EHS} > \sigma$ is an effective hard-sphere diameter, which accounts for the electrostatic repulsion, it is easy to calculate the first few integrals in Eqs. (10–11) by using the known expansions of the two-body hydrodynamic tensors. The result is [18]

$$\frac{H(0)}{D^0} = 1 + \Phi \left[-6x^2 + 1 - \frac{15}{8x} + \frac{9}{64x^3} \right. $$
$$\left. + \frac{75}{256x^4} + O(x^{-5}) \right] \tag{16}$$

and

$$\frac{D^{s,short}}{D^0} = 1 + \Phi \left[-\frac{15}{8x} + \frac{9}{64x^3} + O(x^{-5}) \right] \, , \tag{17}$$

where $x = \sigma^{EHS}/\sigma$. For later use, we have also displayed the result (17) for $D^{s,short}$. There are several ways to

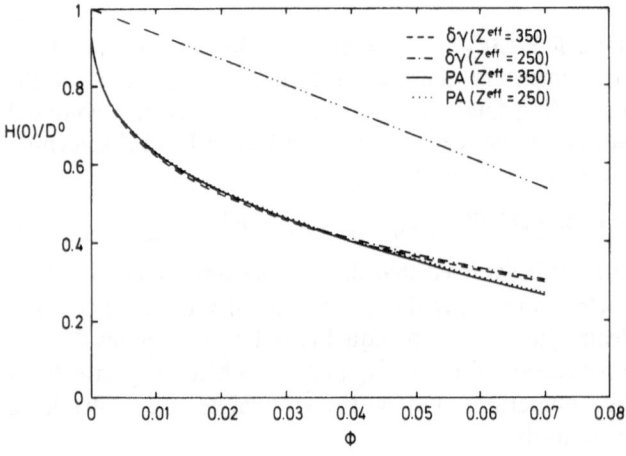

Fig. 1 Volume fraction dependence of the sedimentation coefficient, $H(0)/D^0$, for deionized ($n_s = 0$) suspensions. Comparison of the results obtained from PA-theory and lowest order $\delta\gamma$-expansion, using the RMSA closure with effective valencies $Z^{eff} = 350, 250$. The remaining system parameters are $\sigma = 760$ nm, $L_B = 2.34$ nm. These parameters are typical for suspensions of silica spheres dispersed in ethanol at room temperature, which have been investigated very recently by sedimentation experiments [16]. Dashed-double dotted line: $H(0)/D^0 = 1 - 6.55 \times \Phi$, corresponding to hard-sphere suspensions at small volume fractions [9]

Progr Colloid Polym Sci (1995) 98:117–123
© Steinkopff Verlag 1995

estimate the magnitude of the effective diameter in terms of the parameters of the true system [3, 4]. At any rate, σ^{EHS} can be expected to be significantly larger than σ due to the strong electrostatic repulsion in deionized suspensions. A reasonable choice is $\sigma^{EHS} = r_m$, where r_m is the position of the principal peak of the true $g(r)$. Since x is significantly larger than one ($x > 3$), it is sufficient to consider only the first term in the bracket of Eq. (16), which arises from the long-ranged Oseen tensor. The exponent $b^c = 1/3$ then follows from the sequence of relations $x \propto r_m \propto \bar{r} \propto \Phi^{1/3}$, which hold for deionized suspensions with completely masked hard-core repulsion. Moreover, since $\bar{r}$ is determined by the suspension concentration only, $H(0)$ is expected to be nearly independent of Z^{eff} in accordance with Fig. 1, provided that Z^{eff} is large enough (i.e. > 100) for $r_m \approx \bar{r}$ to be valid. In concluding the discussion of Fig. 1, we mention that $b^c = 1/3$ ceases to hold for extremely low volume fractions ($\Phi \ll 10^{-6}$), usually not resolved by DLS and sedimentation experiments, for in this case is $g(r) \approx \exp[-\beta u(r)]$.

The deviation of $H(0)/D^0$ from its value of one at infinite dilution decreases, when excess electrolyte is added so that the system changes gradually into a hard-sphere like suspension. This fact is demonstrated in Fig. 2, where $H(0)/D^0$ is plotted as a function of Φ for several amounts of added electrolyte. Notice that the results of the PA-theory and $\delta\gamma$-expansion are in good agreement when $\Phi < 0.05$.

For an explanation of this behavior, consider Fig. 3, which shows radial distribution functions calculated in RMSA for suspensions of volume fraction $\Phi = 0.07$ and valency $Z^{eff} = 350$, by varying the amount of added electrolyte from $n_s = 0$ to $n_s = 260 \times 10^{-6}$ mol/l. Notice that

Fig. 2 Resuts for $H(0)/D^0$ as a function of Φ for various concentrations $n_s = 0, 0.08, 5, 20, 260 \times 10^{-6}$ mol/l of excess 1–1 electrolyte. System parameters as in Fig. 1 for $Z^{eff} = 350$, using the RMSA closure. Full lines; PA-theory: dashed lines: lowest order $\delta\gamma$-expansion

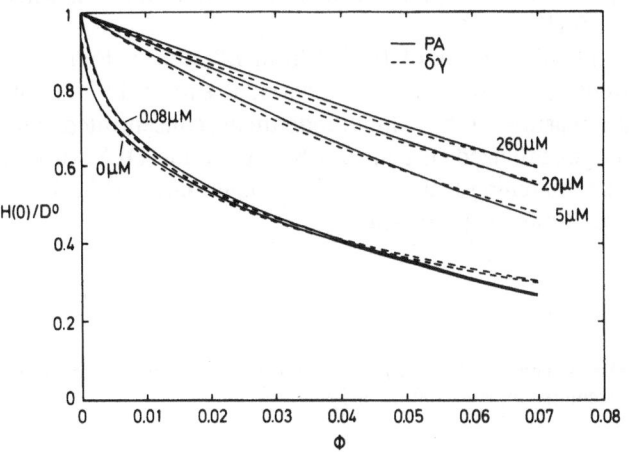

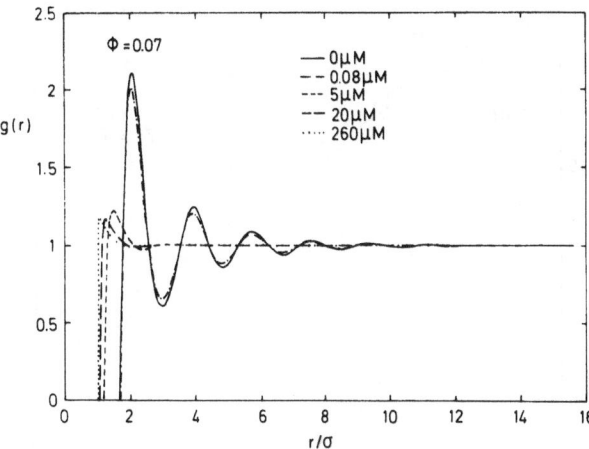

Fig. 3 RMSA radial distribution function $g(r)$ for various concentrations of added 1–1 electrolyte as indicated in the figure. Here, $Z^{eff} = 350$ and $\Phi = 0.07$. All other parameters are as in Fig. 1

the $g(r)$ of the deionized suspension has a strongly developed structure. This structure deteriorates quickly as electrolyte is added, and the position r_m of the principal peak of $g(r)$ is shifted to smaller values. The position q_m of the principal peak of $S(q)$ is shifted towards larger values according to the general rule $r_m \times q_m \approx 2\pi$. The decreasing values of r_m render the leading Oseen contribution to the integrals (11) less significant, since the integrand contains the factor $[g(r) - 1]$ (cf. Ref. [2]). From Eq. (16), we realizes that it is the Oseen contribution which is mainly responsible for the strong deviation of $H(0)/D^0$ from 1. This finding explains why an increasing amount of electrolyte causes $H(0)/D^0$ to increase towards the corresponding hard-sphere value [9].

For the largest amount of added electrolyte, i.e., $n_s = 260\ \mu$mol/l, which corresponds to $\kappa\sigma = 73$ and $\beta\mu(\sigma^+) = 0.27$, the RMSA-$g(r)$ is nearly identical to the $g(r)$ of a suspension of hard spheres at the same volume fraction, as obtained with the Percus–Yevick closure. As a consequence, $H(0)/D^0$ is well described for low Φ by the form $1 - 6.55 \times \Phi$.

It was pointed out that for a large range of volume fractions, $H(0)$ is well described by the parametric form (14), with $b^c = 1$ for hard spheres, and $b^c = 1/3$ for charged spheres in deionized suspensions ($\kappa_s/\kappa_c = 0$). A suspension of charged spheres can be considered as a suspension of hard spheres in the limit of $\kappa_s/\kappa_c \gg 1$. One might try to apply the form (14) also to suspensions with arbitrary amounts of added electrolyte. In this case the ratio κ_s/κ_c, which indicates the relative importance of counterions with regard to added electrolyte ions, is proportional to $\Phi^{-1/2}$. This observation explains why the parametric form (14) becomes now less useful, since the values of a^c and b^c extracted from linear regression may depend crucially on

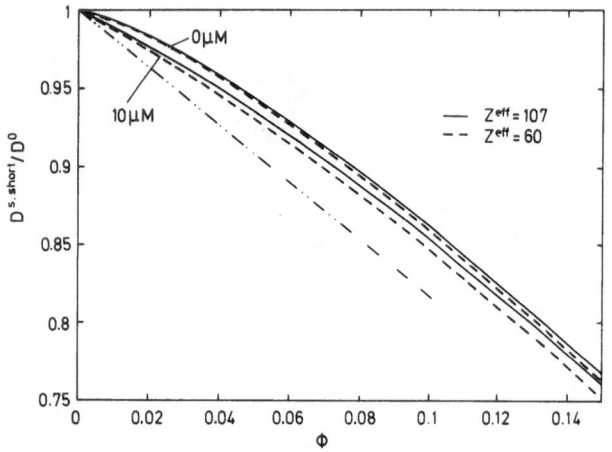

Fig. 4 Normalized short-time self-diffusion coefficient $D^{s,short}/D^0$ versus Φ. Shown is the result of the PA-theory with HNC input for $g(r)$. The system parameters are $n_s = 0$, 10×10^{-6} mol/l; $\sigma = 160$ nm, and $L_B = 5,5$ nm. Full lines: $Z^{eff} = 107$; Dashed lines; $Z^{eff} = 60$; Dashed-double dotted line: $D^{s,short}/D^0 = 1 - 1.831 \times \Phi$, corresponding to hard-sphere suspensions at small volume fractions [19]

the length and the location of the underlying interval of volume fractions.

From Fig. 2, we further observe that the PA-theory ($\delta\gamma$-expansion) result of $H(0)/D^0$ for $n_s = 0$ merges with the result for $n_s = 0.08\ \mu$mol/l when $\Phi > 0.06$. This finding can be attributed to the fact that for $n_s = 0.08\ \mu$mol/l, κ_c becomes larger than κ_s when Φ exceeds a value roughly given by 0.06.

We have analyzed so far the behavior of $H(0)/D^0$ as a function of particle concentration and added electrolyte. In the remaining part, we will focus on the corresponding behavior of the short-time self-diffusion coefficient. Figure 4 displays $D^{s,short}/D^0$ as a function of Φ, as obtained from the PA-theory with HNC input for $g(r)$. Two different values for Z^{eff} are considered. Notice for $n_s = 0$ that $D^{s,short}$ is, similarly, to $H(0)$, nearly independent of Z^{eff}. The system parameters underlying Fig. 4 are characteristic of suspensions of charged silica spheres, which have been studied in great detail both by short-time DLS and static light scattering [1]. We have not included in the figure the result for $D^{s,short}$ as provided by the lowest order $\delta\gamma$-expansion. The reason for this is that the expression for $D^{s,short}$ obtained by Beenakker and Mazur is independent of particle correlations to lowest (i.e., zeroth) order in the renormalized density fluctuations $\delta\gamma$ [3, 5, 6]. Effects due to the direct forces contribute, through $g(r)$, only to order $(\delta\gamma)^2$. The second order contribution in $\delta\gamma$ lead only to minor corrections for hard-spheres suspensions [5]. A similar conclusion cannot be drawn for charge-stabilized suspen-

sions, since in this case $g(r)$ depends not only on Φ, but also rather sensitively on the effective valency Z^{eff} and on the concentration n_s of excess electrolyte. In fact, the PA-theory results shown in Fig. 4 display a modest but non-negligible dependence of $D^{s,short}$ on Z^{eff} and n_s.

The dashed-double dotted line in Fig. 4 is the result $D^{s,short}/D^0 = 1 - 1.831 \times \Phi$, which was obtained for hard spheres to linear order in Φ by including a very large number of terms in the power series of $\mathbf{A}(r)$ [19]. A rather weak but nonetheless observable effect of HI on $D^{s,short}$ is found according to Fig. 4. Moreover, the influence of HI on $D^{s,short}$ is smaller for charged colloidal particles than for uncharged ones. Suspensions with low amount of added electrolyte are particularly weakly influenced by HI. These observations are in agreement with the qualitative conclusions we had drawn from considering the leading terms for $\mathbf{A}(r)$ together with the general behavior of $g(r)$.

It is also interesting to investigate the scaling behavior of $D^{s,short}$ as a function of Φ. In analogy to the discussion given for $H(0)$, let us consider Eq. (17) with the identification $x = r_m$ for the effective diameter. A non-analytic concentration dependence of the form

$$\frac{D^{s,short}}{D^0} = 1 - a^s \Phi^{b^s} \tag{18}$$

with $b^s = 4/3$ is predicted now in case of deionized suspensions of rather low volume fractions, where it is sufficient to consider only the first term in the bracket of Eq. (17). This term arises from the $(a/r)^4$ behavior of the leading contribution to $\mathbf{A}(r)$. An exponent b^s close to the value $4/3$ is indeed obtained for the system of Fig. 4 with no added electrolyte. The value of a^c is determined as $a^c \approx 2.5$. Notice further that it is sufficient here to use the leading term of $\mathbf{A}(r)$ for calculating $D^{s,short}$, even when $\Phi = 0.15$. Including higher order terms, up to the order $(a/r)^{11}$, does not change the numerical result within 2%. This finding also suggests that many-body terms, which first appear in $\mathbf{D}_{11}$ to order r^{-7}, give rise only to minor corrections when $\Phi < 0.15$.

The deviations of $D^{s,short}$ from D^0 seen in Fig. 4 are admittedly rather small. However, it should be possible to measure $D^{s,short}$ for moderately concentrated and charge-stabilized suspensions by combining DLS with an index-matching technique. This would allow to check the predictions of the PA-theory.

Acknowledgement This work has been supported by the Deutsche Forschungsgemeinschaft (SFB 306).

References

1. Philipse AP, Vrij A (1988) J Chem Phys 88:6459
2. Nägele G, Kellerbauer O, Krause R, Klein R (1993) Phys Rev E 47:2562
3. Nägele G, Steininger B, Genz U, Klein R (1994) Physica Scripta T 55:119
4. Nägele G (1994) habilitation thesis, University of Konstanz
5. Beenakker CWJ, Mazur P (1983) Phys Letters A 98:22; Physica A 126:349 (1984); ibid. A 120:388 (1983)
6. Genz U, Klein R (1991) Physica A 171:26
7. Jones RB, Pusey PN (1991) Annu Rev Phys Chem 42:137
8. Kim S, Karrila SJ (1991) "Microhydrodynamics" Butterworth-Heinemann, Boston
9. Batchelor GK (1972) J Fluid Mech 52:245
10. Russel WB, Glendinning AB (1981) J Chem Phys 74:948
11. Cichocki B, Felderhof BU, Hinsen K, Wajnryb E, Blawzdziewicz J (1994) J Chem Phys 100:3780
12. Jeffrey DJ, Onishi YH (1984) J Fluid Mech 139:261
13. Mazur P, van Saarlos W (1982) Physica A 115:21
14. Klein R, Nägele G, Nuevo Cimento, in press
15. Krause R, D'Aguanno B, Mendez-Alcaraz JM, Nägele G, Klein R, Weber R, (1991) J Phys C 3:4459
16. Thies-Weesie D, Philipse AP, Nägele G, Mandl B, Klein R, submitted
17. Hasimoto H (1959) J Fluid Mech 5:317
18. Cichocki B, Felderhof BU (1991) J Chem Phys 94:556; Denkov ND, Petsev DN (1992) Physica A 183:462
19. Cichocki B, Felderhof BU (1988) J Chem Phys 89:3705

Progr Colloid Polym Sci (1995) 98:124–127
© Steinkopff Verlag 1995

J.-C. Bacri
C. Drame
B. Kashevsky
S. Neveu
R. Perzynski
C. Redon

Motion of a pair of rigid ferrofluid drops in a rotating magnetic field

J.-C. Bacri
Université Paris 7 (UFR de Physique)

Prof. J.-C. Bacri (✉) · C. Drame
R. Perzynski · C. Redon
Laboratoire d'Acoustique et Optique
de la Matière Condensée[1]
U.P.M.C.
Tour 13, Case 78
4 Place Jussieu
75252 Paris Cedex 05, France

B. Kashevsky
Heat and Mass Transfer Institute
Belarus Academy of Sciences
15 P. Brovka Street
Minsk 220072, Belarus

S. Neveu
Laboratoire de Physico-chimie Inorganique
URA SRSI
Université Pierre et Marie Curie
Case 63
4 Place Jussieu
75252 Paris Cedex 05, France

[1] Associated with the Centre National de la
Recherche Scientifique

Abstract Ferrofluids are colloidal
suspensions of magnetic particles of
some ten nanometers in size. Such
media are both fluid and magnetic.
We use here a chemically synthesized
system: $CoFe_2O_4$ particles, coated
with a surfactant layer, are dispersed
in a commercial oil based on
hydrogenated terphenyl. Ferrofluid
droplets of 10 μm diameter are
obtained by emulsion in an aqueous
medium. Through direct microscopic
observations, we study the dynamic of
a pair of droplets under a rotating
magnetic field, as a function of its
frequency ω and its amplitude H. At
low frequencies a synchronous regime
is observed up to a frequency
threshold $\omega_c \propto H^2$. After a jerky
regime, the frequency of the
pair rotation reaches a plateau
independent of ω.

Key words Magnetic liquid –
rotational viscosity

Introduction

Physical properties of magnetizable disperse media such as
magnetic fluids, magneto-rheological and electrorheological suspensions and magnetic emulsions are affected by
particle-to-particle interactions.

Investigation of nonequilibrium processes for limited
numbers of interacting particles allows to get useful information for understanding physical properties of the

medium [1]. The simplest case allowing the most precise
investigation is the case of pair particle interaction in
a rotating field, which has been considered theoretically
[2, 3] for magnetic particles and experimentally [4, 5] for
nonmagnetic particles in a magnetic fluid. Hydrodynamic
interactions of particles [3] have a principal role. Because
of the unlimited growth of viscous friction when particles
approach each other, the field action is strongly affected by
the particle surface smoothness and other peculiarities of
surface interaction. For example, it can lead to stable

Progr Colloid Polym Sci (1995) 98:124–127
© Steinkopff Verlag 1995

couples of particles, by their exposition in a constant field. It have been observed in [5].

We present here a new interesting problem, specific of magnetic fluid (MF): in a rotating field, a spin rotation of the magnetic droplets appears which leads for a pair of droplets to a global rotation of the pair through hydrodynamic interaction.

Experiment

Figure 1 shows typical emulsion of ferrofluid droplets with a mean diameter of 10 μm. The MF used here is constituted by cobalt ferrite particles [6] with an average diameter of 10 nm. These particles behave as rigid dipoles. In order to obtain a stable solution in oil base, the particles are coated by a surfactant. The surfactant employed is an anionic surfactant, a phosphorous ester of alkyl phenol; the oil is a commercial one based on hydrogenated terphenyl. This oil base MF is then emulsified in aqueous medium (30% MF, 70% H_2O W/W). The dispersing agent ensuring the stability of the emulsion is an ethylen maleic anhydride copolymer, initially dispersed in water (4% W/W). The emulsion is performed by vigorous stirring (turbine at 24000 t/mn) during 10 mn.

We have focused our measurements on a pair of MF droplets. The measurements are made via an optic microscope and a video camera, and are recorded on video tape. The rotating magnetic field is ensured by two perpendicular sets of two coils supplied by alternating current in quadrature. The frequency of the rotating magnetic field can change from 10^{-2} Hz to 10^3 Hz, with a maximum magnitude of 100 Oe. The MF emulsion fills a glass cell of

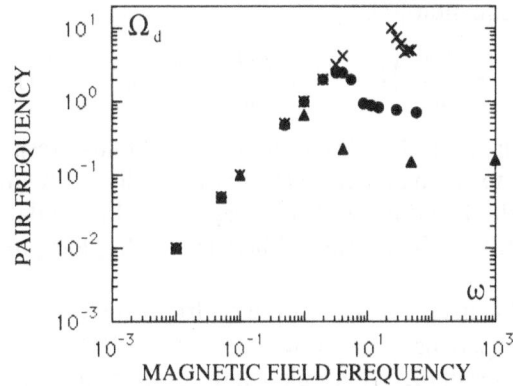

Fig. 2 Rotation frequency of the pair of droplets versus the frequency of the rotating magnetic field for different field amplitude in log–log plot (▲ H = 22 Oe; ● 46 Oe and × 128 Oe)

10 μm thickness. The rotating magnetic field is in the plane of the cell. Figure 2 presents the rotating frequency of a pair of droplets Ω_d as a function of the magnetic field frequency ω, for three different magnetic field amplitudes. A synchronous regime is observed at low frequency, and after a cut-off ω_c depending on the magnetic field amplitude, a jerky domain appears which leads to a small decrease of the mean value of Ω_d. Increasing ω, the pair frequency tends towards a constant frequency slightly smaller than the cut-off one on two or three decades.

Theory

If the viscosity of the magnetic fluid inside the droplets is large in comparison to the surrounding fluid viscosity, the droplet can be considered as a rigid body. The equilibrium droplet magnetization is described by the Langevin law

$$M_0 = M_\infty L(\xi); \quad \xi = mH/kT; \quad L(\xi) = c\tanh\xi - \xi^{-1},$$

where M_∞ is the saturation magnetization of the magnetic fluid, m is the magnetic moment of the MF colloidal particles, k is the Boltzmann constant, and T is temperature. The characteristic time to reach equilibrium magnetization is the Brownian time τ_B:

$$\tau_B = \frac{3v_h\eta_0}{kT},$$

where v_h is the hydrodynamic volume of the particle, η_0 is MF carrier viscosity. Here the value of τ_B is estimated to be of the order of 10^{-4} s.

As the period of magnetic field rotation is large in comparison to τ_B ($\omega\tau_B \ll 1$), the deviation of magnetization from the equilibrium is small and can be described

Fig. 1 Picture of a ferromagnetic emulsion; the mean size is 10 μm

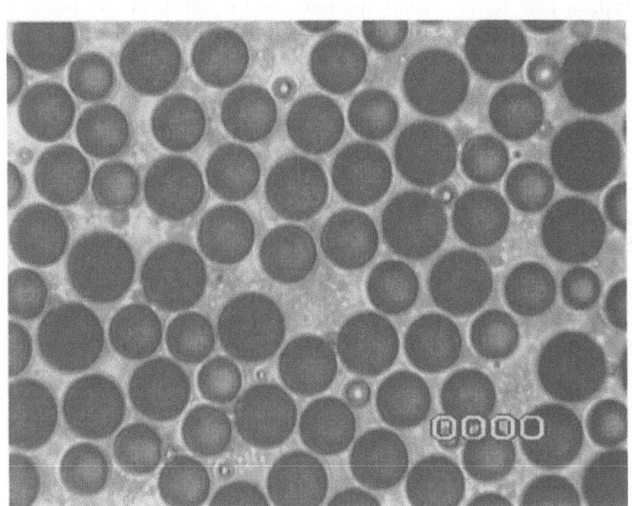

by a linear equation [7]:

$$\vec{M} = \left(M_0 - \tau_{//}\frac{dM_0}{dt}\right)\vec{h} - \tau_\perp M_0\left(\frac{d\vec{h}}{dt} - \vec{\Omega} \times \vec{h}\right), \qquad (1)$$

where $\tau_{//}$ (resp. $\tau_\perp$) is the relaxation time longitudinal (resp. transversal) to the magnetic field; h is the unit vector in magnetic field direction; $\vec{\Omega}$ is the magnetic field angular velocity which can be considered here as the droplet spin velocity.

Due to the magnetization deviation from equilibrium, the droplet experiences a rotation torque, $\vec{T} = V\vec{M} \times \vec{H}$ (V is the volume of the MF droplet). If the droplet moves in the plane of magnetic field rotation, $\vec{\Omega} \cdot \vec{h} = 0$, and:

$$\vec{T} = 4\eta_r V(\vec{\omega} - \vec{\Omega}), \qquad (2)$$

where η_r is the magnetic fluid rotational viscosity, equal to [8]:

$$\eta_r = \frac{3}{2}\eta_0\varphi_h\frac{\xi - \tanh\xi}{\xi + \tanh\xi}, \qquad (3)$$

where φ_h is the colloidal particle hydrodynamic concentration. Under the action of torque (2) a single droplet rotates with a spin frequency determined by the balance of the magnetic torque (2) and the viscous torque $-6V\eta\vec{\Omega}$ (η is the viscosity of the surrounding fluid). So the spin velocity of a single droplet is:

$$\Omega = \frac{2S}{3 + 2S}\omega, \quad \text{with } S = \eta_r/\eta, \qquad (4)$$

where Ω is proportional to the field frequency and depends on the magnetic fluid rotational viscosity. Inside a pair of droplets there are both a magnetic and a hydrodynamic interaction. The magnetic force on the droplet number 1 in the dipole approximation is given by the relation:

$$\vec{F} = \frac{3M_v^2}{r^4}(\vec{\rho} + 2\vec{h}(\vec{\rho}\cdot\vec{h}) - 5\vec{\rho}(\vec{\rho}\cdot\vec{h})^2) \qquad (5)$$

where r is the distance between droplet centers, $\vec{\rho}$ is unit vector from droplet 1 to droplet 2, $M_v = VM$ is the magnetic moment of a droplet. The dipole-dipole interaction force has both a component along the droplet center direction, and along the transversal direction. This means that the pair of droplets is submitted both to a force and to a torque. At low frequency, we can assume that the internal rotation of the magnetic fluid does not affect the movement of rotation of the pair. The pair of droplet is completely described by its angular velocity Ω_d:

$$6\eta K\Omega_d = \frac{3M_v^2}{r^4}\sin 2\theta \qquad (6)$$

where θ is the phase lag between the magnetic field and the axe of the pair of droplets; K is the hydrodynamic volume

of this doublet. The rotation is synchronous up to a critical frequency $\omega_c = M_v^2/6\eta Kr^3$; above this frequency, a jerky regime appears, described by equation (6). But the internal rotation of each droplet has to be taken into account; we assume that a relation exists between the rotation of the doublet Ω_d and the rotation of a single droplet $\Omega = n\Omega_d$, with the coefficient n reflecting the interface interactions between the two droplets; for example, in the case of rolling without gliding $n = 2$. The movement of the doublet is described by the following equation:

$$\Omega_d = \omega_c\frac{1}{(1 + \frac{2}{3}Sn\frac{V}{K})}\sin 2\theta + \frac{\frac{2}{3}S\frac{V}{K}}{1 + \frac{2}{3}Sn\frac{V}{K}}\omega \qquad (7)$$

the first term of the right member leads to the synchronous rotation (slightly modified by the factor $2/3\,SnV/K$) in the regime $\omega < \omega_c$. The second term, preponderant in the regime $\omega \gg \omega_c$, leads to a constant rotation frequency proportional to the magnetic field rotation frequency if the prefactor is frequency independent.

Discussion

Experimentally, a synchronous regime has been found for $\omega \ll \omega_c$ as it is predicted by the theoretical calculation. The dependence of ω_c with H^2 is predicted by the model:

$$\omega_c = \frac{(M_\infty VL(\xi))^2}{6\eta Kr^3} \underset{\xi\to 0}{\simeq} \frac{M_\infty^2 V^2\xi^2}{6\eta Kr^3}. \qquad (8)$$

The initial magnetic susceptibility χ of the magnetic fluid is $\chi = M_\infty\xi/3H$. In our experiment, $4\pi\chi \simeq 1$ and Eq. (8) leads to $\omega_c \simeq 1\ \text{s}^{-1}$, which is in agreement with the experimental results.

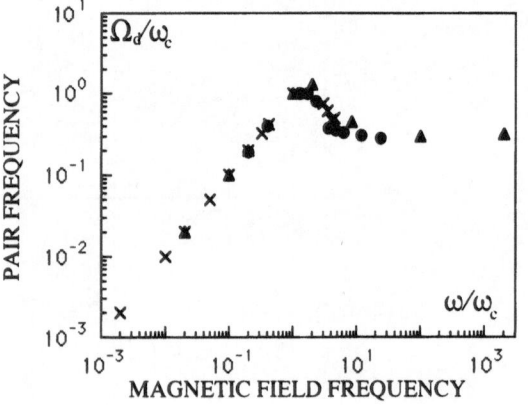

Fig. 3 Log–log plot of the reduced frequency of the rotation of the droplets pair Ω_d/ω_c versus the reduced frequency of the rotating magnetic field ω/ω_c: all the experimental points for different magnetic field amplitudes (see Fig. 2) are on the same master curve

Progr Colloid Polym Sci (1995) 98:124–127
© Steinkopff Verlag 1995

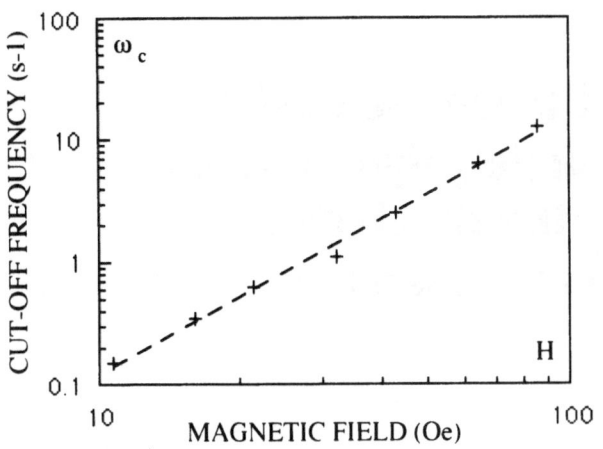

Fig. 4 Cut-off frequency of the rotating doublet versus the magnetic field amplitude ($\omega = 40\,\mathrm{s}^{-1}$)

Figure 3 presents a log–log plot of the reduced frequency of the doublet rotation Ω_d/ω_c versus the reduced frequency of the rotating magnetic field for different magnetic field amplitudes. All the experimental points are on the same master curve which means that a scaling law exists, which is not predicted by expression (7), excepted for synchronous regime.

The unexpected point is that in the high frequency regime, the frequency of the rotation of the doublet tends towards a plateau. The value of this plateau is proportional to H^2 (Fig. 3) as it is predicted by expression (7) because $\eta_r \to H^2$ for $\xi \to 0$. The meaning of this plateau is a subtle compensation between the viscous torque and the magnetic one, resulting in a self-regulation of the doublet rotation in a very large range of the magnetic field frequencies. These phenomena can be used in micro-bio mechanic and for the servo control of micromotors.

Acknowledgement We are greatly indebted to P. Lepert and J. Servais for their technical assistance. One of us thanks the "Ministère de l'Enseignement Supérieur et de la Recherche" of France for a research grant.

References

1. Kashevsky BE (1993) J Mag Mag Mat 122:37–41
2. Kashevsky BE, Novikova AL (1989) Magnetohydrodyanmics 25:304–309
3. Kashevsky BE, Kordonsky WI (1991) Izvestija Akad Nauk SSSR Ser Fiz (in Russian) 55:1110
4. Skjeltorp AT (1987) J Mag Mag Mat 65:195–203
5. Helgesen G, Pieranski P, Skjeltorp AT (1990) Phys Rev Lett 64:1425–1428
6. Massart R, Patents: France 79.18842, 9006484; Germany P 30270123; Japan 98.202/80; USA 4 329241
7. Isaev SV, Kashevsky BE (1980) Magnitnaya Gidrodinamika (in Russian) 4:19–23
8. Shliomis M (1972) Soviet Phys JETP (English translation) 34:1291–1294

Progr Colloid Polym Sci (1995) 98:128–131
© Steinkopff Verlag 1995

W. Liang
G. Bognolo
Th.F. Tadros

Adsorption of macromolecular surfactants on polystyrene latex particles and the rheology of the resulting dispersions

W. Liang · Dr. Th.F. Tadros (✉)
Zeneca Agrochemicals
Jealott's Hill Research Station
Bracknell
Berkshire RG12 6EY, United Kingdom

G. Bognolo
ICI Surfactants
Everslaan 45
3078 Kortenberg, Belgium

Abstract The effect of adsorption of macromolecular surfactants on the stability and rheological properties of latex dispersions has been investigated using adsorption isotherms and rheological measurements. Two latex particles with diameters of 427 and 867 nm were used and two macromolecular surfactants with poly(methyl methacrylate) backbone and poly(ethylene oxide) side chains were studied. It was found that the amount of adsorbed surfactant per unit area was approximately constant for both latex particles. The adsorbed amount increased with increase of temperature as a result of reduction of solvency of the medium for the chains. The steric interaction of the particles at saturation adsorption of macromolecular surfactants was investigated using rheological measurements, namely steady-state shear stress-shear rate and oscillation measurements. The adsorbed layer thickness of surfactants on the latex particle was determined as a function of volume fraction of the dispersions.

Key words Macromolecular surfactants – polystyrene latex – steric interaction

Introduction

Macromolecular surfactants are block or graft copolymer containing both hydrophillic and hydrophobic groups in the molecule. These block or graft copolymers are considered as the best steric stabilizers in practice [1]. By proper choice of the anchor polymer (mostly a hydrophobic chain), e.g. poly(methyl methacrylate), one can ensure that the copolymer is strongly attached to the hydrophobic particle surface. The stabilizing moieties, such as poly(ethylene oxide) can be made of sufficient length and density to provide an effective steric barrier against flocculation.

Macromolecular surfactants are particularly useful for stabilization of aqueous concentrated dispersions. Various factors may affect such stability such as density of adsorbed surfactant layer, particle size, volume fraction of the particles and solvency of the medium. The rheology of the resulting dispersions will also be affected by the above parameters [2].

In this paper, we report the stability of latex dispersions in the presence of macromolecular surfactants using adsorption and rheological techniques.

Experimental

Materials

Polystyrene latex dispersions were prepared using the surfactant-free method developed by Goodwin et al. [3]. The latex was dialysed against distilled water using a Miniton Ultrafiltration System. Basically, a peristaltic tubing pump draws the fluid sample from an un-pressurized reservoir, and pumps it into the Miniton holder at 250 ml/min. Concentrated latex dispersion is returned to the reservoir,

Progr Colloid Polym Sci (1995) 98:128–131
© Steinkopff Verlag 1995

and filtrate is collected in a separate container. The microporous membrane used here was PTHK (OMT 05 polysulfone) with a pore diameter of 0.1 μm. The latex dispersion was diluted to two to three times lower of its original concentration and concentrated after 6 h of filtration. The filtrate was analyzed each time by conductivity measurement. This procedure was repeated 8–10 times until the conductivity of the filtrate became constant. The particle sizes were determined by PCS using a Malvern 4700 PCS instrument. The Z-average particle diameters were found to be 427 and 867 nm respectively and the latices were fairly monodisperse.

Two macromolecular surfactants supplied by ICI Surfactants were used. These surfactants were designated as Atlox 4913 and Hypermer CG-6 (a). Both macromolecular surfactants have the same backbone, namely poly(methyl methacrylate-methacrylic acid) with methoxy-capped poly(ethylene oxide) side chains. However, the CG-6 (a) has a higher proportion of polymethacrylic acid, thus the density of side chains is lower than that for Atlox 4913.

Adsorption isotherms

Latex dispersions (0.100 g) were equilibrated with polymer solutions (30.00 ml) of different concentrations for more than 16 h at fixed temperature (20° or 40 °C). The particles were then removed by centrifugation at 20 000 rpm (25 000 g) for 30 min. The concentration of the polymer in the supernatant was determined colorimetrically using the method originally developed by Balleux [4] and later adapted [5] for other non-ionic surfactants of the ethoxylate type.

Adsorbed layer thickness measurements

The adsorbed layer thickness was determined by rheological measurements. Details of this method has been described previously [6].

Preparation of concentrated latex dispersions for rheological measurements

Copolymer solutions were added to the stock latex dispersion at concentrations corresponding to the plateau of the adsorption isotherm. After equilibration (> 16 h), the latex was concentrated by centrifugation at 10 000 rpm for 30 min in order to cover a wide range of ϕ_s values. Any dilutions were made with the supernatant liquid from the centrifugation.

Rheological measurements

A Bohlin VOR rheometer was used for both oscillatory and steady-state shear stress-shear rate measurements at 25 °C. The details of the measurement has been described before [6, 7].

Results and Discussions

The adsorption isotherm (Γ in mg m^{-2} vs equilibrium concentration C) for surfactant Atlox 4913 on two latex particles studied at 20 °C is shown in Fig. 1. The plateau adsorption amounts on both latex particles are more or less the same in the plateau region, i.e. ~ 1.5 mg/m^2. This implies that the particle curvature has no effect on the adsorption amount per surface area.

For surfactant Hypermer CG-6 (a) adsorbed on two latex particles, similar trends were found as with surfactant Atlox 4913. However, the adsorption amount Γ at the

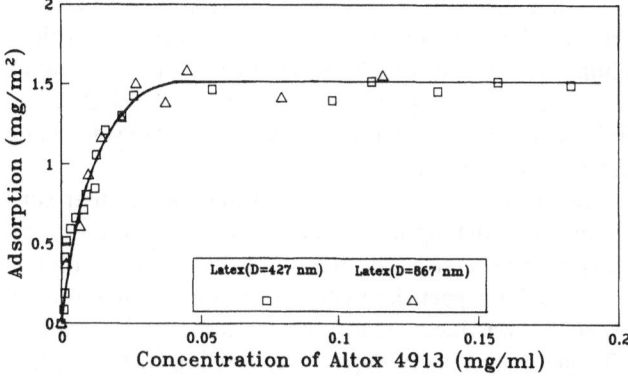

Fig. 1 Adsorption isotherms for Atlox 4913 adsorbed on latices at 20°C

Fig. 2 Adsorption isotherms for Atlox 4913 adsorbed on latex ($D = 867$ nm) at two temperatures

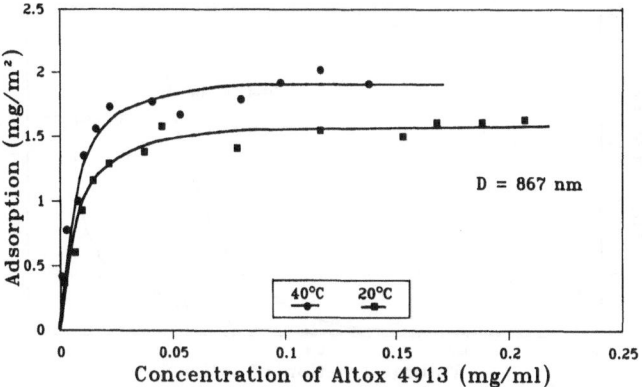

130

W. Liang et al.
Adsorption of macromolecular surfactants

plateau region was slightly lower, i.e. ~ 1.2 mg/m². This is due to the fact that the latter surfactant side-chain density is lower than that for Atlox 4913.

The effect of temperature on the adsorption of macromolecular surfactant Atlox 4913 on two latex particles at 40 °C is shown in Fig. 2. It is clear that with increasing temperature the adsorption amount of the surfactant was increased. This is due to the reduction in solvency for the PEO chains with increase of temperature. Similar results were obtained for surfactant Sypermer CG-6 (a) adsorbed on both latex particles.

Figure 3 shows the plot of yield stress as a function of latex (stabilized by surfactant Atlox 4913) volume fraction. It is clear that at the same volume fractions of particles, the larger the particle, the lower value of yield stress. This is easy to understand if one considers the ratio of the adsorbed layer to particle radius Δ/R [6, 8]. For the small particles Δ/R is relatively larger than that of the larger particles, since the adsorbed layer Δ is more or less the same for both particles. Thus, the effective volume fraction is larger for the smaller particle size latex and hence steric interaction starts at a lower volume fraction when compared to that of the larger latex.

Similar results were obtained for latices stabilized by surfactant Hypermer CG-6 (a). But in this case, the critical volume fraction at which yield stress τ_b shows a rapid increase is much higher than that obtained using Atlox 4913. This may be attributed to the reduction in side-chain density with CG-6 (a).

The adsorbed layer thickness of macromolecular surfactants was determined from the relative viscosity as described before [6]. The results of the effect of volume fraction of latex particles (with diameter of 867 nm) on the adsorbed layer thickness for two macromolecular is listed in Tables 1 and 2. As the volume fraction increases, the

Table 1 The relationship between the volume fraction of latex ($D = 867$ nm + Atlox 4913) and the adsorbed layer thickness.

ϕ	η_r	ϕ_{eff}	Δ (nm)
0.361	5.13	0.407	17.5
0.406	7.33	0.452	15.7
0.465	14.06	0.512	14.2
0.511	36.52	0.566	15.0
0.550	53.59	0.580	7.76
0.553	54.11	0.580	7.02
0.567	85.47	0.593	6.55

Table 2 The relationship between the volume fraction of latex ($D = 867$ nm + CG-6 (a)) and the adsorbed layer thickness.

ϕ	η_r	ϕ_{eff}	Δ (nm)
0.488	16.0	0.529	11.8
0.536	36.9	0.576	10.4
0.555	65.1	0.596	10.4
0.565	70.0	0.598	8.28
0.575	72.7	0.599	5.97
0.592	81.3	0.602	2.44

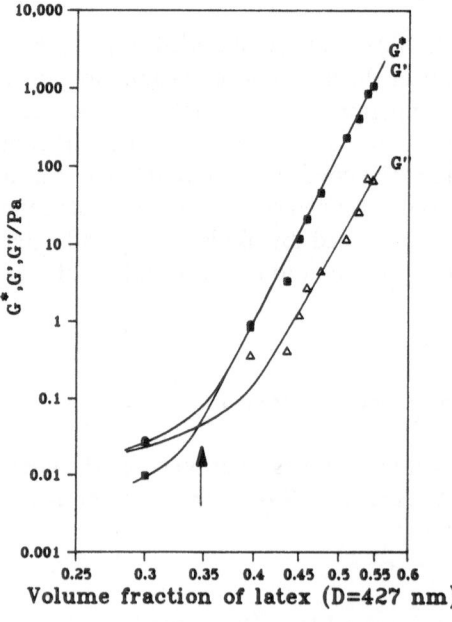

Fig. 4 Plot of moduli as a function of latex ($D = 427$ nm, stabilized by CG-6 (a)) volume fraction

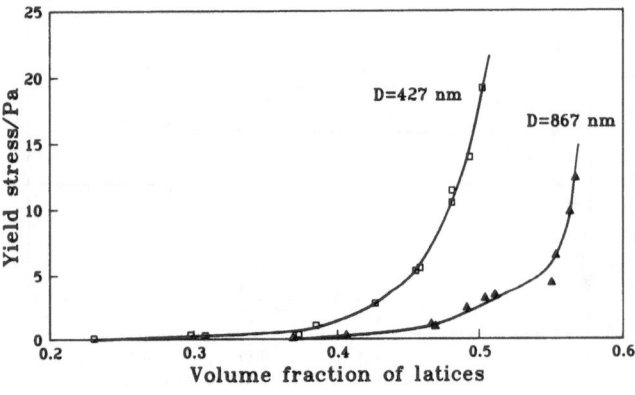

Fig. 3 Plot of yield stress as a function of latices (stabilized by Atlox 4913) volume fraction

adsorbed layer may undergo interpenetration and/or compression, thus the effective adsorbed layer thickness decreases with increase in volume fraction of particles.

Figures 4 and 5 show plots of G^*, G' and G'' as a function of ϕ_s for two particle diameters of 427 and

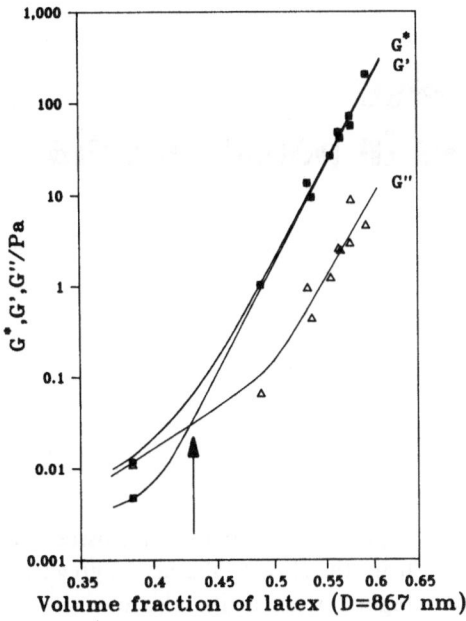

Fig. 5 Plot of moduli as a function of latex ($D = 867$ nm, stabilized by CG-6 (a)) volume fraction

867 nm, respectively. These results are obtained using surfactant Hypermer CG-6 (a) as stabilizer adsorbed on the particle surface at saturation region. It can be seen from Figs. 4 and 5 that at any given ϕ_s, the moduli values are higher for the smaller latex dispersions. The cross-over point at which $G' = G''$ occurs at $\phi_s = 0.35$ for $D = 427$ nm and $\phi_s = 0.43$ for $D = 867$ nm. The cross-over point (ϕ_c) can be taken as the point where the adsorbed layers begin to interfere. Below ϕ_c, $G'' > G'$ and the interaction between the particle is relatively weak. However, above ϕ_c, $G' > G''$, the adsorbed layers undergo interpenetration and/or compression. Under these conditions, the systems becomes predominantly elastic [6].

Similar results are obtained for surfactant Atlox 4913 adsorbed on the same two latex particles. However, the cross-over point, ϕ_c, in this case is lower than those obtained for Hypermer CG-6 (a) as stabilizer. For latex $D = 427$ nm, $\phi_c = 0.26$, and for latex $D = 867$ nm, $\phi_c = 0.42$. The reason for this is that Atlox 4913 has higher graft density of PEO side chains on the backbone than Hypermer CG-6 (a), therefore, the effective adsorbed layer should be thicker than those of Hypermer CG-6 (a).

References

1. Napper DH (1983) In: Polymeric Stabilization of Colloidal Dispersions. Academic Press, London, pp 28
2. Tadros ThF (1993) Adv Colloid Interface Sci 46:1–47
3. Goodwin JW, Hearn J, Ho CC, Ottewill RH (1974) Colloid Polym Sci 252: 464–471
4. Baleux B (1972) C R Acad Sci Ser C 274:1617
5. Tadros ThF, Vincent B (1980) J Phys Chem 84:1575–1580
6. Liang W, Tadros ThF, Luckham PF (1992) J Colloid Interface Sci 153: 131–139
7. Liang W, Tadros ThF, Luckham PF (1993) J Colloid Interface Sci 155: 156–164
8. Tadros ThF, Liang W, Costello B, Luckham PF (1993) Colloids Surfaces A: 79:105–114

Progr Colloid Polym Sci (1995) 98:132–135
© Steinkopff Verlag 1995

Y.D. Yan
M. Borkovec
H. Sticher

Deposition and release
of colloidal particles in porous media

Y.D. Yan · Dr. M. Borkovec (✉)
H. Sticher
Institute of Terrestrial Ecology
Federal Institute of Technology (ETH)
Grabenstraße 3
8952 Schlieren, Switzerland

Abstract The interaction of colloidal carboxylated latex particles with a packed bed of glass beads is studied by means of chromatographic techniques. For these negatively charged surfaces we find strong evidence of particle attachment as well as detachment processes. In particular, the rate constants of the two processes are measured from the particle breakthrough curve of a pulse input. Ionic strength is found to have profound effect on these processes. Upon increasing ionic strength, the deposition rate constant crosses over from the slow to the fast aggregation regime, whereas its release counterpart shows the opposite trend, with their crossing-over concentrations being the same. The DLVO theory with particle-beads interactions is used to explain these findings qualitatively.

Key words Colloids – Deposition and release – particle transport – porous media – column experiments

Introduction

Removal of colloidal particles using packed bed collectors is a basic process in water treatment and wastewater filtration. Similar phenomena also play important roles in subsurface environments such as soils and aquifers. Three underlying particle transport mechanisms leading to particle-collector attachment have been proposed: Brownian diffusion, interception due to fluid stream lines, and gravitational settling (sedimentation) [1]. The interaction forces between the particle and its collector surfaces often play an essential role in determining the rate of deposition.

Equally important is the study of mobilization of those immobilized particles. Its understanding is of particular significance in environmental sciences as colloidal particles can facilitate pollutant transport in groundwater. Depending on the magnitude of the potential energy difference between the primary minimum and the barrier, an adsorbed particle may, due to its thermal motion and the hydrodynamic forces from the mobile suspension fluid, detach from the collector surface and be carried away by the flowing medium. It is, nevertheless, generally believed that particles attached in this primary energy well cannot escape from their collector surfaces without introducing changes in chemical or flow conditions. Recent studies on colloidal particle interactions with glass plates [2] and bacterial transport in packed columns [3, 4] have, however, indicated the reversible nature of the deposition process. There have so far been relatively few studies on the release phenomena in contrast to the extensive research on deposition [5–8].

Here, we present some preliminary results of laboratory column experiments which investigate interactions of carboxylated polystyrene colloidal particles with packed glass beads. In particular, we shall show evidence on the reversibility of our particle attachment and report data on the rate constants of the deposition and release processes.

Progr Colloid Polym Sci (1995) 98:132–135
© Steinkopff Verlag 1995

Experimental

Chemicals

Deionized and filtered water from a Barnstead Nanopure system was used throughout this work. Influent solution of pH 9.6 was obtained by mixing aqueous solutions of 10 mM $NaHCO_3$ (Fluka, p.a.) and 2 mM Na_2CO_3 (Merck, p.a.). The ionic strength was adjusted using NaCl (Merck, p.a.). The actual ionic strength is calculated from the weighted sum of all ionic species in solution.

Glass beads and column packing

Glass beads (Merck) were sieved to a size fraction of 0.40–0.45 mm in diameter and washed by HCl/chromic acid as described by Olson and Litton [9]. The beads were wet packed into a glass chromatography column of 2.4 cm inner-diameter, with a packing height of 31 cm. Figure 1 shows the schematic experimental set-up used in this work. In operation, the influent was passed through a degasser (Erma ERC-3511) and pumped upward through the column by an HPLC-pump (Sykam S 1000). Both the influent and effluent pH values could be measured with a flow-through electrode (Sensorex) and the effluent pH was found to reach that of the influent after 2–3 pore volumes of flushing. Tracers and particle dispersions could, respectively, be introduced into the column inlet through a 100 μL syringe pluse-injection loop. Their concentrations were continuously monitored at the column outlet with flow-through spectrophotometers. The column was equipped with a water-jacket and thermostated to 25.0 ± 0.1 °C. In this study an influent flow rate of 2.82 mL·min^{-1}

was used, which corresponds to a true (interstitial) fluid velocity of 2.55 $\times 10^{-4}$ m·s^{-1}.

UV absorption tests were carried out using molecular tracers NaBr (Fluka, MicroSelect) and $NaNO_3$ (Merck, p.a.) to characterize the packed column. These two tracers showed identical ideal breakthrough behaviour. The measured packing porosity from the tracer breakthrough curves (BTCs) was 0.40. The dispersivity, obtained from the second moment of the ideal tracer BTCs, was 0.26 mm, which is comparable to the size of the glass beads.

Colloidal particles

Fluorescently-labelled, carboxylated polystyrene latex particles (Interfacial Dynamics Corp., Portland, OR) of diameter 306 nm were used. Before use these particles were deionized by using a mixed bed ion exchange resin (AG 501-X8 (D), BIO-RAD). According to the manufacturer, these particles are negatively charged. Their breakthroughs were detected with a spectrofluorometer (Jasco 821-FP).

Results and discussion

Column breakthrough curves (BTCs) reported here are normalized by their respective inlet concentrations measured from separate by-pass pulse experiments by short-circuiting the column with a tubing loop. They are plotted on semi-logarithmic scales as functions of reduced time, namely t/t_0, where t is time and t_0 is the conservative NaBr tracer breakthrough time. As evident from Fig. 2, there are two distinctive differences in the BTCs between the latex particles and the conservative NaBr tracer. Firstly, the breakthrough peak height of the particles is lower than that of the tracer, which reflects the fact that our latex particles did adsorb onto the glass beads surfaces. Secondly, there is significant tailing in the particle BTCs. This clearly resulted from the reentrainment of those previously deposited particles from the beads surfaces, therefore revealing, contrary to most literature findings, the reversible nature of our particle deposition process. Also noticeable in Fig. 2 is that colloidal particles travel through the column faster than the conservative tracer, a few percent in our case. This may be attributed to the size exclusion effect [10, 11]. Ionic strength had significant effect on the deposition and release processes. This can be seen by comparing the particle BTCs of Figs. 2a and 2b. Increasing the solution ionic strength leads to a decrease in the peak height and also slows the decay of the tail.

The convection-dispersion equation coupled with assumed first-order deposition and release kinetics [12]

Fig. 1 Schematic of the experimental set-up used in this work

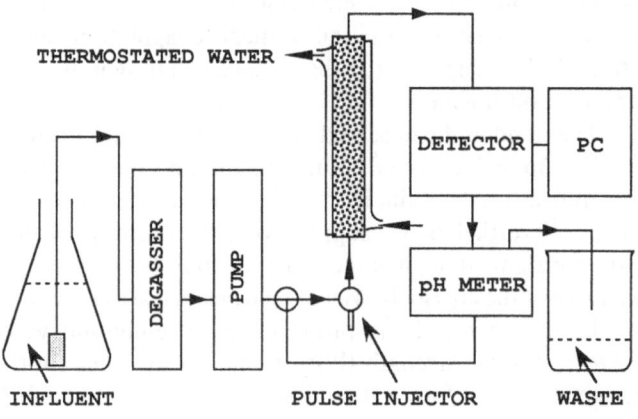

134

Y.D. Yan et al.
Colloidal particle deposition and release in porous media

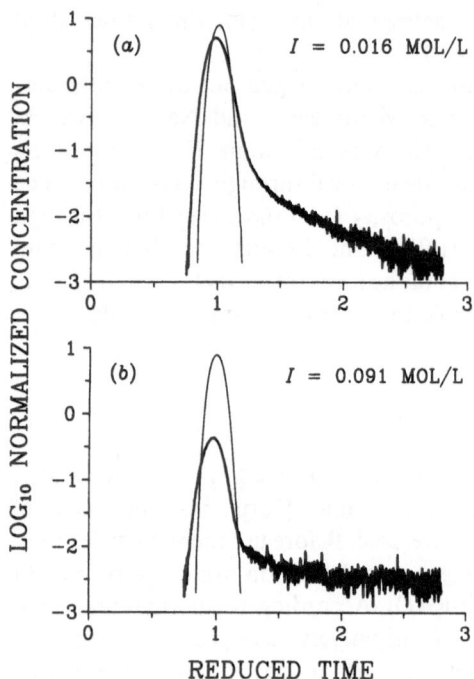

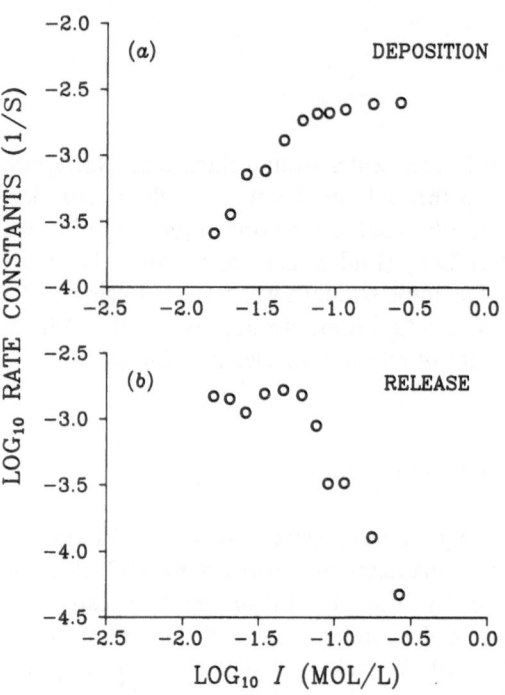

Fig. 2 Column breakthrough curves on semi-log scales as functions of the reduced time, t/t_0, namely, the number of effluent pore volumes. (a) Thin line: conservative NaBr tracer; thick line: latex particles (ionic strength $I = 0.016$ M). (b) Latex particles at $I = 0.091$ M with other conditions the same as in (a). As a reference, the same NaBr tracer as in (a) is also plotted

Fig. 3 Dependence of particle deposition/release rate constants on solution ionic strength. (a) Deposition. (b) Release

presents the basis for extracting the rate constants from our column experiments. According to this equation, the deposition rate constant k_a can be calculated from the area of the non-retarded peak, which originated from the nonadsorbed particles. This area is given by $\exp(-k_a t_p)$, where t_p is the breakthrough time of the corresponding conservative particle tracer, which in our case is close to t_0. This peak area can be accurately estimated by taking twice the area of the BTC up to the peak maximum. On the other hand, if the particle release is a slow first-order rate process, the BTC will decay exponentially for long times. The release rate constant k_d can then be simply obtained from the slope of the release tail, if plotted on a semi-logarithmic scale. Although this assumption of mono-exponential release of particles may be questionable, in this preliminary work, we simply employ the slope of the release tail as the estimate of the particle release rate constant.

Figure 3a shows the measured particle deposition rate constant as a function of the influent ionic strength. This curve is typical of those which can be found in literature [5]. As ionic strength was increased, the heterogeneous aggregation rate became larger (slow regime) and eventually came across the critical coagulation concentration

over to a plateau (fast regime). Our measured plateau values agree reasonably well with those calculated from colloid-filtration theory [1, 5, 13]. Interestingly, the estimated rate of particle release from the beads surfaces shows the opposite trend (Fig. 3b): it remained almost constant at low ionic strength but dropped at high ionic strength with the crossover point corresponding closely to the critical coagulation concentration of Fig. 3a.

The transition of slow to fast aggregation (cf. Fig. 3a) can be qualitatively explained by using the DLVO theory. Since both our latex particles and glass beads surfaces are expected to be negatively charged at the pH values used in our experiments, the repulsive potential energy due to the overlapping of the electric double layers surrounding both surfaces will decrease upon increasing the solution ionic strength. This will effectively reduce the energy barrier for particle deposition. At the critical coagulation ionic strength the energy barrier disappears, resulting in the fast aggregation regime.

The trend in the observed particle release rate constant (cf. Fig. 3b) is less trivial to understand and one needs an accurate estimate of the shape of the total potential energy curve. Nevertheless, we suppose that the decrease of this rate constant at high ionic strength may be due to the increase of the energy barrier for particle release as a result of the deepening of the primary energy minimum with increasing ionic strength (Born repulsion or other short-range structural forces should in this respect be included

Progr Colloid Polym Sci (1995) 98:132–135
© Steinkopff Verlag 1995

into the DLVO theory). On the other hand, the appearance of no significant ionic strength effect on the release rate constant in the slow deposition regime indicates that the energy barrier for particle detachment remains almost constant at low ionic strength. Interestingly, McDowell-Boyer has made a model potential energy calculation (Fig. 2 in ref. [14]) for sphere-plate interactions, which essentially supports the above hypothesis. Additionally, from her calculation one would expect that the release rate constant should go through a maximum and eventually decrease upon further lowering ionic strength. The present interpretation of the release from the primary minimum is endorsed by the order of magnitude of the estimated rate constants, which clearly indicates that the energy barrier for detachment is at least $10\,kT$ (probably much higher) and can hardly originate from a secondary minimum.

In our experiments recovery of the injected particles at the column outlet ranged from ca 90% at the lowest ionic strength used to about 5% at the highest ionic strength reached. The apparent loss of particles was probably due to a combination of two effects: i) a non-exponential decay of the particle release process, and ii) finite detection limit of our detector. One should remark that we have confirmed that the deposition-release process of our particles onto the glass beads was linear, i.e., the response of the column was independent of particle concentration over the range used here. This means that nonlinear processes are unlikely to be the source of the apparent loss of particles. The detailed understanding of the above effects requires additional experiments at different flow rates and particle concentrations, combined with quantitative interpretation of the observed BTCs. These aspects are under further investigation.

Acknowledgements The authors would like to thank Jaro Rička for stimulating discussions. This work was supported by the Swiss COST Project D5.

References

1. O'Melia CR (1980) Environ Sci Technol 14:1052–1060
2. Meinders JM, Busscher HJ (1994) Colloid Polym Sci 272:478–486
3. McCaulou DR, Bales RC, McCarthy JF (1994) J Contaminant Hydrology 15:1–14
4. Kinoshita T, Bales RC, Yahya MT, Gerba CP (1993) Wat Res 27:1295–1301
5. Elimelech M, O'Melia CR (1990) Langmuir 6:1153–1163; (1990) Environ Sci Technol 24:1528–1536
6. Tobiason JE, O'Melia CR (1988) J Amer Water Works Ass 80:54–64
7. Kallay N, Barouch E, Matijević E (1987) Adv Colloid Interface Sci 27:1–42
8. Litton GM, Olson TM (1993) Environ Sci Technol 27:185–193
9. Olson TM, Litton GM (1992) In: Sabatini DA, Knox RC (eds) Transport and remediation of subsurface contaminants: colloidal, interfacial, and surfactant phenomena. ACS Symp Series 491:14–25
10. Gvirtzman H, Gorelick SM (1991) Nature 352:793–795
11. Higgo JJW, Williams GM, Harrison I, Warwick P, Gardiner MP, Longworth G (1993) Colloids Surfaces A 73:179–200
12. Villermaux J (1981) In: Rodrigues AE, Tondeur D (eds) Percolation processes: theory and applications. NASI, Sijthoff and Noordhoff, pp 83–140
13. O'Melia CR (1985) J Environ Eng 111:874–890
14. McDowell-Boyer LM (1992) Environ Sci Technol 26:586–593

Progr Colloid Polym Sci (1995) 98:136–139
© Steinkopff Verlag 1995

P. Van der Meeren
J. Vanderdeelen
L. Baert

On the relationship between dynamic interfacial tension and O/W emulsion stabilizing properties of soybean lecithins

Dr. P. Van der Meeren (✉)
J. Vanderdeelen · L. Baert
University of Gent
Faculty Agricultural and Applied Biological
Sciences
Department of Applied Analytical and
Physical Chemistry
Coupure Links 653
9000 Gent, Belgium

Abstract As the composition of soybean lecithins depends on the growing, storing, and processing conditions of the beans, their functional properties may widely vary. Up to now, confusion exists about the underlying mechanisms responsible for the macroscopically observed differences.

As zeta-potential measurements indicated that the poor O/W emulsion stabilizing effect of powdered soybean lecithin cannot be explained by electrostatic interactions, dynamic surface tension measurements have been conducted.

Thus, it was shown that the adsorption process of powdered soybean lecithin at liquid interfaces occurred in two subsequent steps. The kinetics of the first stage were driven by electrostatic interactions. Comparing two widely different lecithins, it was concluded that the O/W emulsion stabilizing properties of lecithins were directly proportional to the long time interfacial tension.

Key words Lecithins – phospholipids – O/W emulsion – dynamic interfacial tension – zeta-potential

Introduction

As phospholipids are natural surfactants, they are widely used in many pharmaceutical and food emulsion systems [1]. As purified phospholipids are rather expensive, soybean lecithin is mostly preferred. This by-product of the vegetable oil production mainly consists of phospholipids, but includes neutral and glycolipids as well [1, 2]. The composition of lecithin may vary according to the growing, storing, and processing conditions of the soybeans [1, 3], with subsequent variation of the functional properties of the lecithins.

According to Rydhag and Wilton [4, 5], both the emulsifying and the stabilizing effect of lecithins are related to their anionic phospholipid content. However, experiments using commercial lecithins yielded conflicting results, from which it must be deduced that the stabilizing

effect of lecithin cannot be solely explained on the basis of electrostatic interactions. More recently, Kaufman et al. [6] proposed a model containing as much as 20 variables (related to the phospholipid and the fatty acid composition) to predict the functional properties of lecithins. Although the model was quite satisfactory for the 14 commercial lecithins considered, it did not provide a better understanding of the underlying mechanisms because the selection of the variables was merely based on the goodness of fit.

In this contribution, we will try to elucidate the scientific background responsible for the large differences in O/W emulsion stabilizing properties of soybean lecithins. The experiments are mainly focused on the dynamic interfacial tension characteristics of both a deoiled soybean lecithin and a phosphatidylcholine (PC) enriched fraction derived from it.

Materials and methods

Emulsions

All emulsions contained 5% (w/v) soybean oil (Vandemoortele) in a 0.02% aqueous NaN_3 solution, stabilized by 0.6% of phospholipids. Powdered soybean lecithin (Epikuron-100P) originated from Lucas Meyer (Hamburg, FRG); its phospholipid content amounted to about 86%. In addition, a phosphatidylcholine (PC) enriched soybean lecithin sample was used; according to the manufacturer, the Epikuron-200 (Lucas Meyer) contained at least 95% of PC. In all calculations, an average molecular weight of 750 g/mole was assumed for both products.

The procedure for preparing O/W emulsions by microfluidisation, as well as for particle sizing by dynamic light scattering has been described elsewhere [7].

Dynamic interfacial tension measurements

Long-term dynamic interfacial tension measurements of phospholipid solutions in HPLC grade n-hexane (Alltech), reagent grade iso-octane (UCB), and carbon tetrachloride for electronic purposes (UCB) against distilled water or aqueous KCl solutions were performed using the Wilhelmy plate method. The force exerted on a roughened platinum plate, which was directed downward at the interface of hexane or iso-octane and water and upward at the interface of CCl_4 and H_2O, was measured by a Cahn Electrobalance RTL; the analog output signal was either registered on a recorder or converted into a digital signal that was logged into a computer by a PC-Acquisitor (Dianachart Inc.).

Electrokinetic measurements

The zeta-potential of emulsion oil droplets was determined by electrophoretic light scattering (ELS), using a Zetasizer IIc (Malvern). Samples were diluted with 0.02% NaN_3 to a final concentration of about 0.01% (w/v) of oil in water.

Results and discussion

The huge influence of the composition of the lecithin on its functional properties has been described before [7]: particle size analyses of 5% O/W emulsions stabilized by 0.6% of phospholipids revealed that all mixtures of Epikuron-100P and Epikuron-200 were excellent emulsifiers. On the other hand, their stabilizing effect was clearly dependent on the composition of the lecithin used. Mixtures containing at least 50% (w/w) of powdered soybean lecithin were not appropriate to stabilize the formed emulsion against flocculation and subsequent creaming.

Keeping account of the electrostatic interaction mechanism proposed by Rydhag and Wilton [4, 5], the zeta-potential of the emulsion droplets in 0.02% (w/v) NaN_3 was determined by electrophoretic light scattering. In Table 1, it is seen that the zeta-potential becomes less negative as the PC content of the lecithin is increased; this is a logical consequence of the zwitterionic character of PC. Table 1 indicates that the instability of the O/W emulsions towards flocculation cannot be explained from the electrokinetic properties of the emulsion droplets: emulsions containing at least 50% of Epikuron-200 remained stable for several weeks despite their much lower zeta-potentials. Hence, the hypothetical stabilization mechanism postulated by Rydhag and Wilton [4, 5] is at least a severe oversimplification.

In previous experiments, it has also been shown that the observed differences in emulsion stabilizing properties could not be explained on the basis of their phase behavior in the presence of water [7]: a water-swollen lamellar phase was obtained for all phospholipid mixtures under consideration.

In our opinion, the observed phenomena can most probably be explained by the results of the calculations of Hofman and Stein [8]. They demonstrated that the interfacial tension may detrimentally influence the stability of fluid dispersed phases: when the interfacial tension becomes lower than $1 \cdot 10^{-4}$ N/m, there is only a slight additional energy required for two emulsion droplets to assume a deformed state with a flat contact area. Especially at interfacial tension values of $1 \cdot 10^{-5}$ N/m, there is even a substantial potential energy gain to be obtained on deformation. In Fig. 1 the dynamic interfacial tension of solutions of Epikuron-100P in hexane against water are shown. From Fig. 1 it becomes obvious that the interfacial tension becomes very small for concentrations of at least 1 mM. Actually, the values obtained can hardly be discerned from zero, given that the accuracy of the Wilhelmy plate

Table 1 Zeta-potential of O/W emulsion droplets. Zeta-potential of a 5% (w/v) O/W emulsion in 0.02% (w/v) aqueous NaN_3, stabilized by 0.6% of phospholipids; the latter was a mixture of Epikuron-100P and Epikuron-200.

Epikuron-100P content (%)	Zeta-potential (mV)
0	− 28
2	− 32
5	− 37
10	− 44
25	− 58
50	− 69
75	− 74
100	− 76

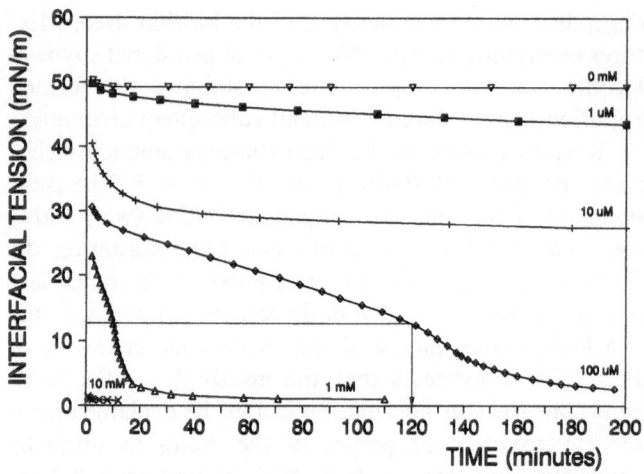
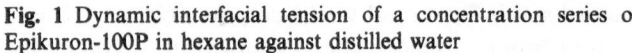

Fig. 1 Dynamic interfacial tension of a concentration series of Epikuron-100P in hexane against distilled water

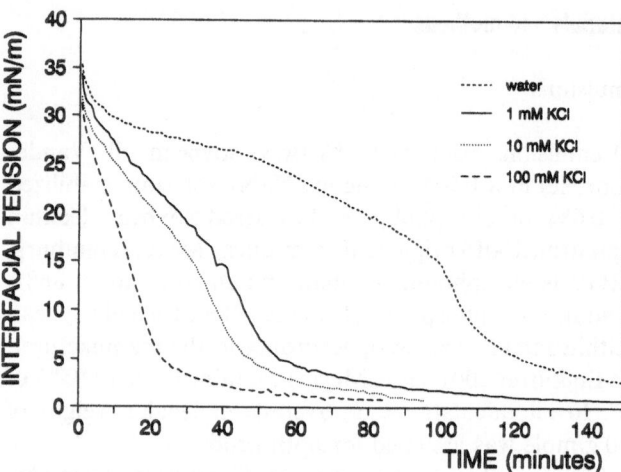

Fig. 2 Dynamic interfacial tension of a solution of 0.1 mg Epikuron-100P per ml of iso-octane against a concentration series of KCl in distilled water

method is limited to about 1 mN/m. Another most striking phenomenon is the presence of two sequential adsorption steps in the overall dynamic behavior. The same phenomenon has been observed by Johnson and Saunders [9] for the dynamic interfacial tension of high concentrations of dimyristoyl phosphatidylcholine in cyclohexane. According to the authors, this so-called anomalous shape is difficult to interpret and may be associated with micelle formation. In addition, Fig. 1 reveals that the transition is taking place at a definite value of the interfacial tension. This two-stage dynamic interfacial tension behavior was also observed for a solution of 0.1 mg of Epikuron-100P per ml of iso-octane against a concentration series of KCl in water (Fig. 2) : a gradual decrease was observed down to about 13 mN/m, at which point the interfacial tension displayed a faster decay. In the latter experiment, the initial part of the decrease of the dynamic interfacial tension with time was steeper as the electrolyte concentration was increased. From this observation, it is deduced that the main barrier for adsorption of phospholipid molecules at the interface during the initial stage of the experiments was due to electrostatic repulsion between the adsorbed layer and the free molecules in the subsurface. On the other hand, the observed interfacial tension decrease during the second stage, starting at about 13 mN/m, was independent from the electrolyte concentration: the different curves all had the same shape but were shifted with respect to time.

For the smallest phospholipid concentration considered, the interfacial tension was linearly related to the square root of the time, which indicated that the adsorption process was diffusion controlled. From the slope of the linear regression, the value of the diffusion coefficient of phospholipids in hexane was estimated to be

$3.62 \cdot 10^{-10}$ m^2/s. This value compared well with the estimated value of PC in decane ($2.0 \cdot 10^{-10}$ m^2/s) as determined by Shchipunov and Kolpakov using the pendant drop method [10]. Hence, for very small concentrations, the transfer of phospholipid molecules from the bulk solution towards the subsurface becomes the limiting step.

For the sake of completeness, it should be mentioned that a minor fraction of the powdered lecithin is insoluble in hexane; similar results were obtained with or without prior filtration of the hazy solution.

Subsequently, a similar experiment using Epikuron-200 was performed. Due to the limited solubility of PC in hexane, carbon tetrachloride was selected as the organic phase. In Fig. 3 it is seen that the interfacial tension of

Fig. 3 Dynamic interfacial tension of a concentration series of Epikuron-200 in carbon tetrachloride against water

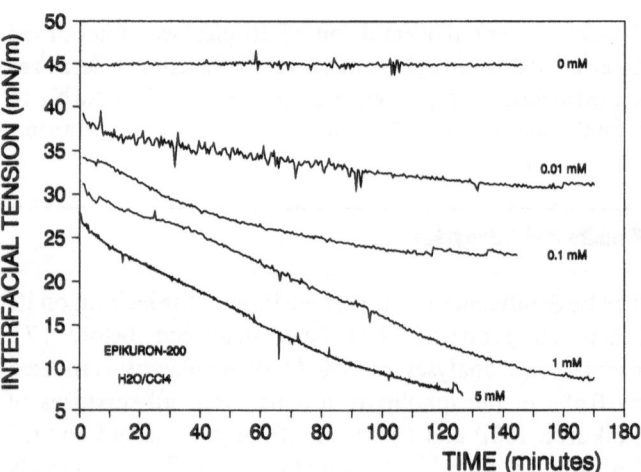

Progr Colloid Polym Sci (1995) 98:136–139
© Steinkopff-Verlag 1995

Epikuron-200 solutions against distilled water was much higher than this of Epikuron-100P solutions: even after several hours, values of over 5 mN/m were obtained. Also, monotonically decreasing curves were obtained for all concentrations studied.

In conclusion, the experimentally determined dynamic interfacial tension values favor the hypothesis of Hofman and Stein [8] stating that the stability of emulsions against flocculation is detrimentally affected by low values of the interfacial tension. In the near future, it will be tried to gather additional evidence using Epikuron-100P fractions with improved functional properties. Thus, it has been described [7] that flocculation may be prevented by a prior selective adsorption of the powdered soybean lecithin in a silica gel dispersion in a 51/41/8 (v/v/v) mixture of hexane, 2-propanol and water. Preliminary experiments indicated that the dynamic interfacial tension behavior was largely affected by this treatment: the second stage of the adsorption process was almost completely ruled out, so that the long-time interfacial tension values were much higher. A more thorough investigation of this phenomenon will be the subject of future research.

Acknowledgements The authors gratefully acknowledge financial support for this study by the Belgian National Fund for Scientific Research (NFWO). Paul Van der Meeren is a senior research associate of the Belgian National Fund for Scientific Research.

References

1. Szuhaj B (1989) Lecithins: sources, manufacture and uses. American Oil Chemists' Society, Champaign, Illinois, 238 p
2. Weber E (1981) J Am Oil Chem Soc 58:898–901
3. Pardun H (1989) Fat Sci Technol 91:45–58
4. Rydhag L (1979) Fette Seifen Anstrichm 81:168–173
5. Rydhag L, Wilton I (1981) J Am Oil Chem Soc 58:830–837
6. Kaufmann P, Olsson U, Herslof B (1990) J Am Oil Chem Soc 67:537–540
7. Van der Meeren P, Stastny M, Vanderdeelen J, Baert L. In: Cevc G, Paltauf F (eds.) Phospholipids – characterization, metabolism and novel biological applications. American Oil Chemists' Society, Champaign (in press)
8. Hofman J, Stein H (1991) J Colloid Interface Sci 147:508–516
9. Johnson M, Saunders L (1973) Chem Phys Lipids 10:318–327
10. Shchipunov Y, Kolpakov A (1991) Adv Colloid Interface Sci 35:31–138

Progr Colloid Polym Sci (1995) 98:140–144
© Steinkopff Verlag 1995

F. Carrique
A.V. Delgado

Effect of the counter- and co-ion valencies on the complex dielectric constant of a colloidal suspension

F. Carrique
Departamento de Fisica Aplicada I
Facultad de Ciencias
Universidad de Málage
29071 Málaga, Spain

Dr. A.V. Delgado (✉)
Departamento de Fisica Aplicada
Facultad de Ciencias
Universitadad de Granada
18071 Granada, Spain

Abstract In this work, we show results based on the model proposed by DeLacey and White concerning the effect of the valency and mobility of ions in solution on the dielectric response of a dilute colloidal suspension. This subject is of interest because it may help to get a better insight into the detailed mechanisms of dielectric response in suspensions; the calculations shown are expected to be found both experimentally and in other theoretical treatments. Our results indicate that an increase in the counterion valency, at constant ζ and κa, leads to larger real and imaginary parts of the dielectric constant. The same result is found for changes in the

coion valency: although the variations are smaller, the assumption of symmetrical electrolyte (neglecting the role of counterions) can be in serious error when dealing with dielectric phenomena. Even if the charge of the ions is the same, different mobilities or drag coefficients also affect the dielectric constant of the suspensions, due to their effects on the polarizability of the double layer.

Key words Dielectric relaxation in colloids – dielectric constant of suspensions – standard electrokinetic model

Introduction

The strong dielectric dispersion shown by colloidal suspensions, in the low frequency range of the applied AC electric field, is well documented, and has been related to the polarization of the electric double layers surrounding the colloidal particles [1–3]. Such polarization is, in general, out of phase with the applied field, this giving rise to a noticeable frequency dependence of the complex dielectric constant of the suspensions; the dielectric behavior of the latter will depend on the characteristics of the particles (size, shape, conductivity), the solution (concentrations, valencies and mobilities of the ions), and the particle/solution interface (surface potential).

The standard model of the dielectric response and the conductivity of dilute colloidal suspensions was elaborated

by DeLacey and White [4], and does not have essential limitations as to the values of κa (κ being the reciprocal double layer thickness, and a the radius of the spherical dispersed particles), zeta potential, ζ, of the interface, or concentrations, valencies or mobilities of ions in solution. The double layer is supposed to be polarizable only in its external or diffuse part, through ionic transport processes taking place in this region.

Since the equations governing the model must be numerically solved, some analytical or semianalytical approximations have been proposed that are valid for either thin double layers ($\kappa a \gg 1$), small zeta potentials, or symmetrical electrolytes composed of ions with equal drag coefficients [5–9]. Rosen et al. have recently elaborated a rigorous generalization of the DeLacey and White's Model (DW hereafter) incorporating mechanisms of adsorption and lateral ionic transport in the inner part of the

Progr Colloid Polym Sci (1995) 98:140–144
© Steinkopff Verlag 1995

double layer (dynamic Stern layer or DSL model). Such generalization gives in many cases a better quantitative agreement with experimental data as compared to DW's predictions [10, 11], and, from a qualitative point of view, it keeps the shape of the relaxation curves and the trends of variation of the dielectric parameters with ζ and κa, or the ionic composition of the liquid medium [12,13]. The thin double layer version of the model has been independently elaborated by Kijlstra et al. [14].

Little has been investigated on the analysis of the effects of counterions and coions of different valencies on the dielectric parameters of the suspensions [4]. In this work, our aim is to quantitatively study such effects, keeping both ζ and κa fixed; we will hence be able to isolate the importance of the ionic composition of the dispersion medium from other influences. Particular emphasis will be placed on the effect of coions, very often neglected when studying colloidal phenomena [15].

Notation

According to the DW model [4], the complex conductivity $K^*(\omega)$ of a dilute colloidal suspension in the presence of an alternating electric field of frequency ω can be expanded in powers of the volume fraction of solids, ϕ, as follows:

$$K^*(\omega) = K_e^*(\omega) + \phi \Delta K^*(\omega) + 0(\phi^2) , \qquad (1)$$

where $K_e^*(\omega)$ is the complex conductivity of the electrolyte, and $\Delta K^*(\omega)$ is the first order coefficient measuring the effect of the particles on the bulk conductivity of the suspension. $K^*(\omega)$ and $K_e^*(\omega)$ can be expressed in terms of their real and imaginary parts:

$$K^*(\omega) = K(\omega) + \omega\varepsilon_0 \varepsilon_r''(\omega) + i\omega\varepsilon_0 \varepsilon_r'(\omega) \qquad (2)$$

$$K_e^*(\omega) = K^\infty + i\omega\varepsilon_{re}\varepsilon_0 , \qquad (3)$$

$K(\omega)$ being the conductivity contribution not related to dielectric loss; $\varepsilon_r'(\omega)$ and $\varepsilon_r''(\omega)$ are, respectively, the real and imaginary parts of the complex dielectric constant of the suspension; ε_0 is the permitivity of a vacuum, ε_{re} is the dielectric constant of the electrolyte solution, and K^∞ its DC conductivity. For dilute suspensions [4]:

$$K(\omega) = K^\infty + \phi \Delta K(\omega) + 0(\phi^2) \qquad (4)$$

$$\varepsilon_r'(\omega) = \varepsilon_{re} + \phi \Delta \varepsilon_r'(\omega) + 0(\phi^2) \qquad (5)$$

$$\varepsilon_r''(\omega) = \phi \Delta \varepsilon_r''(\omega) + 0(\phi^2) , \qquad (6)$$

and, from Eqs. (1–6):

$$\Delta K^*(\omega) = \Delta K(\omega) + \omega\varepsilon_0 \Delta \varepsilon_r''(\omega) + i\omega\varepsilon_0 \Delta \varepsilon_r'(\omega) . \qquad (7)$$

In this work, we will evaluate $\Delta \varepsilon_r'(\omega)$, and $\Delta \varepsilon_r''(\omega)$, for given κa and zeta potential, assuming solutions containing ions

of different valencies and similar equivalent conductances (K^+, Ba^{2+}, La^{3+}), or with the same valency and different mobilities (Cl^-, NO_3^-, $C_2H_3O_2^-$). We have thus the possibility of theoretically analyzing the relative influence, on the above mentioned dielectric quantities, of a wide variety of liquid medium compositions.

Results and discussion

Effect of ionic valencies

Figure 1 shows the frequency dependence of $\Delta \varepsilon_r'$ for fixed κa, in solutions of KCl, BaCl$_2$, and LaCl$_3$ for ζ values of -50 mV (K^+, Ba^{2+}, La^{3+}: counterions; Cl^-: coion), and $+50$ mV (K^+, Ba^{2+}, La^{3+}: coions; Cl^-: counterion). Note that $\Delta \varepsilon_r'$ increases, for a given frequency, with the counterion valency, due to the larger polarizability of the double layer for higher ionic valencies; this fact is a consequence of the existence of a larger charge in the double layer when the valency of the counterions in the latter is increased. As observed, the effect is more important when it is the counterion valency that is increased ($\zeta = -50$ mV), in agreement with the admitted major role played by counterions as compared to coions in most colloidal phenomena. However, the coion effect cannot be

Fig. 1 Real part of the increment of the dielectric constant of a suspension plotted as a function of frequency for the electrolytes indicated. Open symbols: $\zeta = -50$ mV; full symbols: $\zeta + 50$ mV

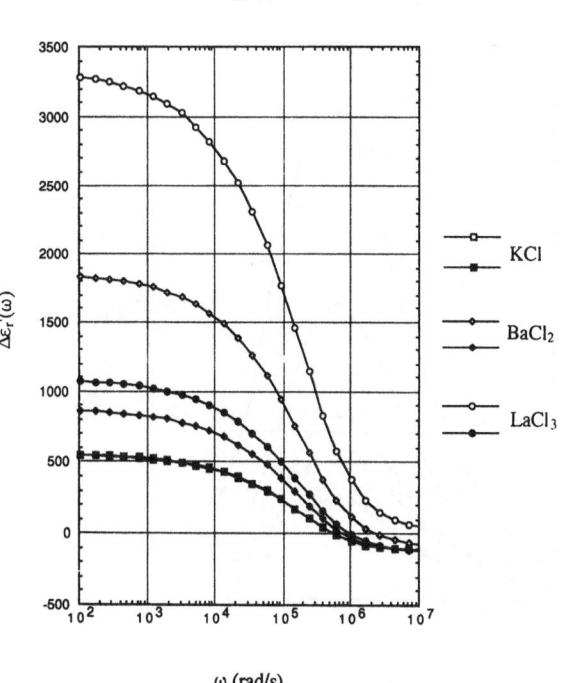

$\kappa a = 10$

KCl

BaCl$_2$

LaCl$_3$

ω (rad/s)

completely neglected, since the variations of $\Delta \varepsilon_r'$ with ω for coions K^+, Ba^{2+}, and La^{3+} are clearly different: the polarizability of the double layer increases with the valency of the coion, in spite of the lower mobile charge in the double layer for higher coion valencies (negative adsorption of coions), for fixed ζ and κa.

The variation of the quantity $\Delta \varepsilon_r''$ with frequency is depicted in Fig. 2, for the same conditions as in Fig. 1; as before, $\Delta \varepsilon_r''$ increases with cation valency as a consequence of the increased polarizability of the double layer, and also in this case the variations are less significant when cations act as coions. It is also worth to note that the frequency observed for the maxima in $\Delta \varepsilon_r'' - \omega$ curves (relaxation frequency) is independent of the cation valency. This can be justified by the following approximate expression for the relaxation frequency, ω_r [4]:

$$\omega_r \simeq \frac{k_B T}{2\pi (a + \kappa^{-1})^2 \lambda} , \qquad (8)$$

where λ is the counterion drag coefficient, and the other symbols have their usual meaning. From Eq. (8), if all the counterions in solution have similar drag coefficients (as in the case of Cl^-, K^+, Ba^{2+}, and La^{3+}), ω_r will also be the same if the value of κ is not changed.

Influence of counterion and coion mobilities on the dielectric response

We will now consider to what extent variations in the drag coefficient of either counter- or co-ions affect the quantities $\Delta \varepsilon_r'$ and $\Delta \varepsilon_r''$, representing the dielectric behavior of the suspensions. We will assume $\kappa a = 10$, with solutions of $BaCl_2$, $Ba(NO_3)_2$, and $Ba(C_2H_3O_2)_2$ for both positive (anions = counterions) and negative (anions = coions) zeta potentials, with $|\zeta| = 50$ mV in both cases. Since the valencies are the same in all such solutions, there will be no differences in the total mobile charge in the double layer between the solutions for each ζ-potential case, and any variation in dielectric behavior must be ascribed to the differences in ionic mobilities (as measured by their limiting equivalent conductances, Λ_i: $\Lambda_{Cl^-} = 75.5$; $\Lambda_{NO3^-} = 70.6$; $\Lambda_{C2H3O2^-} = 40.8$, at 25°C, in units of $Scm^2 eq^{-1}$).

The ω-dependence of $\Delta \varepsilon_r'$ for the six cases considered is shown in Fig. 3. When the zeta potential is positive, $\Delta \varepsilon_r'(\omega)$ almost coincides for Cl^- and NO_3^-, whereas it is somewhat higher (at low frequencies) for acetate counterions. On the other hand, for negative ζ-potential, $\Delta \varepsilon_r'$ is lower for acetate coions for any frequency below 10^7 s^{-1}. As a matter of fact, it is difficult to predict what the effect may be of ion mobilities on the dielectric quantities characterizing a suspension: the drag coefficient or equivalent conduc-

Fig. 2 Imaginary part of the increment of the dielectric constant of a suspension plotted as a function of frequency for the electrolytes indicated. Open symbols: $\zeta = -50$ mV; full symbols: $\zeta = +50$ mV

Fig. 3 Same as Fig. 1, for the electrolytes indicated

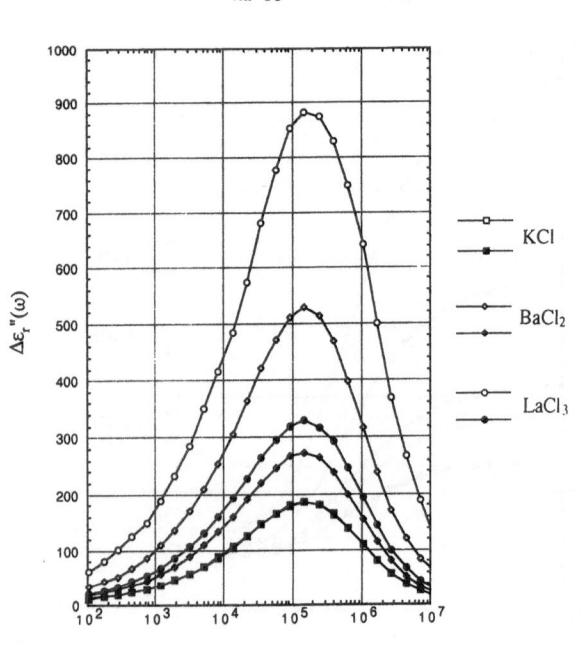

κa=10

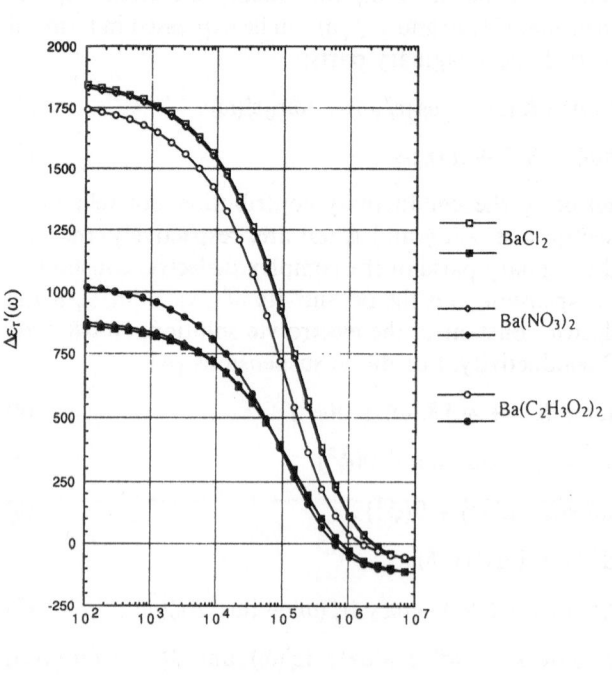

κa=10

Progr Colloid Polym Sci (1995) 98:140–144
© Steinkopff Verlag 1995

tance of each ionic species is involved in a complicated manner in the balance equations of hydrodynamic, electrostatic, and thermodynamic forces [4]. However, some interpretation can be given for the curves shown in Fig. 3.

According to the classical Gouy–Chapman model [15], the densities (number per unit volume) of counterions of equal valency are identical, irrespective of their size and mobility, in the absence of an applied electric field. Let us now consider an applied field of magnitude E and zero frequency (we are interested in the low-frequency dielectric behavior of suspensions) for the case $\zeta = 50$ mV; the field will provoke an unequal distribution of counterions that can be computed by numerical integration of the DW body of equations [16]. The quantity of interest will be:

$$A \left[\frac{\delta n_j(x) - \delta n_{Cl^-}(x)}{\eta_j^0(x)} \right], \qquad (9)$$

where δn_j is the perturbation of the density of the j-th ionic species (relative to the common equilibrium value, $n_j^0(x)$, of all the anions), calculated at a reduced distance $x = \kappa r$ (r: modulus of the radius vector $\mathbf{r}$ with origin in the particle centre). The factor A is given by:

$$A = \frac{\kappa a k_B T \cdot 10^5}{Eae}, \qquad (10)$$

e being the elementary charge. The results are shown in Fig. 4 for $\theta = 0$ and $\theta = \pi$, with θ the angle between $\mathbf{r}$ and the field. As observed, in the limit of low frequencies the ionic perturbation of acetate ion is much larger than that of Cl^-, the difference being almost zero for nitrate anions. These data explain the larger polarizability of the double layer in acetate solutions, and hence the larger values of $\Delta\varepsilon_r'$ (Fig. 3).

The case of negative zeta potential can be understood with reference to Fig. 5, where the perturbations in ionic densities compared to $C_2H_3O_2^-$ coions are plotted vs. the reduced distance x. Note that the perturbations of n_{Cl^-} and $n_{NO_3^-}$ are very similar, and larger in absolute value than that of acetate ions. As a consequence, in this case the polarizability of the system will change in the order $Cl^- \geq NO_3^- > C_2H_3O_2^-$, in agreement with results in Fig. 3.

These arguments are perfectly applicable to the other quantity of interest, $\Delta\varepsilon_r''$, plotted in Fig. 6 for the same cases considered in Fig. 3: the order of variation of $\Delta\varepsilon_r''$ with the anions Cl^-, NO_3^-, and $C_2H_3O_2^-$ is coincident with that found for $\Delta\varepsilon_r'$, both for positive and negative ζ. It is worth to mention that the relaxation frequency is lower for solutions containing acetate than for the other anions.

Fig. 4 Perturbation of the ionic densities (relative to Cl^- perturbation) divided by the equilibrium ion density, for nitrate and acetate ions, calculated in the same ($\theta = 0$) and opposite ($\theta = \pi$) directions to the applied field direction. Abscissa: reduced distance to the particle center ($x = \kappa r$)

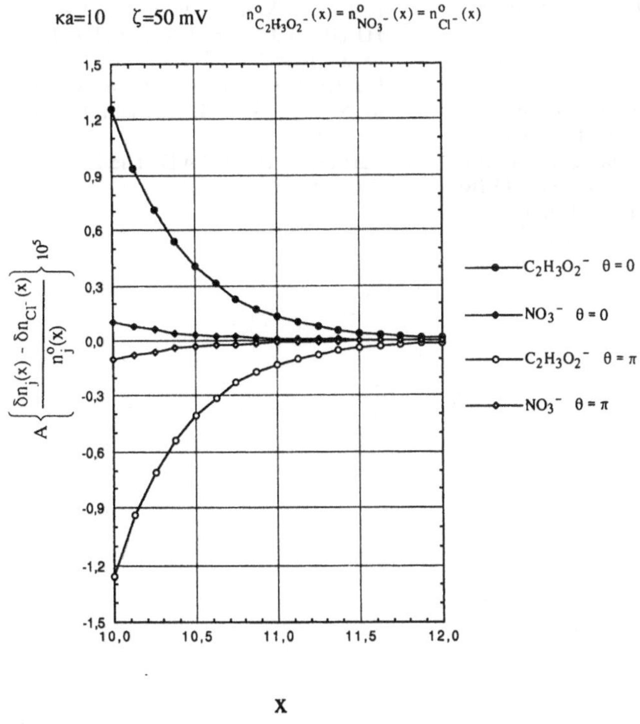

Fig. 5 Perturbation of the ionic densities (relative to acetate perturbation) divided by the equilibrium ion density, for nitrate and chloride ions, calculated in the same ($\theta = 0$) and opposite ($\theta = \pi$) directions to the applied field direction. Abscissa: reduced distance to the particle center ($x = \kappa r$)

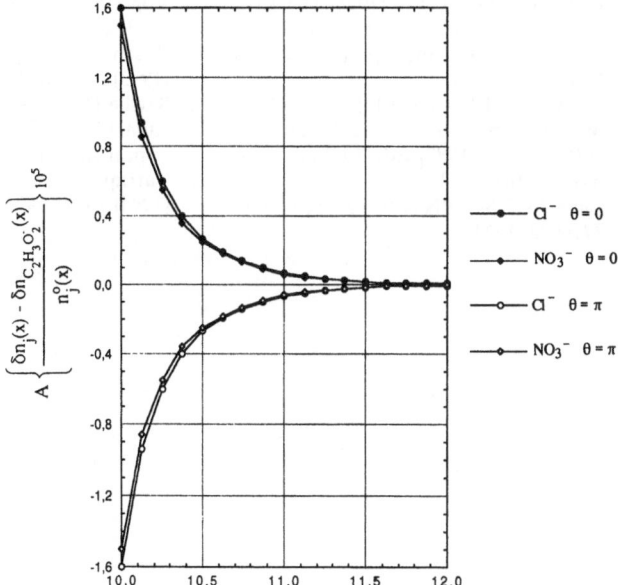

κa=10

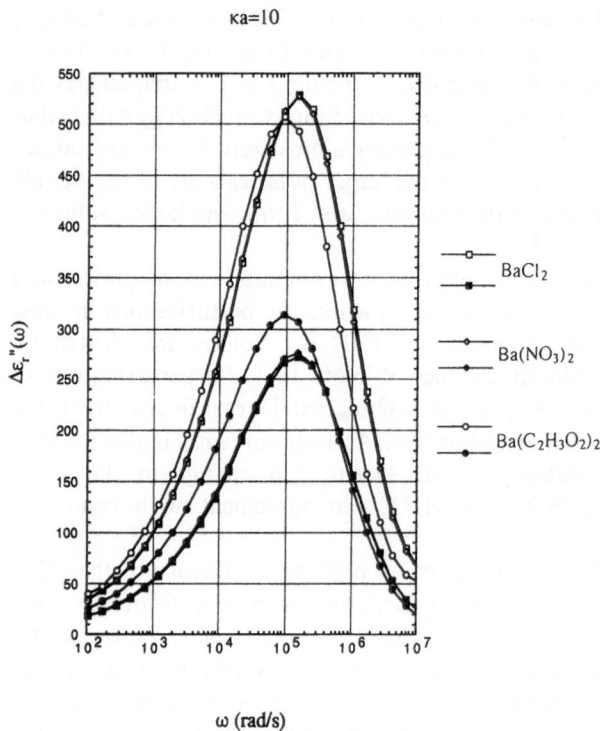

ω (rad/s)

Fig. 6 Same as Fig. 2, for the electrolytes indicated

The fact that the drag coefficient of acetate is larger (or its limiting equivalent conductance lower) than those of Cl^- or NO_3^- can explain such result. This system must show relaxation at lower frequencies, since the relaxation mechanisms are basically diffusive in the electric double layer [3, 4, 7]. This applies to both positive and negative zeta potentials.

Conclusions

The results shown, exploring some aspects of the dielectric relaxation in colloids show that the effect of both counter- and co-ions are significant in the explanation of the dielectric response of a suspension in the presence of AC fields. Both the real and imaginary parts of the dielectric constant are clearly different for different ions (with either equal or opposite charge as the particle surface), affecting the dielectric constant because of their charge and of their mobility.

Acknowledgements Financial support from Fundación Ramón Areces, Spain, and Junta de Andalucía, Spain (Groups No. 6046 and 1168) is gratefully acknowledged.

References

1. Schwarz G (1962) J Phys Chem 66:2636–2642
2. Schurr JM (1964) J Phys Chem 68:2407–2413
3. Dukhin S, Shilov VN (1974) Dielectric phenomena and the double-layer in disperse systems and polyelectrolytes. Wiley, New York
4. DeLacey EHB, White LR (1981) J Chem Soc Faraday Trans 2, 77:2007–2039
5. O'Brien RW (1982) Adv Colloid Interface Sci 16:281–320
6. Chew WC, Sen PN (1982) J Chem Phys 77:4683–4693
7. Fixman M (1983) J Chem Phys 78:1483–1491
8. Chew WC (1984) J Chem Phys 80:4541–4549
9. Vogel E, Pauli H (1988) J Chem Phys 89:3823–3829
10. Rosen LA, Baygents JC, Saville DA (1993) J Chem Phys 98:4183–4194
11. Saville DA (1993) International Symposium Electrokinetic Phenomena '93 (to appear in Colloids and Surfaces)
12. Carrique F, Zurita L, Delgado AV (1994) J Colloid Interface Sci in press
13. Carrique F, Zurita L, Delgado AV (1994) Colloids & Surfaces (submitted)
14. Kijlstra J, Van Leeuwen HP, Lyklema HJ (1992) J Chem Soc Faraday Trans 88:3441–3449
15. Hunter RJ (1987) Foundations of Colloid Science, Vol. 1. Clarendon Press, Oxford
16. Carrique F (1993) Ph.D. Thesis University of Granada, Spain

Progr Colloid Polym Sci (1995) 98:145–150
© Steinkopff Verlag 1995

A. Quirantes
A.V. Delgado

Particle size determinations in colloidal suspensions of randomly oriented ellipsoids

A. Quirantes
Departamento De Física Aplicada
E.U. Politécnica De Jaén
23071 Jaén, Spain

Dr. A.V. Delgado (✉)
Departamento de Física Aplicada
Facultad de Ciencias
Universidad de Granada
18071 Granada, Spain

Abstract The T-matrix (or EBCM) method, which relates the expansion coefficients of the incident and scattered electric fields, has been developed to account for light scattering experiments where the particles are not assumed to be spherical. It is specially suitable for axisymmetric particles, since in this case the T matrix is decoupled into smaller, independent submatrices. A series of experimental measurements of light scattering have been carried out on spherical (polystyrene) and nonspherical (hematite) systems, and the results have been compared to the T-matrix theory. Theoretical fits show good agreement between the size-shape parameters obtained by this method as compared to electron microscopy, thus indicating that the procedure can be used to infer size and shape of suspended particles. Systems ranging from atmospheric aerosols to interstellar clouds can be studied in this way.

Key words Light scattering – colloidal ellipsoids – EBCM theory

Introduction

Light scattering (LS) techniques make an important procedure for size determination of colloidal suspensions since they are nondestructive, simple and able to handle physically inaccesible systems, in situ and in real time. Any LS property (e.g., polarization of scattered light at a certain angle) is a function of several parameters: shape, size, composition, etc. If adequate theories are at hand, geometric features could in principle be found by making multiple fits until the size/shape/composition is found that gives closest values of that property respective to those found experimentally. This would let us characterize a suspension of particles by its LS capabilities.

This approach, valid in theory, is not directly applicable: a fit based on the variation of characteristic parameters (refraction indexes, mean size, nonsphericity) would yield an extremely complex and computationally demanding multidimensional analysis, which in the general case

would be non-univocal (completely different particles can yield identical values of the LS properties considered). For that reason, it is a common norm to fix some of the parameters and vary the others [1–3]. In this work, size and particle/medium refractive indexes have been considered as fixed; intensity – angle variation is the measured LS property, although it is not the only one that can be used.

In order to calculate theoretical LS values, the T-matrix, or EBCM (for Extended Boundary Condition Method) theory has been used [4, 5]. In this theory, both incident $E_i(r)$ and scattered $E_s(r)$ fields are described as an expansion of vector harmonic spherical functions as follows [6]:

$$E_i(r) = \sum_{n=1}^{\infty} \sum_{m=-n}^{n} [a_{mn} Rg\, M_{mn}(kr) + b_{mn} Rg\, N_{mn}(kr)]$$

$$E_s(r) = \sum_{n=1}^{\infty} \sum_{m=-n}^{n} [p_{mn} M_{mn}(kr) + q_{mn} N_{mn}(kr)],$$

(1)

where $k = 2\pi/\lambda$ is the wavenumber corresponding to a wavelength λ, and M_{mn}, N_{mn} are the vector spherical harmonic functions (finite at the origin) whose radial dependences are represented by either spherical Hankel functions $h_n^{(1)}$ (M_{mn}, N_{mn}) or spherical Bessel functions (RgM_{mn}, RgN_{mn}). Linearity of Maxwell's equations means that the relation between the expansion coefficients of the incident (a, b) and scattered (p, q) fields is lineal and is given by a transition matrix, the so-called T matrix:

$$\begin{bmatrix} p \\ q \end{bmatrix} = T \begin{bmatrix} a \\ b \end{bmatrix}. \qquad (2)$$

The elements of the T matrix depend only upon the shape, size parameter and refractive index (relative to the suspending medium) of the scattering particle, as well as upon its orientation respective to a reference system. If the (a, b) coefficients of the incident wave are known and the T matrix is computed for a type of particles, the (p, q) coefficients of the scattered field expansion can be obtained and, therefrom, any property of interest such as cross-sections, scattered intensity at different angles, absorption coefficients, etc. can be calculated. An extension of this method to systems of randomly oriented particles can be made either by numerical integration to every possible orientation angle, or by proper, more subtle analytical developments [4, 7]. Although the T-matrix method can be applied to every shape, it is most efficient in the case of axisymmetric particles, since in this case the transition T matrix divides itself into smaller, independent submatrices [8].

The aim of this work is to analyze the size and shape characteristics of ellipsoidal colloidal particles by means of EBCM computations of the angular dependence of the intensity of light scattered by a suspension of such particles. The accuracy of our size estimations will be checked against electron microscopy data; as a further check, the size distribution of spherical polystyrene particles will also be studied by the same method.

Experimental

Materials

The polystyrene particles were synthesized, in the absence of surfactants, following the method proposed by Goodwin et al. [9]. The dispersion obtained was cleaned by repeated centrifugation-redispersion cycles and serum replacement until the conductivity of the supernatant was equal to that of water (double distilled, deionized and filtered in a Milli-Q Reagent Water System, Millipore). Scanning electron micrographs showed considerable sphericity and monodispersity of the particles, as shown in

Fig. 1, where a histogram of diameters is also included. We assumed a zeroth-order logarithmic distribution of diameters, given by [10]:

$$p(d) = \frac{1}{\sqrt{2\pi}\sigma_0 d_m e^{\sigma_0^2/2}} \exp\left[\frac{-(\ln d - \ln d_m)^2}{2\sigma_0^2}\right], \qquad (3)$$

where d_m and σ_0 are, respectively, the modal diameter and width of the distribution. Fitting of this theoretical distribution to the electron microscopy data yields $d_m = 610$ nm, $\sigma_0 = 0.034$; from these data, the average diameter is found to be 610 ± 20 nm.

Ellipsoidal hematite particles were obtained by precipitation from homogeneous solutions, according to the procedures described by Ozaki et al [11] and Morales et al [12]. The cleaning method used in this case consisted of repeated centrifugation-redispersion in water; the process was considered finished when the isoelectric point of the particles, as determined by electrophoretic mobility measurements (Malvern Zetasizer 2c), coincided with the accepted value for hematite (pH 7–8). Figure 2 is a transmission micrograph of the particles. The two parameters used to describe the ellipsoids are the equivalent-volume-sphere diameter D_{eq}, and the eccentricity or axial ratio $\varepsilon = a/b$, where b is the revolution semiaxis; for prolate ellipsoids, $\varepsilon < 1$. The relation $D_{eq} = 2b\varepsilon^{2/3}$ is straightforward.

Histograms corresponding to the distributions of equivalent-volume diameter and eccentricity are also included. From these data, the hematite sample will be characterized by $D_{eq} = 134 \pm 18$ nm, $\varepsilon = 0.26 \pm 0.02$, or $2b = 330 \pm 50$ nm, $2a = 86 \pm 12$ nm. Although the sample is not completely monodisperse, we will not try to obtain any distribution of size parameters, and will only analyze their mean values.

Intensity vs. angle measurements of these colloidal systems were obtained with a Malvern 4700 PCS analyzer using a 75 mW 2213 Argon Ion laster (Cyonics); for this wavelength (488 nm) the refractive indexes of the particles and the medium are: 1.611 (polystyrene), $3.101 + i\,0.481$ (hematite) and 1.337 (water) [1, 13]. The incident light was vertically polarized. In order to avoid multiple scattering as much as possible, the volume fraction of he suspensions was always kept low ($\phi \leq 10^{-5}$).

Results and discussion

Polystyrene

The use of the EBCM method means a tremendous power in the analysis of light scattering data; even the sphericity of the colloidal particles need not be assumed, but can be

Progr Colloid Polym Sci (1995) 98:145–150
© Steinkopff Verlag 1995

Fig. 1 Scanning electron micrograph and diameter histogram of polystyrene spheres

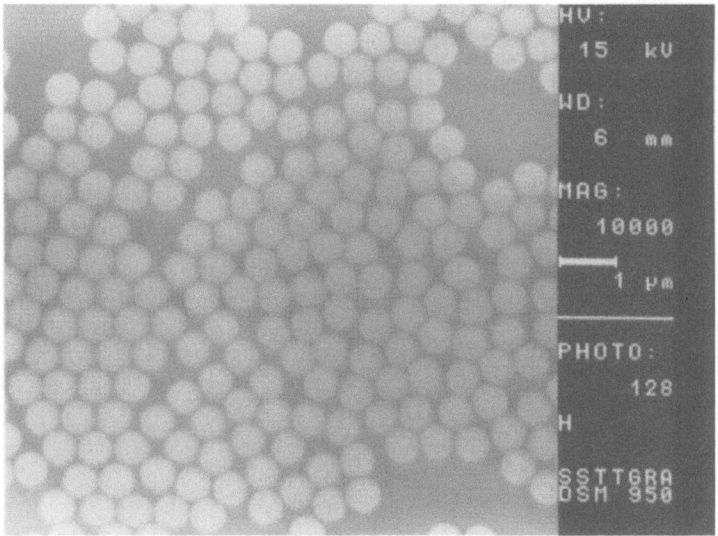

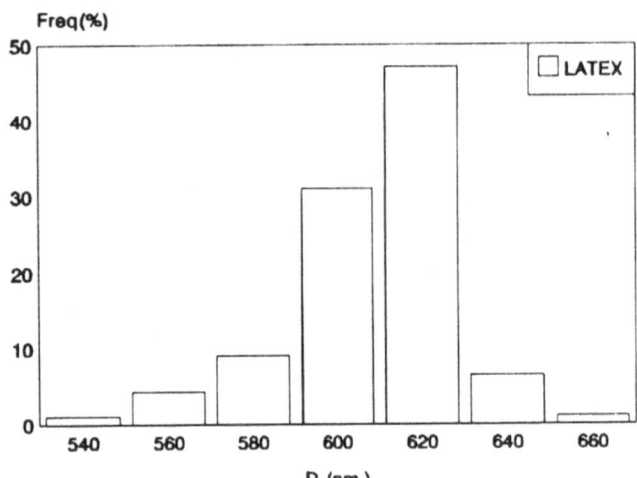

obtained as a result of the scattering pattern of the suspension. However, the computation capacity needed is much higher, and the problem of polydispersity becomes almost untractable. In order to check the capabilities of this formalism, we will perform an analysis of the scattering results obtained with spherical polystyrene using both the classical Mie theory (considering polidispersity) and the EBCM theory (where the mean equivalent diameter and axial ratio will be the only variables).

Figure 3 shows the angular distribution of scattered intensity (relative to the value obtained for 60° scattering angle) for the polystyrene latex sample. The best theoretical fit deduced from Mie theory is also included for comparison; as observed, the agreement between experimental and theoretical data is excellent. The best-fit Mie

curve was calculated with the ZOLD parameters that minimized the ERR parameter defined by:

$$\text{ERR} = \sum_{k=1}^{n} \left[\frac{i_e(\theta_k) - i_t(\theta_k)}{i_e(\theta_k)} \right]^2, \tag{4}$$

where $i(\theta) = I(\theta)/I(60°)$ and the subscripts e, t refer to the experimental and theoretical determinations, respectively. For example, Fig. 4 shows the variation of ERR with d_m for different σ_0 values; an absolute minimum of ERR is found for $d_m = 579$ nm, $\sigma_0 = 0.01$, corresponding to a mean diameter of 580 nm and a standard deviation of 6 nm, in very good agreement with electron microscopy data.

When the same set of scattering data is analyzed with the EBCM formalism, the ERR parameter depends on

Fig. 2 Transmission electron
micrograph and histograms of
equivalent-volume diameter
and axial ratio of ellipsoidal
hematite particles

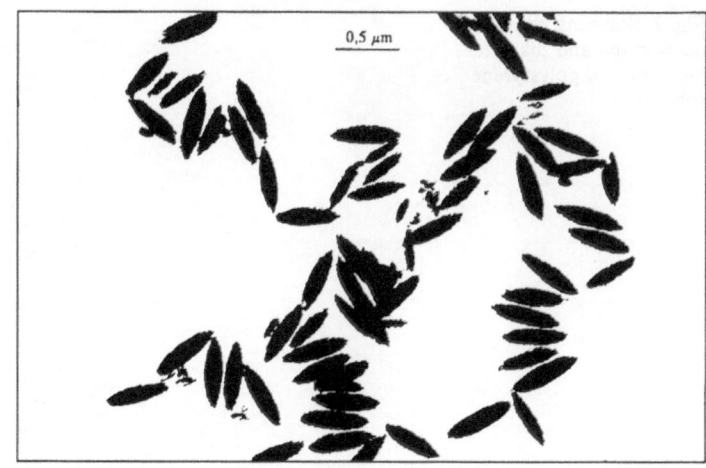

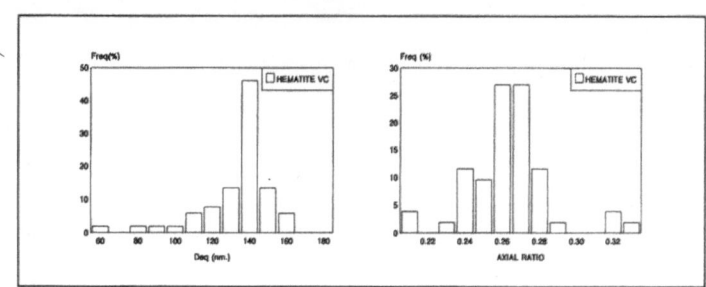

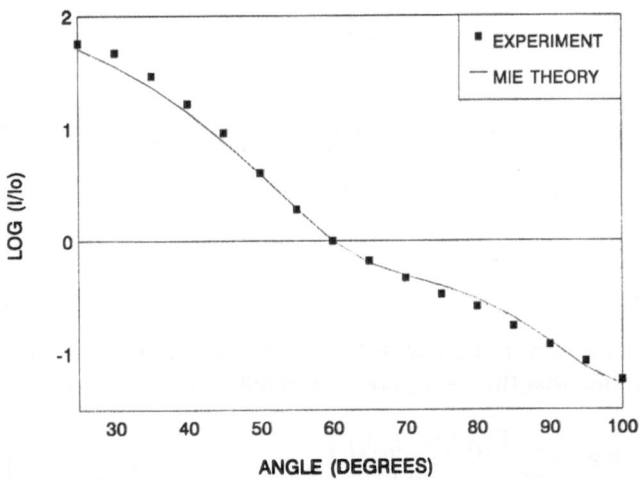

Fig. 3 Experimental and theoretical (Mie) dependence of the intensity scattered by a polystyrene suspensions

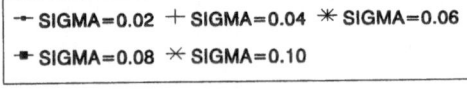

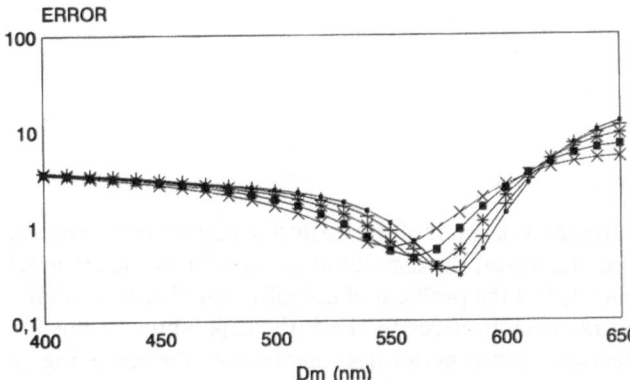

Fig. 4 Values of ERR (see text) for polystyrene suspensions, plotted as a function of modal diameter for different widths of the ZOLD distribution (Mie theory)

the equivalent radius and the axial ratio as shown in Fig. 5. A minimum is again obtained for $\varepsilon = 1.0$ and $D_{eq} = 581$ nm; hence, this data treatment confirms the sphericity of polystyrene spheres and also yields a very reasonable estimation of their average diameter.

Hematite

It is to be expected that the EBCM method will show its capabilities when working with truly ellipsoidal particles of a suitable size. Hematite suspensions prepared as de-

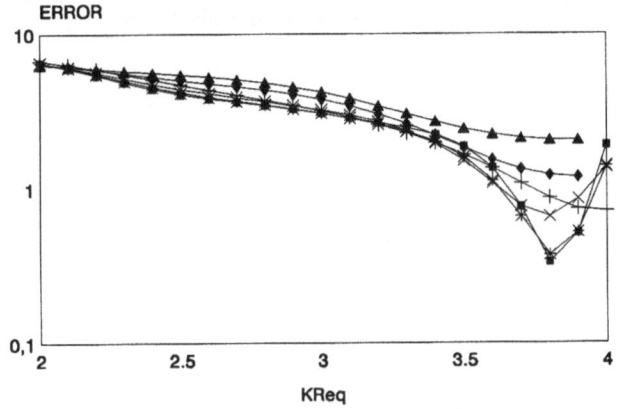

Fig. 5 Values of ERR for polystyrene suspensions as a function of the dimensionless equivalent radius for different axial ratios (EBCM formalism)

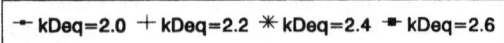

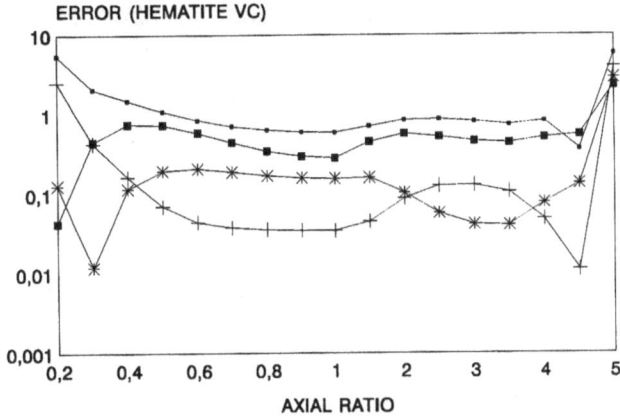

Fig. 7 Values of ERR for hematite suspensions as a function of the axial ratio for several dimensionless size parameters (EBCM formalism)

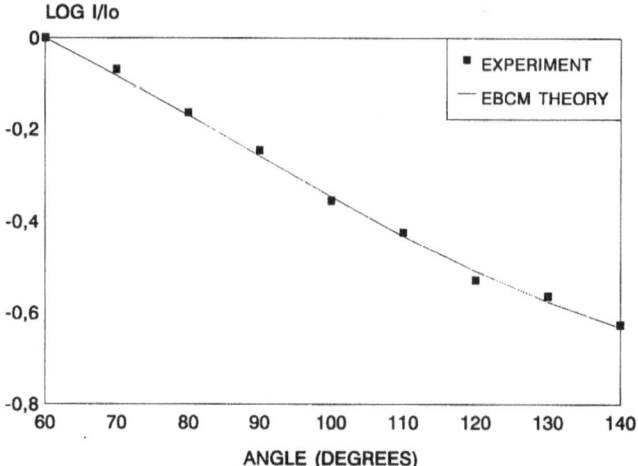

Fig. 6 Experimental and theoretical (EBCM) dependence of the intensity scattered by an ellipsoidal hematite suspension

computed by minimizing ERR as described, an excellent fit being again obtained. In Fig. 7, we have plotted the latter quantity as a function of ε for different values of kD_{eq}; it is interesting to note that ERR reaches relative minima for $D_{eq} = 138$ nm and two essentially reciprocal values of ε, 0.3 and 4.5. This suggests that the characteristics of the scattering of light by prolate and oblate ellipsoids are such that, in a single classical scattering experiment, it is not possible to distinguish between the two geometries. Data obtained with depolarized light scattering [14] allow to carry out such distinction; in our case, the comparison between EBCM predictions and electron micrographs indicates that the particles are, in fact, prolate ellipsoids. Thus, our analysis of the scattering of light by hematite suspensions gives size and shape parameters ($D_{eq} = 138$ nm, $\varepsilon = 0.3$) very close to those directly observed in the microscope ($D_{eq} = 134$ nm, $\varepsilon = 0.26$).

Acknowledgements Financial support for this work by Fundación Ramón Areces, Spain, is gratefully acknowledged.

scribed above appear as an ideal colloidal system in which the achievements of the theory can be checked on a sound basis. Figure 6 shows the angular dependence of scattered intensity for our hematite sample; the continuous line was

References

1. Kerker M (1969) The Scattering of Light and Other Electromagnetic Radiation. Academic Press, San Diego
2. Lettieri TR, Hartman AW, Hembree GG, Marx E (1991) J Res Natl Inst Stand Technol 96:669–691
3. Ray AK, Souyri A, Davis EJ, Allen TM (1991) Appl Opt 30:3974–3983
4. Barber PW, Hill SC (1990) Light Scattering by Particles: Computational Methods. World Scientific, Singapore
5. Waterman PC (1971) Phys Rev D 3:825–839
6. Stratton JA (1941) Electromagnetic Theory. McGraw Hill, New York
7. Mishchenko MI (1991) J Opt Soc Am A 8:871–882

8. Tsang L, Kong JA, Shin RT (1984) Radio Science 19:629–642
9. Goodwin JW, Hearn J, Ho CC, Ottewill RH (1973) Brit Polym J 5:347–353
10. Espenscheid WF, Kerker M, Matijevic E (1964) J Phys Chem 68:3093–3097
11. Ozaki M, Kratohvil S, Matijevic E (1968) J Coll Int Sci 26:62–69
12. Morales MP, González-Carreño T, Serna CJ (1992) J Mater Res 7:2538–2545
13. Kerker M, Scheiner P, Cooke DD, Kratohvil JP (1979) J Coll Int Sci 71:176–187
14. Quirantes A (1994) Ph.D. Thesis. University of Granada, Spain

Progr Colloid Polym Sci (1995) 98:151–154
© Steinkopff Verlag 1995

C. Regnaut
S. Amokrane
P. Bobola

Analysis of the solvent mediated contribution to the effective adhesion of colloidal dispersions using Baxter's model

Dr. C. Regnaut (✉) · S. Amokrane
Laboratoire de Physique des Colloides
Faculté des Sciences et de Technologie
Université de Paris, 12 Val de Marne
Av. du G. de Gaulle
94010 Créteil, France

P. Bobola
Laboratoire de Chimie Physique
URA 176 CNRS
Université Pierre et Marie Curie
11, rue Pierre et Marie Curie
75231 Paris Cedex 05, France

Abstract The influence of the smaller particles in a mixture of asymmetrical adhesive spheres on the structure and stability of the suspension is investigated. From the Ornstein Zernike equations in reciprocal space and the Percus Yevick approximation, an analytical expression of the effective adhesiveness coefficient of the solute spheres is obtained in the asymptotic limit of vanishing small particles/solute size ratio. Otherwise, a numerical solution is obtained at non-zero size ratio. The steric effect, and like-particles stickiness are found to favor suspension instability while hetero-stickiness is found to act in one side or in the opposite depending on the concentration of the small particles. This model is used for discussing the competition between solvent-solvent and solute-solvent stickiness and the resulting solute-solute effective adhesiveness as observed in sterically stabilized suspensions for example.

Key words Adhesive spheres – solvation – depletion forces

Introduction

Baxter's adhesive sphere model has been widely used in the study of colloidal dispersions, i.e., those for which the picture of spherical solute particles with short range effective attractions immersed in a continuous solvent is expected to be reasonable [1]. In spite of the highly idealized form of the sticky potential and the limitations implied by the Percus–Yevick approximation (PYA), such an approach has been found useful for rationalizing the analysis of many experimental data on sterically stabilized suspension and reverse microemulsions such as, for instance, those of SANS or SAXS, rheological and conductivity measurements [2–4]. The aim of such studies is to show that before a phase unstability, gel formation [5] or percolation transition takes place, the modifications occurring in the dispersions when cooling, changing the solvent or adding enzymes [6] can be understood from changes in the effective solute-solute attraction at short range. A drastic representation of the latter by a sticky sphere model allows to discuss the behaviors in terms of the effective stickiness $1/\tau$. All the physico-chemical specificities of the particles (solute and solvent) which determine the effective solute-solute attraction are thus incorporated in this unique parameter. Consequently, important aspects such as the relative weight of specific and non-specific contributions to the observed behavior cannot easily be extracted from such a model and are still subjects of debates. Attempts to go beyond the effective one component model are thus being made.

This work focuses on the examination of some trends in the effective solute-solute attraction which could be expected from solvent specificities such as the size of the solvent, solvent-solvent, and solvent-solute attractions. The simplest model pointing out the influence of the size of the solvent on the structure and stability of a suspension is the binary mixture of large hard spheres (solute) and small (solvent) hard spheres. In spite of the difficulties in the treatment of integral equations for mixtures of large size asymmetry, liquid state theory clearly suggests that an increasing effective adhesion between hard sphere solutes is mediated by the solvent as the solvent/solvent size ratio vanishes at constant packing fractions, eventually producing an instability when this ratio is sufficiently low [7]. An effective adhesion due to such purely steric effects is also predicted when using the PYA but, being underestimated, it does not produce the instability predicted by thermodynamically self-consistent integral equations [7]. However, incorporating a moderate solvent-solvent attraction restores instability [8], so that in this case the PYA may be less disappointing than it stands for pure hard spheres interactions, being able to reproduce, at least qualitatively, the gross structural modifications brought out by the presence of the direct attractions not incorporated in the hard sphere model.

In the following, the dispersion is considered as a solute and discrete solvent mixture with homo and hetero attractions modeled as sticky interactions. We focus here on the trends in the solute-solute structure factors without attempting any detailed thermodynamical analysis. When the solvent to solute size ratio goes to zero, asymptotical expressions are considered in the next section in order to point out which factors compete in the effective adhesion between solutes. In the third section, calculations are performed at finite size ratio for analyzing the temperature dependence of the structure factor of silica particles dispersed in benzene [2].

Model: adhesive sphere mixture in the PYA and analysis of the asymptotic expressions

The sticky potentials $u_{mn}(r)$ of an N-component mixture of adhesive spheres are defined by [9]:

$$e^{-u_{mn}(r)/kT} = \frac{D_{mn}}{12\tau_{mn}} \delta(r - D_{mn-}) + \Theta(r - D_{mn-}),\qquad(1)$$

where $\delta(r - D_-)$ and $\Theta(r - D_-)$ are the asymmetrical δ-distribution and unit step function respectively, τ_{mn}^{-1} is the stickiness parameter, and D_{mn} the mean diameter of two spheres of diameter D_m and D_n. When the PYA is assumed, the partial direct correlation functions can be obtained exactly as:

$$c_{mn}(r) = \Theta(D_{mn-} - r)c_{mn}^R(r) + \frac{\lambda_{mn}D_mD_n}{12D_{mn}} \delta(r - D_{mn-}),\quad(2)$$

where $c_{mn}^R(r)$ is a regular function [9]. The adhesiveness factors λ_{mn} are obtained in terms of D_{mn}, τ_{mn} and the packing fractions η_m by solving the system of $N(N-1)/2$

coupled quadratic equations [9]:

$$\lambda_{mn}\frac{\tau_{mn}}{D_{mn}} = x\frac{D_{mn}}{D_mD_n} + \sum_{k=1}^{N}\left(\frac{\eta_k}{2D_k}\right)$$
$$\times\left[3x^2 - x(\lambda_{mk} + \lambda_{kn}) + \frac{\lambda_{mk}\lambda_{kn}}{6}\right],\qquad(3)$$

where:

$$\eta_k = \frac{\pi}{6}\rho_k D_k^3,\qquad x = \left[1 - \sum_{k=1}^{N}\eta_k\right]^{-1}$$

and ρ_k is the number density.

For a mixture, Baxter's factorization defines the inverse partial structure factors matrix as:

$$\mathbf{S}^{-1} = \mathbf{I} - \mathbf{C} = \mathbf{Q}^\dagger\mathbf{Q},\qquad(4)$$

where $\mathbf{Q}^\dagger$ is the transposed and complex conjugated of the matrix $\mathbf{Q}$ of the Fourier transforms of the Baxter's functions obeying to the general condition det $\mathbf{Q}(0) > 0$ [9]. Expressions of these functions are well known and not reproduced here [8,9]. Let us now consider the special case of a binary mixture with large size asymmetry (1 ≡ solvent; 2 ≡ solute). When a solution of Eq. (3) exists, a careful study of these functions in the limit $D_1/D_2 \to 0$ and η_1, η_2 fixed shows that the structure factor of the large species $S_{22}(q)$ converges towards the expression relative to a one-component system of same packing fraction η_2, but with an effective adhesiveness $\lambda_{22}^{\mathrm{eff}}$ [10]. Introducing $\eta_1^* = \eta_1(1 - \eta_2)^{-1}$, the packing fraction of the small particles relative to the free volume left by the large ones and the following definition: $\lambda_{ij}^* = (1 - \eta_2)\lambda_{ij}$, it can be shown [10] that the effective adhesiveness can be written as the sum of four terms:

$$\lambda_{22}^{*\mathrm{eff}} = \lambda_{22}^* + \lambda_0^{*s} + \lambda_1^{*s} + \lambda_2^{*s}\qquad(5)$$

$$\lambda_0^{*s} = \frac{6\eta_1^*}{1 + 2\eta_1^*}\qquad(5a)$$

$$\lambda_1^{*s} = \frac{9\eta_1^{*2}\lambda_{11}^*}{(1 + 2\eta_1^*)[1 + 2\eta_1^* - \eta_1^*(1 - \eta_1^*)\lambda_{11}^*]}\qquad(5b)$$

$$\lambda_2^{*s} = \frac{\eta_1^*[-6\lambda_{12}^* + (1 - \eta_1^*)\lambda_{12}^{*2}]}{1 + 2\eta_1^* - \eta_1^*(1 - \eta_1^*)\lambda_{11}^*}.\qquad(5c)$$

This limit is, of course, relevant for the study of many colloidal suspensions with large size asymmetry, but due to the limits of the PYA it should not be taken literally as corresponding to an actually very small size ratio. Rather, it should be viewed as a convenient way for analyzing in asymmetric mixtures the qualitative trends which are not critically sensitive to the diameter ratio. As illustrated below, it should provide answers to questions such as, for instance, those relative to the role of the solvent self stickiness or that of hetero-adhesions. Besides these particular questions concerning solvent effects, salient features arising from Eq. (5) and relevant for any asymmetric adhesive spheres mixture are:

i) λ_{22}^* is the direct adhesiveness contribution arizing from the self stickiness of the large particles and adds to the three remaining terms mediated by the small particles.

ii) λ_0^{*s} is always positive and corresponds to the adhesiveness induced by the steric exclusion of the small particles. This effect has been well discussed in [11, 12] and can produce system instability [7].

iii) λ_0^{*s} is always positive and corresponds to the adhesiveness induced by the self stickiness of the small particles which has been discussed in [13]. Such a term can be much larger than λ_0^{*s}. In particular, when the small particles are the solvent ones, that is, $\eta_1^* \approx 0.45$, the self-stickiness $1/\tau_{11}$ can lead to large values of λ_{11}^* and, according to Eq. (5b) a considerable enhancement of the steric adhesion is expected.

iv) λ_2^{*s} is induced by the small particles-large particles stickiness (hetero stickiness). Such a term is either positive or negative depending on the value of η_1^* and the strength of the hetero stickiness. This suggests that the homogeneous mixture is predominantly made of large solvated particles when $\lambda_2^{*s} < 0$ compensates the other terms of Eq. (5), or predominantly made of bridged large particles when $\overline{\lambda_2^{*s}} > 0$ adds to these terms. One illustration of this behavior is shown in the η_1^* vs Ln t_{12}^{-1} diagram (Fig. 1) which

Fig. 1 η_1^* vs hetero stickiness Ln t_{12}^{-1} for a two-component mixture in the asymptotic limit. The dashed curves correspond to the upper and lower limits of t_{12} for which $\lambda_{22}^{*\text{eff}}$ is positive (in region II, $\lambda_{22}^{*\text{eff}}$ is negative). The full curve corresponds to the values $\eta_1^*(t_{12})$ for which a sign inversion occurs in λ_{s2}^*. The dotted curve corresponds to det $\mathbf{Q}(0) = 0$ computed with $\eta_2 = 0.2$. λ_{s2}^* is negative in the solvation regions I_a, I_b and II, and positive in the bridging region III

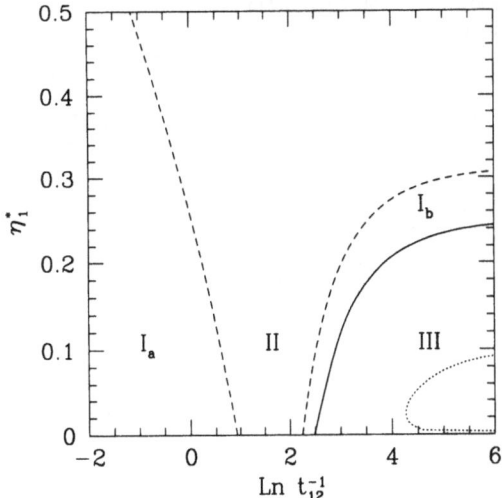

corresponds to the special case of hard spheres mixture with hetero adhesion only (τ_{11} and τ_{22} are infinite so that $\lambda_{11} = \lambda_{22} = 0$). The parameter t_{12} is diameter independent and related to the inverse hetero-stickiness τ_{12} by $\tau_{12} = (D_1 + D_2)/2D_1 t_{12}$ [8].

Comparison with sterically stabilized suspensions

This class of suspensions has been widely studied. One well studied system is a suspension of octadecyl chain-coated silica particles dispersed in benzene [2, 14]. The packing of the grafted chains and the effect of the solvent means that the effective potential between the large particles can be assumed to be hard-sphere-like with, in addition, an attractive short tail [14]. Consequently, Baxter's one-component model has been found to be very convenient for analyzing this system [2, 14]. For instance, there is a good agreement between the values of the effective stickiness parameter obtained by fitting SANS or SAXS structure factors at various temperature with those deduced from viscosity measurements [3].

According to the model in the previous section, let us now consider the packing effects (η_1^*), solvent self stickiness (τ_{11}^{-1}), solvent-solute hetero stickiness (τ_{12}^{-1}), and direct solute-solute stickiness (τ_{22}^{-1}). Calculations with silica particles diameter $D_2 = 48$ nm [2] and benzene mean diameter $D_1 = 0.5$ nm imply a size ratio close to 0.01 for this system. In order to reduce the number of parameters, we fixed $\tau_{22} = 100$. To define the variation of the stickiness $1/\tau_{mn}$ with temperature, we used the second virial coefficient *ansatz* relating the stickiness to square well parameters following:

$$\tau_{mn}^{-1} = 4\left[\left(1 + \frac{\Delta_{mn}}{D_{mn}}\right)^3 - 1 \right]\left[e^{-\frac{E_{mn}}{kT}} - 1 \right], \tag{6}$$

where Δ_{mn} and E_{mn} are the width and the depth of the square well respectively.

For τ_{12} we used a square well potential of fixed range $\Delta_{12} = 0.5D_1$ as suggested in [15] and adjustable depth E_{12}. Representing benzene-benzene interactions as a sticky potential may be questionable due to the large anisotropy of benzene molecule and, correspondingly, the non uniqueness of sphericalized potentials [16]. Here, we fixed $\Delta_{11} = 0.3$ keeping only E_{11} as a free parameter. Finally, the two well depths E_{12} and E_{11} are adjusted at one point of the phase diagram and used to calculate the solute-solute structure factor at any temperature or concentration without further adjustment. Following this procedure the trends observed by varying temperature or solute packing fraction in SANS experiments [2] are correctly reproduced. An example of temperature effect is given in Fig. 2 where the theoretical solute-solute structure

154

C. Regnaut et al.
Analysis of the solvent-mediated adhesion in colloidal dispersions

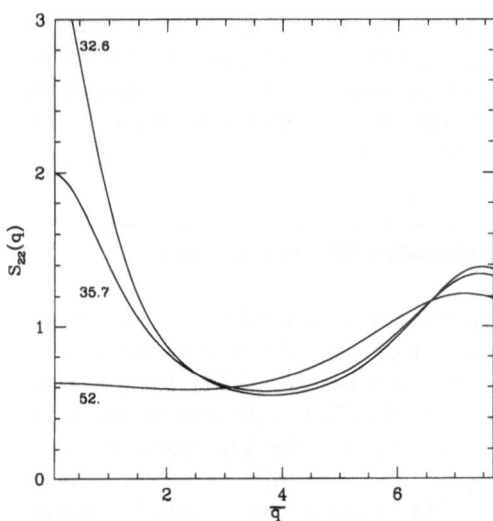

Fig. 2 Theoretical solute structure factors in the two component sticky spheres model at $\eta_2 = 0.19$ and $\eta_1^* = 0.4$ reproducing the structure factors of coated silica particles in benzene. $\bar{q}$ is the reduced wave number qD_2. Temperatures are given in °C

solvent-solvent attraction and in the solvation as the temperature is reduced. Such a sensitive behavior suggests that minute changes in the delicate balance between the solvent-solvent and solute-solvent attractions such as when the temperature or the concentration are changed might have to be considered.

Conclusion

The discrete solvent approach suggests that the stability of some suspensions results from the competition of the following destabilizing and stabilizing factors. Destabilizing factors are the large solute/solvent size asymmetry and mostly the small particles self stickiness. Stabilizing factors are the hetero-adhesions at high packing fraction of the small particles. One possible explanation of the observed decrease with temperature of the effective solute-solute attraction in sterically stabilized suspensions could be the changes in the solvent-solvent attraction and in the solvation. This also suggests that more systematic investigations of the balance between the various effects as well as the use of more realistic potentials would be necessary for a more detailed analysis.

factors at 52°, 35.7° and 32.6 °C are shown. In this example the inverse stickiness τ_{11} decreases from 0.064 to 0.058 while τ_{12} decreases from 0.424 to 0.370. Therefore, the rise of $S_{22}(0)$ appears to be associated to the increase in the

References

1. Baxter RJ (1968) J Chem Phys 49:2770–2774
2. De Kruif CG, Rouw PW, Briels WJ, Duits MHG, Vrij A, May RP (1988) Langmuir 5:422–428
3. Woutersen ATJM, De Kruif CG (1991) J Chem Phys 94:5739–5750
4. Robertus C, Joosten JGH, Levine K (1990) J Chem Phys 42:4820–4831 and ibid 93:7293–7300
5. Grant MC, Russel WB (1993) Phys Rev E 47:2606–2614
6. Pitré F Regnaut C, Pileni MP (1993) Langmuir, 9:2855–2860
7. Biben T, Hansen JP (1991) Phys Rev Lett 66:2215–2218
8. Regnaut C, Amokrane S, Heno Y, J Chem Phys 102:1995 (in press)
9. Perram JW, Smith ER (1975) Chem Phys Lett 35:138–140
10. Heno Y, Regnaut C (1992) Compte Rendus Académie des Sciences, 315:163–168
11. Vrij A, Jansen JW, Dhont JKG, Pathmamanoharan C, Kops-Werkhoven MM, Fijnaut HM (1983) Faraday Discuss Chem Soc 76:19–35
12. Heno Y, Regnaut C (1991) J Chem Phys 95:9204–9208
13. Penders MHGM, Vrij A (1991) Physica A 173:532–547
14. Duits MHG, May RP, Vrij A, De Kruif CG (1991) Langmuir, 7:62–68
15. Dickinson E (1992) J Chem Soc Faraday Trans 88:3561–3565
16. Dows DA, Hsu L (1972) J Chem Phys 56:6228–6233

Progr Colloid Polym Sci (1995) 98:155–159
© Steinkopff Verlag 1995

S. Stölken
E. Bartsch
H. Sillescu
P. Lindner

Spherical polymer micronetwork colloids – A small-angle scattering study

S. Stölken · E. Bartsch (✉) · H. Sillescu
Institut für Physikalische Chemie
Jakob-Welder-Weg 15
55099 Mainz, FRG

P. Lindner
Institute Laue-Langevin
156 X
38042 Grenoble, France

Abstract Deviations of the glass transition dynamics of 1:50 crosslinked polystyrene micronetwork spheres in a good solvent with respect to hard-sphere colloids have been observed [E. Bartsch, V. Frenz, S. Möller, H. Sillescu, Physica A 201, 363 (1993)]. To understand these phenomena, the concentration dependence of particle shape and interactions, the particle form factor, $P(Q)$, and the structure factor, $S(Q)$, have been examined by small-angle neutron scattering (SANS). The radius of gyration R_g was found to be constant over the whole concentration range studied for 1:20 crosslinked spheres. However, it starts to decrease around volume fractions $\phi \sim 0.6$ for a crosslink density of 1:50. The decrease can partially explain the higher glass transition volume fraction of this sample as compared to a hard-sphere colloid ($\phi_g = 0.64$ versus 0.58). 1:80 crosslinked spheres exhibit an even stronger decrease of R_g at all ϕ. While the experimental structure factor $S_{exp}(Q)$ of the 1:20 sample is consistent with hard-sphere behavior, this is not the case for the 1:50 sample as it exhibits a decrease of the peak maximum of increasing volume fraction at high concentration. This behavior may account for the unexpected behavior in the glass dynamics.

Key words Small-angle scattering – micronetwork spheres – crosslink density – structure factor – form factor

Introduction

A new type of colloidal system has been developed [1], consisting of crosslinked polystyrene chains, which are obtained by polymerization in an emulsion employing a crosslinking agent. When placed in good solvents, the polymer particles swell, however, they maintain their spherical shape. Consequently, they form a colloidal system which requires no special surface modification to avoid agglomeration. Another interesting feature of this system is the possibility of varying the particle interactions by changing the crosslink density (i.e., the inverse number of monomer units between crosslinks).

In dynamic light scattering (DLS) studies of concentrated suspensions of micronetwork spheres with a crosslink density of 1:50 it was found that the system undergoes a glass transition at a volume fraction of $\phi_g \cong 0.64$. The dynamics on approaching ϕ_g from the dilute region is qualitatively similar to the behavior of glass-forming molecular liquids [2] and a hard-sphere colloidal system [3], e.g., the density autocorrelation function develops a two-step decay and a critical slowing down [4, 5]. However, some differences with respect to hard-sphere colloids have been observed. The glass transition occurs at a higher volume fraction ($\phi = 0.64$ versus $\phi = 0.58$ for a hard-sphere colloid) and the density fluctuations are not completely arrested even in the glass, but exhibit a remnant

156

S. Stölken et al.
Spherical polymer micronetwork colloids

decay at long times. Specifically, while the dynamics below ϕ_g compares favorably with predictions of the mode coupling theory of the glass transition [4, 5], complete disagreement was stated for $\phi > \phi_g$ [6]. These departures from the hard-sphere behavior could reflect either significant deviations from hard-sphere interaction or changes in particle shape on increasing concentration. Alternatively, they could be due to polydispersity effects. To answer these questions we have studied the concentration dependence of the static structure factor, $S(Q)$, and the particle form factor, $P(Q)$, of the micronetwork spheres as a function of the crosslink density by means of small-angle neutron scattering (SANS). This contribution presents the first results.

Experimental details

Synthesis and characterization

The micronetwork spheres were synthesized by radical copolymerization reaction of styrene and diisopropenyl-benzene (crosslinker) in an emulsion stabilized with ionic surfactant [7]. Micronetwork spheres with crosslink densities 1:20, 1:50, and 1:80 were prepared using protonated as well as deuterated polystyrene such that they swell to about equal size (≈ 20 nm) in a good solvent (toluene). The hydrodynamic radii R_H in the swollen and the unswollen (emulsion) spheres were determined by DLS, yielding the swelling ratio $S = R_H^3(\text{toluene})/R_H^3(\text{emulsion})$.

Small-angle x-ray scattering (SAXS) experiments were performed with a Kratky compact camera ($0.06 \leq Q \leq 5.5$ nm^{-1}) on 1% solutions in toluene in a quartz capillary with 1 mm diameter. The scattering data were corrected for background and solvent. An inverse Fourier transformation algorithm (ITP) derived by O. Glatter [8]

was used to correct the slit-smearing of the data and to evaluate the pair-distribution function $p(r) = \gamma(r)r^2$, where $\gamma(r)$ is the Fourier transform of the desmeared scattering intensity $I(Q)$. The radius of gyration R_g can be extracted from this function by [8]:

$$R_g^2 = \frac{\int_0^\infty p(r)r^2 dr}{2\int_0^\infty p(r) dr} . \tag{1}$$

The results of the SAXS and DLS measurements are summarized in Table 1.

Experiments

Small-angle neutron scattering (SANS) experiments at 20°C were performed on PAXE at the Laboratoire Léon Brillouin (LLB) in Saclay (France) in a Q-range of $0.015 \leq Q \leq 0.46$ nm^{-1} with $Q = \frac{4\pi}{\lambda}\sin\theta$. After radial averaging of the two-dimensional raw data, the empty sample holder, the solvent (deuterated toluene) and the incoherent background were subtracted. The intensities were normalized with respect to a 1-mm-thick water standard and corrected for transmission. The solutions were measured in 2 mm and 1 mm quartz cells, yielding transmissions between 65% and 90%. Multiple scattering effects are thus expected to be negligible.

For a dilute solution the scattering intensity $I(Q)$ in the Guinier regime, $QR_g \ll 1$, is given by [9]:

$$I(Q) = \rho(\rho_p - \rho_s)^2 V_p^2 \exp - [(Q^2 R_g^2)/3] . \tag{2}$$

Here, ρ is the number density of colloidal particles, ρ_i the average scattering length of particle p or solvent s, respectively, and V_p the volume of the particles. R_g is determined by the slope of the Guinier plot, $\ln I(Q)$ versus Q^2. In neutron scattering the large difference between the scattering lengths of hydrogen and deuterium allows variation of the contrast, $(\rho_p - \rho_s)^2$, in order to extract R_g at higher concentrations [10]. Matching the scattering length of deuterated micronetwork spheres to that of the solvent, $(\rho_p - \rho_s)^2$ equals zero. Therefore, concentrated solutions of deuterated micronetwork spheres with a small amount of protonated colloids of the same size scatter as a dilute sample, yielding R_g as function of concentration. Solutions with increasing concentrations of deuterated micronetwork spheres were measured in a solution of deuterated toluene, together with 1% (w/w) protonated micronetwork spheres.

At higher concentrations $I(Q)$ can be written as [9]

$$I(Q) = \rho(\rho_p - \rho_s)^2 V_p^2 P(Q)S(Q) , \tag{3}$$

$P(Q)$ and $S(Q)$ being the particle form factor and the structure factor, respectively. To study the concentration dependence of $S(Q)$, samples of protonated micronetwork

Table 1 Results of the characterization of polystyrene micronetwork spheres in dilute solution, using the good solvent toluene. R_H hydrodynamic radius, which is related to R_g by a factor $\sqrt{3/5}$, determined by dynamic light scattering; S_{DLS}: swelling ratio ($S = R_H^3(\text{toluene})/R_H^3(\text{emulsion})$); σ_{DLS}: polydispersity determined by cumulant analysis of DLS data; data ($\sigma_{DLS}^2 = (\langle R^2 \rangle / \langle R \rangle^2) - 1$). $R_{g\,SAXS}$: radii of gyration as determined via (Eq.1) from small-angle x-ray scattering for the protonated and deuterated species; $R_{g\,SANS}$: radii of gyration as determined via (Eq. [2]) from small-angle neutron scattering.

Crosslink	R_H [nm]	S_{DLS}	σ_{DLS}	$R_{g\,SAXS}$ H	$R_{g\,SAXS}$ D	$R_{g\,SANS}$ [nm]
1:20	20.4	2.84	≤ 0.26	14.5	15.2	15.0
1:50	25.0	4.36	≤ 0.23	16.2	16.2	18.2
1:80	29.0	7.84	≤ 0.26	17.2	17.3	18.9

spheres in deuterated toluene were measured for all three crosslink densities. The experimental structure factor, $S_{exp}(Q)$, is then obtained as the ratio between the intensity of a concentrated sample with concentration c and a dilute sample with concentration c_0 [9]:

$$S_{exp}(Q) = \frac{I(Q, c)}{I(Q, c_0)} \frac{c_0}{c}. \tag{4}$$

Results

For three different crosslink densities R_g, as determined by the Guinier method (Eq. [2]), is shown in Figure 1 (open

squares). The 1:20 crosslink density sample shows no concentration dependence, while the 1:50 sample and the 1:80 sample exhibit a slight and a significant decrease with increasing concentration, respectively. The concentration dependence of R_g has to be taken in account when calculating the volume fraction ϕ, which is defined by $\phi = V_p \cdot S/(V_p + V_s)$. S is connected to R_g in the swollen and unswollen state via $S = R_g^3(\text{toluene})/R_g^3(\text{emulsion})$. Thus, using S values determined for the dilute solution may lead to a significant overestimation of the volume fractions. This effect is demonstrated in Fig. 1, where volume fractions corrected for the concentration dependence of R_g (solid squares), are indicated for comparison.

The concentration dependence of $S_{exp}(Q)$ is displayed in Fig. 2. For atomic liquids $S(Q)$ can be calculated by

Fig. 1 Volume fraction dependence of the radius of gyration R_g of polystyrene micronetwork spheres in toluene for different crosslink densities a) 1:50 and 1:20 and b) 1:80. Open symbols refer to uncorrected volume fractions, while full symbols designate volume fractions corrected for the concentration dependence of R_g (see text). The solid lines indicate the dilute limit of R_g

Fig. 2 Experimental structures factor $S_{exp}(Q)$ for micronetwork spheres with crosslink densities of 1:20 (a) and 1:50 (b) in toluene. The solid lines refer to a polydisperse hard-sphere structure factor with $R_{HS} = 15$ nm, $\sigma = 0.18$ and $\phi = 0.2, 0.3, 0.39$

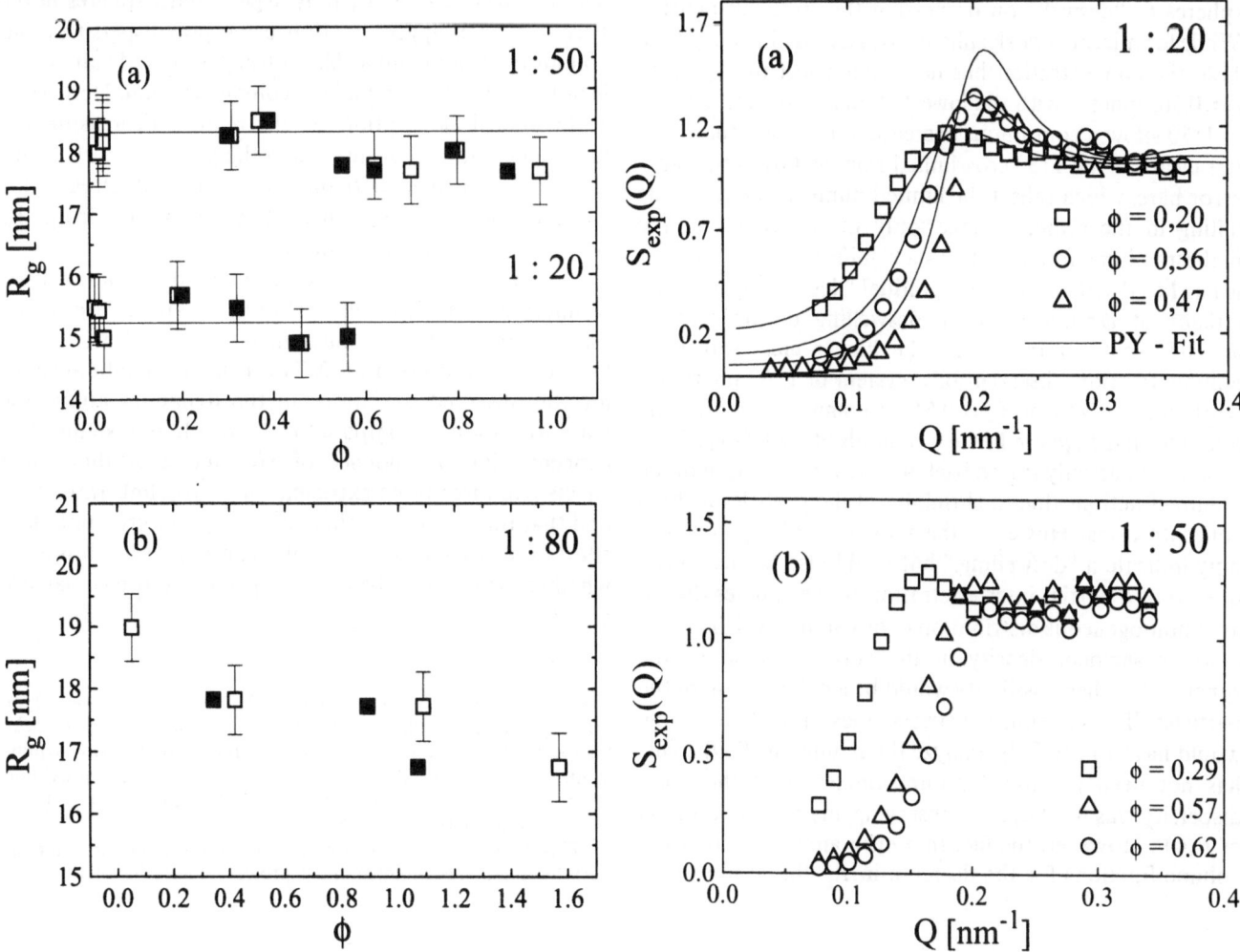

numerically solving the Ornstein–Zernike equation for different potentials using the Percus–Yevick approximation as closure relation [11]. For the hard-sphere potential, analytical solutions are available that have been extended to the case of polydisperse colloidal suspensions [12]. In Fig. 2a results of such calculations for a hard-sphere (HS) system with a radius $R_{HS} = 15$nm, $\sigma = 0.18$ and $\phi = 0.2$, 0.3 and 0.39, are compared to $S_{exp}(Q)$ for the 1:20 crosslinked micronetwork spheres. It can be seen that the experimental data follow the hard-sphere theory quite well up to a volume fraction of $\phi = 0.36$. The 1:50 sample, however, does not show the increase of the maximum expected for a hard-sphere system (Fig. 2b). Thus, a theoretical description with the polydisperse hard-sphere model was not attempted.

Discussion

Figure 1 demonstrates that varying the crosslink density has a significant effect on particle size and shape which can formally be interpreted as an increasing tendency of the spheres to "deswell" on decreasing the crosslink density. While for micronetwork spheres with a crosslink density of 1:20 the concentration has no visible effect on R_g up to $\phi \approx 0.56$, spheres with the lowest studied crosslink density of 1:80 show a continuous decrease of R_g. The situation is less clear for the 1:50 crosslinked sample. Given the large error bars, which reflect the limited number of datapoints falling in the Guinier range, only the datapoint at the highest volume fraction shows a significant effect. Thus, it is hard to decide if there is a gradual "deswelling" starting already at $\phi \approx 0.5$–0.6, or if deswelling occurs only at $\phi \geq 0.8$. In the former case, the higher glass transition volume fraction observed for a system of 1:50 micronetwork spheres [5] of $\phi_g = 0.64$ as compared to the one found for hard-sphere PMMA colloids ($\phi_g = 0.58$) [3] can be at least partially traced back to an overestimation of the volume fractions due to shrinking of the particles at high concentrations. However, the reduction of R_g may not only indicate a "deswelling," but could also be the consequence of a partial interpenetration of the spheres due to an inhomogeneous distribution of crosslinks, which causes a lower segment density in the surface region of the spheres. Another possibility would be a deformation of the particles. To distinquish between these possibilities one would have to carefully analyze the minima in $P(Q)$. This has not been possible for our samples since the polydispersity was too large for observing any distinct minima of $P(Q)$. However, the fact that even after correcting the volume fractions for the decrease in R_g, the 1:50 sample

can be concentrated beyond volume fractions of 0.74, the closest packing obtainable for hard-spheres, indicates that either interpenetration or deformation takes place. In order to distinguish between these effects one would have to repeat the experiments while varying the sphere radius. While a deformation should not depend on overall size, the interpenetration, being a surface effect, should. Some additional clues can be gained from inspecting $S(Q)$. Here, one clearly sees that 1:50 crosslinked spheres do not behave like hard-spheres, since the maxima in the concentrated regime are less pronounced than in the dilute. However, it cannot be excluded that the peak height between $\phi = 0.29$ and 0.57 first increases and then decreases again. Such a behavior has been reported for star polymers [13], which can be thought of as the limit reached for vanishing crosslinking density where the length of dangling ends largely exceeds the size of the solid core of the micronetwork spheres. A slight indication for such behavior can be seen for the 1:20 sample (Fig. 2a) where the peak in $S(Q)$ rises when going from $\phi = 0.2$ to 0.36 and then slightly decreases again towards $\phi = 0.47$. This behavior is consistent with the finding that, while the lower volume fraction data can be described by polydisperse hard-spheres in the Percus–Yevick approximation using polydispersities and volume fractions compatible with experimental values, this is not the case for the highest volume fraction. Here, such a theoretical description would require decreasing the hard-sphere radius or increasing the polydispersity. Thus, the more sensitive $S(Q)$ data indicate that already for crosslink densities as high as 1:20 deviations from the hard-sphere behavior might occur.

In summary, the presented results indicate a tendency of micronetwork spheres to shrink, interpenetrate or deform at high concentrations, which becomes clearly visible for a crosslink ratio of 1:50. This behavior might account for the observed deviations in the dynamics above the glass transition as compared to hard-sphere systems. The concentration dependence of $S(Q)$ indicates that small effects can already be expected for a crosslink ratio 1:20 and that the true hard-sphere limit requires crosslink densities of at least 1:10. It is not clear at present, however, whether the reported behavior depends on overall particle size.

Acknowledgements We are indebted to L. Noirez (LLB, France) for her assistance during the experiments and M. Stamm (Max–Planck-Institut für Polymerforschung, Germany) for use of his SAXS equipment. We thank O. Glatter and A. van Blaaderen for providing copies of their computer programs. The authors also thank A. Dörk and V. Frenz for the preparation of the samples.

This work was supported by the German Bundesministerium für Forschung und Technologie under project number SI3MAI.

Progr Colloid Polym Sci (1995) 98:155–159
© Steinkopff Verlag 1995

References

1. Antonietti A, Bremser W, Schmidt M (1990) Macromol 23:3796–3805
2. Götze W, Sjögren L (1992) Rep Prog Phys 55:241
3. van Megen W, Underwood SM (1993) Phys Rev E47:248–261
4. Bartsch E, Antonietti M, Schupp W, Sillescu H (1992) J Chem Phys 97:3950–3963
5. Bartsch E, Frenz V, Möller S, Sillescu H (1993) Physica A 201:363–371
6. Bartsch E, Frenz V, Sillescu H (1993) J Non-Cryst Solids 172–174, 88–97 (1994)
7. Frenz V (1995) Ph.D. thesis, Mainz
8. Glatter O (1977) J Appl Cryst 10:415–421
9. deKruif CG, Briels WJ, May RP, Vrij A (1988) Langmuir 4:668–676
10. Cotton JP (1991) in Lindner P, Zemb Th (eds) Neutron, X-Ray and Light Scattering: Introduction to an investigative Tool for Colloidal and Polymeric Systems. North-Holland, Amsterdam, pp 1–18
11. Hansen J-P, McDonald IR (1986) Theory of Simple Liquids. Academic Press, London
12. van Beurten P, Vrij A (1981) J Chem Phys 74:2744–2748
13. Willner L, Jucknischke O, Richter D, Farago B, Fetters LJ, Huang JS (1992) Europhys Lett 19:297–303

Progr Colloid Polym Sci (1995) 98:160–168
© Steinkopff Verlag 1995

E. Tombácz
M. Szekeres
I. Kertész
L. Turi

pH-dependent aggregation state of highly dispersed alumina, titania and silica particles in aqueous medium

Prof. Dr. E. Tombácz (✉) · M. Szekeres
I. Kertész · L. Turi
Department of Colloid Chemistry
Attila József University
Aradi Vt. 1
6720 Szeged, Hungary

Abstract Highly dispersed commercial products (Aluminium oxide C, Titanium dioxide P25 and Aerosil 200) of Degussa were investigated. Their pH-dependent surface charge state was determined from acid-base titrations in the presence of KNO_3. The p.z.c. values were measured at pHs 8.0 for Al_2O_3, 5.9 for TiO_2 and ~4 for SiO_2. The calculated intrinsic equilibrium constants for surface charge (σ_0) formation are $\log K_{a1}^{int} = 5.8$, $\log K_{a2}^{int} = 10.2$ for Al_2O_3, $\log K_{a1}^{int} = 3.6$, $\log K_{a2}^{int} = 8.1$ for TiO_2 and $\log K_{a2}^{int} = 8.02$ for SiO_2. The experimental σ_0 vs. pH curves were fitted by using the DDL model of MICROQL-UCR program. The aggregation state in dense suspensions was investigated by means of rheology and SAXS method. The rheological character of flow curves changed from Newtonian to pseudoplastic or to highly thixotropic, depending on the pH of suspensions. The experimental yield values (initial or Bingham) showed maximum at the p.z.c.. Scattering curves of aggregated (at p.z.c.) and well-stabilized (at pH far from p.z.c.) oxide suspensions were determined. The correlation length values were calculated on the basis of theoretical approach of Debye–Bueche which were independent of the solid/liquid ratio of suspensions at p.z.c., indicating that aggregation was controlled by attractive forces between particles, while those increased with increasing dilution in well stabilized systems.

Key words Aqueous oxide suspensions – pH-dependent behavior– rheology – small-angle x-ray scattering – highly dispersed Degussa products

Introduction

The highly dispersed metallic oxides produced by the AEROSIL Process of Degussa are of great practical importance. These fine powders are frequently used as additives in different aqueous suspensions. To obtain defect-free bodies from ceramic slips or smooth thin layers from paint slurries, suspensions are required to be well-dispersed stable systems with low viscosity in the long term. In any aqueous colloidal system of amphoteric ox-

ides, the pH is one of the main factors determining the stability of suspensions. Thus the most frequently used method to obtain well-stabilized suspensions is to work at very high or very low pH values. At these pHs, i.e., far from the pH value of chargeless state (point of zero charge – p.z.c.) of oxides, electrostatic repulsive forces act between the particles originating from the charged surface sites developed due to the specific adsorption of potential-determining ions (H^+ and OH^- for metallic oxides). The pH-dependent surface charge state of oxides is well-known in literature [1–4], and characterized mainly by acid-base

Progr Colloid Polym Sci (1995) 98:160–168
© Steinkopff Verlag 1995

titration method. However, little has been published on the related properties such as colloidal structure, aggregation state, physical network formation in suspension induced by change in pH.

In the present study the pH-dependent properties of the selected metallic (aluminum, titanium, and silicium) oxides are significantly different, but their colloidal features (particle size and shape, polydispersity) are rather similar because of the same manufacturing process. Our objective, beyond the correct characterization of pH-dependent surface charge state of these pyrogenic Degussa products, was to search for correlation between their surface characteristics in aqueous medium and aggregation state of suspensions measured by rheology and small-angle x-ray scattering. Since these pyrogenic products contain small non-porous spheres with 10–20 nm in diameter, they seem to be ideal model materials for rheological and SAXS measurements.

Experimental

Metallic oxides Aluminum oxide C, Titanium dioxide P25, and Aerosil 200 (highly dispersed commercial products of Degussa AG) were investigated. These fine powders are produced by flame hydrolysis of metallic halides ($AlCl_3$, $TiCl_4$, and $SiCl_4$) and are characterized by non-porous structure, high surface area (100 ± 15 m^2/g of Al_2O_3 C, 50 ± 15 m^2/g of TiO_2 P25, and 200 ± 25 m^2/g of SiO_2 A200) and small particle size (13 nm of Al_2O_3 C, 21 nm of TiO_2 P25, and 12 nm of SiO_2 A200). The densities of solid materials are 3.2, 3.7, and ~2.2 g/cm^3, respectively.

The pH-dependent surface charge state was determined from acid-base titrations under CO_2-free condition using indifferent background electrolyte (KNO_3) to maintain the constant ionic strength of aqueous medium over the region 0.001 and 0.1 mol /l. The evaluation of titration data was based on determination of the equilibrium proton concentration which was related to the blank titration data measured at each ionic strength. The surface excess concentrations of H^+ or OH^- were calculated as a function of pH. Since the potential determining ions (p.d.i.) are H^+ or OH^- the surface charge densities can be calculated directly from their surface excess amounts [5]. The intrinsic equilibrium constants for surface charge formation were determined according to the extrapolation method of Stumm and co-workers [6], i.e., the apparent surface association constants (log K_{a1}^{app} for formation of positive and log K_{a2}^{app} for that of negative surface sites) were extrapolated to uncharged state of surface. The experimental data of pH-dependent surface charge densities were fitted by using the DDL model of MICROQL-UCR program [7].

The pH-dependent rheological properties of oxide suspensions with concentration 10 w% ($\phi = 0.034$) for Al_2O_3, 7.5 w% ($\phi = 0.021$) for TiO_2 and 8 w% ($\phi = 0.038$) for SiO_2 were investigated. The flow curves of suspensions were determined by means of HAAKE Rotovisco RV-20, CV-100 apparatus. A Couette type ME-15 measuring cell was used. The shear rate changed between 0 and 300 l/s during 1 min. The up and down curves were measured with increasing and then decreasing shear rate. The temperature was adjusted at 25 ± 0.1 °C.

The aggregated (at p.z.c.) and well-stabilized (at pH far from p.z.c.) oxide suspensions were investigated by SAXS method using a compact Kratky slit-collimation camera attached to a Philips PW1830 x-ray generator ($\lambda(Cu_{K\alpha}) = 0.1542$ nm) under He atmosphere at 25 ± 0.01 °C. Suspensions of different concentrations (0.5, 2.0 and 5.0 g/100 g) were filled into glass capillary tubes of 1 mm diameter. The scattering intensities were measured step-by-step over the distance 0.3 to 25 mm with a position sensitive detector. The appropriate step sizes and duration times were controlled by SDC (Scattering Data Controlling) program. A typical acquisition time was 2 h. The row scattering functions were normalized and corrected by the normalized scattering intensity function of background (capillary filled by water) using PDH program developed by A. Janosi and O. Glatter (Institute of Physical Chemistry, University of Graz). No desmearing correction was applied. The corrected scattering intensity values were plotted in the function of scattering vector ($h = (4\pi \sin\theta)/\lambda$, nm^{-1}).

Results and discussion

The pH-dependence of surface charge (σ_0) of Aluminum oxide C (a), Titanium dioxide P25 (b), and Aerosil 200 (c) solid phases in aqueous medium at different concentrations of indifferent electrolyte (KNO_3) can be seen on the left side of Fig. 1. The point of zero charge (p.z.c.) values can be identified [1–5] as the intersection points of σ_0 vs. pH curves belonging to the different ionic strength. The experimental p.z.c. values are given in Table 1. The p.z.c. values of these oxides are significantly different because of the polarization effect of metallic atoms on chemical bonds to which the OH groups are bound [8]. The surface charge vs. pH functions (left side part of Fig. 1) obviously show that the oxide particles are negatively charged above the p.z.c. and positively below that apart from the silica surface which has relatively low p.z.c. value and the formation of positive charges was not detectable for this sample.

The charges on any amphoteric oxide surface can form due to the specific adsorption of H^+ or OH^-, i.e., potential

Table 1 Surface characterization of some pyrogenic oxides of Degussa in aqueous KNO_3 solutions.

Metallic oxide	S–OH	N_s, nm^{-2}	p.z.c.	$\log K_{a1}^{int}$	$\log K_{a2}^{int}$
Aluminum oxide C	<u>Al</u>–OH	8.0	8.0 ± 0.05	5.8 ± 0.2	10.2 ± 0.2
Titanium dioxide P25	<u>Ti</u>–OH	6.8	5.9 ± 0.02	3.6 ± 0.4	8.1 ± 0.2
Aerosil 200	<u>Si</u>–OH	3.0	~4($\sigma_0 \to 0$)	–	8.0 ± 0.2

Fig. 1 pH-dependence of surface charge (left side) of Aluminum oxide C (a), Titanium dioxide P25 (b), and Aerosil 200 (c) samples measured at ionic strength 0.1 (▲), 0.01 (●) and 0.001 (■) M together with the calculated curves (continuous lines). The DDL model of MICROQL-UCR program [7] was applied using the determined intrinsic equilibrium constants (right side)

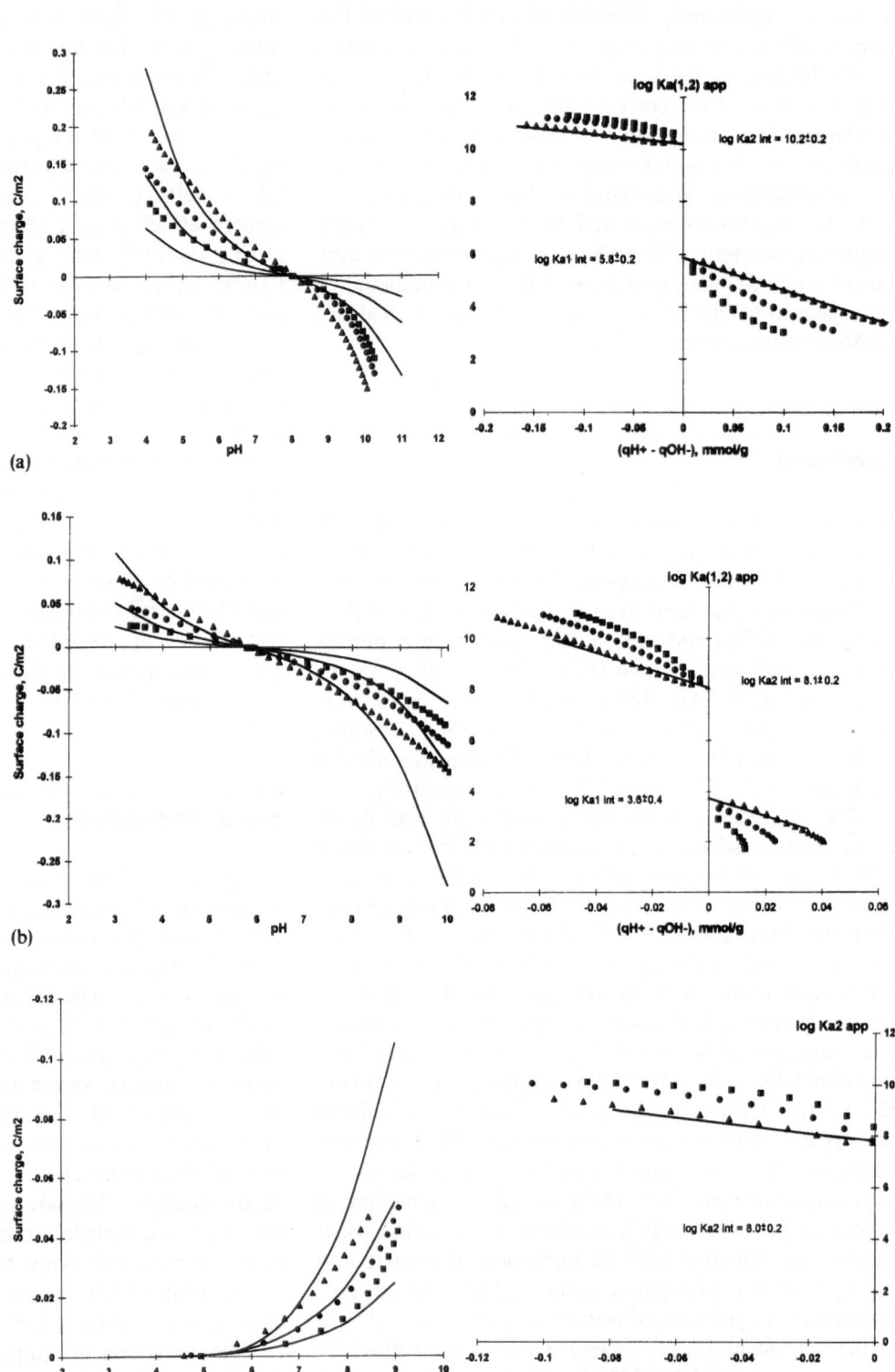

determining ions. The following equilibria can represent the charge formation on oxide surfaces:

$$S-OH + H^+ \rightleftarrows S-OH_2^+$$

$$K_{a1}^{app} = [S-OH_2^+]/([S-OH]_t - [S-OH_2^+])\,[H^+] \tag{1}$$

$$S-O^- + H^+ \rightleftarrows S-OH$$

$$K_{a2}^{app} = ([S-OH]_t - [S-O^-])/H^+][S-O^-], \tag{2}$$

where K_{a1}^{app} and K_{a2}^{app} are the apparent association constants of H^+ and subscript t means the total amount of surface site S–OH. Intrinsic equilibrium constants for surface charge formation were determined by means of an extrapolation method suggested by Stumm and co-workers [6].

$$\lim_{(\sigma_0 \to 0)} \log K_a^{app} = \log K_a^{int} \cdot \tag{3}$$

The apparent equilibrium constants changed monotonously with the surface charge densities of each oxide and the extrapolation to the zero charge could be achieved. The functions are shown on the right side of Fig. 1 and the intrinsic association constants ($\log K_a^{int}$) are summarized in Table 1. The values of intrinsic equilibrium constants are in good harmony with several data in literature [1–6] and, apart from the silica, they correlate well, within the experimental error, with the p.z.c. according to the relation [4, 5, 7]:

$$\text{p.z.c.} = 1/2(\log K_{a1}^{int} + \log K_{a2}^{int}). \tag{4}$$

The experimental points of surface charge vs. pH functions were fitted by using DDL model of MICROOL-UCR program [7]. The total site density data and intrinsic equilibrium constants given in Table 1 were used in the model calculation. The calculated curves are drawn with continuous line in the left side part of Fig. 1.

After the relevant characterization of pH-dependent surface charge state of the highly dispersed alumina, titania and silica particles, the effect of pH on aggregation state of aqueous oxide suspensions was investigated.

The simplest way was the visual observation of test tube series containing dilute oxide suspensions with pHs between 2.5 and 11. The feature of sedimentation showed characteristic differences in the function of pH. At pH values around p.z.c. unstable, coagulated systems formed, the oxide particles stuck closely together because of the absence of electrostatic repulsion between them and settled down with sharp boundary forming loosely packed sediment. Far from this particular pH value in either the acidic or the alkaline region particles could retain their individuality due to the repulsive forces acting between the

charged particles and settled in a diffuse way with primary particle size-dependent velocity. The pH regions for metallic oxides over which the suspensions proved to be unstable systems in the colloid stability point of view were in good harmony with the p.z.c. values. Namely, the alumina (p.z.c. = 8.0) systems were coagulated between pH 6.5 and 9.5, the titania (p.z.c. = 5.9) between pH 4.5 and 8.0, and the silica (p.z.c. ~4) below pH 6.5.

Many more details of pH-induced structural changes in amphoteric oxide suspensions can be obtained from rheological measurements. Significant changes in the flow behavior of stable/unstable systems have been published [9–15]. At the p.z.c. the suspension is coagulated (flocculated) due to the absence of repulsive forces and exhibits high viscosity [9] having pseudoplastic flow character similarly to any coagulated colloidal system [10–15]. Experimental parameters, namely, critical shear rate (D_0) at which the flow curve becomes essentially linear, the Bingham yield value (τ_B) extrapolated back to zero shear rate, and the plastic viscosity (η_{pl}) can be identified from this type of flow curve. As the pH is shifted from the point of zero charge, the particles become charged due to the reaction of surface sites with H^+ or OH^- and a direct electrostatic repulsion comes into action which increasingly hinders the formation of physical network. The viscosity of suspensions decreases as the pH moves away from the p.z.c., the pseudoplastic character may disappear and an ideal viscous behavior (Newtonian flow character) may appear.

The pH-dependent flow properties of alumina suspension are an excellent example for amphoteric behavior. Some characteristic flow curves of alumina suspension (A) can be seen on the left side of Fig. 2. Pseudoplastic systems were observed at the pHs around the p.z.c. and Newtonian flow in the slightly acidic region. The evaluation of the pseudoplastic systems according to the Bingham model leads to a very sharp maximum of the yield value vs. pH function (right side of Fig. 2) at pH ~8, i.e., at the p.z.c. of this solid. Since the yield value has a certain relation to the particle-particle interaction [10–13, 15], it can be concluded that the strongest attractive forces act between the chargeless particles. Although the attraction decreases with increasing pH, a relatively large yield value was still measured at pH 11, which indicates the existence of coagulated structure. The pH-dependent rheological behavior of Alpha-Al$_2$O$_3$ suspensions [9] seems to be very similar to the alumina sample in the present work.

The titania and silica suspensions could not show such significant rheological character as observed in the case of alumina suspensions. Only the very weak physical network could form in the titania suspensions at pHs around p.z.c., which could be broken by very low shear force and the rebuilding of this loose network needed a long time

Fig. 2 Some characteristic flow curves (left side) of 10 w% Aluminum oxide C (a), 7.5 w% Titanium dioxide P25 (b), and 8 w% Aerosil 200 (c) suspensions measured at different pHs, at temperature $2.5 \pm 0.1°C$ and the pH-dependence of the yield values (Bingham or initial) belonging to the coagulated systems (right side)

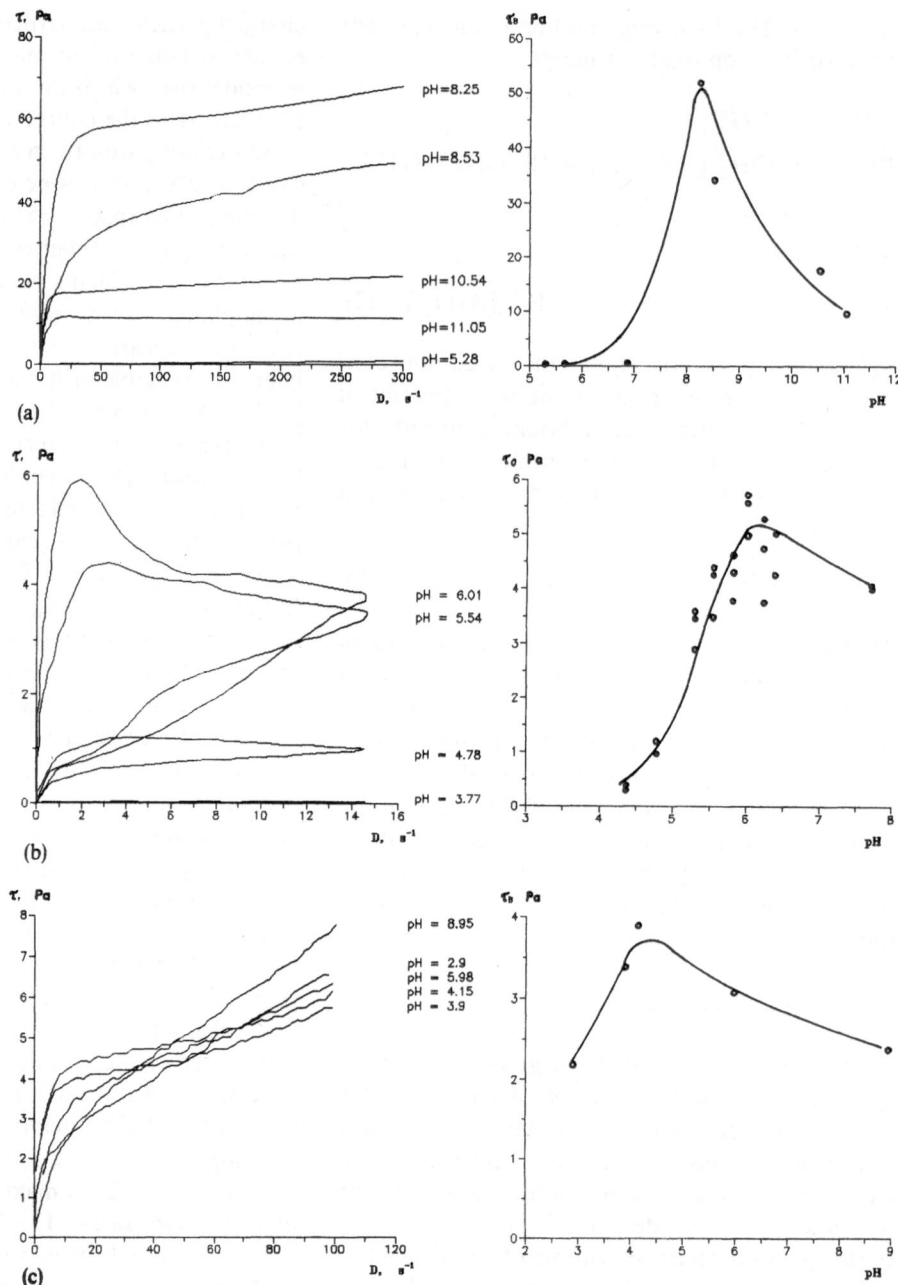

($\sim$ some hours). Huge thixotropic loops were characteristic of the titania suspensions at pHs around p.z.c. (part b on the left side of Fig. 2). This thixotropy was observed when the samples were subjected to either a short ultrasonication or a preliminary shear at higher shear rate ($300\ s^{-1}$) before the flow curves were recorded. The maximum shear stress values of up-curves, to be considered as initial yield values, increased with the volume fraction of solid phase and showed significant change with the pH of suspension (part b on the right side of Fig. 2). Although the change in these shear stresses were not as large as that

in the Bingham yield values of alumina suspensions, but their pH-function shows maximum ($\tau \sim 5.5$ Pa) at the p.z.c. of titania sample. This sort of unusual rheological behavior has not been published for aqueous titania suspensions, e.g., a simple pseudoplastic behavior was observed in the coagulated suspensions of TiO_2 with volume fractions between 0.02 and 0.07 [11].

The flow curves of silica suspensions at different pHs differ slightly from each other (part c on the left side of Fig. 2). Each shows pseudoplastic character. Very small maximum ($\tau_B \sim 3.5$ Pa) of yield value vs. pH function (part

Progr Colloid Polym Sci (1995) 98:160–168
© Steinkopff Verlag 1995

C on the right side of Fig. 2) could be only observed at the charge-free state of silica sample (pH ~ 4).

An equation had been derived in the single particle model [10] related to the flow properties of coagulated colloidal suspensions by which the distance of the closest approach, d_0, can be calculated:

$$d_0 = A \, \Phi^2 / 8\pi^2 \, r^2 \, \tau_B , \qquad (5)$$

where A is the Hamaker constant, ϕ is the volume fraction of particles, r is the particle radius, and τ_B is the yield value. The distances of the closest approach were calculated for the oxide suspensions in which the pH was equal to the individual value of p.z.c., respectively. The Hamaker constant from literature (oxide particle in water $A = 4\,10^{-20}$ J [15]) was used in the calculations. The data and the calculated values are summarized in Table 2. The calculated value of the closest approach of particles in the alumina suspension is essentially smaller than that of the other two examined systems, as could be expected from its strong structure forming at p.z.c.

Numerous geometrical and structural parameters can be obtained from the small-angle x-ray scattering study on colloidal systems. To characterize the pH-induced structural changes in the examined two-phase systems by SAXS method, scattering curves of oxide suspensions being in extremely different (aggregated and well-stabilized) states in the colloid stability point of view were determined.

The normalized and substracted scattering intensities were calculated and plotted in the function of scattering vector ($h = (4\pi \sin\theta)/\lambda$, where θ is the scattering angle (2θ) and λ is wavelength of x-ray). The scattering curves of the well-stabilized Aluminum oxide C, pH ~ 4 (a), Titanium dioxide P25, pH ~ 3.5 (b), and Aerosil 200, pH ~ 8 (c) suspensions with different concentrations in the plot $\log(I_h)$ vs. $\log(h)$ are shown on the left side of Fig. 3. The scattering curves of aggregated systems (the pH of suspensions were adjusted to the p.z.c. of each oxide sample) were situated almost the same lines as the curves of the most concentrated, stable suspensions.

For two-phase systems the final slope of $\log(I_h)$ vs. $\log(h)$ curves toward large angle, i.e., the tail end of the scattering curve, should conform to the asymptotic course of h^{-4}: $I(h) \sim K_p h^{-4}$, were K_p is the Porod or tail end constant [16, 17]. In the case of the smeared intensity curve measured with Kratky camera this relationship

takes the form: $\tilde{I}(h) \sim \tilde{K}_p h^{-3}$, i.e., the final slope of the plot $\log(\tilde{I}_h)$ vs. $\log(h)$ should be equal to -3. The final slopes of the measured scattering curves were between -2.9 and -3.1 apart from the most dilute (0, 5%) systems.

Power-law scattering ($I_h \sim I_0 h^{-\alpha}$, where α is the power-law exponent) to be characteristic of fractal or other disordered systems seems to be non-significant for the investigated suspensions. Only the narrow regions of scattering curves belonging to the more concentrated, unstable systems exist over which the measured curves can be fitted by the power law. The values of mass fractal dimension (D_m) calculated from the power-law exponent ($\alpha = D_m - 1$ for smeared scattering curves) were between 2.8 and 3.0. These values are physically practicable, but too large, considering the aggregated structure of two-phase systems. Among the models which calculate fractal dimension for particle aggregation, the diffusion limited aggregation model (DLA) predicts the largest ($D_m = 2.5$ [18]) mass fractal value for aggregates.

Since Guinier plots for these oxide suspensions proved to be nonlinear, presumably owning to the polydispersity in particle size, further evaluation of the experimental scattering curves should be achieved on the basis of the Debye–Bueche analysis [19]. In this theory the inhomogeneities associated with the domain structure are described by a spatial two point correlation function which for a random two-phase systems should take the form of simple exponential function, $e^{-r/a}$, where a is the correlation length. A correct as well as an easier determination of the zero-intensity (I_0) and the correlation length (l_c) values from the measured data become possible on the basis of the deduced equations even for polydisperse systems [20]. The following equation deducted by A. Jánosi [17] is valid for the smeared scattering curves:

$$\tilde{I}(h) = \tilde{I}_0/(1 + (h \, l_c/2)^2)^{3/2} . \qquad (6)$$

It can be seen that the plot $\tilde{I}(h)^{-2/3}$ vs. h^2 should conform to the straight line with slope $\tilde{I}_0^{-2/3}(l_c/2)^2$ and intercept $\tilde{I}_0^{-2/3}$ at the zero angle.

These plots for the experimental scattering data of different oxide suspensions are shown on the right side of Fig. 3. All the curves proved to be straight lines over the broad region of scattering vector. The calculated correlation length values are summarized in Table 3.

Table 2 Calculation of the distances of the closest approach (d_0) for the different oxide suspensions at p.z.c. on the basis of the single particle model [10].

Metallic oxide	p.z.c.	r, nm	ϕ	τ_B, Pa	d_0, nm
Aluminum oxide C	8.0	6.5	0.034	52.0	0.27
Titanium dioxide P25	5.9	10.5	0.021	$\sim 0.5^*$	4.06
Aerosil A200	~ 4	6.0	0.038	3.5	5.81

*approx. value, the linear part of down-curve was extrapolated.

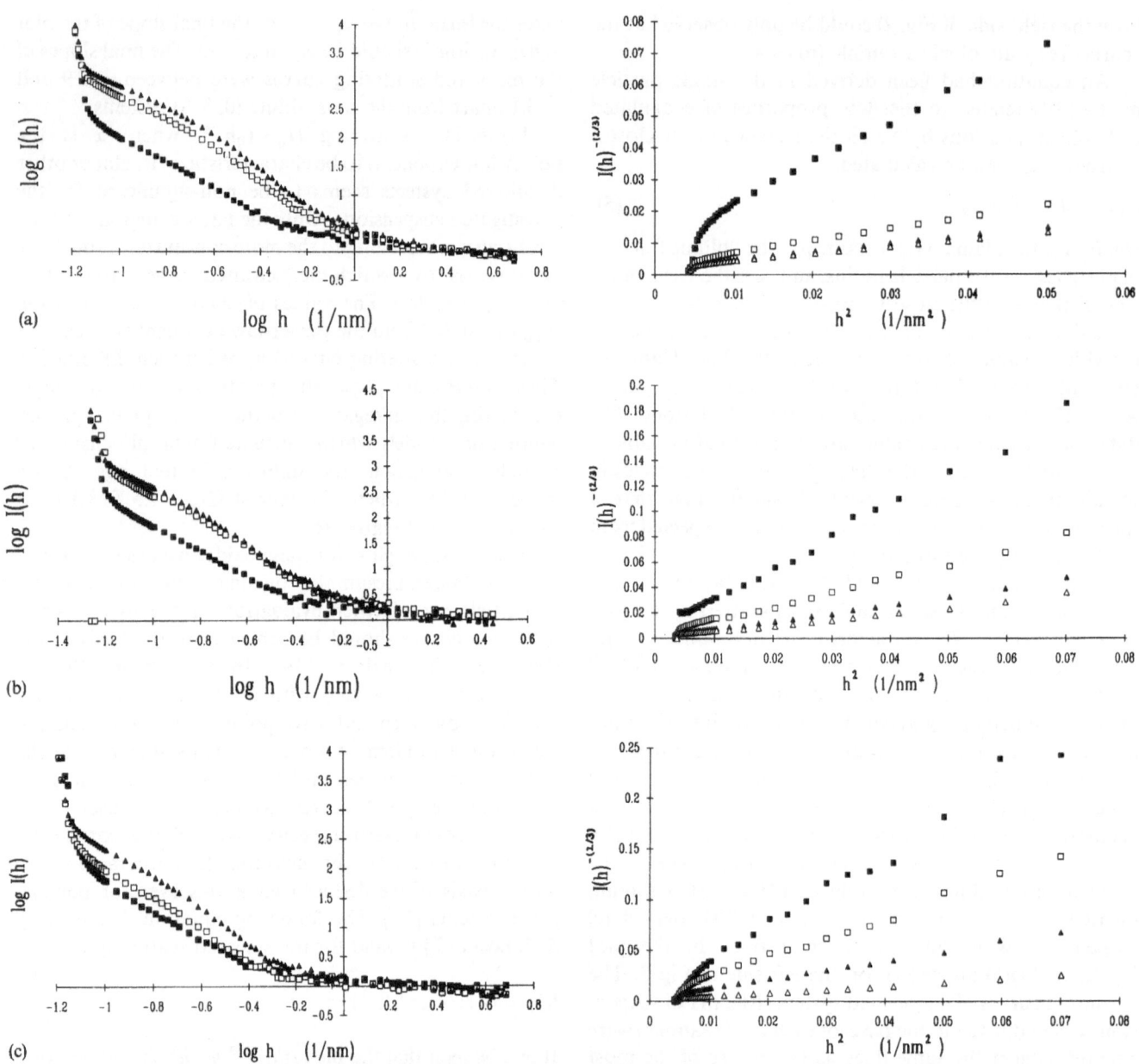

Fig. 3 Scattering curves (left side) of the stable Aluminum oxide C (a), Titanium dioxide P25 (b), and Aerosil 200 (c) suspensions containing 0.5 (■), 2.0 (□) and 5.0 (▲) g solid material in 100 g suspen- sion and Debye–Bueche plots (right side) for the stable (at pH far from p.z.c.) and unstable (at p.z.c., △) suspensions of the same concentration

It can be seen that in all the three oxide suspensions the correlation length values increase significantly with de- creasing concentration of solid particles dispersed well at appropriate pH values far from p.z.c. (pH ~ 4 for alumina, pH ~ 3.5 for titania, and pH ~ 8 for silica) in aqueous phase. However, in the aggregated systems in which the pH was adjusted to p.z.c. (~ 8 for alumina, ~ 5.9 for titania, and ~ 4 for silica) the correlation length values do not show any systematical change with the mass fraction of solid materials, and show certain average values.

Since the correlation length is a parameter (integral breadth) of the correlation function, the corresponding spatial structure is difficult to visualize. For the exponen- tial distribution the correlation length is the double of the reduced chord length or intersection length defined by Porod [ref. 5 in [17]], which is the average length of chords crossing phases 1 and 2 in the arbitrarily chosen direction. Also, for exponential distribution the volume of inhomogeneities can be calculated from the correlation length ($V = \pi l_c^3$).

Table 3 Correlation length values obtained from the Debye–Bueche analysis of scattering curves of aqueous oxide suspensions.

Oxide	Solid conc. (mass fraction)	Correlation length (l_0), nm		
		Well-stabilized	Aggregated suspensions	
Al_2O_3	0.05	20.4	19.2	Av.:
	0.02	23.2	20.0	20.2 ± 1.2
	0.005	54.8	21.4	
TiO_2	0.05	45.6	39.5	Av.:
	0.02	64.2	28.5	34.1 ± 5.6
	0.005	95.2	34.4	
SiO_2	0.05	35.3	28.8	Av.:
	0.02	46.6	28.3	28.2 ± 0.6
	0.005	71.6	27.4	

It can be supposed that the calculated coherence length values correspond with the average distance between the centers of gravity of particles. In the well-stabilized suspensions the particles move away from each other as far as possible due to the repulsive forces acting between them. The distance between them is limited only by the number of particles in unit volume of system. The change in the value of the calculated coherence lengths shows just this tendency (see Table 3). For the sake of comparison the average distance between the centers of particles were calculated from the known particle size, density and mass fraction data. The coherence length values corresponded well with these distances, especially for the silica suspensions. In the aggregated systems independently on the concentration of solid phase up to a certain level, the particles are situated at the closest distance determined by the resulting attraction [10, 12] and the thickness of hydration layers in aqueous medium. Considering the average values of the calculated correlation lengths for the different oxide suspensions (see data in Table 3), it can be concluded that the equilibrium distances between the particles are obviously different, although the surface layer of oxides are in chargeless state and their Hamaker constants, being characteristic of attractive interaction, cannot differ considerably. It has to be supposed that the thickness of hydration layer on these oxide surfaces is different. Substracting the particle diameter from the coherence length and dividing by two, the thickness of hydration layer can be estimated. These thickness values are ~3.5 nm for alumina, ~6.5 nm for titania, and ~7 nm for silica particles. These values for the hydration layer thickness seem to be too large. Similar large value (5 nm) for fused silica plates have been reported [21, 22].

Comparing the experimental results for the aggregated structure of different oxide suspensions derived from rheology and small-angle x-ray scattering, it can be stated on the one hand that the distances of the closest approach calculated from the rheological data and the thicknesses of hydration layer from SAXS evaluation related to the different oxide samples are in harmony, especially in the case of titania and silica samples, and, on the other hand, that a much stronger physical network can form in the alumina suspensions due to the poorer hydration of solid surface in aqueous medium at p.z.c. than in titania and silica systems under corresponding conditions.

Acknowledgements This work has been supported by the Hungarian National Scientific Research Foundation (OTKA I/5 No. T006077), PHARE ACCORD (No. H 9112–153) and by the NATO Collaborative Research Grant No. 920785.

References

1. Parks GA, de Bruyn (1962) J Phys Chem 66:967–973
2. Healy TW, White LR (1978) Adv Colloid Interface Sci 9:303–345
3. Westall J, Hohl H (1980) Adv Colloid Interface Sci 12:265–294
4. James RO, Parks GA (1982) in:Matijevic E(ed) Surface and Colloid Science Vol 12, Plenum Press, New York, pp 119–216
5. Lyklema J (1987) In:Tadros ThF(ed) Solid/Liquid Dispersions, Academic Press, London, pp 63–90
6. Hohl H, Sigg L, Stumm W (1980) In:Kavanaugh MC, Leckie JO(eds) Particulates in Water. Adv Chem Ser No 189, ACS, Washington, DC, pp 1–31; Huang CP, Stumm W (1973) J Colloid Interface Sci 43:409–420; Hohl H, Stumm W (1976) J Colloid Interface Sci 55:281–288
7. Coves J, Sposito G (1989) MICROQL-UCR:A Surface Chemical Adaptation of the Speciation Program MICROQL, University of California, Riverside
8. Sparks DL (ed) (1986) Soil Physical Chemistry. CRC Press, Boca Raton, Florida, pp 70–77
9. Nikumbh AK, Schmidt H, Martin K, Porz F, Thümmler F (1990) J Mater Sci 25:15–21
10. Firth BA, Hunter RJ (1976) J Colloid Interface Sci 57:248–256

11. Firth BA (1976) J Colloid Interface Sci 57:257–265
12. Firth BA, Hunter RJ (1976) J Colloid Interface Sci 57:266–275
13. Hunter RJ, Frayne J (1979) J Colloid Interface Sci 71:30–38
14. Hunter RJ (1982) Adv Colloid Interface Sci 17:097–211
15. Wiese RG, Healy TW (1975) J Colloid Interface Sci 51:427–433
16. Kratky O, Laggner P (1987) Encyclopeadia of Psysical Science and Technology
17. Jánosi A (1993) Monatscefte für Chemie 124:815–826
18. Schaefer DW, Martin JE (1984) Physical Review Letters 52:2371–2374
19. Debye P, Bueche A (1949) J Appl Phys 20:518–525
20. Russel TP, Lin JS, Spooner S, Wignall GD (1988) J Appl Cryst 21:629–638
21. Peschel G, Belouschek P, Müller MM, Müller MR, König R (1982) Colloid Polym Sci 260:444–451
22. Israelachvili JN (1985) Intermolecular and surface forces. Academic Press, London pp 194–208

Progr Colloid Polym Sci (1995) 98:169–172
© Steinkopff Verlag 1995

C. Dicharry
B. Mendiboure
G. Marion
J.L. Salager
J. Lachaise

Contribution to the modelization of the emulsification in a colloid mill

C. Dicharry · B. Mendiboure
G. Marion · Dr. J. Lachaise (✉)
L.T.E.M.P.M.
Centre Universitaire de Recherche
Scientifique
Avenue de l'Université
64000 Pau, France

J.L. Salager
L.F.I.R.P.
Universidad de los Andes
Merida, Venezuela

Abstract We propose a stochastic model to forecast the droplet size distributions of oil in water emulsions generated in a colloid mill. The model lays on breakup sequences of the oil droplets induced by the shear field existing within the mill. One single set of the three fitting parameters is sufficient to account for the variations of the droplet size distributions in function of the surfactant concentration, the surfactant nature, the rotation speed of the rotor.

Key words Colloid mill – emulsification – droplet size distribution – shear-field – viscous dispersed phase – oil-in-water emulsion

Introduction

Emulsions can be produced by a turbulent agitation in closed vessels. As long as the turbulent agitation is isotropic, the emulsions exhibit lognormal unimodal droplet size distributions. This is the case for little agitated volumes and not very viscous liquids. Recently, we proposed a stochastic model based on the competition between the breakups and the coalescences of the droplets [1, 2].

However, when the agitation is anisotropic, the lognormality and, sometimes, the unimodality of the droplet size distributions disappear. This is the case for emulsions where one phase of which is viscous, such as bitumen in water emulsions. To obtain lognormal unimodal distributions in these systems it is necessary to use colloid mills. Widely used in industrial processes [3–10], such apparatus allows to produce in-line large amounts of emulsions.

If the influence of some parameters on the properties of the emulsions thus generated is correctly described [11], there is no model to estimate their droplet size distributions.

This paper is an attempt to model the emulsification process in a colloid mill. After presenting the essentials of the model, we test it on bitumen in water emulsions.

The permanent breakup model

Several mathematical studies on solid grinding in colloid mills have already shown that breakup sequences of solid fragments lead to lognormal particle size distributions [12, 13]. Furthermore, it has been recently claimed that in a colloid mill, liquid droplets are subjected to permanent breakups [14]. Thus, we assume that the liquids which compose the emulsion are sufficiently viscous to validate this claim.

As in the *breakup coalescence model* of the emulsification in a turbulent stirring, the droplets of the emulsion are ranked in N classes, numbered from 1 to N between the limits d_{min} and d_{max}. The droplets contained in the class i are represented by the diameter d_i, geometric mean of the two limits $d_{i\,min}$ and $d_{i\,max}$ of the class.

During a time lag Δt, each droplet is submitted to a random transition: it has the probability P_b to break into two droplets of half volume, and the probability P_{iv} to remain unaltered (Fig. 1). Of course, the two probabilities satisfy the relation:

$$P_b + P_{iv} = 1 . \tag{1}$$

At the time t, the Markov chain theory allows to represent the droplet size distribution by a line vector $V(t)$, the elements of which are the volume percentages of the N classes. During the time lag Δt, taken as the step of the calculus, the random transitions are represented by a matrix $[N, N]$, the elements of which are the two transition probabilities of all the classes. The elements are updated after each step to take account of both the creation of the oil/water interface and the corresponding surfactant adsorption. The distribution at the time $t + \Delta t$ is derived from the product of the matrix $[N, N]$ by the vector line $V(t)$. The initial distribution is located in the four last classes with equal volume percentages. d_{max} is assumed to be equal to the gap width e of the colloid mill; we fix $d_{min} = 0.1$ µm to be sufficiently lower than the finest droplets formed.

When the random transitions are applied to this distribution, it is possible to follow the evolution of the transient droplet size distribution until the time t_p which corresponds to the resident time of the liquids in the mill. t_p is proportional to the free space volume in the gap, and inversely proportional to the flow of the emulsion in the mill.

The breakup probability P_b is built from an expression based on hydrodynamics [15]:

$$P_b = \beta . \dot{\gamma} . \Delta t . \exp\left[- \alpha \frac{E_{coh}}{E_{def}} . \frac{d_{max} - d_i}{d_i - d_{min}} \right] . \tag{2}$$

In this expression the shear rate $\dot{\gamma}$ is inversely proportional to the breakage time of the droplets, and the exponential

term is proportional to the fraction of droplets of size d_i which are broken by units of time. α and β are coefficients of proportionality.

$\dot{\gamma}$ has the form:

$$\dot{\gamma} = \frac{\pi D \omega}{60 e} , \tag{3}$$

D is the distance between the center of the mill and the middle of the gap; ω is the rotation speed of the rotor (rpm)

The time lag Δt is assumed to be an increasing function of the resident time in the mill and a decreasing function of the shear rate:

$$\Delta t = \sqrt{\frac{t_p}{\dot{\gamma}}} . \tag{4}$$

Thus, a sufficient number of steps in the calculus is ensured.

The exponential term depends upon the competition between the cohesive energy E_{coh} which tends to keep the droplets unchanged, and the deformation energy E_{def} which tends to break them.

The cohesive energy E_{coh} is given by [16]:

$$E_{coh} = \left(\frac{\sigma}{d_i} X_i + \eta_d \frac{dX_i}{dt} \right) . \Delta S_i . \tag{5}$$

In this expression σ is the oil/water interfacial tension. It is calculated step by step as a function of the surfactant adsorption on the droplet surfaces [17]. This calculus involves the knowledge of surfactant characteristics such as critical micellar concentration (CMC), molecular weight (M) and molecular area (a_0) at the oil/water interface. $X_i = C_1 \dot{\gamma} d_i \Delta t$ is the droplet stretching, and $\frac{dX_i}{dt} = C_2 \dot{\gamma} d_i$ is the stretching speed. η_d is the dynamic viscosity of the dispersed phase and ΔS_i is the increase of the droplet surface under the effect of the droplet stretching. It has the form:

$$\Delta S_i = C_3 \pi (2 d_{i-1}^2 - d_i^2) = C_3 \pi (2^{1/3} - 1) d_i^2 . \tag{6}$$

The ratio $\frac{d_{max} - d_i}{d_i - d_{min}}$ imposes the breakup probability to be maximum for the largest droplets and minimum for the finest droplets, while C_1, C_2, C_3 are proportionality coefficients.

The deformation energy E_{def} is defined by [18]:

$$E_{def} = C_4 . \eta_c . \dot{\gamma}^2 . d_i^3 . \Delta t . \tag{7}$$

In this expression C_4 is a proportionality coefficient and η_c is the dynamic viscosity of the continuous phase.

Fig. 1 Evolution of the simulated droplet size distribution

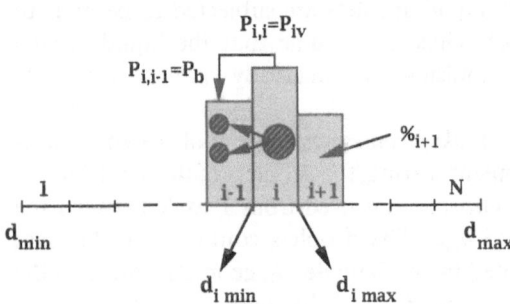

$P_{i,i} = P_{iv}$

$P_{i,i-1} = P_b$

$\%_{i+1}$

d_{min} d_{max}

$d_{i\,min}$ $d_{i\,max}$

Noting $\frac{C_2 C_3}{C_4} = \alpha_1$ and $\frac{C_2}{C_1} = \alpha_2$, the breakup probability can be written in the form:

$$P_b = \beta.\dot{\gamma}.\Delta t.\exp\left[-\alpha_1 \frac{\left(\frac{\sigma}{d_i}\dot{\gamma}\,d_i\Delta t + \alpha_2\,\eta_d\dot{\gamma}\,d_i\right)\pi(2^{1/3}-1)d_i^2}{\eta_c.\dot{\gamma}^2.d_i^3.\Delta t}\cdot\frac{d_{max}-d_i}{d_i-d_{min}}\right];$$

(8)

α_1, α_2 and β are three adjustment parameters which can be determined by fitting the measured Sauter mean diameters of two emulsions.

Test on bitumen in water emulsions

We have tested the permanent breakup model on bitumen in water emulsions generated in an Atomix C colloid mill. The experimental data were kindly supplied by the Groupement de Recherche de Lacq (Artix, France).

The characteristics of the Atomix C were $D = 59.5$ mm, $e = 1$ mm, $\omega_0 = 6600$ rpm (maximum rotation speed of the rotor). The flow of the emulsion in the mill gave $t_p = 0.15$ s. The values of the precedent parameters gave for the maximum shear rate 20600 s^{-1} and for the minimum time lag 0.0027 s.

The rotation speed of the rotor varied from 60% up to 100% of ω_0.

The oil phase was the ESSO 180–220 bitumen; its viscosity was 200 cP at 140 °C, the temperature used for the oil. The volume fraction of the oil phase was fixed equal to 0.60.

Three different cationic surfactants were used:

- Dinoram S (CMC = 1.53 10^{-3} mol/l, $M = 355$ g/mol, $a_0 = 115$ Å^2)
- Monomethyl Dinoram S (CMC = 1.73 10^{-3} mol/l, $M = 369$ g/mol, $a_0 = 150$ Å^2)
- Stabiram MS 3 (CMC = 2 10^{-3} mol/l, $M = 474$ g/mol, $a_0 = 195$ Å^2)

The surfactant concentrations were varied from one to four times the CMC.

The droplet size distributions are determined by laser diffractometry with a Cilas granulometer. All were found to be lognormal.

The fitting of the measured Sauter mean diameters with the simulated Sauter mean diameters of two emulsions generated with the Dinoram S surfactant gave the following adjustment parameters:

$$\alpha_1 = 3\ 10^{-5} \quad \alpha_2 = 1.7\ 10^{-2} \quad \beta = 0.80 .$$

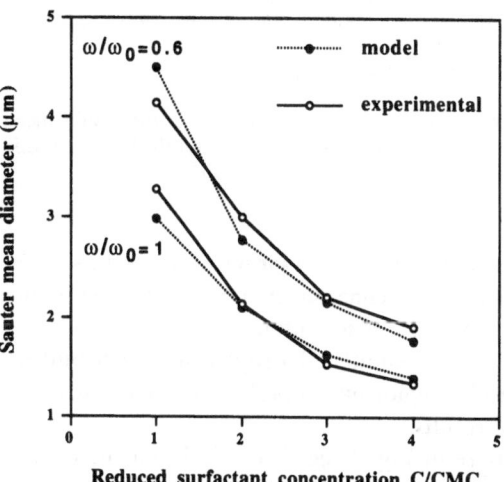

Fig. 2 Experimental and theoretical variations of the Sauter mean diameter versus the surfactant concentration of Dinoram S

Fig. 3 Experimental and theoretical variations of the Sauter mean diameter versus the surfactant concentration of Monomethyl Dinoram S

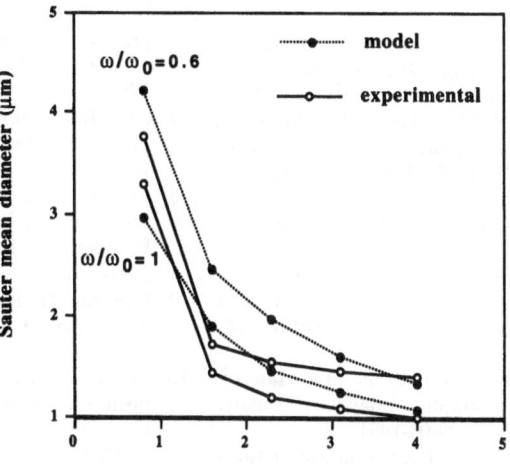

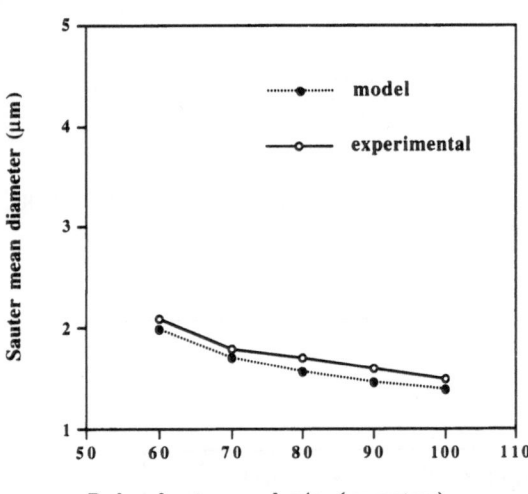

Fig. 4 Experimental and theoretical variations of the Sauter mean diameter versus the rotation speed of the rotor, for the Stabiram MS3 used at the concentration of 1.6 times the CMC

This triplet remained unchanged for the whole of the simulations on the surfactant concentration, the surfactant nature and the rotation speed of the rotor.

In all the cases simulated droplet size distributions were found to be lognormal, in perfect agreement with the experimental results.

We compare in Fig. 2 the theoretical and the experimental variations of the Sauter mean diameters for different Dinoram S concentrations, and for two rotation speeds of the rotor. The agreement between theory and experiment is excellent.

We compare in Fig. 3 the theoretical and experimental Sauter mean diameters obtained for different Monomethyl

Dinoram S concentrations, and for two rotation speeds of the rotor. The agreement between theory and experiment is good considering the degree of approximation of the model.

We compare in Fig. 4 the theoretical and experimental Sauter mean diameters obtained for different rotation speeds of the rotor, at a Stabiram MS 3 concentration of 1.6 times the CMC. The agreement between theory and experiment is again excellent.

Thus, the triplet of adjustment parameters defined from two emulsions made with the Dinoram S allow to predict the influences of the surfactant concentration, the surfactant nature and the rotation speed of the rotor.

Conclusion

The *permanent breakup model* that we propose accounts for the lognormality of the droplet size distributions of bitumen in water emulsions generated in a colloid mill. It has three adjustment parameters which can be easily determined from the fitting of theoretical and experimental Sauter mean diameters of two emulsions. One single set of these parameters is sufficient to account for the variations of the droplet size distributions in function of the surfactant concentration, the surfactant nature, the rotation speed of the rotor.

The model will be used successfully each time the viscosity of the system will be sufficiently high to avoid coalescences of the droplets within the colloid mill.

Acknowledgements The authors gratefully acknowledge the *Elf Aquitaine Society* for financial support and for permission to publish this work.

References

1. Mendiboure B, Dicharry C, Marion G, Morel G, Salager JL, Lachaise J (1993) Progr Colloid Polym Sci 93:307–311
2. Lachaise J, Mendiboure B, Dicharry C, Marion G, Bourrel M, Chenevière P, Salager JL, Colloids and Surface, to be published
3. Labofina SA (1981) Use of elastomer bitumens, France
4. Colas SA, Composite binders, their manufacture and their application in road construction, France, C04B007-00, 25-09-1991
5. Ciba-Geigy AG, Photographic Material and Photographic Images, Germany, G03C005-54, 09-08-1981
6. Eastman Kodak Co, Precipitated Photo-

graphic Materials Having Increased Activity, England, G03C007-32, 16-04-1991
7. CPC International Inc., Sugar-shortening Emulsions, England, A21D, 13-09-1970
8. Yissum Research Development Co, Emulsions of Essential Oils for Citrus Beverages, England, A23L002-00, 30-04-1984
9. Hagstam H, Barium Sulfate Emulsion for x-ray Contrats, Swed, A61K, 23-01-1968
10. Toray Silicone Co, Ltd, Cosmetic compositions containing microemulsions of dimethylpolysiloxane, England, A61K007-06, 01-06-1988

11. Durand G, Poirier JE, Richard F, Lenfant M (1993) Proceedings of the First Mondial Congress of the Emulsion (Paris), vol 1, EDS edition
12. Epstein B (1947) Ind and Eng Chem, 40 (12): 2289–91
13. Epstein B (1947) JFI, 471–477
14. Ito R (1978) Kagaku Kogaku Rombunshu, 4 (6): 629–633
15. Coulaloglou CA, Tavlarides LL (1977) Chem Eng Sci 32:1289–1297
16. Arrai K, Konno M, Matunaga Y, Saito S (1977) J Chem Eng Japan 10 (4), 325
17. Mendiboure B, Thesis (1992) University of Pau
18. Tavlarides LL, Stamatoudis M (1981) Adv in Chem Eng 11:199–273

Progr Colloid Polym Sci (1995) 98:173–176
© Steinkopff Verlag 1995

Properties of emulsion mixtures

J.-L. Salager
M. Ramirez-Gouveia
J. Bullon

Prof. J.-L. Salager (✉)
M. Ramirez-Gouveia · J. Bullon
Lab. FIRP
Ingeniería Quimica
Universidad de Los Andes
Mérida 5101, Venezuela

Abstract Emulsions are systems which exhibit some memory of their manufacturing process; thus, the same surfactant-oil-water system can result in different emulsions. This paper deals with the mixing of emulsions with the same formulation, but different drop-size distributions.

The two base emulsions are prepared in different vessels according to a standard procedure to attain fine or coarsely dispersed systems, with wider or narrower size distribution. Then, they are poured together and blended with a gentle stirring.

Experimental evidence indicates that the characteristics of the drop-size distribution of the emulsion mixture has a strong influence on the properties of the emulsion mixture, such as its viscosity and its stability. A considerable viscosity reduction can be attained by mixing two emulsions with identical formulation but different size distributions. The features which are found to promote the viscosity reduction are a strong bimodality of the combined distribution, a deep gap between modes, and a certain degree of polydispersity.

Key words Emulsion – mixture – inversion – bimodal – distribution

Introduction

The viscosity of a dispersed system, and particularly of an emulsion, is known to depend on its external phase viscosity, its internal phase ratio, and its drop size [1, 2]. The effect of the drop size is twofold: on the one hand, the smaller the average diameter, the higher the viscosity. On the other hand, the more polydispersed the drop size distribution, the lower the viscosity. This paper deals with the second effect, when the drop-size distribution exhibits two modes, i.e., two maxima, as found when different drop-size emulsions are mixed [3]. Most previous publications on the size distribution of mixed dispersions dealt with solid suspensions [4–6]; however, they can be taken as a guideline.

Experimental procedures

The surfactant-oil-water systems are pre-equilibrated at room temperature (21–24 °C) for at least 48 h prior to emulsification. Used surfactants are a detergent grade dodecylbenzene sulfonate sodium salt symbolized as DDBS, and a commercial alpha-olefin sulfonate (Stepanflow 30) manufactured by Stepan Chemicals. All systems contain 1 wt.% hydrophilic surfactant, so that all emulsions are of the O/W type; the aqueous phase contains 1 wt.% NaCl; the oil phase is either an adjusted viscosity mixture of a low viscosity kerosene cut (1.3 cP) and a 4000 cP lubricating oil base, or a viscous (>10000 cP) heavy crude oil from the Hamaca field.

The emulsion viscosity is measured with a Contraves Rheomat 30 apparatus equipped with a Couette cell, at 25 °C.

The internal phase proportion is 70 vol.% oil in all cases, that is a high internal phase ratio, for which the effect of the size distribution is enhanced [6].

Mixed emulsions always correspond to the same original formulation and composition, and they only differ by their drop-size distribution.

Different average drop sizes are attained by changing the stirring energy input (time, rpm) with an Ultraturrax T4558 turbine blender; drop-size distribution can be altered also by spacial and time programming techniques reported elsewhere [7–10].

The drop-size distribution is determined with a Malvern Master Sizer laser diffraction apparatus with a 0.1–500 μm range and 2% accuracy. O/W emulsions are diluted in two steps to the required water-to-oil ratio (oil content about 1/1000) with a 0.5 wt% sodium pyrophosphate solution. The average drop size is the so-called D(V, 0.5), i.e., the diameter corresponding to 50% of the drop-size distribution in volume. Note that in a Gaussian distribution, this average diameter matches the mode.

Figure 1 shows the drop-size distribution of two 70/30 O/W emulsions which happen to exhibit essentially the same viscosity (220 and 230 cP at 25 °C). These two emulsions are prepared by very different methods, and it is worth noting that the difference in average drop size is almost tenfold; as a consequence, the smaller drop size emulsion might be expected to be much more viscous than the coarse one; however, this is not the case, and the lower than expected viscosity of the fine emulsion is attributed to

two features exhibited by the drop-size distribution: polydispersity and bimodality.

Lets now look at these characteristics in detail.

Viscosity of emulsion mixtures

The simplest way to produce a bimodal emulsion is to mix two emulsions that are prepared separately. The resulting emulsion mixture size characteristics depend upon the characteristics of the two base emulsion distributions and the relative amount of each emulsion.

Figure 2, leftside, shows the variation of the viscosity of emulsion mixtures containing different proportions of two base emulsions; the size distributions of these two base emulsions are plotted in Fig. 2, rightside. The fine emulsion is the more viscous, both because it has smaller drops and because its distribution is more monodispersed; note that the opposite occurs for the coarse emulsion. Figure 2, rightside, indicates that the two distributions are overlapping only slightly in the 10–20 μm range.

Figure 2, leftside, shows the viscosity of emulsion mixtures as a function of the volume proportion (φ) of the fine emulsion. The data show that mixtures containing up to 50% of the fine emulsion exhibit a viscosity that is equal or even slightly lower than the viscosity of the coarse emulsion. In this case, the viscosity minimum seems to occur for a mixture containing about 25% of the fine emulsion.

Figure 3 shows the same type of data for emulsions made with a heavy crude oil. In this case, both the fine and coarse emulsions are very polydispersed (see Fig. 3, rightside), with log-normal characteristics that come from the stirring process. Because of this polydispersity, the two base emulsion distributions overlap over a wide range, e.g., 10 to 60 μm. Nevertheless, the viscosity of the emulsion mixture still goes through a minimum (Fig. 3, leftside) that this time occurs for a composition at about 50–60% of fine emulsion.

Other results (not shown) indicate that the overlapping of the two distributions tends to reduce and even

Fig. 1 Drop-size distributions of two emulsions exhibiting the same viscosity, but produced from a same formulation by different stirring methods

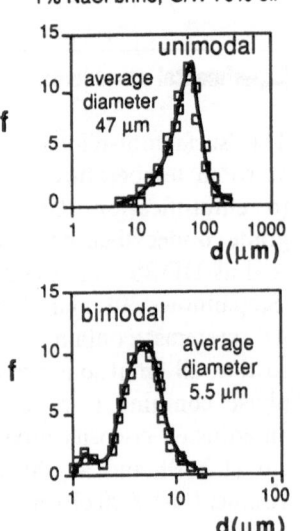

Fig. 2 Left: variation of the viscosity of emulsion mixtures as a function of the vol. proportion (φ) of the fine emulsion. Right: drop-size distribution of the two base emulsions

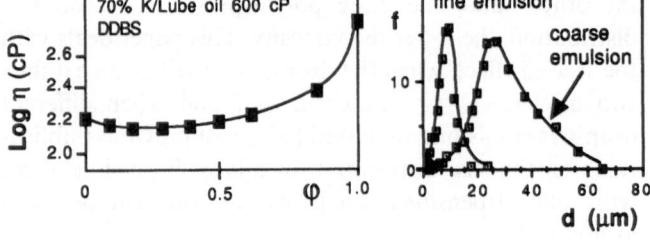

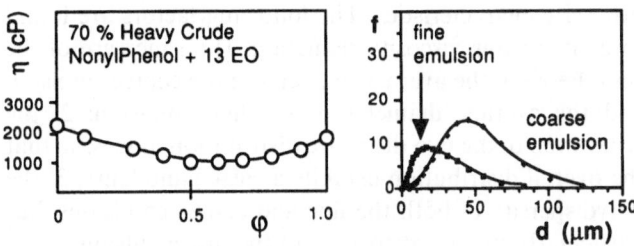

Fig. 3 Left: variation of the viscosity of emulsion mixtures as a function of the vol. proportion (φ) of the fine emulsion. Right: drop-size distribution of the two base emulsions

Fig. 5 Drop size distributions of the 50% fine emulsion mixture for different binary cases. Same systems as in Fig. 4

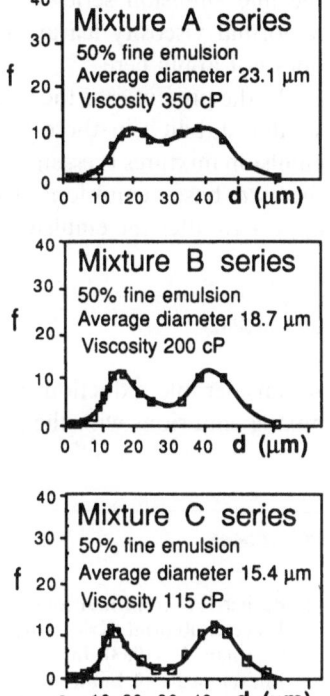

eliminate the viscosity minimum [7, 8]. The case shown in Fig. 3 is rather exceptional, in the sense that the fine emulsion can be considered to be extremely polydispersed. Indeed, it is so polydispersed that its viscosity is lower than the viscosity of the coarse emulsion, in spite of a much lower average diameter.

Figure 4 shows the influence of the difference between the two base emulsions. This difference is measured as the ratio between the average diameters of the coarse and fine emulsions.

Mixtures of the A series are produced by mixing a fine emulsion (average diameter 20 μm) and a coarse emulsion (average diameter 42 μm), with an average diameter ratio = 2.1. The two distributions overlap to a great extent, as shown in Fig. 5-A, which indicates the overall size distribution of a 50–50% mixture (arrow in Fig. 4, curve A). The resulting viscosity is rather high (350 cP), and no viscosity minimum is apparent over the whole composition range.

Curve B in Fig. 4 indicates the viscosity variation of mixtures containing the same coarse emulsion (diameter

Fig. 4 Variation of the viscosity of emulsion mixtures as a function of the vol. proportion (φ) of the fine emulsion, for different binary cases

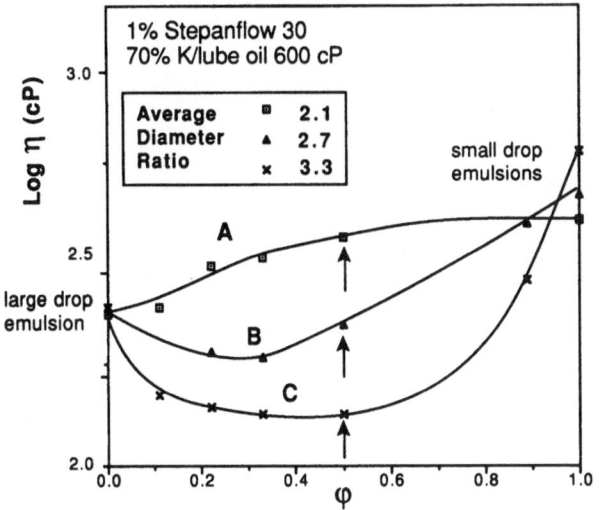

42 μm) and a lower diameter fine emulsion (diameter 15.5 μm), so that the average diameter ratio increases up to 2.7. The curve exhibits a clear minimum in viscosity at about 30% of the fine emulsion. Figure 5-B shows that the overall size distribution of a 50–50% mixture exhibits a strong bimodality and a narrower region of overlapping. The viscosity of this mixture (200 cP) is substantially lower than in case A, in spite of a smaller average drop size.

Cases C in Figs. 4 and 5 indicate the same results for mixtures of the same coarse emulsion with an even lower diameter fine emulsion (diameter 12.7 μm), so that the average diameter ratio is 3.3. In this last case, the viscosity minimum is enhanced both in width and depth. Figure 5-C indicates a better separation of the two distributions, although the difference in contrast to Fig. 5-B is not extremely apparent on the distribution shape. It can be said only that the gap between the two peaks is slightly deeper than in the previous case. Actually, the drop in viscosity from 200 cP to 115 cP seems to be the most sensitive measurement of the change.

The ratio between the average diameters of the coarse and fine emulsions seems to be of first importance in determining the possibility of a minimum in viscosity when two emulsions are mixed.

This is true if the two emulsions exhibit the same kind of polydispersity, and it is worth noting in Fig. 4 that the viscosity of the fine emulsion component in the mixture ($\varphi = 1$) increases steadily when its average drop diameter increases. This is an indication that the polydispersity of

176
J.-L. Salager et al.
Properties of emulsion mixtures

the fine emulsion series is fairly constant, and that the abnormal viscosity feature mentioned in the Fig. 3 case does not apply here.

In the present case the two base emulsions exhibit very similar stability; in these conditions, the stability of the emulsion mixtures is essentially the same as the stability of the two bases emulsions, so that the mixing does not appear to alter the emulsion stability in this case.

Conclusions

A considerable reduction of viscosity can be attained by mixing emulsions with the same formulation but different drop-size characteristics. The following factors are found to enhance this viscosity reduction: 1) an increase of the ratio between the average diameter of the coarse emulsion and the average diameter of the fine emulsion; 2) the separation of the two drop-size distribution modes, so that the overall distribution exhibits a clear bimodality; 3) the polydispersity of both the fine and coarse emulsions, but particularly the polydispersity of the fine emulsion.

Acknowledgements The Lab. FIRP research program at Universidad de Los Andes, is sponsored by the University Research Council CDCHT and the Lab. FIRP Industrial Sponsor Group: Corimon, Hoechst de Venezuela, INTEVEP and Procter & Gamble de Venezuela.

References

1. Becher P (1977) Emulsions: Theory and Practice, Reprint, Krieger R, New York
2. Sherman P (1983) In: Becher P (ed) Encyclopedia of Emulsion Technology. Dekker M, New York, vol I, p 405
3. Rivas H, Nuñez G, Dalas C (1993) Vision Tecnológica 1:18
4. Parkinson C, Matsumoto S, Sherman P (1970) J Colloid Interface Sci 33:150
5. Sengun M, Probstein P (1989) Rheological Acta 28:382, 28:394
6. Poslinski A, Ryan M, Gupta K, Seshadri S, Frechette F (1988) J Rheology 32:751
7. Ramirez-Gouveia M (1992) Informe Técnico FIRP #9210, Universidad de Los Andes, Mérida-Venezuela
8. Bullón J (1993) III Jornadas Facultad de Ingeniería, Universidad de Los Andes, Mérida-Venezuela, October 10–15
9. Ramirez M, Bullón J (1994) 10th Int Symposium on Surfactants in Solution, Caracas-Venezuela, June 26–30
10. Salager JL, Pérez-Sanchez M, Garcia Y, Physico-chemical parameters influencing the emulsion drop size (this volume)

Progr Colloid Polym Sci (1995) 98:177–179
© Steinkopff Verlag 1995

M. Miñana-Pérez
A. Graciaa
J. Lachaise
J.-L. Salager

Solubilization of polar oils in microemulsion systems

Dr. M. Miñana Pérez (✉) · J.-L. Salager
Lab. FIRP
Ingeniería Química
Universidad de Los Andes
Merida 5101, Venezuela

M. Miñana-Pérez · A. Graciaa
J. Lachaise
Lab. TEMPM CURS
Université de Pau
64000 Pau, France

Abstract A new type of amphiphile that contains both conventional surfactant and lipophilic linker features in a single molecule was designed and tested. In these so-called extended surfactants, a polypropylene oxide chain of variable length is inserted in between the conventional polar and apolar groups.

With this type of surfactant, it was possible to produce for the first time a middle phase microemulsion in alcohol-free systems with long chain ($C_{10}–C_{18}$) synthetic and natural triglyceride oils. High molecular weight hydrocarbons were solubilized as well.

The reported solubilization at optimum is found to depend upon both the propylene oxide chain length and the oil structure. The solubilization parameter of polar oils is found to attain quite remarkable values, in the range of several milliliters of oil per gram of extended surfactant, an interesting feature as far as the applications are concerned.

Key words Extended surfactant – lipophilic linker – microemulsion – optimum formulation – optimum solubilization parameter

Introduction

Semiempirical correlations for the attainment of an optimum formulation have been proposed for both anionic and nonionic surfactant systems containing hydrocarbon oils and aqueous brines [1, 2]. An optimum formulation is attained when the system exhibits a three-phase behavior with equal amounts of oil and brine solubilized in the microemulsion middle phase [3]. The guidelines to improve the interactions between the surfactant and its physico-chemical environment have been known for some time [4].

Recently [5, 6], it has been shown that a very lipophilic amphiphile, the so-called lipophilic linker, may substantially improve the solubilization in properly formulated surfactant-oil-water microemulsions. All reasoning

concerning the solubilization in microemulsion has been handled easily thanks to the original concepts proposed by Winsor 40 years ago [7].

Because of its molecular structure, this amphiphilic compound is able to induce some orientation in the oil phase near the interface, thus extending the reach of the surfactant "tail" deeper into the oil phase and providing additional interactions; moreover, this situation probably creates a gradient of polarity in the vicinity of the interface.

In the present study the same polarity gradient concept is attained by using a new type of surfactant structure, the so-called extended surfactant. Such a label comes from the fact that this type of amphiphile contains an intermediate polarity "extension" in between a conventional hydrophilic group (sulfate) and the typical $C_{12}–C_{18}$ lipophilic hydrocarbon chain. The extension is provided by poly-propylene oxide chain of variable length, which can result in extra molecular interaction between the surfactant and the polar oil.

This paper reports the phase behavior and microemulsion formation with extended surfactants. It is shown that the variation of the number of propylene oxide units is a simple way to adjust the intermediate polarity feature, so that the solubilization of different alkanes and synthetic or natural polar oils can be achieved. Optimum salinity (S*) is used to characterize the relative hydrophilic-lipophilic balance of the extended surfactants for brine and a given oil. Optimum solubilization parameter (SP*), defined as the volume of solubilized oil or water per unit mass of surfactant uptake in the microemulsion middle phase, is being used as a yardstick of the surfactant efficiency.

Materials and experimental procedure

The extended surfactants are alkyl polypropoxylated ethoxy sulfates, containing two ethylene oxide units to insure the link between the polypropylene oxide chain and the sulfate group; they were synthesized and provided by SEPPIC. Table 1 indicates the different labels used for extended surfactants with different alkyl chain (12 or 18 carbon atoms), and different average propylene oxide numbers (PON = 6, 10 or 14). They were used as received from the manufacturer.

The hydrocarbons are purum grade Merck reagents. The polar oils, synthetic as well as natural, are pharmaceutical grade products manufactured by Laserson-Sabetay, Hüls, Aldrich Chimie, and Touzart-Matignon.

The phase equilibrium experiments are carried out according to a standard procedure [1–2, 8], by scanning a single formulation variable. Water-to-oil ratio (WOR) is kept equal to 1. Oil, brine, and surfactant are poured into graduated test tubes in the required proportions. The test tubes are sealed and placed in a thermostated bath, gently shaken twice a day over 1 week, and then left to equilibrate until the volume of the phases in equilibrium becomes constant. Then, the phase behavior is noted and the phase volumes are recorded to compute the optimum solubilization parameter value.

Optimum salinity (S*) as a function of PON and oil structure

Optimum three-phase systems are produced with alkanes as well as polar oils, sodium chloride brine, and 1.25 wt.% extended surfactant. The temperature is kept constant at 35 °C, to simulate body conditions. Salinity scans are carried out with every extended surfactant-oil pair.

Figure 1 shows the variation of the logarithm of the optimum salinity (Ln S*) as a function of either the Alkane Carbon Number (ACN) for hydrocarbons (circle dots) or the Equivalent Alkane Carbon Number (EACN[4–6]) for polar oils. The latter were evaluated from experimental optimum salinity and subsequent graphical extrapolation. As the number of propylene oxide groups (PON) increases (for a given oil), the interaction energies of the surfactant with the oil phase A_{CO} increases, and it has to be compensated by a decrease in salinity, in order to similarly increases A_{CW}.

When comparing data for the same surfactant and different oils, a similar reasoning applies; for instance, the increase in optimum salinity in switching from ethyl oleate to soya oil is expected to be a direct consequence of the increase in oil lipophilicity; in other words, the increase in

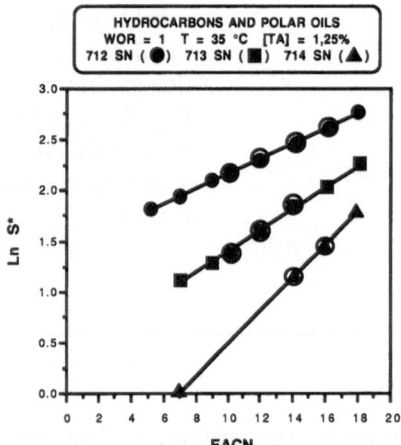

Fig. 1 Optimum Salinity (Ln S*) variation as a function of Alkane Carbon Number (ACN) for Hydrocarbons (circled) or Equivalent Alkane Carbon Number (EACN) for polar oils (filled dots)

Table 1 Characteristics of the Extended Surfactants.

Surfactant label	Lipophile length (Cn)	PON
712 SN	12	6
713 SN	12	10
714 SN	12	14
978 SN	18	14

Table 2 Characteristics of alkanes, Hydrocarbons and Polar Oils.

Hydrocarbons	ACN	Polar oils	EACN
Decane	10	Ethyl myristate (ester)	5
Dodecane	12	Ethyl oleate (ester)	7
Tetradecane	14	Miglyol 840 (diglyceride C10–C12)	9
Hexadecane	16	Miglyol 812 (triglyceride C10–C12)	14
Paraffin	18	Soja oil (triglyceride C18)	18

Progr Colloid Polym Sci (1995) 98:177–179
© Steinkopff Verlag 1995

A_{CW} would be compensated by an increase of oil EACN. An important feature of this plot is the reproducibility of the EACN values with the three surfactants.

As expected from the anionic nature of the extended surfactants, the LnS*-ACN plots are straight lines, and the slope is consistent with the reported one for alkyl sulfates [1, 4]; however the slope depends upon the PON, i.e., upon the nonionic nature of the surfactant.

Solubilization parameters (SP*) as a function of PON and oil structure

Figures 2 and 3 data show the variation of the solubilization as a function of ACN (respectively EACN). Each surfactant is kept at its optimum salinity. As can be seen, the solubilization either decreases as the oil molecular weight increases, or it exhibits a maximum, depending on the surfactant intermediate chain length and the type of oil. The shortest surfactant SN712 (C_{12} and 6 PO) allows the micro emulsification of alkanes, but the solubilization

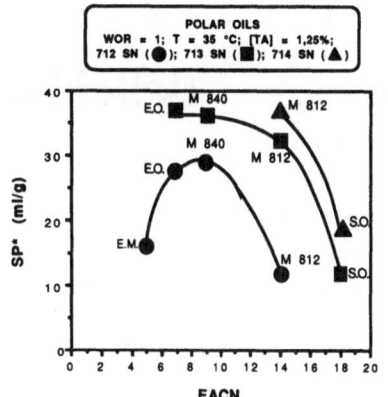

Fig. 3 Variation of Optimum Solubilization Parameters (SP*) as a function of the Equivalent Alkane Number (EACN) of polar oils

is decreasing with increasing ACN or reaching a maximum when polar oils are being solubilized.

Only hexadecane and synthetic as well as natural triglycerides could be solubilized with the longest surfactant SN 714 (C_{12} and 14 PO). On the other hand, neither the shortest alkanes ($< C_{14}$) nor the shortest (C_8-C_{10}) synthetic di- and triglycerides could be solubilized. The first addition of four propylene oxide units (6 PO → 10 PO) is more significant than the second one (10 PO → 14 PO).

Conclusions

It has been shown that the so-called extended surfactants allow the solubilization of long chain alkanes, single chain polar oils, as well as di- and triglyceride oils in alcohol-free microemulsion systems.

The general trends to select the best surfactant structure as a function of the oil to be solubilized are reported.

The experimental solubilization parameters reach extremely outstanding values, e.g., 10–20 ml/g for triglycerides and 30–40 ml/g for single chain polar oils.

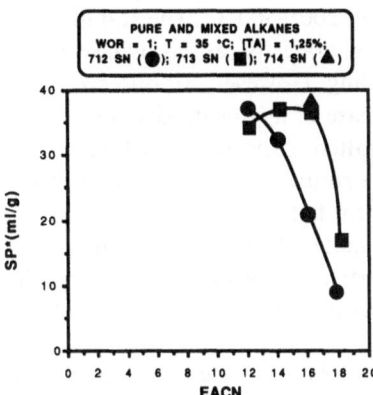

Fig. 2 Variation of Optimum Solubilization Parameters (SP*) as a function of Alkane Carbon Number (ACN) or Equivalent Alkane Carbon Number (EACN)

References

1. Salager JL, Morgan J, Schechter RS, Wade W (1979) Soc Pet Eng J 19:107
2. Bourrel M, Salager JL, Schechter RS, Wade W (1980) J Colloid Interface Sci 75:451
3. Reed RL, Healy RN (1977) In: Shah DO, Schechter RS (eds) Improved Oil Recovery by Surfactants and Polymer Flooding. Academic Press, New York
4. Bourrel M, Schechter RS (1988) Microemulsions and Related Systems. Dekker M, New York
5. Graciaa A, Lachaise J, Cucuphat C, Bourrel M, Salager JL (1993) Langmuir 9:669
6. Graciaa A, Lachaise J, Cucuphat C, Bourrel M, Salager JL (1993) Langmiur 9:3371
7. Winsor P (1954) Solvent Properties of Amphiphilic Compounds, Butterworth, London
8. Miñana-Pérez M (1993) Doctoral Thesis, University of Pau, France

Progr Colloid Polym Sci (1995) 98:180–183
© Steinkopff Verlag 1995

V. Guille
D. Espinat
L. Barré
J.C. Ravey
J. Lambard
Th. Zemb

Colloidal macrostructure of crude oil

Dr. V. Guille (✉) · D. Espinat · L. Barré
Institute Francais du Pétrole
1-4 Avenue de Bois Préau
92 506 Rueil Malmaison, France

J.C. Ravey
Laboratoire de Physico Chimie
des Colloides
UA 406
54506 Venoeuvre les Nancy, France

J. Lambard · Th. Zemb
CEN Saclay
Service de Chimie Moléculaire
91121 Gif sur Yvette Cedex, France

Abstract Asphaltenes, defined as the fraction insoluble in *n*-heptane, are responsible for flocculation of crude oil. We consider both crude oil and asphaltene solutions as dynamical colloidal systems containing particles that may form different aggregates upon changing the temperature and polarity of the solvent. For example, resins, the polar molecules present in crude oils, constitute a good solvent for asphaltenes.

We used ultra small-angle x-ray scattering (USAXS) in order to investigate the evolution of the macrostructure of asphaltenes, resins, and crude oils. This technique enables to investigate particles of the size range 10–10000 Å.

We have studied a simplified system composed of asphaltenes and resins in a toluene solution, as well as the vacuum residue that contains asphaltenes, resins, aromatics, and saturated species.

An important finding is the presence of large-size fluctuations of concentration in the temperature region 200°–300 °C. The stability of agglomerates is caused by covalent bonding. However, we observe the decrease of aggregate size for asphaltenes and resins at higher temperatures. The rheology results confirm flat shape of the asphaltene's aggregates. The effect of temperature is particularly pronounced for small aggregates of asphaltene molecules.

Key words Asphaltenes – aggregation – temperature – x-ray scattering

Introduction

Crude oil is a complex colloidal system. It contains a variety of molecules: asphaltenes, resins, aromatics, and saturated hydrocarbons. The heaviest fraction contains asphaltenes, responsible for the flocculation phenomena that occur during oil production in the field.

Yen et al. [1] proposed a structural model for the aggregation of asphaltene molecules as stacks of a few aromatic sheets surrounded by aliphatic chains. These sheets further self-assemble into micelles [2] that form flat, large aggregates [3]. Their size depends on a number of factors: i) the presence of heteroatoms such as nitrogen, sulphur, and oxygen [4], ii) environment via the polarity of the solvent [5, 6], and iii) the presence of transition metal atoms such as nickel or vanadium. Cohesion between the micelles is ensured by hydrogen bonds [7], van der Waals forces, and π-acceptor-π-donor (covalent) bonds. The evolution of molecular aggregation in asphaltenes has been previously studied using neutron scattering [8] and structural information over a limited size range has been obtained. In order to extend this range USAXS was employed in the present study. However, the low electronic density contrast between the toluene and asphaltenes limits the feasibility of this technique at low concentrations.

Progr Colloid Polym Sci (1995) 98:180–183
© Steinkopff Verlag 1995

In this report, we present data on the influence of temperature on the aggregation properties of asphaltenes and resins in toluene solution as well as in pure VR. The highest temperature used was 300 °C, just below the cracking temperature. The studies versus concentration of asphaltenes and resins in toluene solution have been performed at room temperature.

The asphaltenes were obtained from Safanya Vacuum Residue by n-heptane precipitation using the AFNOR T60-115 procedure. The resins were separated by liquid chromatography. The density at 20 °C was 1.19 g/cm³ for asphaltenes and 1.08 g/cm³ for resins.

Theoretical

Small-angle x-ray scattering experiments are performed by measuring the intensity scattered by the specimen at a small angle 2θ. The scattered intensity (differential cross-section) $I(Q)$ is given for centrosymmetric systems by the following general equation:

$$I(Q) = (\rho_1 - \rho_2)^2 \int_0^\infty \gamma(r) \frac{\sin(qr)}{qr} 4\pi r^2 \, dr \qquad [I]$$

where $\gamma_0(r)$ is the characteristic function introduced by Porod. This allows determination of an area per unit volume, as well as a characteristic size of the dispersion of two homogeneous media.

If we can use a further approximation, the scattering may be reduced to a product of a structure factor by an average form factor in the following expression, only valid for spheres in thermodynamic equilibrium:

$$I(Q) = c_i n_e v (1 - n_e v) (\rho_1 - \rho_2)^2 F(Q) S(Q), \qquad [II]$$

where $Q = (4\pi/\lambda)\sin\theta$, $C_i = 7,9010^{-26}$ cm², ρ is the electronic density of solvent (1) and colloidal particle (2), n_e is the number of particles per unit volume and v is the volume of a single particle. The quantity $(\rho_1 - \rho_2)^2$ is called the contrast term, $F(Q)$ is the form factor, and $S(Q)$ is the structure factor that depends on interparticular interactions.

For diluted systems (i.e., $1 - n_e v \cong 1$), it is assumed that there are no interactions between the particles and, consequently, $S(Q) = 1$. Moreover, for the scattering vector equal to zero the form factor is equal to the volume v of the particle. Relation [I] thus becomes:

$$I(Q = 0) = C_i n_e v^2 (\rho_1 - \rho_2)^2. \qquad [III]$$

This scattered intensity is therefore proportional to the molecular weight. Using the above relations, we determine the molecular weight and characteristic aggregate dimensions for each scattering curve. We assume a disk model for the form factor $F(Q)$.

Experimental

Two experimental arrangements were used for x-ray measurements. For the ultra small Q values, we employed a triple-axis Bonse–Hart camera at CEN, Saclay, with Q-range 5×10^{-4} to 5×10^{-2} Å⁻¹ [9]. For Q values in the range 10^{-2} to 0.5 Å⁻¹, we used a conventional Guinier camera at the French Institute of Petroleum. For high-temperature experiments (up to 300 °C) the sample was introduced into a high-temperature cell equipped with 25 μm thick kapton windows. For the studies of asphaltenes and resins at ambient conditions glass capillaries were used.

Results

Asphaltenic systems

Asphaltenes Safanya were diluted in toluene at 20 °C at concentrations 2 wt.% and 6 wt.%. The corresponding scattering curves are presented in Fig. 1. The two curves are similar and, as expected, the scattered intensity for the more concentrated solution is three times larger. There is no modification to the structure of aggregates with concentration. The intrinsic viscosity of the Safanya asphaltenes is 6.3, which indicates the presence of flat aggregates.

For both 2 wt.% and 6 wt.% solutions, we obtain the same aggregate characteristics: diameter 110 Å, thickness 15 Å and molecular weight 170000. At small Q values the intensity follows a power law with a slope equal to -2.4. It reveals the existence of fluctuations of concentration on a size scale up to 1000 Å. The heterogeneities in solution are caused by the occurrence of regions rich and poor in asphaltene; the solution looks like a two-phase system.

Fig. 1 X-ray scattering for Safanya asphaltenes 2 wt.% (1) and 6 wt.% (2) in toluene at 20 °C

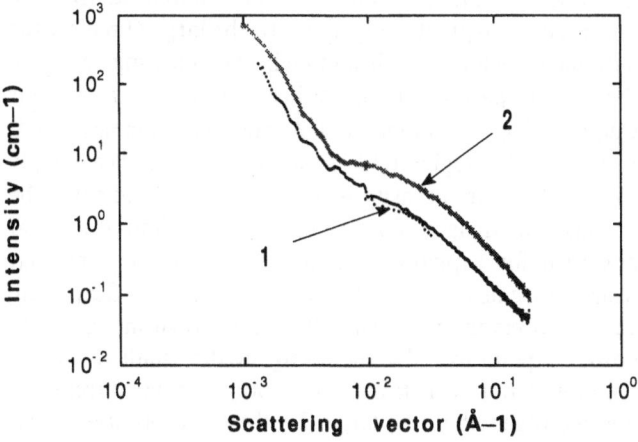

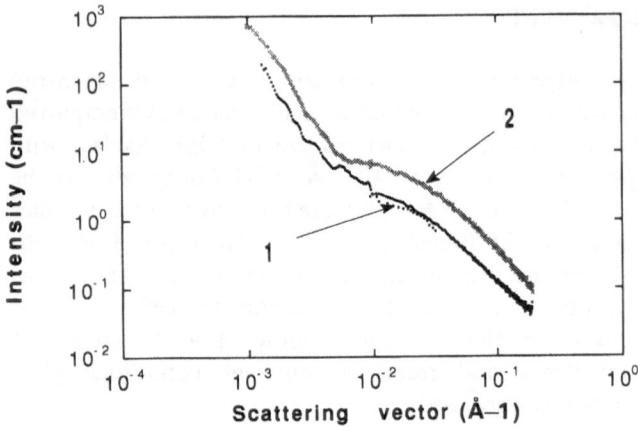

Fig. 2 X-ray scattering for Safanya asphaltenes 6 wt.% in toluene at 20 °C (1) and 80 °C (2)

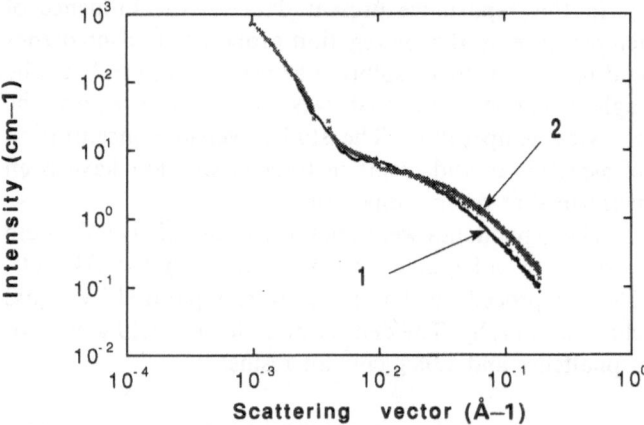

Fig. 3 X-ray scattering for Safanya resins 6 wt.% (1) and 10 wt.% (2) in toluene at 20 °C

In Fig. 2, we present the effect of temperature on the scattering of asphaltenes. The scattering curve can be separated into two regions. At large Q values the intensity decreases slowly, which indicates presence of small particles. At small Q values the scattering curves for 20° and 80 °C are identical and very steep, which indicates the existence of large, thermally stable aggregates. Upon the increase of temperature some bonds break which results in decrease of the small particle size, but on the large scale the concentration fluctuations are still present. We have measured viscosity versus shear rate for 2 wt.% and 6 wt.% solutions and a Newtonian regime was observed. Our small-Q scattering results strongly suggest the presence in solution of some large, thermally stable agglomerates.

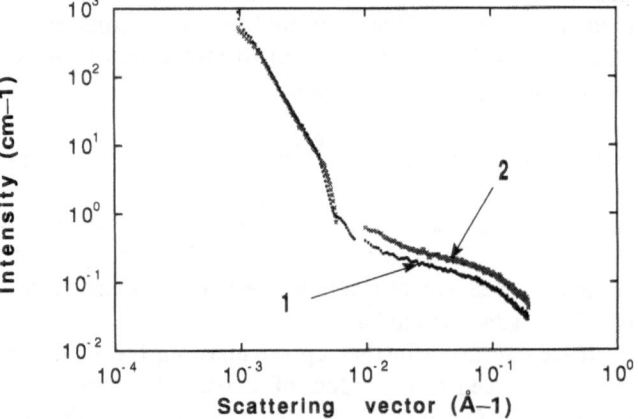

Fig. 4 X-ray scattering for Safanya resins 10 wt.% in toluene at 20 °C (1) and 80 °C (2)

Resin systems

The scattering curves for two different concentrations of resins in toluene solution are shown in Fig. 3. The shape of the scattered intensity curve is very similar to the one observed for asphaltene systems. In the large Q region the data have been fit to a disk model. The following values of the fitting parameters have been obtained: molecular weight 8000, disk diameter 30 Å, and disk thickness 10 Å. In the small Q region (Q less than 4×10^{-3} Å^{-1}) a steeply rising scattering intensity is observed, which indicates the presence of density fluctuations of size similar to that observed for asphaltene solutions despite the fact that resin molecules are smaller than asphaltenes. Scattering curves observed at 20° and 80 °C are shown in Fig. 4. At both temperatures the scattering in the small Q region remains strong, which indicates good thermal stability of the heterogeneities. At large Q values, we observe much

weaker Q-dependence of the scattering curves, which indicates presence of smaller aggregates. These small aggregates seem to be affected by the increased temperature.

Vacuum residue

The effect of temperature on the x-ray scattering spectra of the Safanya Vacuum Residue is presented in Fig. 5. As for the other systems discussed above, we observe different behavior in the small Q and large Q regions. The small density fluctuations that dominate the scattering intensity for Q values from 10^{-3} to 9×10^{-3} Å^{-1} seem to be thermally stable up to 200 °C. On the other hand, the strong scattering observed in the small Q region ($Q < 10^{-3}$ Å^{-1}) at 20 °C (indicating particle size about 8000 Å) markedly decreases at elevated temperatures. This may be brought about by the presence of paraffins in the industrial VR. We

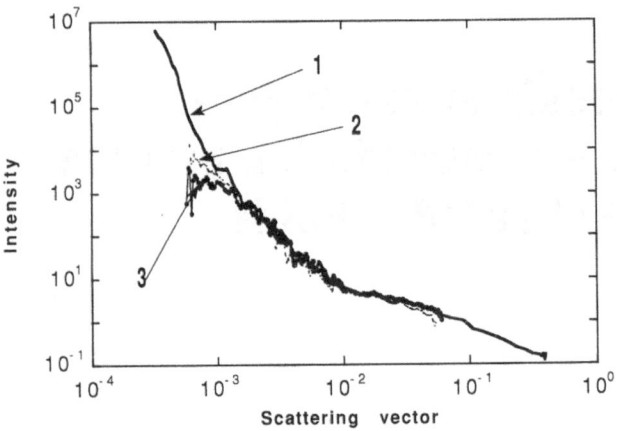

Fig. 5 X-ray scattering for Safanya Vacuum residue at 20 °C (1), 90 °C (2) and 200 °C (3)

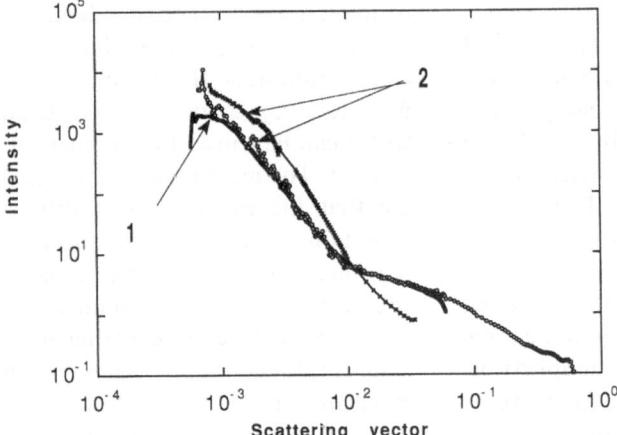

Fig. 6 X-ray scattering for Safanya Vacuum residue at 200 °C (1) and 300 °C (2)

also beleive that at room temperature the small domains aggregate into larger particles. These larger particles mostly dissociate upon the increased temperature, although some heterogeneities are thermally stable. Given the extent of this thermal stability, we conjecture that the structure is strongly connected in a gel-like manner. Covalent bonds ensure the cohesion of larger domains. Therefore, we suggest that at elevated temperatures van der Waals type interactions become unimportant and the stability of large domains is ascertained by covalent bonding between the smaller organic structures, in full analogy to polymer gels.

In the temperature region 200°–300 °C the scattering results for VR may be irreproducible. This is illustrated in Fig. 6, where two different experimental curves registered at 300 °C are shown. In one case the structure does not change between 200° and 300 °C, whereas in the other, we observe an increase of scattering intensity in the Q region between 10^{-3} and 10^{-2} Å^{-1}. The irreproducibility is probably caused by the inhomogeneity of the VR and in some samples we may observe the onset of the cracking process. Upon the increase of temperature some organic molecules may be transformed by breaking the carbon-carbon covalent bonds, which results in a locally less dense region.

Conclusion

The USAXS technique offers an important insight into the macrostructure of petroleum products. Using a conventional Guinier-type camera, one can obtain only a local description for sizes up to 500 Å. Using USAXS, we have been able to demonstrate the presence of large-scale density fluctuations for vacuum residue as well as for asphaltene and resin solutions. These fluctuations are thermally stable up to 300 °C, probably due to covalent bonding similar to that reported in gel-like structures.

We propose a model for asphaltene and resin aggregates in toluene solutions, which may also apply to vacuum residue. According to this model, the structure responsible for the x-ray scattering pattern is held together by a small percentage of covalent bonds forming three-dimensional cages that contain the electronically dense regions. Smaller aggregates composed of a variety of small molecules interacting via the van der Waals forces form independently of those large aggregates. The size of those small aggregates is temperature-dependent. Newtonian flow behavior of the solutions studied here indicates that only a small proportion of large covalently-bound agglomerates exists.

Acknowledgements Cotton and J. Teixeira from Leon Brillouin Laboratory, Saclay, France, for their involvement in the experimental stage of this work.

References

1. Yen TF, Erdmann JG, Pollack SS (1961) Analyt Chem 33:1587
2. Dickie JP, Yen TF (1967) Analyt Chem 39:1847
3. Ravey JC, Ducouret G, Espinat D (1988) Fuel 67:1560
4. Ignasiak T, Bimer J, Samman N, Montgommery DS (1979) ACS Div Petr Chem (1979) 24:1001
5. Maruska HP, Rao BML (1987) "The Aggregation of Asphaltenes" Marcel Decker Inc.
6. Moschopedis SE, Fryer JF (1979) Speight JG, Fuel 55:227
7. Moschopedis SE, Speight JG (1976) Fuel 55:187
8. Espinat D, Ravey JC, Guille V, Lambard J, Zemb Th, Cotton JP (1993) Journal de Physique IV 3:C8
9. Lambard J, Zemb Th (1991) 24:551–561

Progr Colloid Polym Sci (1995) 98:184–188
© Steinkopff Verlag 1995

S. Durand Vidal
J.P. Simonin
P. Turq
O. Bernard

Acoustophoresis of electrolyte solutions and sedimentation profiles of screened charged colloids

S. Durand Vidal · J.P. Simonin
P. Turq (✉)
Laboratoire des Propriétés Physico-
Chimiques des Electrolytes
URA CNRS 430
Université Pierre et Marie Curie
Bât. 74, Boite courrier 51
4, Place Jussieu
75252 Paris Cedex 05, France

O. Bernard
Laboratoire d'Electrochimie Analytique
et Appliquée
URA CNRS 216
Ecole Nationale Supérieure de Chimie
de Paris (ENSCP)
11, rue P. & M. Curie
75231 Paris Cedex 05, France

Abstract Acoustophoresis consists of
applying an ultrasonic wave to
a charged solution and measuring the
induced electric field, originating from
the local separation of charges. This
effect, predicted by Debye in 1933, has
been studied for electrolytes (Ionic
Vibration Potential, IVP) and for
colloids (Colloidal Vibration
Potential, CVP).

In the first part of this work, we
analyze the effect of interionic forces
on IVP within the Mean Spherical
Approximation, at the same level of
description as for conductance of
electrolytes. This analysis is required
by the fairly high concentrations used
in IVP experiments. Experimental
data found in the literature are used
to find the "acoustophoretic mass" of
cations, which involves solvation
water. The values obtained for alkali
and alkali-earth cations are in good
agreement with other cation solvation
numbers.

In the second part, a method
for the calculation of equilibrium

sedimentation profiles of charged
colloids, in the presence of added salt,
is presented. It is found that the
concentration of the colloidal
particles obeys a non-linear equation,
which can be solved by a numerical
method. The main assumptions used
are that the electrical neutrality
condition is verified locally and that
the colloids interact through a hard-
sphere potential. The results are
compared with recent experimental
data. It is found that the agreement is
excellent in the case of high screening
(excess of salt), but becomes poorer
when the salt concentration is
decreased. The salt-free case is also
approached.

We finally present our projects
that describe acoustophoresis applied
to colloidal suspensions where the
two previous studies are used.

Key words Acoustophoresis – mean
spherical approximation – hydration
numbers – sedimentation – charged
colloids

Acoustophoresis of electrolyte solutions

Previous works

Acoustophoresis was originally described by Debye in
1933 [1]. Basically, the phenomenon can be pictured as

follows: when an ultrasonic wave is applied to a liquid, the
dilatation/compression induces a motion of the solvent. In
a 1:1 electrolyte solution for instance, the two ions move
differently in the solvent's velocity field, because they have
different masses and frictional coefficients. This effect
causes local heterogeneities of charge, which create a

---

[1] to whom any correspondence should be addressed.

macroscopic electric field. Associated potential differences can be measured in the solution.

Since 1933, acoustophoresis has been extensively studied from the experimental point of view, in the case of electrolyte solutions [2, 3] and colloids [4]. Also, commercial apparatus have been devised for the purpose of measuring surface electric charges of colloids (ζ-potential).

Since the basic Debye treatment was based on an ideal solution approximation, workers attempted to take into account the various departures from ideal behavior [5] of Debye. Unfortunately, in the case of electrolytes, these theories do not agree with experiments. Therefore, it appeared to us necessary to revisit completely the theory of acoustophoresis, starting with IVP. Besides, at a practical level, a precise calculation of IVP is useful because IVP can contribute appreciably to the acoustophoresis signal when the mixture of a colloidal suspension with a salt is studied. The study of the acoustophoresis of collids (CVP) is based on sedimentation calculation techniques and will be described subsequently.

Here, the Mean Spherical Approximation (MSA) is used to describe departures from ideality in IVP experiments. This theory was previously applied to conductance [6] and to self diffusion [7] in electrolyte solutions.

In the first section, we present the main features of our treatment. In the second section we compare our results with experimental values.

MSA treatment

Debye treatment leads to a plateau for the potential Φ_0, at high c. However, it is observed experimentally that some salts do not exhibit any plateau, but rather a continuously decreasing pattern. The following treatment is an attempt to explain this behavior from the consideration of non-ideal terms, namely: relaxation effect and hydrodynamic corrections.

We find for the relaxation force:

$$\delta \mathbf{k}_i^{rel} = e_i \, \alpha_k \mathbf{E} + i \, e_i \, \omega \, \beta_k \mathbf{v}_s \,, \tag{1}$$

where

$$\alpha_k = \frac{\Re(\kappa_q^2)}{3} \left(i_0(\kappa_q \sigma) - \frac{\varepsilon k_B T}{e_i e_j} \kappa_q \sigma^2 \cdot i_1(\kappa_q \sigma) \right)$$

$$\times \frac{\kappa_q B_{ij} \exp(-\kappa_q \sigma)}{\kappa_q^2 + 2\Gamma \kappa_q + 2\Gamma^2 (1 - \exp(-\kappa_q \sigma))} \tag{2}$$

$$\beta_k = -\frac{4\pi \sum_{i=1}^{2} \bar{c}_i e_i m_i D_i^0}{3\varepsilon k_B T \sum_{i=1}^{2} D_i^0} \left(i_0(\kappa_q \sigma) - \frac{\varepsilon k_B T}{e_i e_j} \kappa_q \sigma^2 \cdot i_1(\kappa_q \sigma) \right) \tag{3}$$

$$\times \frac{\kappa_q B_{ij} \exp(-\kappa_q \sigma)}{\kappa_q^2 + 2\Gamma \kappa_q + 2\Gamma^2 (1 - \exp(-\kappa_q \sigma))} \,, \tag{4}$$

and

$$B_{ij} \simeq \frac{e_i e_j}{\varepsilon k_B T (1 + \Gamma \sigma)^2} \tag{5}$$

$$\kappa_q^2 = \frac{4\pi}{\varepsilon} \frac{n_1 e_1^2 a_1 + n_2 e_2^2 a_2}{D_1^0 + D_2^0} + i \frac{\omega}{D_1^0 + D_2^0} \tag{6}$$

$$a_i = (\zeta_i + i \, m_i \omega)^{-1} \,, \tag{7}$$

and where $\Re(z)$ is the real part of the complex z, Γ is the MSA parameter, κ the inverse Debye screening length, σ the mean-closest approach distance, n_i is the mean number density, D_i^0 is the self-diffusion coefficient, ζ_i is the friction coefficient, e_i is the charge, m_i is the mass, $\bar{c}_i$ is the mean concentration of the ion of species i and k_B is the Boltzmann constant, T is the temperature, ω is the pulsation of the acoustic wave and

$$i_0(\kappa_q \sigma) = \frac{\sinh(\kappa_q \sigma)}{\kappa_q \sigma} \tag{8}$$

$$i_1(\kappa_q \sigma) = \frac{\cosh(\kappa_q \sigma)}{\kappa_q \sigma} - \frac{\sinh(\kappa_q \sigma)}{\kappa_q^2 \sigma^2} \,. \tag{9}$$

For hydrodynamic correction, the velocity increment $\delta \mathbf{v}_j^{hyd}$ induced on a particle j by the motion of the surrounding particles i can be found by the use of the Oseen tensor. The result reads

$$\delta \mathbf{v}_j^{hyd} = \sum_{i=1}^{2} n_i \int_r h_{ji}^0(\mathbf{r}) \, \mathcal{T}(\mathbf{r}) \zeta_i (\mathbf{v}_i - \mathbf{v}_s) \, d\mathbf{r} \,, \tag{10}$$

where $\mathcal{T}$ is the Oseen tensor.

We finally find:

$$\delta \mathbf{v}_i^{hyd} = e_i \alpha_v \mathbf{E} + i \, e_i \omega \beta_v \mathbf{v}_s \,, \tag{11}$$

with

$$\alpha_v = -\frac{\kappa}{6\pi\eta(1 + \Gamma\sigma)^2} \tag{12}$$

$$\beta_v = \frac{2}{3\eta\varepsilon k_B T(1 + \Gamma\sigma)^2} \sum_{i=1}^{2} \bar{c}_i e_i m_i / \kappa \tag{13}$$

and the electrophoretic force is given by:

$$\delta \mathbf{k}_j^{hyd} = \zeta_j \, \delta \mathbf{v}_j^{hyd} \,. \tag{14}$$

Introducing these corrections, we obtain a new expression for the potential

$$\Phi_0 = \frac{4\pi g_s v_{sO}}{\omega \varepsilon k_B T} \left(\frac{\left[\sum_{i=1}^{2} \bar{c}_i e_i m_i \left(\Re(\omega_i) - \frac{k_B T D_i^0}{g_s^2 m_i} \right) \right]^2 + \left[\sum_{i=1}^{2} \bar{c}_i e_i m_i \Im(\omega_i) \right]^2}{\left[1 + \frac{4\pi}{\varepsilon k_B T \omega} \sum_{i=1}^{2} \bar{c}_i e_i^2 \Im(\omega_i) \right]^2 + \left[\frac{4\pi}{\varepsilon k_B T \omega} \sum_{i=1}^{2} \bar{c}_i e_i^2 \Re(\omega_i) \right]^2} \right)^{1/2} \,, \tag{15}$$

186

S. Durand Vidal et al.
Acoustophoresis of electrolyte solutions

with

$$\omega_i = D_i^0 \, (1 + \alpha_k) \left(1 - \frac{k_B T \kappa}{6 \pi \eta \, D_i^0 (1 + \Gamma \sigma)^2} \right), \qquad (16)$$

where $\mathfrak{I}(z)$ is the imaginary part of the complex z, ε is the dielectric constant of the medium, g_s is the sound velocity and v_{s0} is the solvent velocity amplitude which depends on the power of the acoustic ultrasonic wave.

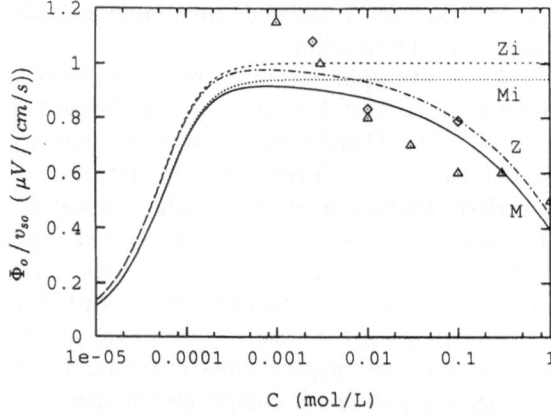

Fig. 1 Acoustophoretic potential divided by the solvent velocity amplitude against NaCl salt concentration. Triangles ($\triangle$): Zana's experimental results. Z: fitted curve with a mass of Na^+ of 70 g/mol. Zi: corresponding ideal curve. Diamond ($\diamond$): Millner's experimental results. M: fitted curve with a mass of Na^+ of 71 g/mol. Mi: corresponding ideal curve

Results and discussion

We have focused our attention on the alkali and alkali-earth chlorides which were studied by two main groups: Zana et al. and Millner et al..

Equation (15) contains three unknowns: the closest-approach distance σ, the mass of the solvated anion, and that of the solvated cation.

The closest-approach distance σ is calculated by fitting the conductance data of a given salt by the expression for the equivalent conductivity; this fitting is rather good. Concerning the masses of the ions, it is commonly believed [8, 11] that the chloride ion is weakly solvated. As a consequence, in the following we make the approximation that its solvation number is zero. With these two approximations, we can calculate the masses of the solvated cations by fitting the experimental data with Eq. (15) and also the cation's hydration number. The case of LiCl, KCl, RbCl, CsCl, CaCl$_2$, SrCl$_2$, and BaCl$_2$ were studied. As an example, the plots of the scaled signal Φ_O/v_{sO} are given in Fig. 1 as a function of the NaCl concentration. Concerning this salt, the experimental measurements found by the two authors are consistent; it is seen that they decrease with the concentration (no plateau value is found) and that the signal measured is rather weak. A good fitting is obtained with $M_+ \approx 70$ g/mol for the molar mass of Na^+. At 1 M, the relative contributions from the relaxation and the hydrodynamic corrections to the ideal plateau value are about 10% and 50%, respectively.

In addition, it is found that the hydration numbers N_Z and N_M found from acoustophoresis measurements per-

formed respectively by Zana and Millner on 1:1 electrolytes and 2:1 are consistent and, as expected, they decrease with the ionic radius of the cation; although pressure-gradient effects were not allowed for in the present treatment, our results are in rather good agreement with those from the literature (Table 1).

One of the major results of the present study is to account for the decrease of the IVP in a wide concentration range.

Calculation of equilibrium sedimentation profiles of screened charged colloids

Introduction

The sedimentation of charged colloidal suspensions has been the subject of recent interesting experimental studies, e.g. [12]. Equilibrium profiles have been obtained [13]. However, it was found in this latter work that the data were well fitted by taking an effective mass which was

Table 1 Hydration Number N of the cations deduced from the fitted mass: N_Z, obtained from Zana's experimental results; N_M, from Millner's results; N_L, from the literature (the hydration number of the chloride ion is supposed to be zero).

Salts	BaCl$_2$	SrCl$_2$	CaCl$_2$	CsCl	RbCl	KCl	NaCl
N_Z	–	–	–	0.6	1.1	1.2	2.7
N_M	9.3	8.5	10.1	− 0.6	0.0	1.1	2.7
N_L	7.7[a]	10.7[a]	12[a]	0.7[b]	1.2[a]	1.9[a]/1.3[b]	3.5[a]/2[b]
$c^c(M)$	0.1–1.8	0.1–1.8	0.1–1.4	1.2	0.1–1.5	0.1–4/1.2	0.1–5/1.2

[a]From ref. [8, 9], [b]from ref. [8, 10], c^c concentration range relative to N_L (c in mol/L).

smaller than the expected mass of the particles, as calculated from the known particle parameters. Decreasing the relative amount of added salt led to a mass reduction which could reach ca. 40%. The authors interpreted this phenomenon by supposing that there was an additional force acting on the particles, likely arising from a residual electric repulsion between the colloid spheres.

A theoretical treatment [14, 15] of the calculation of density profiles of colloidal particles in sedimentation equilibrium has been developed on the basis of the minimization of the free energy functional of the system. Here, we present a method which, although similar, is based on a simpler treatment.

Theoretical

We consider that salt, e.g., NaCl, is added to a suspension of charged colloids and their small counterions. We assume that the size of the colloid spheres is much smaller than the gravitational length (characteristic scale of sedimentation). This assumption allows a macroscopic treatment involving the local concentrations of each species.

We first express that each species is in equilibrium in the gravitational field and that the departures from ideality arise mainly from the hard-sphere repulsion between the colloid spheres, as calculated from the Carnahan-Starling equation of state [16]. Denoting by $C(x)$ the colloid concentration, $C_+(x)$ that of the small positive ions and $C_-(x)$ that of the small negative ions, and neglecting the masses of the small ions, we find

$$eE/k_BT = \partial_x \ln C_+ = -\partial_x \ln C_- = -\frac{1}{Z}\partial_x \ln(\gamma Ce^{qx}), \quad (17)$$

with E the local electric field, $q = mg/k_BT = l^{-1}$, (m is the buoyant mass of the colloid particles), $-Z$ the charge of the colloid, and γ the activity coefficient of the colloid particles.

As a closure relation it is assumed that the local electrical neutrality condition [17, 19, 20] prevails in the bulk of the solution, that is,

$$C_+ - C_- - ZC = 0. \quad (18)$$

This equation expresses the strong electric attraction between positive and negative species. It does not contradict the dynamic fundamental relationship, in which the electric field E appears, for this field originates from a gravity-induced charge separation existing at a microscopic scale. When averaged on a larger scale, it may be assumed that the charge density nearly vanishes, yielding Eq. (18).

After some elementary algebra, an algebraic equation verified by $C(x)$ is obtained, which can be solved by a numerical method.

Results

The present treatment has been applied to the experiments described in ref. [13], in which work PTFE copolymer latex spheres were used (about 1300 ionizable sites, radius $R = 73$ nm, gravitational length $l \sim 0.0235$ cm; $2R/l \sim 6 \; 10^{-4}$).

Figures 2 and 3 illustrate the results obtained from our treatment for experiments A (excess of salt) and E (salt-free).

The agreement found in case A between the hard-sphere result and the experimental points is excellent, and

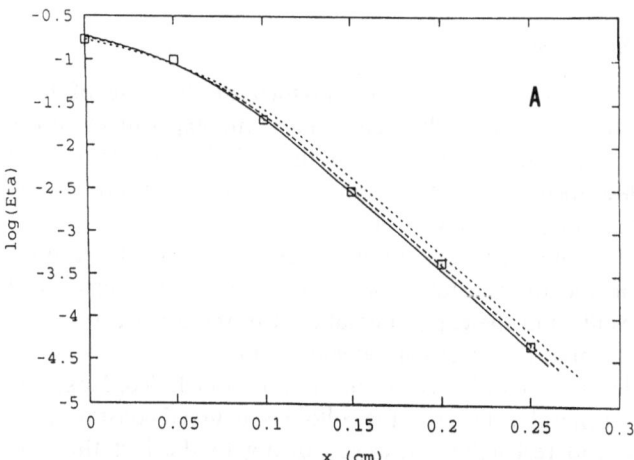

Fig. 2 Experiment A. Decimal logarithm of the packing fraction of the colloid as a function of the position x in the container. Squares (□▷: experimental data. Continuous curve: theoretical result for the fluid of hard spheres. Dashed curve: result for $Z = 400$. Dotted curve: result for $Z = 700$

Fig. 3 Experiment E. Decimal logarithm of the packing fraction of the colloid as a function of the position x in the container. Squares (□): experimental data. Full line e: experimental barometric line. Continuous curve: result for the uncharged hard-sphere fluid. Curve 1: result for $Z = 0.5$. Curve 2: $Z = 2$. Curve 3: $Z = 3$. Curve 4: $Z = 15$

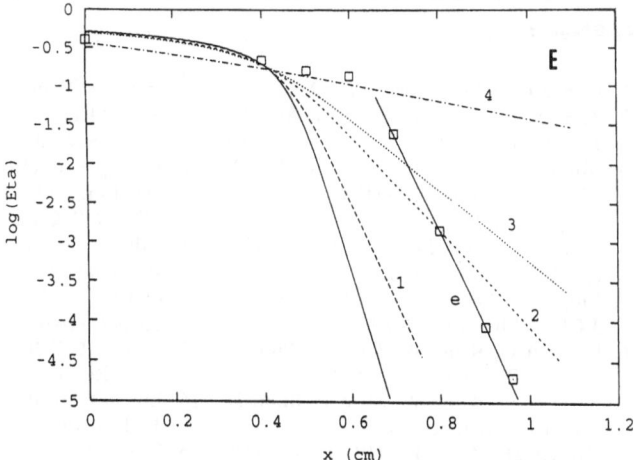

weakly sensitive to the charge of the colloid. In case C (lower screening, not shown) it is seen that the theoretical results are more sensitive to the effect of the particle charge. At the same time, the effective mass, as deduced from the present treatment, is found to remain equal to the buoyant mass, in contrast with the experimental findings.

As regards experiment E, it can be shown that the apparent mass should be $m_{app} = m/(Z + 1)$, if Z is not too small. In Fig. 4 the experimental data are compared with the theoretical results for various charges. The strong influence of the charge of the colloid is shown by plotting the results for $Z = 2, 3$, and 15. This latter value yields a satisfying agreement with the data for $x < 0.6$ cm.

Discussion

The excellent agreement obtained in the case of high screening (exp. A) between theory and experiment shows that the repulsion between the colloid spheres is well described by the hard-sphere volume exclusion, at the Carnahan–Starling level.

Tentative explanations can be put forward to account for the discrepancies observed between the theoretical results and the experimental data in the other cases.

First, the electrostatic short-range repulsion between charged colloid particles has not been included. Next, the fact that the charge of the colloid is assumed constant may be too restrictive. However, owing to the fact that the colloid concentration is very low in the barometric region, it is likely that the slopes of the plots in that region would be unchanged.

The last possible explanation is connected with the time relaxation of the colloid profile. Starting from an initial homogeneous suspension, one may wonder how long one would have to wait before equilibrium is reached. As a preliminary study, we have considered the ideal system, without departures from ideality. The characteristic relaxation times of the system can be calculated using the normal-mode technique [18–20]. The result of this simplified treatment is that, in all cases and for any charge Z, the relaxation of the present system may be governed by the relaxation process of the salt. This behavior would be owed to the low mass of the salt ions; though the small ions hardly sediment, the motion of the colloid particles is even so coupled to their motion through the electrical neutrality condition. In conditions of moderate screening, the time to be elapsed before equilibration may be very long.

Conclusion

We are working presently on the theoretical aspects of the acoustophoresis of micelles and colloids. Colloidal suspensions yield significant signals, typically on the order of 1 mV. The acoustophoresis of colloids has become a widespread tool in the investigation of their surface charge.

The main problems lie in the calculation of the various departures from ideality: relaxation effect, hydrodynamic interactions and interactions between the large particles and their counterions. So far, results have been obtained on these questions using mainly the well-known cell model combined with Debye–Hückel theory. However, such treatments are limited to cases where the double-layer thickness is much smaller than the particle size and the surface charge is rather low.

A preliminary study has led us to the conclusion that the MSA may also be applied to the acoustophoresis of colloidal suspensions. This theoretical frame work would permit studies in a much wider range of surface charges and colloid concentration.

Also, we are working on the application of the electroneutrality condition (used in the section sedimentation) to CVP. First results show that the method seems valid at high ionic strength.

References

1. Debye P (1933) J Chem Phys 1:13
2. Zana R, Yeager E (1982) Mod Aspects Electrochem 14:1
3. Millner R (1961) Z Elektrochem 65:639; Millner R, Müller H-D (1966) Ann Phys (Leipzig) 17:160
4. Marlow BJ, Fairhurst D, Pendse HP (1988) Langmuir 4:611
5. Bugosh J, Yeger E, Hovorka F (1947) J Chem Phys 15:592–597
6. Bernard O, Kunz W, Turq P, Blum L (1992) J Phys Chem 96:3833
7. Bernard O, Kunz W, Turq P, Blum L (1992) J Phys Chem 96:398
8. Hinton JF, Amis ES (1971) Chemical Reviews 71 num 6:627
9. Stokes RH, Robinson RA (1948) J Amer Chem Soc 70:1870
10. Principles of electrochemistry, edited by MacInnes DA (Reinhold, New York, NY, 1939, Chapter 4).
11. Salmon PS, Lond PB (1992) Physica B 182:421–430
12. Okubo TJ (1994) Phys Chem 98:1472–1474
13. Piazza R, Bellini T, Degiorgio V (1993) Phys Rev Lett 71:4267–70
14. Biben T, Hansen JP, Barrat JL (1993) J Chem Phys 98:7330–44
15. Biben T, Hansen JPJ Phys Condens Matter 6:A345–A349
16. Carnahan NF, Starling KE (1969) J Chem Phys 51:635
17. Robinson RA, Stokes RH (1959) Electrolyte Solutions 2nd ed.; Butterworths; London
18. Turq P, Orcil L, Chemla M, Mills RJ (1982) Phys Chem 86:4062. Simonin JP, Turq P, Soualhia E, Michard G, Gaillard JF (1989) Chem Geol 78:343
19. Mills R, Perera A, Simonin JP, Orcil L, Turq P (1985) J Phys Chem 89:2722–25
20. Simonin JP, Gaillard JF, Turq P, Soualhia E (1988) J Phys Chem 92:1696–1700

Progr Colloid Polym Sci (1995) 98:189–192
© Steinkopff Verlag 1995

L. Motte
F. Billoudet
J. Cizeron
M.P. Pileni

Synthesis "in situ" in reverse micelles of silver sulfide semiconductors

L. Motte · F. Billoudet · J. Cizeron
M.P. Pileni
Université Pierre et Marie Curie
Laboratoire S.R.S.I.
BP 52
Bat 74
4 Place Jussieu
75005 Paris, France

L. Motte · J. Cizeron · M.P. Pileni (✉)
C.E.N. Saclay
DRECAM-SCM
Bat 522
91191 Gif sur Yvette, France

Abstract Functionalized reverse micelles are used to control the size of silver sulfide, Ag_2S, nanosized particles. The size of the crystallites varies linearly with the water content from 2 to 10 nm. The particles have been coated by dodecanethiol and extracted from micelles. The size of the particles dispersed in heptane has been determined by SAXS experiments and compared to those obtained by TEM. A good agreement between these two technics is obtained. A drop of the particles previously dispersed in heptane is dried on a carbon grill. A network of the particles forming monolayers of crystallites in a hexagonal distribution appears.

Key words Nanosized crystallites – Ag_2S – reverse micelles – crystallites monolayers

Introduction

Reverse micelles are droplets of water in oil stabilized by monolayer of surfactant. With AOT as surfactant, micellar system presents two properties very important for co-precipitation reactions: i) the droplet size increases linearly with the water content, $w = [H_2O]/[AOT]$; ii) due to Brownian motion, some collisions between droplets are efficient and an exchange process between water pools occurs.

In this paper, it is shown that the silver sulfide nanosize semiconductors have been synthesized in reverse micelles. The semiconductor size is controlled by the size of the water content and varies from 2 to 10 nm. The particles have been coated by dodecanethiol and extracted from micelles. The sizes of such particles have been characterized by transmission electron microscopy (TEM) and small-angle x-ray scattering (SAXS). By TEM, we observed a network of nanosize particles forming monolayers of crystallites in a hexagonal distribution. The size of particles dispersed in heptane has been determined by SAXS experiments. A good agreement between these differents techniques is obtained.

Experimental section

Products

Sodium di(2-ethylhexyl) sulfosuccinate, usually called NaAOT, was obtained from Sigma; isooctane was from Fluka, sodium sulfide Na_2S from Janssen, dodecanethiol, heptane from Merck, and ethanol from Prolabo. Silver di(2-ethylhexyl) sulfosuccinate, AgAOT, was prepared as described previously [1].

Synthesis of Ag$_2$S nanocrystallites

The preparation is achieved by mixing the reverse micellar system with an aqueous solution containing sodium sulfide Na$_2$S. Reverse micellar solution is formed by solubilizing in isooctane an AOT micellar solution containing silver and sodium AOT. For a 0.1 M total concentration of AOT surfactant, the silver [Ag$^+$] and sulfide [S^{2-}] concentrations are $4 \cdot 10^{-4}$ M.

Size-selective precipitation and formation of ordered monolayer of Ag$_2$S nanocrystallites

Pure dodecanethiol is added to reverse micellar system containing Ag$_2$S nanocrystallites. After evaporation at 60 °C, the precipitate is washed with ethanol and filtrate. The nanocrystallites coated by dodecanethiol are dispersed in heptane, forming an optically clear solution.

Apparatus

Optical absorption spectra were collected at room temperature on a UVIKON 931 spectrophotometer.

A Jeol electron microscope, model JEM 100CX II, is used to image nanocrystallites. The samples were prepared by placing a drop of solution on a surface of a copper grid coated with amorphous carbon film.

The SAXS experiments were performed at LURE (Orsay, France), on the D22 diffractometer. The treatment of the experimental data has been described previously [2].

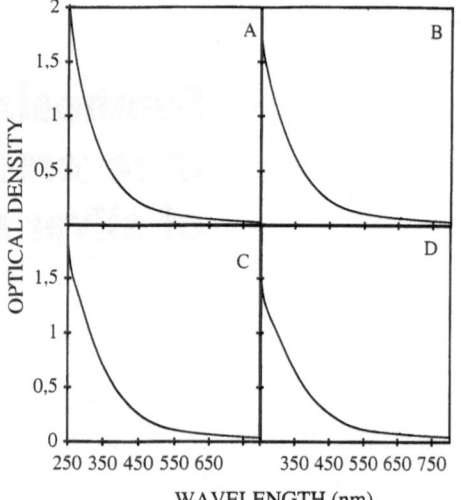

Fig. 1 Absorption spectra of Ag$_2$S synthesized in reverse micelles at various water content. A: $w = 5$, B: $w = 10$; C: $w = 20$; D: $w = 40$; system AOTNa/AOTAg/Isooctane/Water, [AOT] = 0.1 M, [Ag$^+$] = [S^{2-}] = $4 \cdot 10^{-4}$ M

Results and discussion

Synthesis in reverse micelles

The absorption spectrum of Ag$_2$S synthesized in AOT(Na)-AOT(Ag)-water-isooctane reverse micelles recorded at various water constent, w, is given in Fig. 1. No change is observed in the absorption spectra with the water content, w. However, a blue shift is obtained compared to the optical band edge of bulk silver sulfide which is well known to be at 1240 nm (1eV). In contrast to what has been observed with other semiconductors, such as CdS[3], CdSe[4], ZnS[5], etc., the absorption spectra obtained at various water contents are not structured and no peak or bumps characteristics to excitonic peaks are observed.

Electron microscopy pictures have been performed after making Ag$_2$S synthesis at various water contents. Figure 2 shows histograms and pictures obtained by transmission electron microscopy. The size of the Ag$_2$S particles increases linearly with the water content, indicating a control of the crystallite sizes by the water content, Fig. 3.

Nanosize particles coated by dodecanethiol

The silver sulfide nanosize particles synthesized in reverse micelles are coated by dodecanethiol as described above and are redispersed in heptane. Absorption spectra of various size coated Ag$_2$S particles are similar to those given in Fig. 1.

By SAXS, the size of the particles has been determined. Figure 4 shows the simulated and experimental data obtained on the Porod plot representation for particles prepared at various water content. A well-defined maximum and minimum can be observed, indicating a low polydispersity in size. From the maximum and the minimum, the radius of the particle is deduced and reported in Table 1. Table 1 indicates an increase in the size with that of the water droplet in which the synthesis has been made before coating the particles.

Figure 5 shows histograms and pictures of dodecanethiol capped Ag$_2$S particles obtained by TEM. Ordered Ag$_2$S colloid monolayer in the form of hexagonally close packed colloid particles is observed. The interparticle spacing and the hydrocarbon length of dodecanethiol are

Progr Colloid Polym Sci (1995) 98:189–192
© Steinkopff Verlag 1995

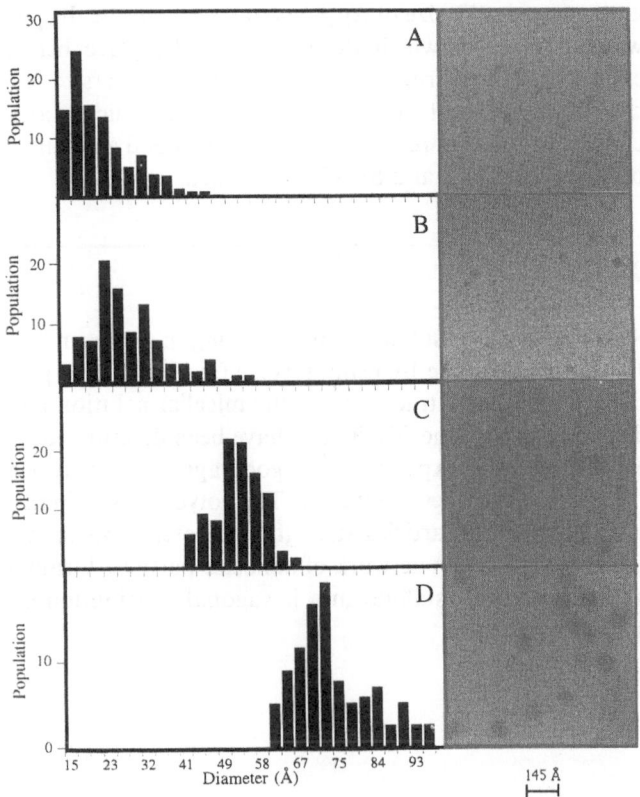

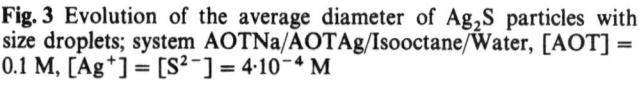

Fig. 2 Size distributions and electron microscopy patterns of Ag_2S synthesized in reverse micelles at various water content. A: $w = 5$, B: $w = 10$; C: $w = 20$; D: $w = 40$; System AOTNa/AOTAg/Isooctane/Water, [AOT] = 0, 1 M, $[Ag^+] = [S^{2-}] = 4 \cdot 10^{-4}$ M

Fig. 3 Evolution of the average diameter of Ag_2S particles with size droplets; system AOTNA/AOTAg/Isooctane/Water, [AOT] = 0.1 M, $[Ag^+] = [S^{2-}] = 4 \cdot 10^{-4}$ M

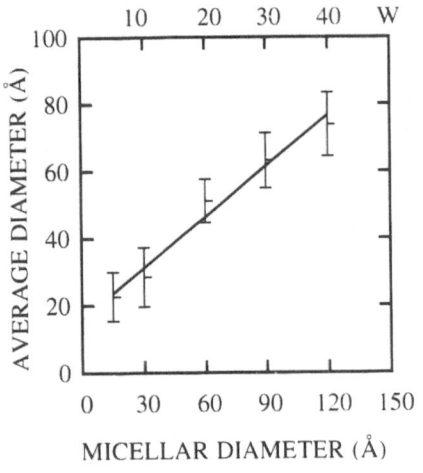

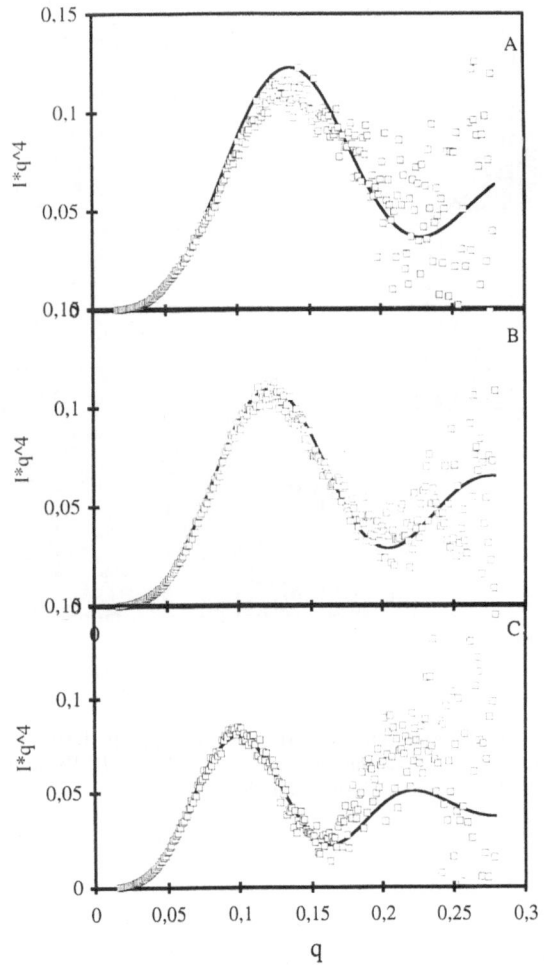

Fig. 4 Porod representations obtained from SAXS experiments, + + + experimental data: Ag_2S particles synthesized in reverse micelles at $w = 5$ (A), $w = 10$ (B) and $w = 20$ (C) and extracted with dodecanethiol, — simulated form factor for sphere

Table 1 Crystallites radius and polydispersity deduced from SAXS analysis.

w	R_{Porod} (Å) max	min	$R_{simulation}$ (Å)	Polydispersity
5	21, 5	–	20	14%
10	23	22	22	14%
20	27, 5	29	27, 5	14%

on the same order of magnitude $1 \approx 18$ Å. The hydrocarbon length of dodecanethiol has been estimated by using the empirical form derived by Bain et al. [6]: $1 = 2.5$ Å + $1.27 \cdot n$ Å where n is the number of CH_2 groups and 2.5 Å takes into account the thiol and methyl terminating functional groups particles. We observed, as in reverse micelle,

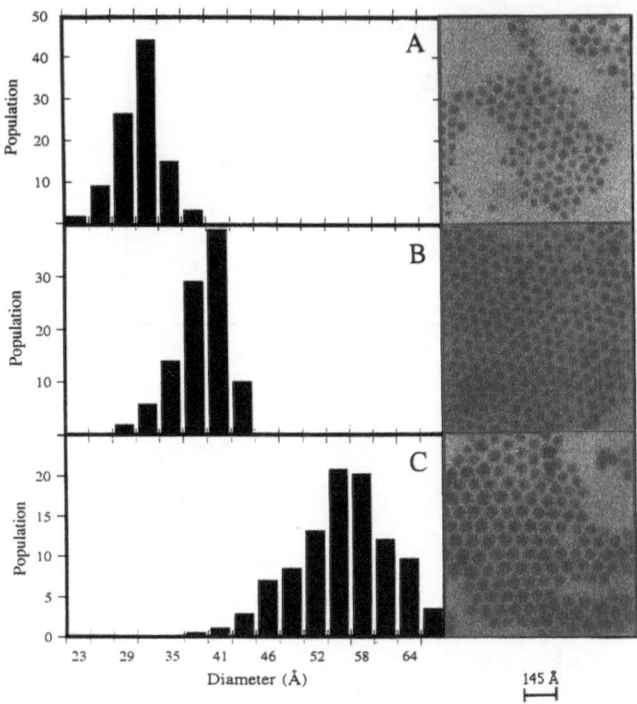

Fig. 5 Size distributions and electron microscopy patterns of Ag_2S synthesized in reverse micelles at $w = 5$ (A), $w = 10$ (B) and $w = 20$ (C) and extracted with dodecanethiol

an increase in the size of Ag_2S crystallites with the droplet water content in which the synthesis take place before extraction. Furthermore, the polydispersity in size decreases compared to that obtained by electronic microscopy histograms in reverse micelles, and a good agreement between the TEM and SAXS is obtained.

Conclusion

Synthesis of silver sulfide particle varying in size from 2 to 10 nm can be made by using reverse micelles. These particles have been extracted from the micellar solution and dispersed in heptane. Their sizes have been determined by SAXS and TEM experiments. A good agreement between these two techniques is obtained. The polydispersity in size of the extracted particles strongly decreases. This favors the appearance of a network of nanosize particles forming monolayers of crystallites in a hexagonal distribution on a solid support.

References

1. Petit C, Lixon P, Pileni MP (1993) J Phys Chem 97:12974–12983
2. Pitre F, Regnaut C, Pileni MP (1993) Langmuir 9:2855–2860
3. Pileni MP, Motte L, Petit C (1992) Chem Mater 4:338–345
4. Murray CB, Norris DJ, Bawendi MG (1993) J Am Chem Soc 115:8706–8715
5. Cizeron J, Pileni MP, manuscript in preparation
6. Bain CD, Evall J, Whitesides GM (1989) J Am Chem Soc 111:7155–7164

Progr Colloid Polym Sci (1995) 98:193–196
© Steinkopff Verlag 1995

Radiometry of colloid systems

I. Krznaric
R. Despotovic

Dr. I. Krznaric (✉) · R. Despotovic
Ruder Boskovic Institute
Colloid Chemistry Laboratory
P.O. Box 1016
Bijenicka 54
41001 Zagreb, Croatia

Abstract It is well known that by using the same surfactant and the same inorganic sol particles essentially different colloid phenomena can be observed. If the identical surfactant is present under different conditions, added in various concentrations or mixed at various conditions, i.e., depending on the preparation conditions, the resulting colloids exhibit specific properties. In order to obtain highly reproducible quantitative characteristic data (critical flocculation concn. c_f, flocculation concn. f_c, critical stabilization concn. c_s, stabilization concn. s_c) radiogravimetric methods were applied. As a model system the AgI^-_{solid} suspension in an aqueous solution of NaI + n-dodecyl-ammonium nitrate was used with carrier free $^{131}I^-$ as a radioactive indicator. Changes in radionuclide distribution between solid and liquid phase were followed during the solid/liquid phase separation, using high-speed centrifuge, and the radiogravimetric data obtained were used to calculate c_f, f_c, c_s, and s_c. The radiogravimetric data show interesting phenomena closely connected with the characteristic of bulk colloid system properties. Colloid stability as a function of surfactant concentration, suspension age, potential determining ion concentration appears as determining factors, that can be quantitatively determined by means of radiogravimetry. The measured radiogravimetric data are highly reproducible, simple and sensitive even for determination changes in very dilute systems like in various waste waters.

Key words Colloids – inorganic suspension – radiometry – surfactants – critical concentrations

Introduction

Fine dispersed inorganic sol prepared in surfactant solution show a strong dependence on the chemical nature and concentration of the surfactant present in the precipitation solution. There is a voluminous literature with various results related to the inorganic sol particles/surfactant interactions [1].

Unfortunately, much quantitative published data cannot be compared because of noncoherent, nonsensitive, or nonselective methods applied. On the other hand, radiometry method applied in colloid chemistry examination (radionuclide diffusion, heterogeneous exchange of radionuclides, Ostwald ripening, phase separation, adsorption-desorption equilibria, etc.) appears to be extremely sensitive, selective, and accurate [2]. The stabilization and destabilization of various dispersions are very important from both practical and theoretical points of view. With the ultimate goal to determine coagulation concentration c_c, stabilization concentration s_c, critical coagulation concentration c_k, and critical stabilization concentrations C_s, the radiogravimetry method was developed and applied using very stable negative silver iodide sol in n-dodecylammonium nitrate, following the mechanisms determining the mutual interactions between cationic surfactant and anionic counter-ion labelled by radioactive $^{131}I^-$. The same radiogravimetry method was applied in order to explain time-dependent adsorption/desorption process in dilute surfactant solution. Because of growing interest in waste water purification processes [3] the interactions of fine inorganic suspension particles with dilute surfactant solutions is of great importance, and the quantitative and selective methods appear also to be important.

Experimental

Materials

n-Dodecylammonium nitrate was prepared from purified n-Dodecylamine (Fluka) and nitric acid (Merck p.a.). The surfactant was purified by recrystallizing four times ethanol-petrol ether mixture. Recrystallized n-Dodecylammonium nitrate was vacuum-dried and stored in a desiccator. Analar grade chemical silver nitrate and sodium iodide (Merck) were used throughout the experiments. The water used was double distilled. Radioiodine was applied as a solution of carrier free $^{131}I^-$ in the NaI form. The solutions used were prepared and standardized in the usual ways.

Systems

System A; n-Dodecylammonium nitrate + NaI($^{131}I^-$)

Samples were prepared by mixing equal $DDANO_3$ and NaI($^{131}I^-$) solutions. All the samples were thermostated at 293 K and aged for $t_a = 600, 1800, 3600, 6000, 9600, 86400$, and 604800 s. Initial radioactivity A_0 was determined for the A system without surfactant added. Using a superspeed Sorval RC2B centrifuge the solid liquid separation was carried out in order to count A_t radioactivity of the clear supernatant at the t_a. Radioactivities of clear supernatants A_L (systems in water), $A_{L(EtOH)}$ (systems in 10% EtOH) and/or $A_{L(PropOH)}$ (systems containing 10% PropOH) were calculated as the mean of seven measurements for the same samples. All A_L radioactivity values correspond to $A_L = 100A_t/A_0$. Clear supernatants were used for A_0 and A_t counting by means of Nuclear Enterprises ST6

with well-type NaI(TII) scintillation crystal. The background values were always less than 1.3% of total counting values.

System B

The negative charged silver iodide sols were prepared by adding 0.0020 mol/dm^3 silver nitrate solution to an equal volume of 0.0040 mol/dm^3 sodium iodide solution labeled with NaI($^{131}I^-$) and containing n-Dodecylammonium nitrate or various concentrations; 0.0000010 mol/dm^3 to 0.0030 mol/dm^3. The systems for t_a at 293 K contain 0.0010 mol AgI/dm^3 + 0.0010 NaI/dm^3 + n-Dodecylammonium nitrate. 100 cm^3 of labeled suspension divided into five equal samples was aged 100 min (t_a). After t_a minutes and centrifugation fo 5 min, 0.20 cm^3 of supernatant over sediment from the precipitated sol was pipetted off in order to determine the radioactivity A_t. Centrifugation was carried out at different relative centrifugal forces R.C.F. from a to f (Fig. 1) corresponding to 1000 G up to 32000 G. 0.20 cm^3 of homogenized suspension was used for determining the A_0 radioactivity. The fraction of sedimented sol f_s can be derived as $f_s = (A_0 - A_t)/A_0$.

Results and discussion

The general Težak scheme of the formation of colloids in aqueous media points to several mesophase substructures as transition substructures between equilibrium phase structure. Using radiometry, such concept was verified quantitatively [4]; application of radiometry has led to several selective and simple solutions in the investigation of solid/liquid equilibria, giving us the possibility for quantitative interpretation of mesophase transformation [5]. On the same level of the fine structure study in a complex polycomponent colloid systems containing surfactants, the most important four characteristic values (c_c, S_c, c_k, C_s) can also be determined using radiogravimetry. The experimental data (system B) in the determination of c_c, S_c, c_k, and C_s illustrate (Fig. 1) the simple possibility for quantitative determination of various mutual interactions between S^+ and inorganic sol particles; in contrast to relatively simple Coulombic interactions between inorganic sol particles and inorganic ions, surfactants cause significant complex and substantially different phenomena, especially with respect to the colloid stability of sols depending on their chemical nature. Following radiogravimetry, the low concentration slope of f_s versus log c_{S^+} curve corresponds to classical adsorption, occurs at the lowest S^+ concentration at which colloid stability

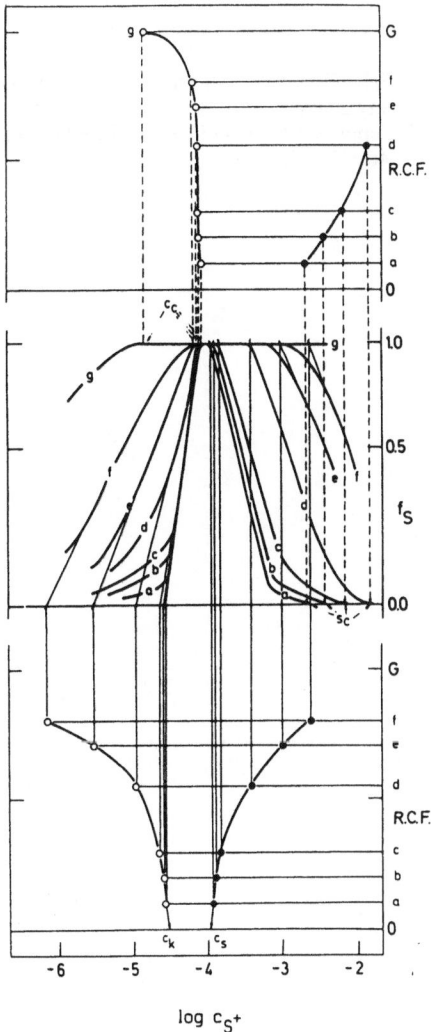

Fig. 1 System: 0.0010 mol AgI_{sol}^-/dm^3 + 0.0010 mol $NaI(^{131}I^-)/dm^3$ + 0.0000010 to 0.010 mol n-dodecylammonium nitrate S^+/dm^3. Relative centrifugal force R.C.F., and the factor of sedimentation f_S, are plotted versus S^+ concentration (log scale). Using graphical extrapolation, arrows show coagulation concentration C_C, stabilization concentration S_c, critical coagulation concentration C_k, and critical stabilization concentration C_S

begins to decrease, and is labeled as the critical coagulation concentration c_k. The second point is where total sedimentation of the inorganic sol particles occurs ($f_S = 1$), and corresponds to the coagulation concentration c_c. The second slope f_S versus log c_{S^+} corresponds to different interactions between inorganic sol particles and surfactant species of higher concentration. With an increase in surfactant concentration, deflocculation or redispersion begins to increase at the critical stabilization concentration or critical deflocculation concentration c_S, reaching the stabilization concentration s_c, i.e., the lowest surfactant concentration at which all the inorganic particles are suspended or deflocculated. Thus, different interactions

between S^+ and AgI_{sol}^- are probably caused by different structures of submicellar S^+ associates, depending on the surfactant concentration. For various surfactant concentrations, S^+ associates are probably of various structures, with different electrostatic capacity for each submicellar structure [6], and with different corresponding adsorption capacity. In order to verify such model, radiogravimetry method was applied, and radiometry data (Tables 1, 2) show very interesting dependence of S^+ adsorption capacity on the system aging. For all followed samples ($t_a = 600$ to $604\,800$ s) the radioactivity A_L (Table 1) increases by the aging of colloid system from $A_L = 88.0$ to 96.0%, indicating the decrease of adsorption capacity of aged associates. At the higher counter-ion concentration the A_L is higher and relative palisade adsorption [7] is lower, probably because of a strong influence of counter-ion concentration on associate structure growth, like in the Carpineti–Giglio model [8]. By using 10% alcohol EtOH and/or PropOH solutions, all the measured A_L values are slightly higher (Table 2), indicating an influence on the acceleration of associate aging processes [6]. It is of interest to point out the possibility of following the time-dependent processes in dilute solutions of submicellar structures for $-C_{12}-$ derivatives, that is to say, for most important aliphatic derivatives in waste waters [3].

Table 1 Systems: n-dodecylammonium nitrate S^+ of 0.050 mol/dm^3 + $NaI(^{131}I^-)$ of $c_1 = 0.000010$, $c_2 = 0.00010$, $c_3 = 0.0010$ mol/dm^3. Radioactivity A_L of liquid phase after centrifugation for the systems aged throughout t_a seconds

t_a	600	1800	3600	6000	9600	86400	604800
A_L							
c_1	88.0	88.5	88.5	88.5	90.0	96.0	96.0
c_2	89.6	89.6	90.4	91.0	91.5	96.0	98.0
c_3	92.4	92.8	95.2	96.8	96.8	97.0	96.5

Table 2 Systems: n-dodecylammonium nitrate S^+ of 0.0050 mol/dm^3 + $NaI(^{131}I^-)$ of $c_1 = 0.000010$, $c_2 = 0.00010$, $c_3 = 0.0010$ mol/dm^3 in 10% EtOH and/or 10% PropOH. Radioactivities $A_{L(EtOH)}$ and $A_{L(PropOH)}$ of liquid phase after centrifugation for the samples aged throughout t_a seconds

t_a	600	1800	3600	6000	9600	86400	604800
$A_{L(EtOH)}$							
c_1	88.2	88.0	91.6	92.8	93.5	98.6	98.8
c_2	90.8	92.8	93.2	93.0	93.4	98.6	98.5
c_3	95.5	95.5	97.0	97.0	96.6	98.0	98.0
$A_{L(PropOH)}$							
c_1	92.4	92.4	93.4	93.0	95.0	98.4	98.8
c_2	94.6	94.8	95.0	95.2	95.2	96.0	98.8
c_3	98.4	98.8	98.0	97.8	98.0	98.8	99.0

References

1. Parfitt GD (1964) Annu Rep Chem Soc 64:125–176
   Ottewill RH (1969) Ibid. 69:183–234
2. Despotović R (1978) Croat Chem Acta 51:113–132
3. Röhl W, Rybinski W, Schwuger MJ (1991) 84:206–214
4. Despotović R, Subotić B (1976) J Inorg Nuclear Chem 38:1317–1319
5. Despotović R, Subotić B (1982) Powder Technology 31:63–73
6. Salaj Obelić I (1994) Ph D Thesis, University Zagreb
7. Sepulveda L (1974) J Colloid Interface Sci 46:372–375

Progr Colloid Polym Sci (1995) 98:197–200
© Steinkopff Verlag 1995

BIOCOLLOIDS

L. Cantu
M. Corti
E. Del Favero
N. Maurer

Spontaneous vesicle formation: transition between single- and bicomponent system

L. Cantu · E. Del Favero
Dipartimento di Chimica e
Biochimica Medica
Università di Milano
via Saldini 50
20133 Milano, Italy

Prof. M. Corti (✉)
Deipartimento di Eletronica
Università di Pavia
via Abbiategrasso 209
27100 Pavia, Italy

N. Maurer
Institute of Physical Chemistry
Karl-Franzen-University Graz
Heinrichstraße 28
8010 Graz, Austria

Abstract Vesicles of the ganglioside GM3 form spontaneously in water. Their average diameter is about 500 Å. Vesicles are in equilibrium with a very small number of much larger non-spherical aggregates. This single component vesicle system is quite "frustrated", since the two monolayers of the bilayer have spontaneous curvatures with the same magnitude, but opposite sign. This situation is dramatically changed if a second amphiphile GM1, of the same type of the ganglioside GM3 but with larger headgroup, is added. In this case, spontaneous curvature readjustments of the two monolayers via demixing can give rise to a finite spontaneous curvature of the bilayer, energetically favoring vesicles towards other structures. Experimentally, it is observed that, by adding increasing proportions of GM1, the large aggregates gradually diminish in number until a pure vesicle solution is obtained for a GM1 to GM3 ratio of 35 to 65. Observations are performed by static and dynamic, both polarized and depolarized, laser light scattering.

Key words Glucosidic amphiphiles – vesicles – light scattering

Introduction

Spontaneous self-assembly of amphiphilic molecules into vesicles is an interesting thermodynamic problem [1, 2]. It has been shown both theoretically and experimentally [3, 4] that mixing two ionic surfactants with oppositely charged headgroups can lead to spontaneous formation of stable unilamellar vesicles in aqueous solutions. In this case vesicles are stabilized energetically, which means that the vesicle phase can have lower free energy than the competing lamellar phase, even in the limit of large curvature elastic modulus. This can happen because the mixing of two such amphiphiles can be non-random, due to some sort of interaction between them, allowing different spontaneous curvatures of the two monolayers. The bilayer can have, therefore, a finite spontaneous curvature, since it is the sum of the two monolayer spontaneous curvatures. Such vesicles have a well defined size, dictated by their spontaneous curvature, and their distribution is fairly monodisperse.

Recently, it has also been shown experimentally [5, 6] that vesicles can form spontaneously in a water solution of a single amphiphile. In this second case, vesicle stabilization cannot be explained with energetic considerations, but entropic effects have to be considered. In fact, the two monolayers of the bilayer, being equal, have the same spontaneous curvature and therefore flat lamellar structures are energetically favored. Some energy cost has to be necessarily paid to curve the bilayer into a closed vesicle and the system is said to be "frustrated". However, if the bending modulus is low with respect to the thermal energy $k_B T$, the gain in entropy of mixing due to the formation of small vesicles can easily compensate this energy cost. The system frustration may also be connected with another interesting feature, which has been found experimentally in the single component system: vesicles are in thermodynamic equilibrium with a very small amount of larger aggregates of lamellar type. Indeed, also the small population of lamellar fragments is not energetically favored against infinite lamellar sheets due to the unfavorable edge energy; but, here again, it can be stabilized entropically. In other words, energetically unfavored single-component vesicles and lamellar fragments can be found in water solution because of entropic effects, and their coexistence arises from the fact that their energy cost is equivalent.

These two pictures of spontaneous vesicle formation are not conflicting at all. Indeed, in the following, it will be shown that it is possible to observe a gradual transition from an entropically-stabilized single-component vesicle system to an energetically-stabilized two-component system. In fact, it has been found experimentally that the addition of a second amphiphile to the single-component vesicle solution makes mixed vesicle energetically stable. This becomes evident by observing that the addition of the second amphiphile destroys the coexistence of vesicle and lamellar fragments. The lamellar fragments gradually disappear at increasing proportions of the second amphiphile which interacts sterically with the first. The disappearance of the lamellar fragments in favor of mixed vesicles indicates that mixed vesicles are in a lower energy state than the single-component vesicles.

Experimental

Single-component vesicles have been found to form spontaneously in water solutions of the ganglioside GM3, which is a biological amphiphilic molecule [7, 8] normally embedded in cell membranes. Like phospholipids, gangliosides are a family of molecules which have a double-chain hydrophobic part. The hydrophilic head is an oligosaccharide chain. GM3 has three saccharide rings, one of which is a sialic acid, attached to a ceramide with a 20 carbon sphingosine and an 18 carbon fatty acid. GM3, prepared as sodium salt, was dissolved at room temperature in a 30 mM NaCl water solution at a concentration of 0.1 mM. The GM3 molecular weight is 1195. NaCl was added to shield Coulomb interactions among vesicles. Aggregates in solution were studied by static and dynamic, both polarized and depolarized, laser light scattering [9, 10]. GM3 vesicles are found to have a diameter which slightly decreases with temperature, that is, from 530 Å at 7 °C to 490Å at 50 °C. No hysteresis was observed on heating and cooling the sample. Data were insensitive to scan rate variations between 100 s/°C and 1000 s/°C.

These features are not at all usual in the common phospholipid vesicles, which are produced with the supply of some form of external energy.

Vesicles were found to be in equilibrium with a small fraction of large aggregates, mostly of lamellar type. Equilibrium is verified by the fact that large aggregates reform at the expense of vesicles immediately after microporous filtration. At 7 °C only about 4% of the GM3 in solution contribute to the formation of large aggregates, while at 50 °C this amount is even reduced by a factor of 2. Figures 1 and 2 (filled squares) show the angular distribution of the polarized and depolarized scattered intensity, respectively, for the GM3 solution. The large intensity values at low angles for the polarized data indicate the presence of large objects in the solution, while the existence of the depolarized intensity of Fig. 2 reveals the lamellar nature of these aggregates [10].

The addition of a second amphiphilic molecule, the ganglioside GM1 of molecular weight 1560, to GM3 leads to a gradual decrease of the amount of large aggregates in solution, as shown in Figs. 1 and 2. GM1 has a packing parameter smaller than GM3, that is, a larger ratio of the head-to-tail cross-sections. GM1 has the same hydrophobic part of GM3 but a larger headgroup with five sugar rings instead of three. Indeed GM1 is capable, by itself, to form micelles with a small radius of curvature in solution. The large polarized scattered intensity at low angles, which is a clear indication of the existence of large aggregates in solution, disappears when GM1 is added. Besides, the scattered depolarized intensity is observed to drop down completely, which is a further verification that lamellae are

Fig. 1 Polarized scattered intensity versus momentum transfer for the pure GM3 solution (filled squares), 90:10 GM3-GM1 solution (circles), 75:25 GM3-GM1 solution (diamonds), 70:30 GM3-GM1 solution (triangles), 65:35 GM3-GM1 solution (open squares), 60:40 GM3-GM1 solution (crosses)

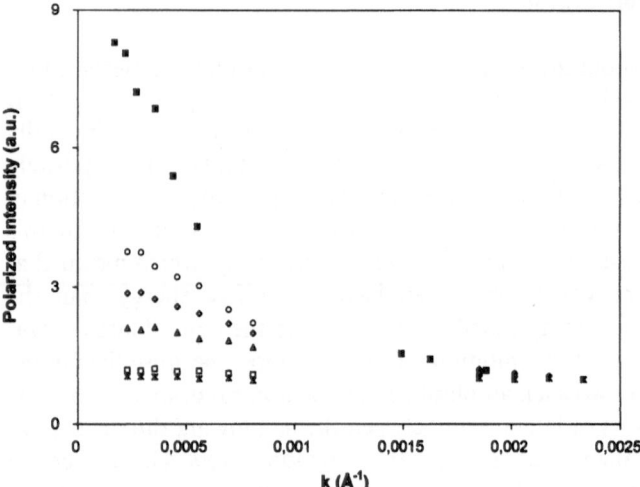

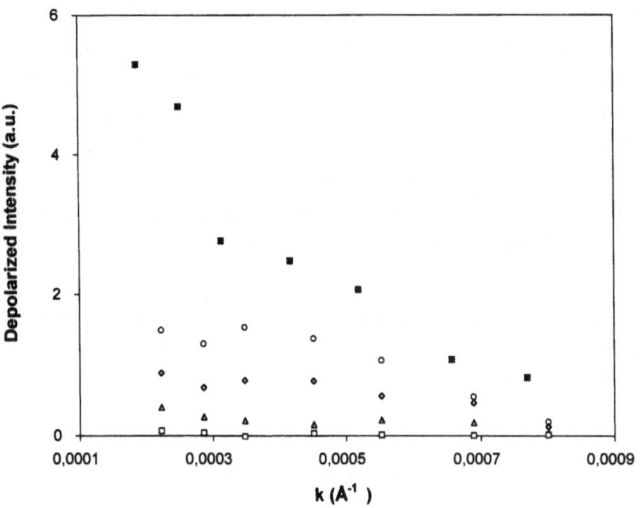

Fig. 2 Depolarized scattered intensity versus momentum transfer for the pure GM3 solution (filled squares), 90:10 GM3-GM1 solution (circles), 75:25 GM3-GM1 solution (diamonds), 70:30 GM3-GM1 solution (triangles), 65:35 GM3-GM1 solution (open squares), 60:40 GM3-GM1 solution (crosses)

no more present in solution. All measurements were performed at 25 °C. GM3-GM1 solutions, all at a total concentration of $1.5 \, mg/cm^3$, were prepared in the desired molar ratios first in chloroform-methanol, so that gangliosides are dissolved in monomeric form, dried under vacuum and then dissolved in water.

At a 65:35 GM3:GM1 molar ratio only vesicles are found in solution. The mixed vesicle diameter is 410 Å, which is about 20% smaller than the pure GM3 one. Interesting enough, these mixed vesicles are found to increase their diameter with temperature, as usually happens with normal phospholipids vesicles.

Discussion

The addition of a second amphiphile, the ganglioside GM1, to GM3 eliminates the coexisting large aggregates and only vesicles are present in solution. This means that the additional degree of freedom of the composition of each monolayer of the bilayer allows the formation of an equilibrium phase of vesicles with an effective negative bending energy, so the vesicles have lower energy than the lamellar phase. In fact, it can happen that the lowering of the free energy by the outer layer matching its curvature to the spontaneous curvature can compensate the cost due to frustration of the inner monolayer, mostly when vesicles are not too large like in the present case.

The mixed GM3-GM1 system is qualitatively similar to the mixed copolymer system which make vesicular

solutions as recently described theoretically [11]. The model predicts that curvature energy effects favor an inhomogeneous distribution of the two components among the layers. In fact, ideal mixing of the two components would leave the bilayer in a flat geometry, with both monolayers equally frustrated with respect to their preferred curvature. Demixing, when possible, can relieve this frustration by modifying the spontaneous curvature of each monolayer, so that the spontaneous curvature of the bilayer is then finite. The balance between mixing and demixing energies determines the molar fraction at which a given mixture forms stable vesicles.

Theoretically, it has been stressed that stabilization of vesicles by surfactant mixtures only occurs when interactions of the surfactants is considered. Indeed, gangliosides with different oligosaccharide headgroups have been shown to mix non-ideally [12]. For instance, in a GM3-GM1 vesicle, GM3 should be more abundant in the inner layer where headgroup lateral compression is higher. For the present case, interactions can be of steric nature and not necessarily electrostatic as they are for the mixing of anionic and cationic surfactants [4].

The fact that single component vesicles of the ganglioside GM3 form spontaneously in solution is a clear indication that the energy cost to form a vesicle is rather small, of the order of thermal energy. In turn, this means that the bending elasticity of the GM3 bilayer should be quite small, compared to the usual bending elasticity of phospholipids. This is not surprising, if only for simple geometrical considerations. The oligosaccharide chain of gangliosides has a lateral extension which is larger than the corresponding one of the phospholipids headgroups, phosphatidylcholine for example. On the other hand, the hydrophobic dimensions are about the same. It is reasonable, therefore, to think that in the bilayer the sugar headgroups play a much more important role in determining the response of the layer to lateral compression than the choline groups, which do not exceed the hydrocarbon chain lateral dimension. This can reduce the bending elasticity, as the restoring force due to the sugar headgroup compression upon bending is smaller than the one coming from the hydrocarbon chains, which are known to have a quite high compression elasticity modulus. Entropic stabilization of single component GM3 vesicle is therefore rather plausible.

The small bending energy of the GM3 bilayer also allows large thermal undulations [13], which may make vesicles form and break quite easily. Indeed, for pure GM3 solutions, the fast equilibration times of the vesicle distribution as temperature is varied and when large aggregates reform after filtration is an interesting indication that monomer exchange is not the mechanism responsible for the equilibrium among the different aggregates, but rather it is the exchange of membrane fragments. It is known, in fact, that monomer exchange in ganglioside micellar solution is an extremely slow process [14]. Indeed, an accurate deconvolution of dynamic light scattering data seems to show that these fragments exist in solution.

The reduction of the GM3 average vesicle-diameter with temperature can than be understood as a readjustment of the vesicle distribution to higher curvatures which are allowed by the well known reduction of the effective bending elasticity modulus with temperature [13].

Mixed GM3-GM1 vesicles, instead, increase their dimension with temperature, and hysteresis effects have been observed. This is an indication that these vesicles have a more rigid structure, like normal phospholipids. The increase of vesicle size with temperature, may have the usual meaning of mass conservation of the vesicle bilayer with a reduction of its thickness, due to the increased fluidity of the hydrocarbon chains. Indeed, the stability of the mixed vesicle system is perfectly consistent with an energy stabilization process of vesicles.

Acknowledgements This work has been partially supported by CNR progetto Speciale "Complex Fluids" and by ECC Project nr. ERBCHR × CT920019.

References

1. Degiorgio V, Corti M (eds) (1985) Physics of Amphiphiles: Micelles, Vesicles and Microemulsions, North Holland, Amsterdam
2. Israelachvili JN (1991) Intermolecular and Surface Forces, Academic Press, London
3. Safran SA, Pincus PA, Andelman D, MacKintosh FC (1991) Phys Rev A 43:1071
4. Kaler EW, Murthy AK, Rodriguez BE, Zasadinski JAN (1989) Science 245: 1371
5. Cantù L, Corti M, Musolino M, Salina P (1990) Europhs Lett 13:561
6. Cantù L, Corti M, Del Favero E, Raudino A (1994) J Phys II (France) 4:1587
7. Tettamanti G, Sonnino S, Ghidoni R, Masserini M, Venerando B (1985) In ref.1 p 607
8. Corti M, Cantù L (1995) In: Barenholtz E, Lasic D (eds) Non Medical Applications of Liposomes, CRC Press Inc, Boca Raton, FL, USA to appear
9. Cantù L, Corti M, Lago P, Musolino M (1991) In Photon Correlation Spectroscopy: Multicomponent Systems, SPIE Vol.1430 p. 144
10. Cantù L, Mauri M, Musolino M, Tomatis S, Corti M (1993) Progr Colloid Polym Sci 93:30
11. Dan N, Safran SA (1993) Europhys Let 21:975
12. Cantù L, Corti M, Degiorgio V (1990) J Phys Chem 94:793
13. Helfrich H (1986) J Physique 470:321
14. Cantù L, Corti M, Salina P (1991) J Phys Chem 95:5981

Progr Colloid Polym Sci (1995) 98:201–205
© Steinkopff Verlag 1995

BIOCOLLOIDS

G. Förster
O. de la Cruz Rodríguez
G. Bendas
P. Nuhn

The influence of single-chain-anchored galactose on the polymorphism of resuspended lyophilized dipalmitoylphosphatidylcholine liposomes

Dr. G. Förster (✉)
O. de la Cruz Rodríguez
Institute of Physical Chemistry
Martin Luther University Halle/Wittenberg
Mühlpforte 1
06108 Halle, FRG

G. Bendas · P. Nuhn
Department of Pharmacy
Martin Luther University Halle/Wittenberg
Weinbergweg 15
06120 Halle, FRG

Abstract Three glycolipids consisting on a hexadecyl chain and galactose headgroup without or with different ethylenoxide spacers (EO) were studied in 1:2 mixtures with DPPC by means of calorimetry (DSC) and X-ray diffraction. Despite the high content of the glycolipid in the system without spacer and with two EO-units, miscibility was found. Only the glycolipid with three EO-units in the headgroup shows a demixing in the sample. It occurs in subgel and gel phases, but in the liquid crystalline phase the glycolipid solves itself in the DPPC matrix. In dependence on the headgroup structure of the glycolipids two different polymorphisms were detected. The consequence of mixing glycolipids to DPPC is that an interdigitation of the molecules occurs in gel phases, but in the liquid crystalline phases the situation is more complicated. In the $L\alpha$ phase a shortening of the bilayer distance was observed as well as its increase with respect to the gel phase. Nevertheless, a predominate influence of glycolipids on DPPC is obvious in liquid crystalline phases which should be studied further with relevant methods.

Key words X-ray diffraction – dipalmitoylphosphatidylcholine – liposomes – glycolipid – polymorphism

Introduction

Lyophilization is an employed method for the physical stabilization of phospholipid (PL) vesicles [1]. From investigations of natural organism it is known that sugar plays a protective role in its interaction with the dehydrated membrane [2, 3]. The first investigation at model membrane vesicles with diverse glycolipids has shown that glycolipids play only restricted its protective role during the lyophilization [4].

The aim of this paper is to investigate the whole polymorphism of some selected phospholipid – glycolipid – lyophilizates with physico-chemical methods. Synthetic DPPC was used as model membrane because of the higher information content about the miscibility behavior of glycolipids in the subgel, as well as gel and liquid crystalline phases.

Material and Methods

Lipids

1,2-dipalmitoyl-sn-glycero-3-phosphocholine (DPPC) was obtained from SIGMA and used without further purification. The galactopyranosides hexadecylgalactose C_{16}-Gal, diethylenglycolgalactosylhexadecylether C_{16}-(EO)$_2$-Gal, triethylenglycolgalactosylhexadecylether

C_{16}-$(EO)_3$-Gal were synthesized as described elsewhere [5].

Vesicle preparation

Large unilamellar vesicles (LUV) were prepared by sonication of multilamellar vesicles (MLV). Briefly, a lipid film (30 μmol) was suspended above the phase transition temperature in 1 ml phosphate buffered solution (PBS, pH 7.4) containing the required amounts of glycolipids. The MLV-dispersions were sonicated with a Bandelin Sonoplus HF 70 for 10 min [4].

Lyophilization and resuspension

Aliquots of 250 μl LUV-dispersion were rapidly frozen in a methanol bath at $-45\,°C$ for 15 min. Then, the frozen product was lyophilized for 20 h (Christ Beta 1–8 K freeze dryer, Osterode). The resulting lyophilization cakes were in the state of monohydrates according to a Karl–Fischer-titration (Karl–Fischer-Titrator Aqua 2000, Analysentechnik Beringer GmbH). In all mixtures the relation DPPC:glycolipid was taken 2:1. The lyophilized lipids were resuspended to 50 wt% mixtures at room temperature with distilled water.

Differential Scanning Calorimetry (DSC)

50 wt% lipid suspensions were equilibrated to full hydrated samples and contain excess of water in all phases. Heating and cooling scans were performed with a Perkin-Elmer DSC-2 fitted with a sub-ambient accessory. The temperature and enthalpy calibration were achieved using indium standards. A scanning rate of 5 °C/min was employed. The calorimetric scan was recorded and planimetrically evaluated.

X-ray diffraction

The low- and wide-angle studies were carried out with a horizontal goniometer HZG 4 (Freiberger Präzisionsmechanik GmbH). Patterns from monochromized CuKα radiation (35 kV, 60 mA) were obtained as difference of Ni and Co filtered patterns. The X-ray samples were prepared in 50 wt% vortexed mixtures and sealed in thin-wall glass capillaries of diameter 1.5 mm. The samples were studied in 8-h steps at controlled temperatures ($\pm 0.1\,°C$) in the interval $-25\,°C...65\,°C$.

Results and discussion

Differential Scanning Calorimetry (DSC)

In resuspended lyophilized LUVs of DPPC the well known pretransition at 34 °C and the main transition at 41 °C are seen with almost unchanged parameters [7]. A polymorphism $L\beta' - P\beta' - L\alpha$ is found without any thermal treatment to grow the subgel phase structure. The pretransition and the main transition temperatures in the mixture DPPC:C_{16}-Gal (2:1) decrease of about two degrees, respectively. The shape of the heating curve implies an unchanged polymorphism of pure DPPC.

One very broad maximum at 41 °C without pretransition was found in the mixture of DPPC:C_{16}-$(EO)_2$-Gal (2:1). The character of the transitions changes, but the temperature stays in the same range as in pure DPPC. The mixture DPPC:C_{16}-$(EO)_3$-Gal (2:1) has a similar broad form, but the maximum decreases to 38 °C. In both cases an extended two-phase region is expected because of the broadening of the peaks.

X-ray diffraction

In order to obtain the origin of the different transitions seen in the thermal measurements, X-ray diffraction measurements were made under similar conditions. Series of diffraction patterns obtained in the interval from $-25\,°C$ to $+65\,°C$ are displayed in the Figs. 2–4 in order to illustrate the transition of the sample from one to another phase. Both regions at low and wide angles are shown separately and amplified in different scales.

DPPC:C_{16}-Gal (2:1) liposomes

Figure 1 shows diffraction patterns obtained from resuspended lyophilized DPPC:C_{16}-Gal (2:1) liposomes. The following variations in the diffraction patterns indicate that the C_{16}-Gal is solved in the DPPC matrix and does not separate itself in a redundant phase. At low temperatures a subgel phase Lc is found, which in pure DPPC appears only after special treatment [7]. In Fig. 1 the subgel phase Lc is characterized by several wide-angle reflections, which are similar to those described for DPPC by Füldner [7]. As only one set of long spacings with four orders is seen in the small-angle region, a homogeneous packing in bilayers for the mixture is assumed. At $-25\,°C$ sharp crystalline reflections appear from frozen water.

Fig. 1 Review of diffraction patterns on heating mixtures of DPPC:C_{16}-Gal (2:1) in excess of water. Left: wide-angle region, right: small-angle region, rescaled

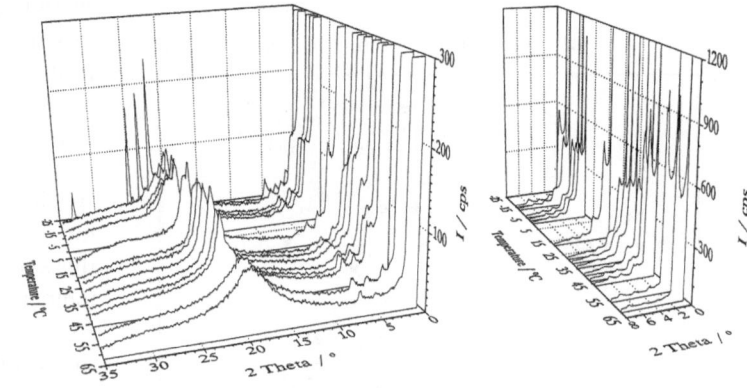

Fig. 2 Review of diffraction patterns on heating mixtures of DPPC:C_{16}-$(EO)_2$-Gal (2:1) in excess of water. Left: wide-angle region, right: small-angle region, rescaled

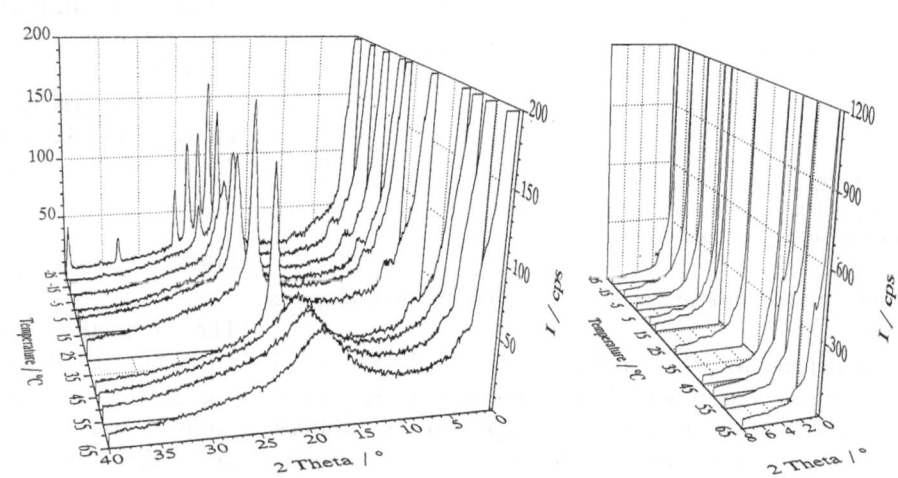

Fig. 3 Review of diffraction patterns on heating mixtures of DPPC:C_{16}-$(EO)_3$-Gal (2:1) in excess of water. Left: wide-angle region, right: small-angle region, rescaled

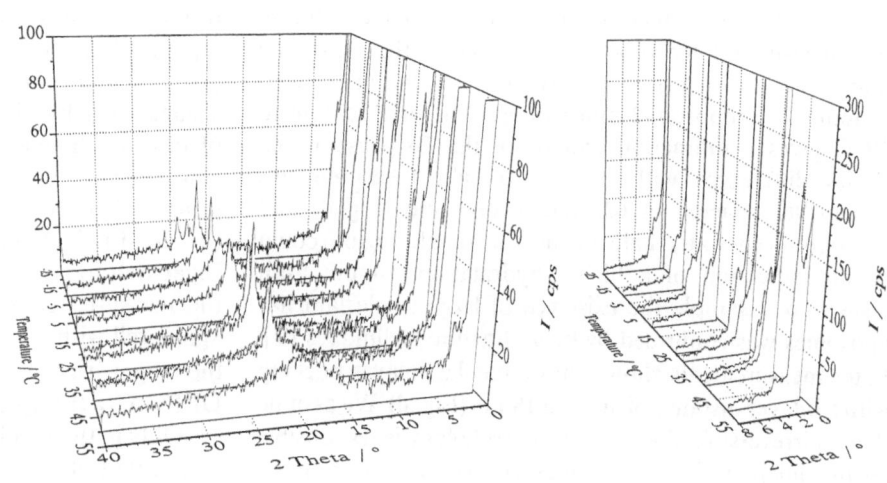

Between 1° and 35°C a wide-angle scattering is observed which is in agreement with the gel phase $L\beta'$ analyzed by Tardieu et al. [6]. In addition, the pretransition $L\beta'$–$P\beta'$ is deduced from the temperature dependence of the long spacings (Fig. 4). During the main

transition an extended two-phase region $P\beta'$ – $L\alpha$ appears. In the liquid crystalline phase $L\alpha$ an unusual decreasing of the long spacings is observed between 40° and 55°C, which ends at an average value of 5.5 nm for this phase. The polymorphism Lc + ice – Lc – $L\beta'$ – $P\beta'$ – $L\alpha$ was

204
G. Förster et al.
The influence of single-chain galactose on liposomes

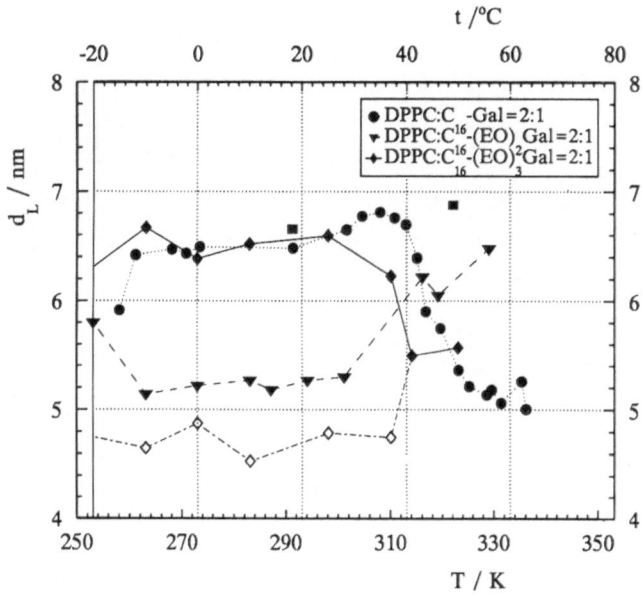

Fig. 4 Temperature dependence of the bilayer distance of resuspended lyophilized mixtures of dipalmitoylphosphatidylcholine with different glycolipids (2:1) in excess water

found in the mixture which is similar to that of pure DPPC.

A dependence of the polymorphism on the history of the samples was observed. A comparison of respective diffraction curves in the phases $L\beta'$ and $L\alpha$ in fresh resuspended state and in the course of systematic investigations (Fig. 1) shows characteristic differences. The $L\beta'$ phase as the initial state after resuspension gives a bilayer distance which is comparable to that which also was measured later (Fig. 4, ■, ●). Only the scattering intensity of the long spacings is smaller, which is interpreted in terms of a liposomal morphology. The first heating to 49 °C gives a diffraction pattern of the $L\alpha$-phase which shows differences with respect to later measurements. A lamellar spacing of 6.87 nm is observed (Fig. 4) and a pronounced water halo in the wide-angle region could be an indication of a higher hydratation. In the fresh resuspended sample the existence of large single lamellar liposomes was confirmed by freeze fracture techniques [9]. Later measurements show a more regular bilayer packing with a bilayer distance of about 5.18 nm (Fig. 4). We assume that an irreversible change in the morphology is responsible for the different phases structures. It takes place during the measurement in the $L\alpha$ phase with its high mobility.

DPPC:C$_{16}$-(EO)$_2$-Gal (2:1) liposomes

Figure 2 shows the diffraction curves of resuspended lyophilized DPPC:C$_{16}$-(EO)$_2$-Gal (2:1) liposomes. The scattering in the wide-angle region is different from the mixture discussed above. The gel phase shows one sharp peak which becomes very sharp at high temperatures. It is interpreted as the diffraction from a two-dimensional (2D) hexagonal packing of untilted chains and named β. A transition takes place at lower temperatures. The small spacing splits into the peaks (11) and (20) which are typical for orthorhombically packed hydrocarbon chains. From this subgel phase, named ρ, water becomes frozen at lower temperatures and the bilayer distance decreases (Fig. 4). After an evaluation of the long spacings it is obvious that the whole polymorphism takes place in bilayers with a very small thickness. Therefore, an antiparallel arrangement of the molecules in lamellar gel and subgel phases is assumed. On the other hand, the bilayer distance increases continuously on heating the liquid crystalline phase $L\alpha$.

In this system in which two EO spacers were introduced in the glycolipid headgroup the polymorphism $L\rho_a + ice - L\rho_a - L\beta_a - L\alpha$ was observed.

DPPC:C$_{16}$-(EO)$_3$-Gal (2:1) liposomes

The diffraction curves of resuspended lyophilized DPPC:C$_{16}$-(EO)$_3$-Gal (2:1) liposomes are shown in Fig. 3. It is remarkable that the overall scattering of the sample is weak compared to the other mixtures prepared under the same conditions. Indeed, there is a physical reason which can explain this fact. In the low-angle region two sets of long spacings appear in the gel phase. One set of long spacings shows two orders, and additionally one separate long spacing is seen. This points to a demixing phenomenon in the mixture into a lipid-rich fraction and a sugar-rich fraction, respectively. Of course, this takes place in small areas and disturbs the morphology with the consequence of a weak scattering intensity from the samples. The two bilayer distances, $d_L = 6.0 \pm 0.3$ nm in the lipid-rich fraction and $d_L = 4.8 \pm 0.2$ nm for the sugar-rich fraction, are compared in Fig. 4 with those of the measured substances. The lipid-rich fraction values agree well with the data of DPPC:C$_{16}$-Gal (2:1), whereas the sugar–rich fraction is smaller than the data of DPPC:C$_{16}$-(EO)$_2$-Gal (2:1).

Only in the $L\alpha$ phase is the scattering in the small-angle region that of a homogeneous phase deducible from only one set of long spacings with two orders. Therefore, a mixing of components occurs in the liquid crystalline phase with fluid chains. But the resulting bilayer distance is smaller compared to the mixture with the glycolipid with two EO units as spacers.

In the wide-angle region of a demixed system also spacings of two different phases should appear. But both phases consist of packings of aliphatic chains, and the superposition of their scattering is hardly to decompose. Approximately, it can be seen that the same short spacings appear in the scattering curves as in the system DPPC: C_{16}-$(EO)_2$-Gal (2:1). However it has not been elucidated in which of the demixed phases this polymorphism $L\rho_a$ + ice $- L\rho_a - L\beta_a - L\alpha$ exists or if it is the same in both phases.

Conclusions

Three glycolipids consisting of a hexadecylchain ether connected with galactose and without or with different EO spacers are mixed in the ratio of 1:2 with DPPC. Their history was freeze-drying of large single lamellar vesicles (SLV) followed by resuspension in excess water.

An influence on the history of the phase structure in the initial state in relation to later studies was found only in the system DPPC:C_{16}-Gal which has a polymorphism like DPPC. Irrespective of the high contents of the glycolipid, a miscibility behavior was found in two of the systems. Only the glycolipid with a spacer of three EO-units in the headgroup shows a demixing in the sample. It occurs in subgel and gel phases, but in the liquid crystalline phase the glycolipid solves itself in the DPPC matrix.

In dependence on the headgroup structure of glycolipids two different polymorphisms were detected.

The polymorphism Lc + ice $- Lc - L\beta' - P\beta' - L\alpha$ was found, surprisingly, in the mixture DPPC:Gal-C_{16} (2:1) which is similar to that of pure DPPC. As almost the structure parameters of DPPC are unchanged, it is assumed that the large sugar headgroup of the glycolipid is packed in a pocket of DPPC headgroups, which was recently proposed for lecithins which have a short-chain branching in the C2 chain [8].

The introduction of EO spacers changes the polymorphism to $L\rho_a$ + ice $- L\rho_a - L\beta_a - L\alpha$. In the mixture of DPPC:C_{16}-$(EO)_2$-Gal (2:1) with two EO units a bilayer with reduced thickness appears in which the molecules should be arranged antiparallely. If the spacer is one EO longer a demixing into a lipid-rich-fraction and a sugar-rich-fraction occur with different bilayer thicknesses in the coexisting gel phases.

The consequence of mixing of glycolipids to DPPC is that in gel phases an interdigitation of the molecules is induced, but in the liquid crystalline phases the situation is more complicated. In the $L\alpha$ phase a shortening of the bilayer distance was observed as well as its increasing in respect to the gel phase. Nevertheless, a predominate influence of glycolipids on DPPC is obvious in liquid crystalline phases which further should be studied with relevant methods like spectroscopy, for instance.

Acknowledgments This work was supported by the Deutsche Forschungsgemeinschaft (SFB 197 "Bio -und Modellmembranen"). The authors would like to thank F. Wilhelm for the synthesis of the galactopyranosides.

References

1. Crowe JH, Crowe LM, Chapman D (1984) Science 223:701–703
2. Higa LM, Womersley CS (1993) J Exp Zool 267:120–129
3. Madin KAC, Crowe JH (1975) J Expl Zool 193:335–342
4. Engel A, Bendas G, Wilhelm F, Mannova M, Ansban M, Nuhn P (1994) Int J Pharmac 107:99–110
5. Ogawa T, Beppu K, Nakabayashi S (1981) Carbohydr Res 93:C6–C9
6. Tardieu A, Luzzati V, Reman FC (1973) J Mol Biol 75:711–733
7. Füldner HH (1981) Biochemistry 20:5707–5710
8. Rattai B, Brezesinski G, Dobner B, Förster G, Nuhn P (1995) Chem Phys Lipids 75:81–91
9. Meyer HW, Richter W, unpublished results

Progr Colloid Polym Sci (1995) 98:206–211
© Steinkopff Verlag 1995

BIOCOLLOIDS

S. Beugin
C. Grabielle-Madelmont
M. Paternostre
M. Ollivon
S. Lesieur

Phosphatidylcholine vesicle solubilization by glucosidic non-ionic surfactants: a turbidity and x-ray diffraction study

S. Beugin (✉)
C. Grabielle-Madelmont
M. Paternostre · M. Ollivon · S. Lesieur
Equipe "Physicochimie des Systèmes
Polyphasés"
CNRS URA 1218
Université Paris Sud
92296 Chatenay-Malabry Cedex, France

Abstract The solubilization mechanism of egg phosphatidylcholine (PC) vesicles by octyl glucoside (OG) and hecameg (HG) was examined at 25 °C by turbidity and small-angle x-ray scattering (SAXS) analysis. Turbidity was recorded upon continuous surfactant addition to PC vesicles in the highly diluted region of the lipid-surfactant-water phase diagram. SAXS analysis was performed on more concentrated samples with the same surfactant and lipid compositions as in the diluted domain studied by turbidity. Very similar vesicle-to-micelle transition mechanisms are observed for OG and HG. The solubilization process involves four steps, the limits of which univocally correspond to precise surfactant-to-PC molar ratios in the aggregates and surfactant concentrations in the aqueous continuum. However, HG shows a better efficiency to dissolve PC vesicles than OG. These limits correspond to the boundaries of distinct structural states observed in the more concentrated region. With increasing surfactant-to-lipid ratios, these states are successively: 1) a lamellar structure characterized by a main repeat distance close to that observed for the PC L_α phase, beside one or more other apparently lamellar structure(s) of smaller periodicity, 2) a unique lamellar phase, 3) coexisting lamellar and micellar assemblies, and 4) interacting micelles.

Key words Vesicle-to-micelle transition – liposome – egg phosphatidylcholine – octyl glucoside – hecameg

Introduction

Functional and structural studies of membrane proteins often involve the dissolution of their native membrane, by the addition of surfactant, before their reconstitution into simplified lipid assemblies such as liposomes. These last are formed by removal of the surfactant from mixed lipid-protein-surfactant micelles. The understanding of the micelle-to-vesicle transition or reverse process of vesicle micellization thus appears essential to control solubiliza-tion and reconstitution experiments [1]. Egg phosphatidyl-choline (PC) vesicles associated with octyl glucoside (OG), a non-ionic surfactant which presents the advantage of having a high critical micelle concentration (CMC) and a low-denaturing effect on proteins, are frequently chosen to modelize the vesicle-to-micelle transition [2–5]. Recently, a glucosidic analogue of OG, hecameg (HG), has been synthesized and used for membrane protein reconstitution [6].

The aim of this work is to compare the solubilization processes of PC vesicles induced by OG and HG,

Progr Colloid Polym Sci (1995) 98:206–211
© Steinkopff Verlag 1995

respectively. On a supramolecular level, the rearrangements of the mixed PC-surfactant aggregates are followed by turbidity measurements performed in the highly diluted region of the lipid-surfactant-water phase diagram. In parallel, structural information is obtained by small-angle x-ray diffraction analysis of more concentrated samples, with the same surfactant-to-lipid ratios as those investigated for the vesicle solubilization experiments.

Materials and methods

Products

Egg phosphatidylcholine (PC, $MW = 760.1$), octyl β-D-glucopyranoside or octyl glucoside (OG, $MW = 292.4$), and 6-O-(N-heptylcarbamoyl)-methyl-α-D-glucopyranoside or hecameg (HG, $MW = 335.4$) were purchased from Avanti, Sigma, and Vegatec, respectively. All samples were prepared in a 10 mM Hepes and 145 mM NaCl buffer (pH = 7.4). In this buffer and at 25 °C, the critical micelle concentrations of both surfactants, OG and HG, are 21.6 and 19.5 mM, respectively.

Sample preparation for x-ray diffraction

A dry lipid film of about 15 mg was first prepared, by evaporation of a chloroformic PC solution under nitrogen and then by lyophilization (12 h). The surfactant powder (0 to 20 mg, depending on the sample) was then added and the lipid-surfactant mixture was dissolved in 2–3 ml chloroform. Then, the mixed lipid-surfactant film was obtained as above, hydrated with 30 μl buffer and sonicated until homogenization of the suspension.

Vesicle preparation

Small unilamellar PC vesicles were prepared by sonication, according to Lesieur et al. [7]. They were characterized by quasi-elastic light scattering (QELS) using a Nanosizer (N4, Coultronics), and by gel exclusion high performance liquid chromatography according to the protocols set up by Ollivon et al. [8] and Lesieur et al. [9, 10]. QELS analysis indicates a bimodal distribution centered at 30 and 120 nm, respectively. Besides, gel exclusion chromatograms only show a symmetric elution peak corresponding to a particle size of 35 nm. These results make evident that the unit size of the vesicles is 35 nm, the particle population at 120 nm corresponding to vesicle aggregates [10].

X-ray diffraction

X-ray diffraction analysis was performed on the D24 line of the DCI Synchrotron ($\lambda = 1.499$ Å) at L.U.R.E. (Orsay, France) using the initial set up of P. Vachette and C. Bourgaux. The different samples were introduced into thin-walled glass capillary tubes (ext. diameter < 1.5 mm) (GLAS, Germany) by successive centrifugations and placed in a sample holder thermostated at 25 °C. Data were collected using a linear 1024-channel position-sensitive detector. Two different sample-to-detector distances were used, 30 cm and 120 cm, in order to screen s values (s = reciprocal space co-ordinate $= 2\sin\theta/\lambda$, where 2θ is the scattering angle) in the range 0–0.30 Å^{-1} ($q = 2\pi s = 0$–2 Å^{-1}) and then more precisely between 0 and 0.1 Å^{-1} ($q = 0$–0.6 Å^{-1}).

Solubilization

The solubilizations were performed, at 25 °C on stirred samples, as previously described [7]. The HG solution was continuously added to the vesicle suspension placed in a quartz cell, at a rate of 13.9 μl/min using a 1 ml glass precision-syringe pushed by a syringe-pump. During the solubilization, turbidity was continuously recorded at 350 nm using a double-beam spectrophotometer (Lambda 2, Perkin Elmer). The HG concentration in the syringe was chosen low enough (60 mM) to get accurate correspondence between turbidity recording and surfactant concentration in the cell.

Results and discussion

Solubilization of PC vesicles by HG

Solubilization experiments were performed by continuous addition of HG solution to PC vesicles of initial concentrations ranging from 0.5 to 3 mM. The recording of the optical density (OD) variations as a function of the surfactant concentration characterizes the evolution of the aggregates morphology. A typical solubilization curve is shown in Fig. 1. This curve can be divided into four parts separated by three break points noted A, B, and C, by analogy with the study of the PC-OG system [2]. Before break point A, the addition of HG only causes a slight OD increase. Break point A marks the first drastic change in turbidity, and break point B corresponds to maximum OD values ranging from 0.1 to 0.7 while increasing the initial PC concentrations from 0.5 to 3 mM. Beyond B, OD first decreases, then shows a significantly higher peak (noted

208

S. Beugin et al.
Phosphatidylcholine vesicle-to-micelle transition by glucosidic surfactants

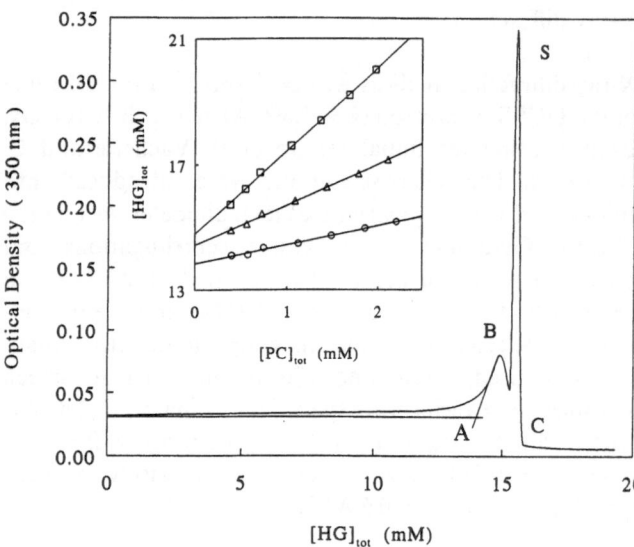

Fig. 1 Variations of the optical density during the solubilization of PC vesicles by HG (initial [PC] = 0.5 mM). Inset: Total HG concentration $[HG]_{tot}$ as a function of total PC concentration $[PC]_{tot}$ at break points A (○), B (△) and C (□)

Table 1 Non lipid-associated HG concentration $[HG]_{bulk}$ and HG-to-PC molar ratios in mixed aggregates $(HG/PC)_{agg}$ at the solubilization curve break points A to C. Comparison with the similar solubilization parameters calculated from literature data relative to OG [2].

	$[OG]_{bulk}$ (mM)	$(OG/PC)_{agg}$	$[HG]_{bulk*}$ (mM)	$(HG/PC)_{agg*}$
A	14.3	1.1	13.9	0.61
B	15.2	2.1	14.5	1.26
C	15.7	3.0	14.9	2.58

(∗): the relative error deduced from two sets of experiments is equal to 2.5%.

S in Fig. 1) reaching intensities between 0.35 and 1.2 in OD. Complete solubilization is depicted by the drastic decrease in turbidity (break point C) towards very low OD values.

The successive events observed during the solubilization of PC vesicles by HG (this work) and by OG [2–5] seem to be very comparable. However, the higher turbidity peak (S) observed between B and C, and presumably due to a macroscopic phase separation, is also observed with OG, but only for initial PC concentrations higher than 2.4 mM [2, 3].

From the different solubilization curves and at each break point "i" (i = A, B or C), both total surfactant concentration $[HG]_{tot,i}$ and total lipid concentration $[PC]_{tot,i}$ were determined. The inset of Fig. 1 reports the plots of $[HG]_{tot,i}$ versus $[PC]_{tot,i}$. Linear regression analysis allowed to calculate 1) the concentration of HG molecules which are not associated with the lipids ($[HG]_{bulk,i}$) from the extrapolation to zero of the total lipid concentration, 2) the surfactant-to-lipid mole ratio in the aggregates ($(HG/PC)_{agg,i}$) from the slopes of the lines. $[HG]_{bulk,i}$ and $(HG/PC)_{agg,i}$ are independent of the initial vesicle concentration and characterize precisely the system at each break point. These parameters, which were calculated on the basis of two series of experiments, are summarized in Table 1 together with the $[OG]_{bulk,i}$ and $(OG/PC)_{agg,i}$ values determined using the same method [2]. The comparison of the two systems shows that, at all steps of the solubilization process, $[HG]_{bulk,i}$ is lower than

$[OG]_{bulk,i}$, reflecting the solubility properties of surfactant monomers in buffer ($CMC_{HG} < CMC_{OG}$). In addition, the fact that surfactant-to-PC ratios in the aggregates are lower for HG at each stage of the vesicle-to-micelle transition indicates a better efficiency for HG to destabilize PC bilayers. Particularly, complete micellization (break point C) is reached in the case of HG for 2.6 surfactant molecules per PC in the aggregates, instead of 3 in the case of OG.

By analogy with a recent study of the OG-dipalmitoylphosphatidylcholine-water system [12], and to understand the surfactant behavior, the modifications of concentrated PC lamellae induced by OG and HG additions were examined as an approach of the solubilizing effect of the surfactants on PC vesicle bilayers.

Modifications of the PC lamellar phase L_α induced by the addition of OG or HG

The influence of both OG and HG addition onto the PC lamellar phase L_α was studied by small-angle x-ray scattering (SAXS), for total surfactant-to-PC molecular ratios ($(OG/PC)_{tot}$ or $(HG/PC)_{tot}$) ranging from 0 to 2.7 (precise ratios are listed in Table 2) and constant PC-to-buffer weight ratio equal to 30:70. It is interesting to note that the total surfactant-to-PC mole ratios indicated in Table 2 are close to the effective surfactant-to-PC ratios in the lipidic structures, since the surfactant monomer proportion dissolved in water is negligible in such concentrated samples.

Samples containing OG (or HG) and PC at $(OG/PC)_{tot}$ ratios ranging from 0 to 1.53 (or $(HG/PC)_{tot}$ from 0 to 0.97) present a turbid white lower phase and a clear upper phase. The latter is very likely constituted of excess water since PC L_α phase reaches its maximum swelling around 50% water (w/w) and then over 50% rejects the excess water [13]. In the followings, only the structure of the lower phase is described. At ratios of 2.01 for OG and 1.73 for HG, samples show an opalescent lower phase and a cloudy upper phase. The samples of ratios equal to 2.75 for OG and 2.33 for HG are uniformly cloudy. An HG-to-PC ratio of 2.66 gives a clear sample.

Small-angle x-ray diffraction patterns representative for the different surfactant-to-PC ratios investigated are presented in Figs. 2 and 3, and the Bragg spacings deduced from the reflections are reported in Table 2. The scattering bump observed below 0.01 $Å^{-1}$ is only due to the rings of the direct beam.

The SAXS pattern of the sample containing only PC and buffer shows low-angle reflections in the ratios 1:1 and 2:1, characteristic of a one-dimensional lattice. The observed lamellar repeat distance of 63.7 Å (Figs. 2a, 3a, and Table 2) agrees with the thickness of PC lamellae (one lamella including a PC bilayer and a water layer) at their maximum hydration (above 50%) [13]. In addition, a broad scattering band centered at $(4.5 \text{ Å})^{-1}$ is observed at high angles and provides the average distance between the lipid chains in a plan parallel to the lamellae. The width and the position of this band indicate the disordered

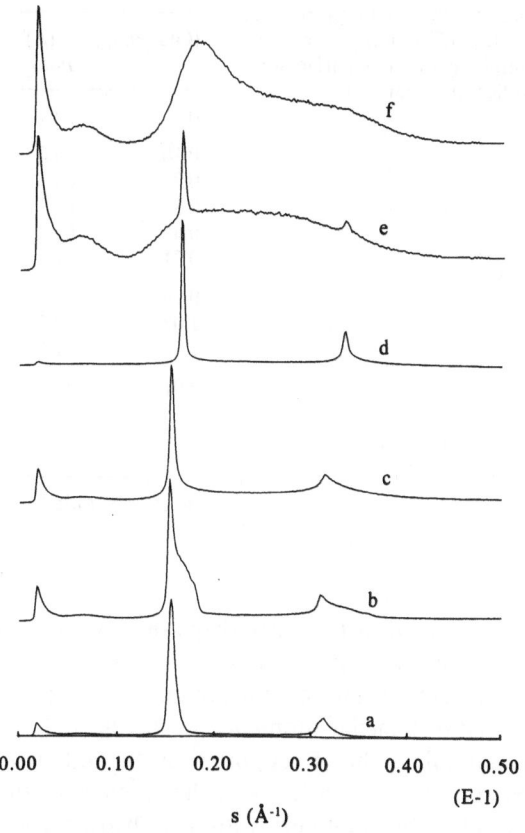

Fig. 3 SAXS patterns obtained for samples of total HG-to-PC ratios equal to: 0 (a), 0.28 (b), 0.97 (c), 1.73 lower phase (d), 1.73 upper phase (e), 2.66 (f)

state of these chains at 25 °C and is not significantly modified upon surfactant addition.

At surfactant-to-lipid ratios in the range 0.10–0.71 for OG and 0.10–0.62 for HG, diffraction patterns still show two main lamellar reflections (Figs. 2b and 3b). Each main reflection is accompanied by one (OG) or several (HG) shoulders on its high angle side. With increasing surfactant proportions, the lamellar periodicity decreases from 64.0 Å to 62.5 Å for OG, and from 64.5 Å to 63.4 Å for HG. However, it should be pointed out that the shift of the lamellae repeat distance can be enhanced or simply only caused by the shoulder rise. In the case of OG, the shoulder reflections show increasing intensity with increasing OG concentration. Although these reflections are not totally deconvoluted, their respective positions likely correspond to a lamellar structure of periodicity around $(57 \text{ Å})^{-1}$. Incorporation of low amounts of HG induces similar structural changes; however, shoulders observed in the $(55–60 \text{ Å})^{-1}$ range are significantly broader and not resolved. The main lamellar repeat distances can be attributed to a structure of lipid bilayers

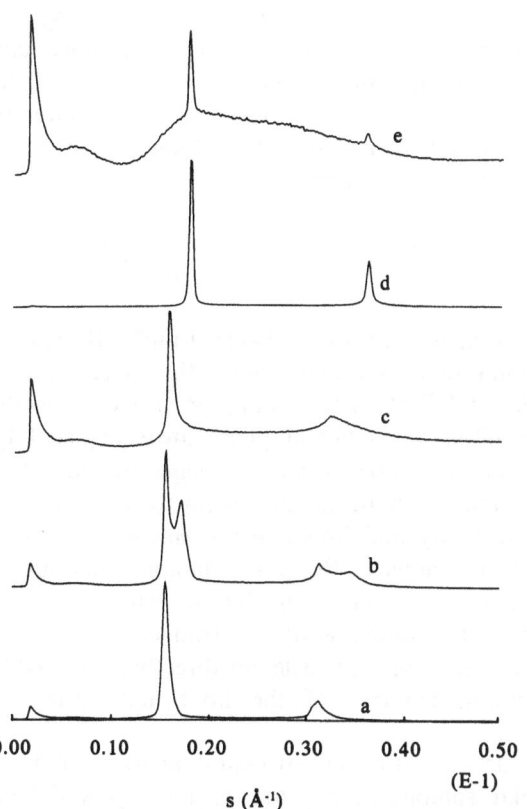

Fig. 2 SAXS patterns obtained for samples of total OG-to-PC ratios equal to: 0 (a), 0.45 (b), 1.53 (c), 2.01 lower phase (d), 2.01 upper phase (e)

Table 2 Bragg spacings deduced from the small-angle x-ray diffraction patterns of lipid-rich samples containing surfactant and PC at different molar ratios.

$(OG/PC)_{tot}$	$d_1(Å)$ Peak	$d_2(Å)$ Shoulder	$(HG/PC)_{tot}$	$d_1(Å)$ Peak	$d_2(Å)$ Shoulder
0	63.7	–	0	63.7	–
0.10	64.0	sh.*	0.10	64.5	sh.
0.21	64.1	sh.	0.19	64.2	sh.
0.35	63.2	56.1	0.28	64.1	sh.
0.45	63.3	57.6	0.40	63.8	sh.
0.71	62.5	sh.	0.62	63.4	sh.
0.99	62.3	–	0.84	63.5	–
1.08	62.2	–	0.97	63.5	–
1.53	61.6	–			
2.01	55.0 + mic	–	1.73	59.5 + mic	–
2.75	51.4 + mic	–	2.33	56.0 + mic	–
			2.66	mic	–

(*) sh. = shoulder maximum not resolved

essentially constituted of PC, since they are very close to the spacing of the pure PC L_α phase. The significantly shorter repeat distances indicate the presence of one (OG) or several (HG) lamellar domains probably rich in surfactant molecules. This is supported by the following simple geometrical approach: in a lipid bilayer, the replacement of a PC molecule with two hydrocarbon chains of average length corresponding to a C16 chain by a surfactant molecule composed of one C8 chain, leads to a decrease of the bilayer thickness estimated to 9.1 Å $(6(CH_2–CH_2) + 1(CH_2–CH_3) = 6(1.27) + 1.5)$ [13]. The modification in the observed lamellar periodicity close to 7 Å is in agreement with this estimation; the difference of 2 Å can reflect the influence of the polar head group (replacement of a phosphatidylcholine by a glucosidic group) on the water layer thickness. For surfactant-to-PC molar ratios ranging from 0.10 to 0.71 for OG, and from 0.10 to 0.62 for HG, surfactant appears not uniformly distributed in the lipid matrix: the PC L_α phase, probably saturated with very few surfactant molecules, coexists with another lamellar structure richer in surfactant.

At higher ratios, in the 0.99–1.53 (OG) and 0.84–0.97 (HG) ranges, a single set of sharp lamellar reflections is observed and corresponds to repeat distances from 62.3 Å down to 61.6 Å (OG) and of 63.5 Å (HG), intermediate between the preceding values (Figs. 2c and 3c). The formation of a unique lamellar structure may indicate a homogeneous surfactant and lipid distribution.

At ratios of 2.01 (OG) and 1.73 (HG), the macroscopically separated phases of each sample were analyzed. Diffraction patterns of the lower phase present sharp and very intense lamellar reflections of periodicity 55 Å (OG) and 59.5 Å (HG) (Figs. 2d and 3d). The upper phases (Figs. 2e and 3e) exhibit, between 0.015 and 0.04 Å^{-1}, a broad scattering band, ascribable to mixed PC-surfactant micelles. The superimposed diffraction lines are related to a part of the lower phase not separated by centrifugation. Samples with ratios of 2.75 (OG) and 2.33 (HG) show diffraction patterns similar to those of the previous upper phases, indicating that the same two phases are still coexisting, but could not be separated here. In this overall concentration range, these results support the coexistence between lamellar and micellar structures, already mentioned for diluted PC-OG system [2–5].

At HG/PC ratio of 2.66 (Fig. 3f), reflections have totally vanished and only a dissymmetric scattering band (intensified around 0.02 Å^{-1}) is observed. At this ratio, the system is thus only composed of mixed micelles, probably in strong sterical interaction.

Conclusions

From the comparison between Tables 1 and 2, the surfactant-to-lipid ratios in the aggregates at the break points of the vesicle solubilization curves appear very close to the ratios delimiting the different phase areas depicted by x-ray diffraction patterns. As a consequence, the successive stages observed by both methods seem to be correlated, although turbidity and SAXS measurements have been performed in regions of the phase diagram differing in their dilution state and in the bilayer curvature. This suggests that the supramolecular rearrangements induced by surfactant addition on vesicles are directly connected to the structural behavior of the lipid and surfactant association.

Both turbidimetric and structural analyses indicate very similar solubilization processes for OG and HG.

Progr Colloid Polym Sci (1995) 98:206–211
© Steinkopff Verlag 1995

However, HG shows a better efficiency than OG to solubilize PC vesicles, likely in relation with OG and HG different polar head groups (conformation, hydration).

The understanding of the molecular mechanisms involved during the vesicle-to-micelle transition deserves complementary studies. Samples with intermediate compositions should be analyzed, first to determine more exactly the phase boundaries and their structure parameters, and then to confirm the correlation between highly diluted and more concentrated regions of the ternary phase diagram.

Acknowledgements We thank C. Bourgaux for allowing these preliminary experiments to be performed using D24 line and help in experimental set-up, F. Lavigne for writing the computer programs, and M. Dahim for advice concerning sample preparation.

References

1. Walter A (1992) In: Gaber BP, Easwaran KRK (eds) Biomembrane Structure and Function. The State of the Art. Adenine Press, p 21
2. Ollivon M, Eidelman O, Blumenthal R, Walter A (1988) Biochemistry 27(5):1695–1703
3. Paternostre M-T, Roux M, Rigaud J-L (1988) Biochemistry 27:2668–2677
4. Almog S, Litman BJ, Wimley W, Cohen J, Wachtel EJ, Barenholz Y, Ben-Shaul A, Lichtenberg D (1990) Biochemistry 29:4582–4592
5. Vinson PK, Talmon Y, Walter A (1990) Biophys J 56:669–681
6. Plusquellec D et al. (1989) Anal Biochem 179:145–153
7. Lesieur S, Grabielle-Madelmont C, Paternostre M-T, Moreau JM, Handjani-Villa RM, Ollivon M (1990) Chem Phys Lipids 56:109–121
8. Ollivon M, Walter A, Blumenthal R (1986) Anal Biochem 152:262–274
9. Lesieur S, Grabielle-Madelmont C, Paternostre M-T, Ollivon M (1991) Analytical Biochem 192:334–343
10. Lesieur S, Grabielle-Madelmont C, Paternostre M-T, Ollivon M (1993) Chem Phys Lipids 64:57–82
11. Ueno M (1989) Biochemistry 28:5631–5634
12. Dahim M, Grabielle-Madelmont C, Lesieur S, Paternostre M, Ollivon M, in preparation
13. Small DM (1986) In: The Physical Chemistry of Lipids. Plenum Press, New York and London, chaps. 3 and 12

Progr Colloid Polym Sci (1995) 98:212–214
© Steinkopff Verlag 1995

BIOCOLLOIDS

H. Hermel
U. De Rossi

Polypeptide β-sheet detection by cyanine dyes

Dr. H. Hermel (✉)
Max-Planck-Institut for Colloid and
Interface Research
Rudower Chaussee 5
12489 Berlin, FRG

U. De Rossi
Federal Institute for Materials
Research and Testing
12200 Berlin, FRG

Abstract Mixtures of oxacarbo-cyanine or thiacarbocyanine dye and protein- or polypeptides aggregates show a shifted absorption band for the dye compared to its aqueous solution in the case that all or most of the protein- or polypeptide aggregates have β-sheet structure. The shift can be a red or a blue of 10 nm to 25 nm, depending on the kind of cyanine dye. In contrast are mixtures with protein or polypeptides with helical or random coil conformations, where no change in the position of absorption maxima of dye was observed. That is, if possible, also a way to analyze different secondary structures in thin polypeptide and protein adsorption layers and membranes.

Key words Polypeptides – secondary structure – cyanine dyes

Introduction

The adsorption of proteins and polypeptides at fluid and solid interfaces is of great interest for live sciences and technological processes. Our aim is to detect structural differences of the protein secondary structure, i.e., α-helix and β-sheet. Poly-L-lysine (**PL**) is an excellent model system for such investigations because it can exist either in the α-helix or in the β-sheet conformation at the same environmental conditions (pH, ion strength, temperature) [1]. We have investigated the interaction between different **PL** secondary structures and cyanine dyes which represent a very sensitive system: the equilibrium between the monomer (M), dimer (D) and the higher aggregated forms (H and J aggregates) depends in an aqueous emulsion and suspension on the nature of the fluid/fluid and solid/fluid interface [2, 3].

Materials and methods

Poly-L-lysine-hydrobromide (**PL**) was obtained from SIGMA. $M_w = 288400$ D was used to produce the α-helix type and $M_w = 61000$ D to produce the β-sheet type and the random coil type [4]. It was dissolved in water to a 7.2×10^{-3} M (amino acid residues) solution. In this solution the secondary structure was obtained using the method of Greenfield [5, 6].

- α-helix type: pH = 11.5, $T = 22\,°C$
- β-sheet type: pH = 11.5, solution heated for 25 min at 50 °C and then cooled down to 22 °C
- random coil type: pH = 4.5, $T = 22\,°C$

The formation of these different structures was checked by circular dichroism measurements.

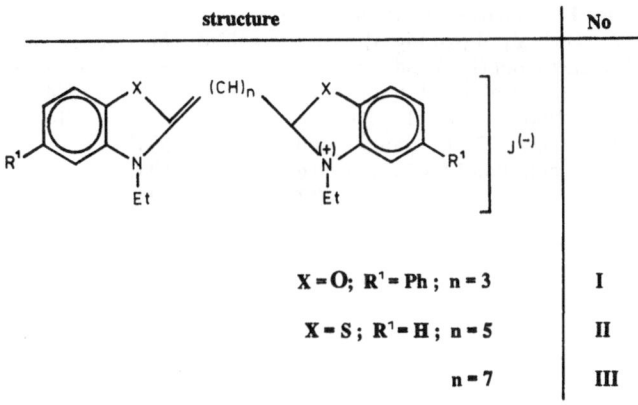

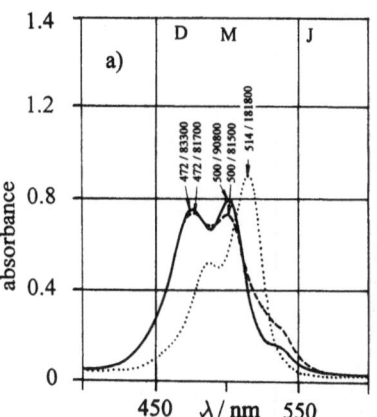

Fig. 1 Oxacarbocyanine and thiacarbocyanines which we have used

The dyes used in our experiments (Fig. 1) were dissolved in methanol to a 1.5×10^{-4} M solution. Protein and dye stock solutions were mixed in the ratio **PL** solution/cyanine dye solution = 9/1. At this water/methanol ratio the cyanine dye monomer molecules and aggregates possess a good solubility. That means: within 1 h after dissolution no flocculates or coatings were observed. Moreover, using this relatively high water content also the **PL** secondary structure was stable. This was demonstrated by circular dichroism measurements.

Results and discussion

The following results are obtained from a UV-Vis-analysis:

1) Without **PL** in the dye solution λ_{max} of the absorption bands independent of the pH-value (pH = 11.5 and pH = 4.5). Different at both pH-values is only the absorptivity.
2) This is valid also in the presence of **PL** α-helix or random coil in the solution. In the presence of α-helix and random coil the J-aggregation increases at the cost of the M and D bands. However, no shift of the M band compared to aqueous solution can be observed (Fig. 2).
3) In the presence of **PL** β-sheet the behavior is different: one observes a strong red-shift of the M band for the oxacyanine and the thiacyanine ($n = 5$) (14 nm and 26 nm, respectively) and a blueshift of 10 nm for the thiacyanine with the long methinbridge ($n = 7$) (Fig. 2).
4) This effect is not restricted to the **PL** β-sheet. The same spectroscopic behavior is observed with other polypeptides containing a high amount of β-sheet structure, such as ovalbumine and β-lactoglobuline (Fig. 3).

A second condition for a shifted M band is a sufficiently high polypeptide β-sheet concentration leading to aggregate formation similar to micelles in the case of surfactants.

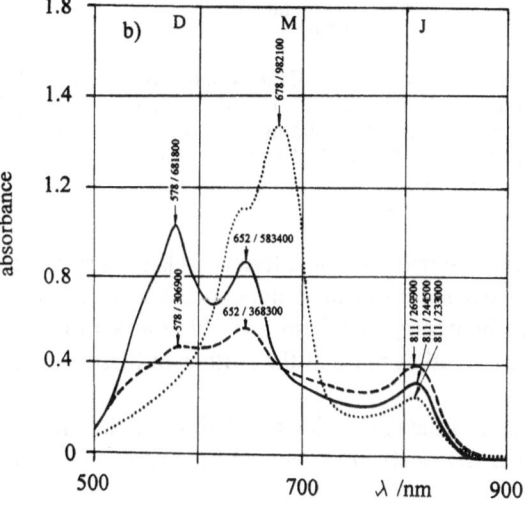

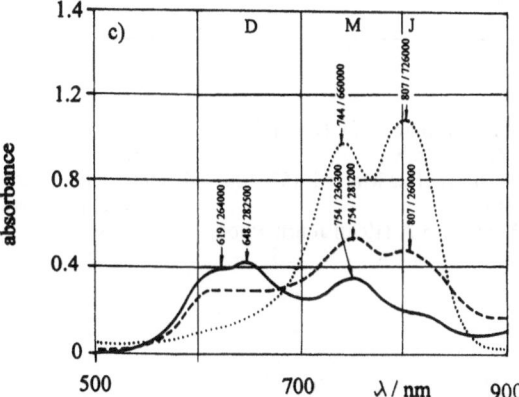

Fig. 2 UV/Vis absorption spectra of the cyanine dyes I, II and III (see Fig. 1) in water/methanol = 9/1, pH = 11.5 at 22°C without (——) and with poly-L-lysine in the α-helix (– – – –) and in the β-sheet (·····) conformation. Band maxima λ_{max}/nm and the absorptivity $\varepsilon / 1 \times mol^{-1} \times cm^{-1}$ are indicated. a) cyanine dye I, b) II, c) III

For **PL** β-sheet we found that the concentration must be higher than 1.5×10^{-5} M (Table 1) which is higher than the "critical aggregation concentration" (cac = 9.55×10^{-7} M) determined from the adsorption isotherm of **PL**

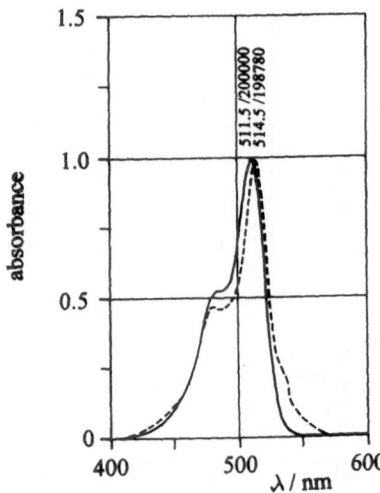

Fig. 3 UV/Vis absorption spectra of the cyanine dye I (see Fig. 1) in water/methanol = 9/1, pH = 7 at 22 °C with ovalbumin (——) and β-lactoglobulin (————) in the solution. Solution concentration of ovalbumin is 1.12×10^{-4} M and of lactoglobulin is 1.6×10^{-4} M. λ_{max} and ε are indicated

Table 1 Comparison between the cyanine dye I (see Fig. 1) monomer band (λ_{max}) and the concentration of poly-L-lysine β-sheet. Water/methanol = 9/1, pH = 11.5, $T = 22$ °C, cyanine dye concentration 1.5×10^{-5} M

$c_{\beta-sheet} \times 10^5$ M	M-band	λ_{max}/nm
3.24		514
1.62		514
1.08	500	
0.65	500	
0.32	500	

β-sheet at the water/dodecane interface [4]. That means in our cyanine dye solutions the polypeptide β-sheet molecules exist in the aggregated form and the interaction with the cyanine dye takes place at the surface of these aggregates.

The origin of the effects described above is the interaction between the electron systems of the cyanine dye and the polypeptide backbone which is greatest for the β-sheet conformation. Dependent on the cyanine dye both hypsochromic and bathochromic shifts are possible. Further results will be published elsewhere.

Conclusion

The interaction between the surface of β-sheet polypeptide aggregates and the cyanine dyes depicted in Fig. 1 leads to a strong shift of the M band. The possibility to work in an aqueous system opens new ways for recognizing β-sheet structures in biological and other systems containing polypeptides with cyanine dyes as sensor. But additional investigations are necessary for conceivable applications in the field of polypeptid membranes and LB-layers.

References

1. Davidson B, Fasman GD (1967) Biochemistry 6:1616–1629
2. Dietz F (1973) J Inf Rec Mater 1: 157–180, 237–252, 381–382
3. Hermel H, Seeboth A (1990) Journ Phot Sci 38:70–72 (1993) Thin Solid Films 223:371–374
4. Hermel H, Miller R (1995) Colloid Polym Sci 273: (in press)
5. Greenfield N, Davidson B, Fasman GD (1967) Biochemistry 6:1630–1637
6. Greenfield N, Fasman GD (1969) Biochemistry 8:4108–4116

Progr Colloid Polym Sci (1995) 98:215–218
© Steinkopff Verlag 1995

BIOCOLLOIDS

J.S. Pedersen
S. Hansen
R. Bauer

Aggregation behavior of zinc-free insulin studied by small-angle neutron scattering: analysis by use of a thermodynamic equilibrium model

Dr. J.S. Pedersen (✉)
Department of Solid State Physics
Risø National Laboratory
4000 Roskilde, Denmark

S. Hansen · R. Bauer
Department of Mathematics and Physics
Royal Veterinary and Agricultural
University
Frederiksberg, Denmark

Abstract The aggregation behavior of zinc-free insulin has been studied by small-angle neutron scattering as a function of protein concentration, pH, and ionic strength of the solution. The analysis of the data shows that the weight-averaged molecular mass and the z-average radius of gyration varies systematically with the experimental conditions. They increase with decreasing pH (decreasing charge) and with increasing ionic strength (increasing charge screening). The radius of gyration scales as a power law of the weight-average mass with the exponent 0.44. A similar scaling is found for a set of oligomer structures based on the crystal structure of zinc-free insulin. The scattering data were fitted by a model based on these oligomer structures and on the equilibrium model recently introduced by Kadima et al.. The model takes into account the variation of the effective charge of the monomer with pH and ionic strength. The model has only three parameters which were fitted simultaneously to the neutron scattering data for 12 different experimental conditions.

Key words Zinc-free insulin – aggregation – small-angle neutron scattering – equilibrium model

The hormone insulin is a small protein with a molecular weight of 5778 Daltons. When used in diabetes therapy insulin is injected at high concentrations, but the protein functions as a monomer at very low concentration. Thus, studies of the dependence of the aggregation behavior of insulin on protein concentration, pH, and ionic strength are of fundamental interest. The system also serves as a biochemical model system for studies of protein-protein interaction and as a physico-chemical system for studies of self-association at thermodynamic equilibrium.

The aggregation of insulin has been investigated by several experimental techniques (see the references in [1, 2]). The most recent works are an extensive static and dynamic light-scattering study [1] and a small-angle neutron scattering (SANS) study [2, 3]. The charge repulsion between the insulin oligomers gives rise to strong interparticle interference effects which hampers the interpretation of the light scattering data [1]. It was therefore necessary to invoke a thermodynamic equilibrium model which includes these effects. Due to the relatively small size of the insulin oligomers no variation in the scattering intensity due to the particle form factor can be observed in the light-scattering experiments. This is in contrast to the SANS technique [2] which allows a direct measurements of the average form factors of the oligomers. However, at small q, the scattering curves are also influenced by the inter-oligomer charge repulsion. In the present paper, we describe some results and aspects of the SANS study [2, 3]. We also present the results from fitting a thermodynamic equilibrium model to the neutron scattering data.

The sample preparation procedure is described in [2]. We have performed the following four series of measurements:

1) Concentration series for $C = 1, 2, 5, 10$ mg/ml with 10 mM NaCl at pH 11;
2) Concentration series for $C = 1, 2, 5, 10$ mg/ml with 100 mM NaCl at pH 8;
3) pH variation for pH 8, 9.5, 11 at $C = 10$ mg/ml with 10 mM NaCl;
4) pH variation for pH 8, 9.5, 11 at $C = 10$ mg/ml with 100 mM NaCl.

The SANS measurements were carried out at the DR3 reactor at Risø National Laboratory in Denmark. In the experiments the scattering patterns were recorded in the q range 0.013–0.500 Å^{-1}, using two instrumental settings. Throughout the data analysis the smearing of the ideal cross-section by the instrumental resolution was included as described in [4].

The SANS spectra are displayed in Fig. 1. In Fig. 1(a) it can be observed that the charge repulsion of the oligomers gives rise to a reduction of the intensity at small q. If this effect is neglected, the intensity at $q = 0$ is proportional to the weight-average molecular mass and the width of the curve is inversely related to the z-average radius of gyration. The weight-average molecular mass is significantly larger at pH 8 and 100 mM NaCl (Fig. 1(b)) than at pH 11 and 10 mM NaCl (Fig. 1(a)). In Fig. 1(a) the mass seems nearly independent of the concentration, whereas

a concentration induced increase is observed in Fig. 1(b). A comparison of Fig. 1(c) and 1(d) shows that the mass is larger at the highest NaCl concentration. At 10 mM NaCl (Fig. 1(c)) the interparticle interference effects are most pronounced. This is due to the long Debye–Hückel screening length ($r_d = 30$ Å) at 10 mM NaCl compared to $r_d = 10$ Å at 100 mM.

The qualitative aspects of the aggregation behavior can be explained in terms of the net charge of the insulin monomer and the screening of the electrostatic interactions by the ions in the solution. The net charge Z_0 of the insulin molecules in 100 mM KCl as a function of pH has been determined by proton titration [5]. At pHs 8, 9.5, and 11 it is respectively -2, -3.6, and -6.3. Thus, the large charge at pH 11 and the corresponding charge repulsion hinder the aggregation. The paramount importance of the charge is also emphasized by the observed changes for the variation of pH at constant NaCl concentration (Fig. 1(c, d)).

The indirect Fourier transformation method [6] can be used for obtaining the distance distribution function $p(r)$, which describes the distance histogram of the particles. The results for the 12 samples are given in [2, 3]. For systems with particles of different size and shape the $p(r)$ function is an average one. A comparison between the results for the distance distribution function and the function calculated for the monomer and dimer in the crystal structure of zinc-free insulin [7] shows that insulin is present as a dimer at high pH and low ionic strength. For

Fig. 1 SANS spectra. a) Concentration series at an ionic strength of 10 mM NaCl and at pH 11. b) Concentration series at an ionic strength of 100 mM NaCl and at pH 8. From top: 10, 5, 2, 1 mg/ml. c) pH variation series at 10 mg/ml and 10 mM NaCl. From top: pH 8, pH 9.5, pH 11. d) pH variation series at 10 mg/ml and 100 mM NaCl. From top: pH 8, pH 9.5, pH 11. The curves are the fits from the equilibrium model described in the text. The curves are dotted in the low-q regions which were not included in the fit

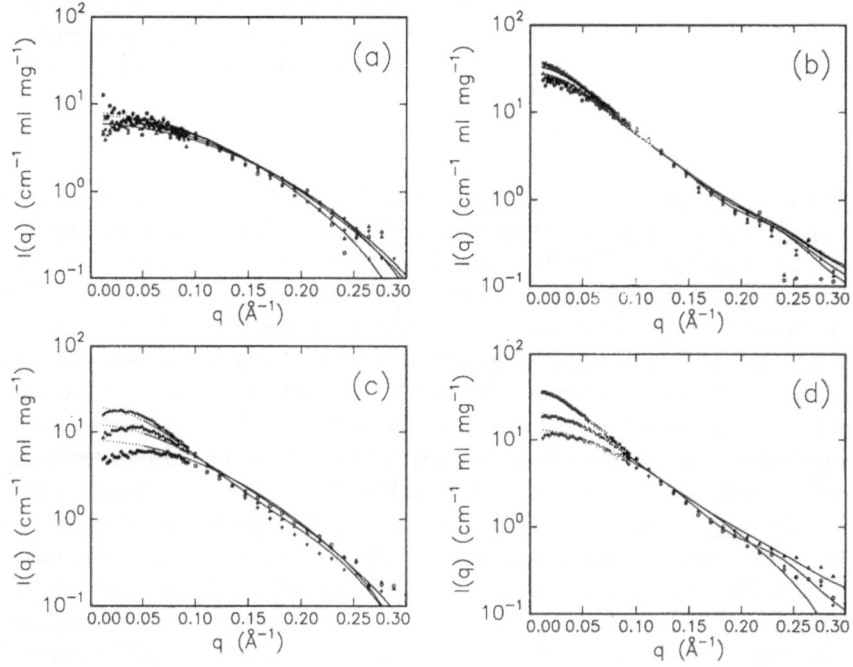

lower pH and higher ionic strength the distance distribution functions show that oligomers of much larger size than the dimer are present. The functions also show concentration-induced aggregation under these conditions.

The z-averaged radius of gyration R_{gyr} and the weight-average molar mass M can be calculated from the distance distributions functions. However, this can only be done at infinite dilution where the functions are not influenced by interparticle interference effects. The SANS spectra for the samples at low ionic strength are influenced by these effects at low q. The effects are caused by characteristic minimum distances between the particles and therefore they occur for $q \ll 1/R_{gyr}$, where R_{gyr} is the Guinier radius. If this part of the spectra is omitted from the data analysis $p(r)$ functions free of correlation effects can be obtained. The indirect Fourier transformation method [6] automatically provides an extrapolation to $q = 0$. Estimates R_{gyr} and M can be derived from the $p(r)$ functions obtained in this way.

The dependence of R_{gyr} on M approximately follows the scaling relation $R_{gyr} \propto M^\alpha$, where $\alpha = 0.44$. For $\alpha = 1/3$ one has compact objects, whereas $\alpha = 1/2$ corresponds to the open structure of a random Gaussian coil. Therefore the observed exponent corresponds to relatively open structures, however, one should bear in mind that changes in polydispersity might influence the value of α.

The free energy of the oligomers is strongly influenced by the specific interactions between the molecules. It can therefore be expected that the arrangement of the molecules in the oligomers is similar to the arrangement in the crystal structure. We have estimated a possible set of structures [2, 3] by taking a subset of the monomers/dimers in the crystallographic unit cell [7]. There are several ways of selecting, for instance, a hexamer from the crystal structure, but for simplicity we will only consider one particular structure for each oligomer in which the monomers have a high number of contacts. The radius of gyration calculated for the chosen set of oligomers [2, 3] follows approximately the same scaling behavior as the experimental (average) values. For the model structures $\alpha = 0.46$, however, the two power laws have different prefactors. The difference can be explained by the polydispersity which gives rise to a larger shift of R_{gyr} as compared to M due to the respective z-average and weight-average of the two parameters. The model structures and the observed results are compatible with a mass distribution with a constant ratio between the width and the center of the distribution. This is, in fact, what the thermodynamic equilibrium model by Kadima et al. [1] predicts.

By use of the model oligomer structures the mass distributions have been determined [2, 3]. Insulin is known to have a strong affinity for dimer formation, and therefore only the monomer and the oligomers with an even aggregation number were taken into account in the analysis. The scattering intensity is given as a linear combination of the form factors and the coefficients are the mass fractions of the individual oligomers. They were determined by a least-squares method with non-negativity and smoothness constraints. The mass distributions determined by this procedure were in good agreement with the mass

Fig. 2 Mass distribution f_i obtained by fitting the equilibrium model described in the text. a), b), c) and d) refer to the same conditions as they do in Fig. 1

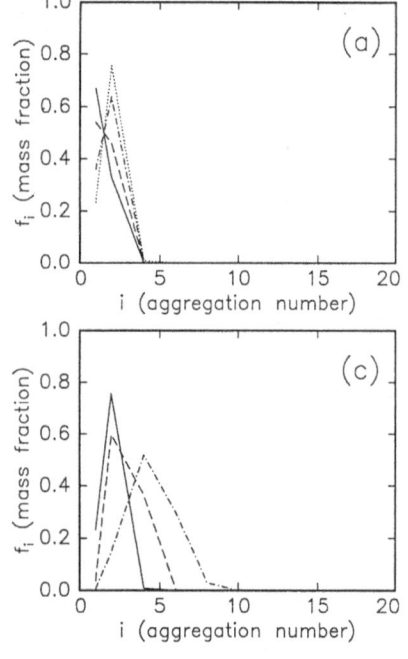
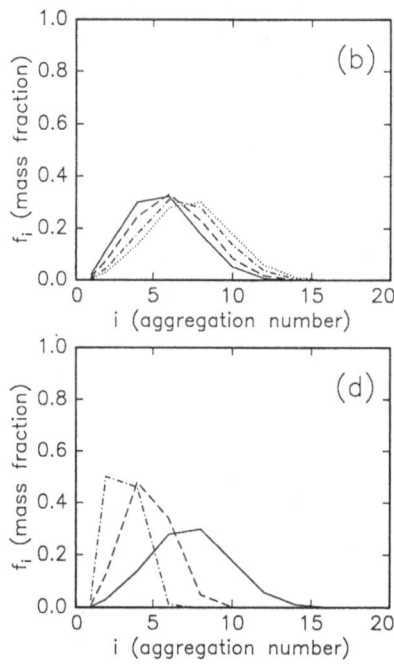

distribution determined indirectly by the equilibrium model from the light-scattering experiments [1].

The good agreement between the mass distribution determined from the SANS data and by the equilibrium model suggests that one can fit the equilibrium model directly to the SANS data. The model by Kadima et al. [1] is based on the law of mass action. The equilibrium constants are in a conventional approach related to the standard Gibb's free energy change of the oligomers formation. The expression for Gibb's free energy is composed of a term which takes into account the energy gain by bond formation and of a term which describes the charge repulsion. The first term is assumed to scale as i and the second one as $i(i - 1)/2$, where i is the aggregation number of the oligomer. The charge repulsion term is also proportional to the square of the net charge (Z_0) of the monomer. In order to get a dependence on the ionic strength it is assumed that Na^+ ions binds to Z_0 specific binding sites on the monomer with an equilibrium constant K_{Na}. This gives an effective charge of the monomer which depends on the Na^+ concentration.

The model was fitted simultaneously to the 12 SANS data sets obtained for the different conditions taking the oligomer form factors from the crystal structure [2, 3]. The model has three parameters to describe the variation of the aggregation: The prefactor ΔG^- of bonding term, ΔG^+ of $Z_{eff}^2 i(i - 1)/2$ in the charge repulsion term, and K_{Na} the binding constant of the cations. These parameters were optimized by a conventional least-squares method and the following values were obtained: $\Delta G^- = 0.47$ kJ/mole, $\Delta G^+ = -34$ kJ/mole and $K_{Na} = 88$ mM. These values agree reasonably with the values obtained by Kadima et al. [1] in the analysis of the light scattering data. The fit to the experimental data is shown as curves in Fig. 1. Only the parts of the spectra which are not influenced by the interparticle interference effects have been fitted. The agreement between the data and the fit is very good. The mass distributions obtained by the fit are shown in Fig. 2. They agree well with those obtained from the previous free-form analysis. It can therefore be concluded that the relatively simple equilibrium model is able to describe the aggregation behavior of zinc-free insulin.

Acknowledgments The insulin powders were kindly provided by Novo Nordisk A/S. The financial support from the Danish Natural Science, Technical, and Veterinary and Agricultural Research Councils, and the Swedish Natural Science Research Council is gratefully acknowledged.

References

1. Kadima W, Øgendal L, Bauer R, Kaarsholm N, Brodersen K, Hansen JF, Porting P (1993) Biopolymers 33:1643–1657
2. Pedersen JS, Hansen S, Bauer B (1994) Eur Biophys J 22:379–389 and (1994) Eur Biophys J 23:227–229
3. Pedersen JS, Hansen S, Bauer B (1995) Il Nouvo Cimento, to appear
4. Pedersen JS, Posselt D, Mortensen K (1990) J Appl Crystall 23:321–333
5. Kaarsholm N, Havelund S, Hougaard P (1990) Arch Biochem Biophys 283:496–502
6. Glatter O (1977) J Appl Crystall 10:415–421
7. Badger J, Harris MR, Reynolds CD, Evans AC, Dodson EJ, Dodson GG, North ACT (1992) Acta Crystall B47:127–136

Progr Colloid Polym Sci (1995) 98:219–223
© Steinkopff Verlag 1995

C. Otero
L. Robledo

Lipase activity in anionic reverse micelles. Inhibition effect of the system

Dr. C. Otero (✉) · L. Robledo
Instituto de Catálisis
CSIC
Campus de la Universidad Autonoma
Cantoblanco 28049, Spain

Abstract The inhibition effect of AOT reverse micelles in the p-nitrophenol ester hydrolysis, has been compared with lipases from *Pseudomonas sp.*, and isoespecies A and B from *Candida rugosa*. These anionic micelles induced a displacement of the optimal pH of lipase in opposite direction to that observed with other hydrophilic non surface-active enzymes. A relation between the hydrophobic character of each lipase and a blue displacement of their fluorescence emission in micelles was found. Also, a decay of the emission fluorescence intensity was detected under experimental conditions at which

lipases showed maximal activity. Lipases located in a more hydrophobic environment were less inhibited by the AOT system.

Key words Candida rugosa – Candida cylindracea – lipase – lipase PS – reverse micelles – AOT – fluorescence

Abbreviations AOT, Sodium bis-(2-ethylhexyl) sulfosuccinate; pNPB, p-nitro-phenyl butyrate; $\omega_0 = [H_2O]/[surfactant]$; $k_2^0 = kcat/Km$, app; CCL, *Candida cylindracea or rugose* lipase; PSL, *Pseudomonas sp.* lipase (lipase PS).

Introduction

Lipolytic enzymes suffer an interfacial activation at the water/oil interface. This distinguishes lipases from esterases and other hydrolases. Lipases whose structures have been elucidated to date, present their catalytic center covered by an amphophilic lid which is opened after interfacial activation [1]. The movement of the lid in presence of an interface stabilizes the interaction with the substrate [1, 2].

Isoenzymes A and B from *Candida rugosa* were purified and characterized previously [3–5]. Both lipases have the same molecular mass, similar aminoacid content (79% identical, [4]) and N-terminal sequences (with only the fifth aminoacid of the ten determined being different: Thr in Lipase B and Lys in Lipase A). But they differ in their hydrophobic interactions [3, 4]. cDNA clones for these

lipases have been isolated and their nucleotide sequences determined [6]. These isolipases are highly homologous, but the lid or flap is the most variable structure of these isolipases [7].

Reverse micelles of AOT have been frequently used to study the enzyme action into them. Although these systems present a high interfacial area (about 100 m²/ml), the hydrolysis rate of different esters catalyzed by lipases in the AOT systems proves to be always slower than in water. However, under certain conditions other enzymes like chymotrypsin or LADH [8] present superactivity in these media.

In this work, the role of the molecular properties of the enzyme and the physico-chemical properties of the micellar media in the inhibition effect of the lipase activity are discussed. The results obtained with lipase PS from AMANO and those obtained with two lipase isoforms from *Candida rugosa* have been compared.

Materials

Lipase PS from *Pseudomonas sp.* was purchased from AMANO. Lipase type VII from *Candida rugosa*, p-nitrophenyl butyrate, AOT, and the buffers Epps, Mes, Bes were from Sigma (St Louis, MO, USA). Sodium bicarbonate was from PANREAC. *n*-Heptane (Scharlau, HPLC grade) was dried over a type 4A molecular sieve prior to use.

Methods

Lipase purification

The purification of two extracellular lipases from *Candida rugosa* (lipase A and lipase B) was carried out as previously described [4].

Enzyme assays

The hydrolytic activity of lipase PSL was measured by following the accumulation of p-nitrophenol in a Kontron 930 UVIKON spectrophotometer equipped with thermostated cells at 30 °C. The monitoring wavelength was selected at the isosbestic point (λ_{iso}) of the nitrophenol/nitrophenolate couple. The isosbestic wavelengths for p-nitrophenol in reverse micelles of AOT were collected in Table 1. When necessary, small corrections were made for the non-enzymatic-catalyzed rates of hydrolysis.

In water solution, the assays were measured at 346 nm (isosbestic point of p-nitrophenol) considering the molar extinction coefficient as 4800 M-1 cm-1 [9].

The initial rate was obtained in less than 2 min. Thus, a possible influence of enzyme stability was avoided. Concentration units for the enzyme and reactants were referred to the total volume of the micellar system.

Fluorescence study of CCL and PSL lipases

Flourescence spectrum of PSL in 1 mM sodium bicarbonate buffer, pH = 8.5, was compared with the spectra in micellar systems of AOT at ω_0 values of 5 and 20, at its optimal pH (Fig. 1B).

Fluorescence spectra of purified lipases A and B in 1 mM Sodium phosphate buffer, pH = 7.2, were also compared with the spectra in micellar systems of AOT at ω_0 values of 5 and 20, at the optimal pH of lipases A and B, 5.6 and 6.1, respectively [5].

Protein concentration in the cuvette was 0.02, mg/ml, 0.017 mg/ml and 0.067 mg/ml of lipases PSL, CCL A and B, respectively. Spectra were made at 26 °C at 280 nm excitation and 280–450 nm emission, in a Perkin Elmer fluorescence spectrometer LS 50B. Spectra were uncorrected for the instrument sensitivity, but Raman emission of the solvent – buffer or micellar solution – was subtracted.

Results and Discussion

Dependence of the reaction rate on the pH at distinct ω_0

The influence of the pH on the hydrolysis rate catalyzed by PSL was determined using pNPB as substrate in bulk water and micellar systems of $\omega_0 = 2.2$–50. The pH value was referred to that of the water before solubilization.

Results are shown in Fig. 1. In aqueous solution (Fig. 1A) the hydrolysis rate increased with pH, and reached a plateau at pH ≥ 8.5. The AOT system (Fig. 1B) shifted the optimal pH of PSL to lower values, except at the highest droplet size ($\omega_0 = 50$). Shifts on the optimal pH of 0.5–1 units in the acid direction were found for lipases CCL A and B [5], and for *Chromobacterium viscosum* lipase [10] in AOT/*n*-heptane/water systems. This phenomenon is frequently attributed to the physical state of water in reverse micelles and its subsequent effect on the pKa values of aminoacid residues in the active site of the enzyme [10, 11]. At low ω_0 values, it has qualitatively different properties from those of bulk water such as extent of hydrogen-bonding, effective dielectric constant, viscosity, etc. [12].

Non-surface-active enzymes like α-chymotrypsin [12], tyrosinase [13], alkaline phosphatase [14], etc., show shifts to the basic side in a similar AOT system, especially at low droplet size. The opposite shift in the optimal pH value of lipases might be due to the different localization of their active center into the micelles: lipases at the interface, and non-surface-active enzymes in the water core of the droplets. Thus, changes in the dielectric constant of the environment of the active center should produce different shifts in their pKa values.

These results show an optimal ω_0 value ($\omega_0 = 5$, pH = 5.5) for PSL. From data of Fig. 1B a bell-shaped

Table 1 Isosbestic wavelengths (λ_{iso}) and the extinction coefficients of the nitrophenol/nitrophenolate couple in reverse micelles of AOT at 30 °C

ω_0	λ_{iso} (nm)	$\varepsilon(M^{-1}cm^{-1})$
2.2	327	6825
3.0	329	6361
5.0	331	5600
≥ 7.0	333	5212

Fig. 1 Dependence of the initial rate of pNPB hydrolysis catalyzed by PSL on the pH, at 30 °C. A) In aqueous medium ([Buffer] = 0.5 M, [pNPB] = 0.12 mM, [PSL] = 2.5 mg/ml). B) In micelles ([AOT] = 0.1 M, ω_0 = 2.2–50, [pNPB] = 2.4 mM, [Buffer] = 2 mM, [PSL] = 5 mg/ml). (■, ω_0 = 2.2; ▲, ω_0 = 3.0; ●, ω_0 = 5.0; ■, ω_0 = 7.0; ▽, ω_0 = 10; ♦, ω_0 = 15; △, ω_0 = 30; ○, ω_0 = 50)

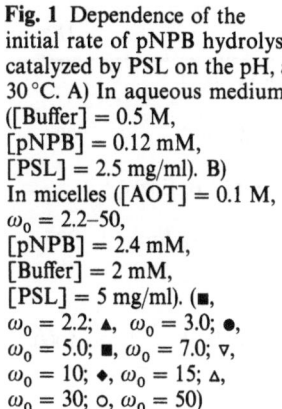

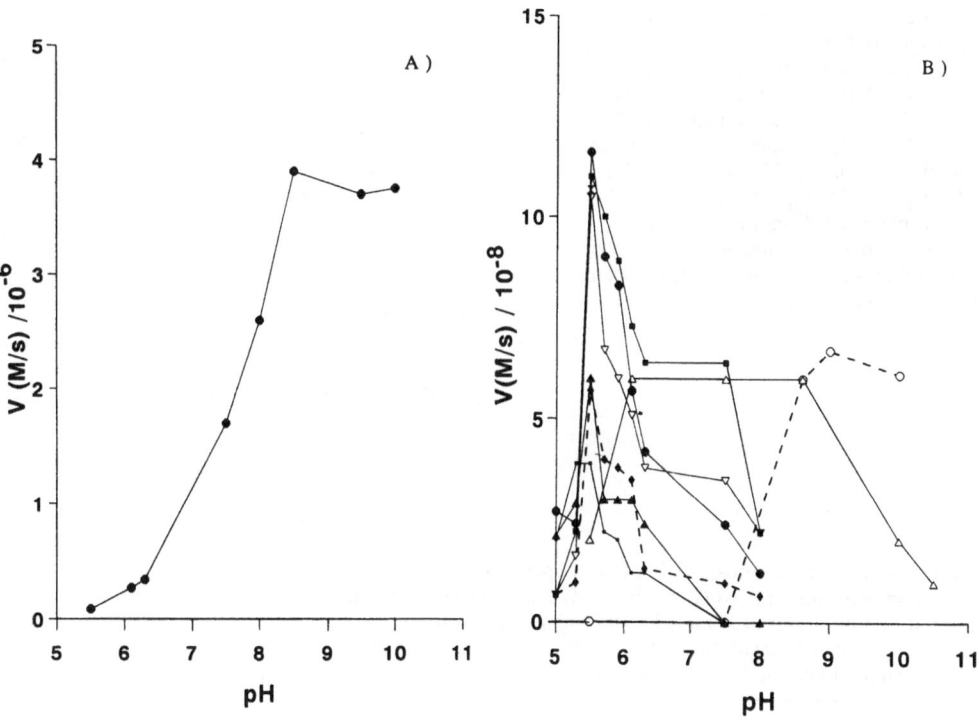

curve was obtained for the variation of initial rate vs ω_0 (not shown). Our results fitted to an enzyme which expresses the highest activity in the bound water, or in the surfactant apolar tail domains of the micellar system, according to the theoretical model developed by Bru et al. [15]. Similar behavior was reported for *Rhizopus delemar* lipase in a lecitin/isooctane/water system [16] and bilirubin oxidase in a CTAB/chloroform-*n*-heptane/water system [17], which also acts on interfacially bound substrates.

Kinetic parameters in bulk water and in reverse micelles

The kinetic parameters for the pNPB hydrolysis catalyzed by PSL were obtained in aqueous and micellar media at 30 °C. The assays in aqueous system were carried out in 0.5 M bicarbonate buffer (pH 8.5, the optimal pH in Fig. 1A), and in micelles at ω_0 = 5, with 2 mM Mes buffer, pH = 5.5. The obtained values of Km, app were 0.35 mM and 20.0 mM; kcat values were 3320 10^{-7} and 792 10^{-7} Ms^{-1}/(g Enz L^{-1}), in aqueous and micellar systems, respectively; and k_2^0 values are given in Table 2.

A similar study was done with CCL A and B at 26 °C (Table 2). The assays were carried out in 0.5 M Mes buffer (pH 7.2) and in AOT systems at ω_0 = 5, their optimal ω_0 values (unpublished work).

The micellar system resulted to be a non-favorable medium for the lipase catalyzed hydrolysis, especially for lipase B. The specificity constant, k_2^0, for lipase B was 1000 times lower in reverse micelles than in water, 237 times lower for PSL, and 30 times lower in the case of lipase A.

The substrate, PNPB, distributes in the micelle between the interface and the organic bulk. Substrate partition decreases its accessibility to the enzyme. Consequently, reaction rates in micelles lower than in water can be expected. But, in this case the effect of pNPB distribution must be the same for these three lipases at ω_0 = 5. Thus, a distinct inhibition effect of the anionic micelles (*I.F.*, in Table 2), might be due to a different ability of these lipases to change to the active conformation [1, 2, 18] in the micellar interface (see fluorescence study).

Fluorescence study of lipases in aqueous and micellar systems

Results of fluorescence study of PSL, and isoenzymes A and B in aqueous and micellar systems (ω_0 = 5 and 20) are shown in Table 3. Fluorescence spectra of these lipases are dominated by tryptophan's absorbance and emission.

A blue displacement of the emission band was found for the more hydrophobic lipase A in low and high ω_0 micelles, but not for lipase B (Table 3). Once again, the

Table 2 Kinetic parameters of PSL, lipase A and lipase B from *Candida rugosa*, for the hydrolysis of *p*-nitrophenyl butyrate in aqueous solution and in optimal conditions of AOT/heptane/water system. Conditions for PSL: 30 °C; in aqueous system: 0.5 M bicarbonate buffer, pH = 8.5; in micelles: [AOT] = 0.1 M, $\omega_0 = 5$ with 2 mM Mes buffer, pH = 5.5. Conditions for lipases A and B: 26 °C; in aqueous system: 0.5 M Mes buffer, pH = 7.2; in the AOT system, [AOT] = 0.1 M, $\omega_0 = 5$

Enzyme	System	k_2^0	IF*
Lipase A	aqueous	$6.90\ 10^{+6}\ M^{-1}s^{-1}$	
	micellar	$0.22\ 10^{+6}\ M^{-1}s^{-1}$	30
PSL	aqueous	$950\ 10^{-3}\ L\ s^{-1}\ (g\ Enz)^{-1}$	
	micellar	$4.00\ 10^{-3}\ L\ s^{-1}\ (g\ Enz)^{-1}$	237
Lipase B	aqueous	$6.45\ 10^{+6}\ M^{-1}s^{-1}$	
	micellar	$6.45\ 10^{+3}\ M^{-1}s^{-1}$	1000

* Inhibition factor: k_2^0 (aqueous)/k_2^0 (micelles)

Table 3 Fluorescence emission of PSL and lipases A and B from *C. rugosa* in aqueous and AOT systems. Maxima wavelengths and intensities. Conditions: 0.02 mg/ml, 0.017 mg/ml and 0.067 mg/ml of PSL, CCL A and B, respectively. Excitation at 280 nm, emission at 280–450 nm. Other conditions in the text.

System	λ_{max} (nm) CCL A	I_{max}	λ_{max} (nm) PSL	I_{max}	λ_{max} (nm) CCL B	I_{max}
Aqueous	333	101	344	151	333	790
$\omega_0 = 5^a$	324	047	316	056	333	700
$\omega_0 = 20^b$	325	111	342	362	333	625

[a] optimal value for pNPB hydrolysis catalyzed by the three lipases.
[b] optimal value for lipase B with C_8 and C_{12} ester substrates [5].

PSL behavior resulted to be intermediate between lipases A and B, and the blue displacement was only observed at low ω_0. The emission intensity of lipase A at $\omega_0 = 20$ increased with respect to its spectrum in buffer solution. However, in case of lipase B the intensity decreased at this ω_0 more than at $\omega_0 = 5$. The emission intensity of the three lipases increased, with respect to the intensity in water, at ω_0 values where these lipases showed their minimum activity, and decreased at ω_0 values of maximum activity (lipases A and B showed maxima at $\omega_0 = 5$, and 5 and 20, respectively, [5].

This distinct spectral behavior of the three lipases may be due to their distinct environment in the micellar system (most probably distinct penetration degree of their active center/amphiphatic lid region into the micellar interface), which may induce different conformational changes of these proteins in the micelle. These two options could be expected for isolipases CCL A and B considering the higher interaction of lipase A on hydrophobic chromatography [4], and for lipases in general which are susceptible to conformational changes in interfaces [1, 2].

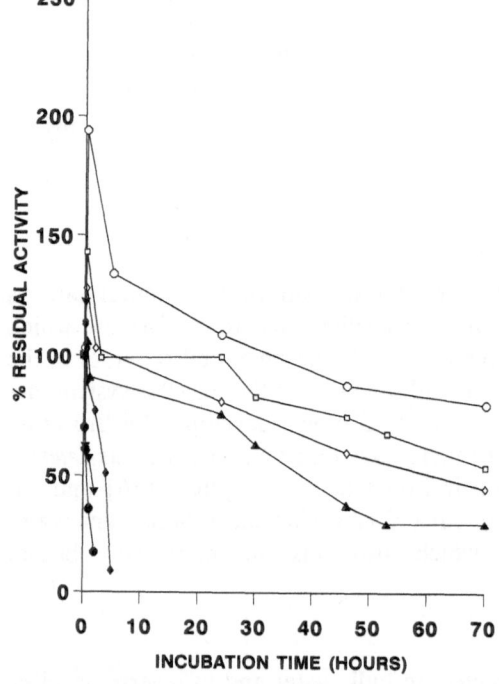

Fig. 2 PSL stability in aqueous medium and in reverse micelles of AOT of distinct ω_0. (▲, 0.5 M buffer, pH = 8.5; ●, $\omega_0 = 5$ and pH = 5.5; ▼, $\omega_0 = 10$ and pH = 5.5; ◆, $\omega_0 = 20$ and pH = 5.5; ◇, $\omega_0 = 30$ and pH = 7.5; □, $\omega_0 = 50$ and pH = 8.5; ○, $\omega_0 = 70$ and pH = 8.5). Conditions: [AOT] = 0.1 M, [Buffer] = 2 mM, [PSL] = 45 mg/ml, 30 °C

Stability of lipases A and B in AOT/*N*-heptane/water systems

The stability of PSL was studied in aqueous medium (2 mM buffer, pH = 8.5), and in reversed micelles of

$\omega_0 = 5, 10, 20, 30, 50$ and 70) containing 2 mM Mes buffer (pH 5.5, except $\omega_0 = 30$ at pH = 7.5, and $\omega_0 = 50$ at pH = 8.5, the optimal values for activity in Fig. 1B). After incubation of the micellar solutions containing the enzyme at 30 °C, aliquots were taken at the indicated time and the residual activity was measured (Fig. 2).

The results obtained showed lower stability at low ω_0 than in the aqueous medium, but higher at $\omega_0 \geq 30$. PSL resulted to be more stable in reversed micelles than in water, but only when it was highly hydrated (absence of blue displacement of Trp's emission). However, it has a low stability in conditions of maximum activity in micelles (at $\omega_0 = 5$ and a more hydrophobic environment; see kinetic and fluorescence studies in Fig. 1B and Table 3).

Inhibition/activation behavior of lipases in anionic reverse micelles

The comparative study of physico-chemical, molecular and kinetic properties of lipases PSL, CCL A and B, allow us to suggest that a factor that governs the lipase activation in a water-tensioactive-organic medium is the hydrophobicity of lipase, which determines its penetration degree at the interface. Therefore, the interfacial activation of lipases could be unfavored in the presence of ionic surfactant, AOT, in the micellar interface.

Acknowledgements We thank V.M. Fernández for his help in the manuscript edition. This work has been financed by Spanish CICYT (No. PB92–0495), and a grant (PFPI) of Spanish Ministerio de Educación y Ciencia.

References

1. Bradly L, Brzozowski AM, Derewenda ZS, Dodson E, Dodson G, Tolley S, Turkenburg JP, Christiansen L, Huge-Hensen B, Norskov L, Thim L, Menge U (1990) Nature 343:767–770
2. Derewenda U, Brzozowski AM, Lawson DM, Derewenda ZS (1992) Biochemistry 31:1532–1541
3. Rúa ML, Díaz-Mauriño T, Otero C, Ballesteros A (1992) Ann NY Acad Sci 672:20–23
4. Rúa ML, Díaz-Mauriño T, Fernández VM, Otero C, Ballesteros A (1993) Biochem Biophys Acta 1156:181–189
5. Otero C, Rua ML, Robledo L (1995) FEBS Letters 360:202–206
6. Longhi S, Fusetti F, Grandori R, Lotti M, Vanoni M, Alberghina L (1992) Biochim Biophys Acta 1131:227–232
7. Lotti M, Tramontano A, Longhi S, Fussetti T, Brocca S, Pizzi E, Alberghina L (1994) Protein Engineering 7(4):531–535
8. Khmelnitsky YuL, Kabanov AV, Klyachko NL, Levashov AV, Martinek K, in Pileni MP, Troyanowsky C (eds) Structure and Reactivity in Reverse Micelles Elsevier. Amsterdam
9. Fletcher PDI, Robinson BH, Freedman RB, Oldfield C (1985) J Chem Soc Faraday Trans 81:2667–2679
10. Oldfield C (1987) Ph.D. University of Kent (UK)
11. Castro MJM, Cabral JMS; Enzyme Microb Technol 11:6–11
12. Kuntz ID, Kauzmann W (1974) Adv Protein Chem 28:239–345
13. Bru R, Sanchez-Ferrer A, García-Carmona F (1989) Biotech Bioeng 34:304–308
14. Ohshima A, Narita H, Kito M (1983) J Biochem 93:1421–1425
15. Bru R, Sanchez-Ferrer A, García-Carmona F (1989) Biochem J 259:355–361
16. Schmidli PK, Luisi PL (1990) Biocatalysis 3:367–376
17. Oldfield C, Freedman RB (1989) Eur J Biochem 183:347–355
18. Grochulski P, Li Y, Schrag JD, Bouthillier F, Smith P, Harrison D, Rubin B, Cygler M (1993) J Biol Chem 268:12843–12847

Progr Colloid Polym Sci (1995) 98:224–227
© Steinkopff Verlag 1995

Shape transformations in biological mixed surfactant systems: from spheres to cylinders to vesicles

S.U. Egelhaaf
Labor for Elektronenmikroskopie
ETH Zürich
8092 Zürich, Switzerland

J.S. Pedersen
Department of Solid State Physics
Riso National Laboratory
4000 Roskilde, Denmark

PD Dr. P. Schurtenberger (✉)
Institut für Polymere
ETH-Zentrum
Universitätsstraße 6
8091 Zürich, Switzerland

Abstract Aqueous mixtures of lecithin and bile salt form aggregates of various structures depending on their mixing ratio and concentration. Upon a reduction of the bile salt-to-lecithin ratio they exhibit a characteristic sequence spherical micelle – flexible cylindrical micelle – vesicle – lamellar sheet in the aggregate morphology. Static (SLS) and dynamic (DLS) light scattering as well as small-angle neutron scattering (SANS) experiments were performed to investigate the structural properties of the different types of aggregates as well as the transitions between them.

Special attention was paid to the dependence of the size and flexibility of the wormlike mixed micelles on the micellar composition, the coexistence of mixed micelles and vesicles, and the occurrence of large sheetlike structures. The experimental results were interpreted within the framework of current theoretical models for the shape transformations in surfactant systems.

Key words Shape transformation – micelle – vesicle – light scattering – neutron scattering

Introduction

Considerable attention has been devoted to theoretical and experimental investigations [1–3] of the shape and structure of surfactant aggregates and the shape transitions observed upon a change of control parameters such as surfactant or cosurfactant concentration, ionic strength, pH or temperature. A transition of the aggregates from spherical micelles to cylindrical micelles, bilayers, and lamellae is predicted by these theories if, for example, the spontaneous curvature is decreased. The theoretical models are based on different physical concepts such as curvature elasticity, spontaneous curvature or geometrical constraints which were related to the chemical nature of the surfactant.

In our experiments we used mixtures of lecithin and bile salt in buffer. These solutions not only serve as model systems, but are also of relevance in biology and physiology as well as in pharmaceutical applications [4–10]. The solutions were prepared by diluting a mixed micellar stock solution with buffer (for details see [11]). Due to the much higher monomer solubility of the bile salt, dilution results in a change of the aggregate composition (a reduction of the bile salt-to-lecithin ratio) and induces transitions in the shape and structure of the aggregates. The solutions were investigated by static (SLS) and dynamic (DLS) light scattering as well as small-angle neutron scattering (SANS) experiments. We then aimed for a self-consistent description of the data using concepts from colloid and polymer physics. Quantitative information on the dependence of the structural properties of the aggregates upon composition is vital for any attempt to derive a thermodynamic model for the shape transformations in these systems.

Results

The experimental details are given elsewhere [11, 12]. Figure 1 summarizes the concentration dependence of the apparent molar mass, M_{app}, of a lecithin-bile salt solution with a lecithin-to-bile salt molar ratio of 0.9 as obtained by SLS (open circles). With increasing dilution M_{app} increases, reaches a maximum and decreases again. The radii of gyration, R_g, and the hydrodynamic radii, R_h, as determined in SLS and DLS experiments (data not shown) exhibit a similar dependence on dilution.

Mixed micelles

At low dilutions small mixed micelles are present. The shape of these micelles could not be determined due to the limited range of scattering vectors Q accessible by light scattering. At higher dilutions we observed an extended micellar growth. The data are consistent with the presence of polydisperse, flexible cylindrical micelles. A quantitative description can be achieved based on the wormlike chain model, which is frequently used to characterize static and dynamic properties of semiflexible polymers [13]. The model parameters are the contour length L, the persistence length l_p, which describes the flexibility of the chain, and the polydispersity σ. An iterative procedure was used to simultaneously fit the SLS and DLS data. Educated guesses of the parameters were initially made, and a step-wise procedure was then applied for refinement until convergence was obtained. Based on this procedure a self-consistent interpretation of the normalized scattering intensity (i.e., excess Rayleigh ratio) $\Delta\Re(Q)$, and the hydrodynamic radius $R_h(Q)$ could be obtained as shown in Fig. 2A. The model not only provides $\Delta\Re(Q)$ and $R_h(Q)$, but correctly describes the distribution of relaxation times

Fig. 1 Experimentally determined apparent molar mass M_{app} of a lecithin-bile salt stock solution ($L/BS = 0.9$, $C_{tot} = 50$ mg/mL, $T = 25\,°C$) as a function of dilution. The molar mass M_0 (dashed line) and the apparent molar mass $M_{c,app}$ (solid line) are calculated based on the different structural models. In addition, the phase boundary for the onset of vesicle formation is shown as the dashed-dotted line. No contributions from wormlike micelles could be detected for dilutions beyond the dotted line

Fig. 2 Micellar solution (1:20.25). A) Excess Rayleigh ratio $\Delta\Re(Q)$ (●) and hydrodynamic radius $R_h(Q)$ (△) as functions of the scattering vector Q as obtained by light scattering. The curves are calculated on the basis of the wormlike chain model. B) Normalized scattering intensity $I(Q)$ as function of the scattering vector Q as determined by SANS. Individual symbols represent data obtained with different combinations of neutron wavelength and sample-detector distances (note that these curves are slightly shifted due to resolution effects). The line represents the result from a fit with a two-shell model for the cylindrical mixed micelles and corresponds to the intensity smeared by the instrumental resolution (see [12] for details)

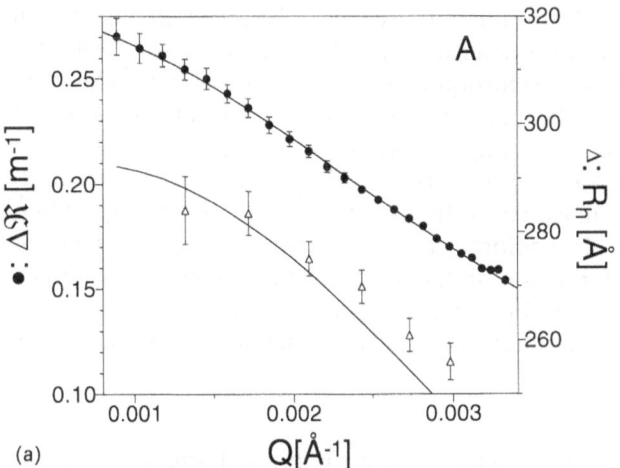

(a)

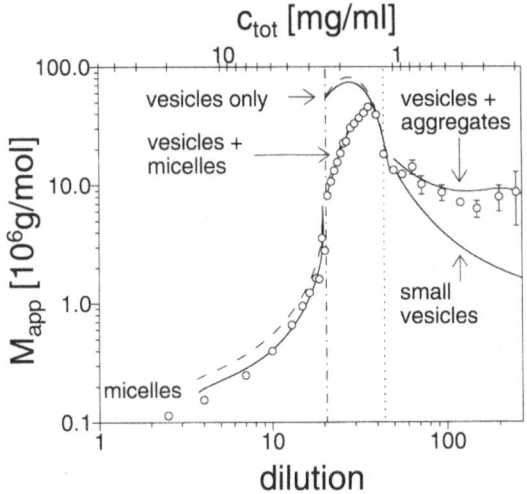

(b)

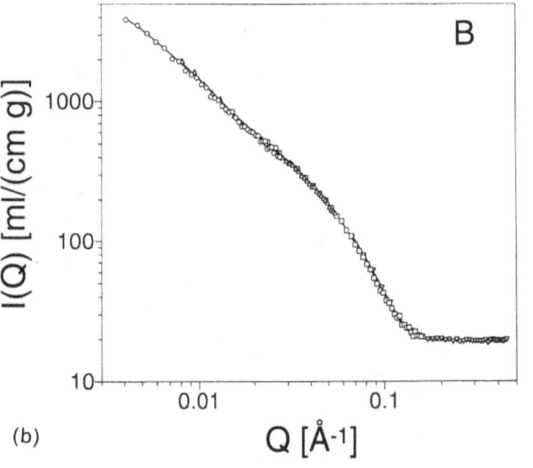

226

S.U. Egelhaaf et al.
Shape transformation in surfactant systems

measured in a DLS experiment, if internal modes of the polymerlike micelles are taken into account.

The results from the SLS and DLS experiments are summarized in Fig. 3, where the contour length L and the persistence length l_p of the cylindrical micelles is shown as a function of dilution, which corresponds to different compositions of the mixed micelles. At the lowest dilution $R_h \approx 59.5$ Å is comparable to the diameter of the micelles ($d \approx 54$ Å) and therefore indicates a nearly spherical shape of the micelles. On the other hand, close to the phase boundary highly elongated polymerlike micelles ($L \approx 8700$ Å) are present.

Whereas the length scale of light-scattering experiments ($1/Q > 300$ Å) is appropriate to determine the overall size of the mixed micelles, the determination of the persistence length is only achieved indirectly by a combination of SLS and DLS. In order to test this procedure, we examined a sample close to the phase limit with SANS, which allows a direct determination of the persistence length due to the more extended Q-range (Fig. 2B). In the fitting procedure, we used an expression for the scattering cross-section of polydisperse wormlike micelles which is capable to quantitatively describe the scattering intensity over an extended range of scattering vectors and which allows the incorporation of the overall size, polydispersity, flexibility and local structure of the micelles in a self-consistent way (for details see [12]). This measurement yielded a persistence length $l_p = 150$ Å in perfect agreement with the value obtained by SLS and DLS ($l_p \approx 140$ Å).

In addition, the SANS experiment yields important information on the local structure of the mixed micelles. We can, for example, determine the linear number density λ_L of lecithin molecules in the micelles. This value can then

be used to test the applied structural model on an absolute scale: From the Q-dependence of $\Delta\Re(Q)$ and $R_h(Q)$ as obtained in the SLS and DLS experiments the contour length L and the polydispersity σ was determined, which together with λ_L yields the molar mass M_0 of the micelles. If we include interaction effects on the level of a second-order virial expansion for excluded volume interactions between semiflexible polymer coils [13], we can then calculate an apparent molar mass $M_{c,app}$. This can directly be compared with M_{app} determined in the SLS experiment. The results are shown in Fig. 1 (dashed line: M_0, solid line: $M_{c,app}$, open circles: M_{app}).

Coexistence

Beyond the micellar phase boundary, where a spontaneous micelle-to-vesicle transition can be observed, $\Delta\Re(Q)$ and $R_h(Q)$ as determined by SLS and DLS are consistent with a single phase of vesicles, but the experimentally determined M_{app} differs significantly from the values which were calculated based on the parameters obtained by the fit to $\Delta\Re(Q)$ and $R_h(Q)$ (Fig. 1). A self-consistent description of $\Delta\Re(Q)$ and $R_h(Q)$ as well as M_{app} can only be obtained, if we allow for coexisting cylindrical micelles and vesicles (Fig. 1). The relative amount of vesicles increases with increasing dilution. Whereas the observation of the coexistence based on SLS and DLS measurements is rather indirect, SANS measurements allow for an unambiguous identification of coexisting micelles and vesicles due to the larger Q-range. The SANS data shown in Fig. 4

Fig. 4 Normalized scattering intensity $I(Q)$ as a function of scattering vector Q for a sample beyond the micellar phase limit as obtained by SANS. The lines represent the results from fits with a model of vesicles only (dashed line) and coexisting micelles and vesicles (solid line), respectively. They correspond to the intensity smeared by the instrumental resolution (see [12] for details)

Fig. 3 Contour length $L(\bullet)$ and persistence length $l_p(\triangle)$ of the mixed micellar solutions as a function of dilution

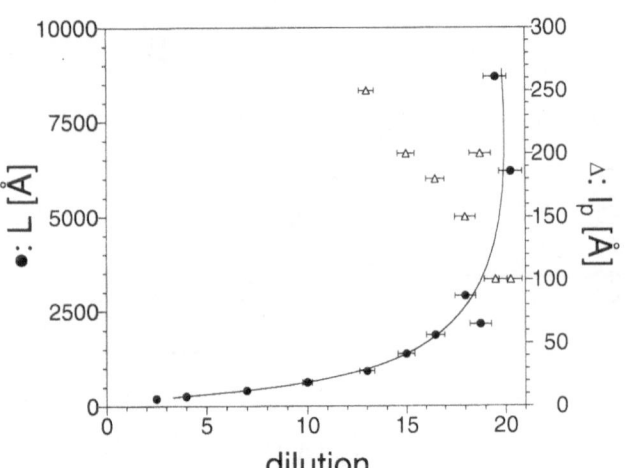

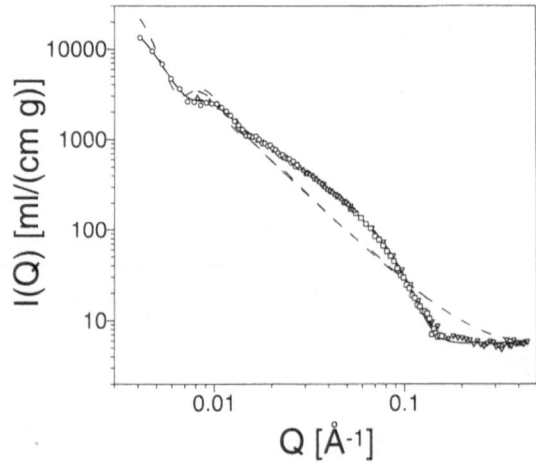

demonstrate that the normalized scattering intensity $I(Q)$ measured from samples in the coexistence region are not consistent with a population of vesicles only, and that a coexistence of vesicles and wormlike micelles has to be assumed. Coexistence of cylindrical micelles and vesicles was observed in the relatively narrow region indicated in Fig. 1. It is interesting to note that the data in this region is consistent with a constant size of micelles and vesicles, and that only the relative amount changed.

Vesicles

At still higher dilutions the angular dependence of $\Delta\mathfrak{R}(Q)$ and $R_h(Q)$ are consistent with a single phase of relatively monodisperse vesicles. The size of the vesicles decreases upon dilution in agreement with previous findings and the prediction of a simple equilibrium model that relates the local bilayer composition to the overall size of the vesicles [7, 14]. At very high dilutions, we finally observed a significant increase of the polydispersity, and the light-scattering data at low values of Q indicate the presence of an additional component of very large aggregates. The two populations are well separated in size and hence affect different Q-ranges, which allows us to estimate the relative mass fractions. However, we cannot unambiguously determine the shape of the large aggregates (see [11] for details). It is feasible that fragments of the lamellar phase are formed in agreement with the proposed phase diagram [15] and the theoretical models [3].

Conclusions

We were able to show that the data obtained by SLS, DLS and SANS measurements from aqueous mixtures of lecithin and bile salt at concentrations above the mixed micellar phase limit can self-consistently be interpreted using a structural model of wormlike mixed micelles. Upon dilution the shape of these micelles changes from nearly spherical micelles to elongated, flexible cylindrical micelles (Fig. 3). This is caused by the decrease of the bile salt-to-lecithin-ratio in the aggregates due to dilution, which appears to lower the spontaneous curvature of the mixed micelles and forces them to avoid endcaps and therefore grow. Furthermore, the decrease of the bile salt-to-lecithin-ratio decreases the relative charge density of the micelles and hence allows for a greater flexibility. At concentrations below the micellar phase limit, mixed lecithin bile salt vesicles form spontaneously. Close to the phase limit they coexist with cylindrical micelles. At very high dilutions, we observed an additional second population of very large aggregates, which most likely have a sheetlike structure. By changing the composition of the aggregates, i.e., the bile salt-to-lecithin-ratio, we therefore passed from spheres to elongated, cylindrical micelles to vesicles to lamellae, in agreement with the phase diagram [15] and current theories of surfactant self-assembly [3].

References

1. Israelachvili JN, Mitchell DJ, Ninham BW (1976) J Chem Soc Faraday Trans II 72:1525–1568
2. Porte G, Marignan J, Bassereau P, May R (1988) J Phys France 49:511–519
3. Safran SA (1992) In: Chen SH, Huang JS, Tartaglia P (eds) Structure and dynamics of strongly interacting colloids and supramolecular aggregates in solution. Kluwer Academic Publishers, Dordrecht, pp 237–263
4. Mazer NA, Benedek GB, Carey MC (1980) Biochemistry 19:601–615
5. Hofmann AF (ed) (1984) The physical chemistry of bile in Health and Diseases. Hepatology 4
6. Barbara L, Dowling RH, Hofmann AF, Roda E (eds) (1985) Recent advances in bile acid research. Raven Press, New York
7. Schurtenberger P, Mazer N, Känzig W (1985) J Phys Chem 89:1042–1049
8. Hjelm RP, Alkan MH, Thiyagarajan P (1990) Mol Cryst Liq Cryst 180A:155–164
9. Hjelm RP, Thiyagarajan P, Alkan-Onyuksel H (1992) J Phys Chem 96:8653–8661
10. Long MA, Kaler EW, Lee SP, Wignall GD (1994) J Phys Chem 98:4402–4410
11. Egelhaaf SU, Schurtenberger P (1994) J Phys Chem 98:8560–8573
12. Pedersen JS, Egelhaaf SU, Schurtenberger P (1995) J Phys Chem 99:1299–1305
13. Yamakawa H (1971) Modern Theory of Polymer Solutions. Harper and Row, New York
14. Israelachvili JN, Mitchell DJ, Ninham BW (1977) Biochem Biophys Acta 470:185–201
15. Small DM (1984) Hepatology 4: 180S–190S

Progr Colloid Polym Sci (1995) 98:228–232
© Steinkopff Verlag 1995

M. Winterhalter
K.-H. Klotz
D.D. Lasic
R. Benz

Electric field-induced breakdown in lipid membranes

M. Winterhalter · K.-H. Klotz
Biozentrum der Universität Basel
Biophysical Chemistry
Klingenbergstr. 70
4056 Basel, Switzerland

D.D. Lasic
MegaBios Corp.
871-J Industrial Road
San Carlos, CA 94070, USA

Dr. R. Benz (✉)
Lehrstuhl für Biotechnologie
Biozentrum der Universität Würzburg
Am Hubland
97074, Würzburg, FRG

Abstract We present a new and fast method to obtain information on macromolecule-lipid interaction. Lipid bilayers were formed without and in presence of macromolecules. Mechanical rupture of the membranes was induced by applying high electric fields. The kinetics of pore formation during irreversible breakdown is sensitive to the binding affinity of macromolecules and the lipid composition of the membrane. We apply this finding to investigate the so-called stealth property of recently introduced PEG-grafted lipids used in liposomal drug carrier systems.

Key words Bilayer lipid membrane – electrical breakdown – adsorption on membranes – protein-lipid interaction – stealth lipids

Abbreviations DPhPC, diphytanoyl phosphatidylcholine; Stealth lipid, DSPE-PEG, distearyl phosphatidylethanolamine poly-(ethyleneglycol); EYPC, egg yolk phosphatidylcholine; HDL, high density-lipoproteins;

External electric fields deliver a unique tool for cell manipulation. Several advantages in comparison to mechanical or chemically mediated pertubation are evident: easy and possible automatic regulation from the outside, fast and specific application, as well as fast switch off. Many applications are commonly used nowadays. Especially in biotechnology [1, 2] electrophoresis, electroinjection of macromolecules into living cells or electrofusion are standard methods. Recently, pulsed electric fields found a promising application in medicine when used to permeabilize tissue to increase the efficiency of antitumoral drugs [3]. Although electric field techniques have been applied for quite a long time, the exact mechanism of various methods is still a matter of investigation [4, 5]. Most of the experimental accessible parameters are found in a phenomenological way. More information is required to improve their efficiency.

In the present work, we present a new application to study adsorption or incorporation of macromolecules to membranes. External electric fields were used to induce mechanical breakdown of the lipid film. The dynamics of the widening of the defects after the initial process is for lipids in the fluid phase sensitive to surface tension as well as film thickness [4, 5]. These two factors are altered by incorporation or adsorption of surface-active substances into the membrane. As an example, we demonstrate this in case of protein adsorption on biosurfaces.

It was found that the therapeutic efficiency of many drugs can be significantly improved by using drug-delivery systems, such as liposomes [6–8]. The main problem of these drug carriers is, however, their biological instability. For instance, liposomes are, upon intravenous administration, either destabilized by plasma lipoproteins and/or rapidly taken up by the cells of the immune system. Destabilization by lipoproteins requires close approach and possibly adsorption of proteins onto liposomes while the uptake is a two-step process: immunoglobulins or proteins of the complement cascade adsorb onto foreign particles and this marks them for the subsequent uptake by macrophages. The first part of this process is called opsonization. One can expect that the presence of a steric barrier would reduce the rate of these two interactions. A clear example is the so-called stealth lipids (DSPE-PEG) which consist of a lipid, a stearic acid derivative, and a covalently bonded polyethylene glycol (PEG) polymer of different chain length.

The adsorption of lipoproteins was modeled by studying the adsorption of HDL onto bilayers in the presence and absence of DSPE-PEG [6]. HDL contains apolipoprotein I which is known to strongly adsorb onto liposomes [9].

External electric fields are acting on membranes in different ways. Short electric field pulses across planar or almost planar membranes are inducing surface charges at the lipid/water interface. This causes so-called Maxwell stresses which are increasingly quadratic with the applied electric fields. These externally induced stresses act in addition to the underlying mechanical forces [10].

The exact process of pore formation is not yet known. We assume that thermal fluctuation will cause statistical defects, also called pores, in the membrane. The stability of a pore with radius a can be described by two opposite energy contributions,

$$E_{pore} = 2 \cdot \pi \cdot a \cdot \Gamma - \pi \cdot a^2 \cdot \sigma \; . \tag{1}$$

The first term describes the energy needed to form the edge per unit length with Γ as the edge energy. The second term is the energy gained by the widening of the pore area due to the surface tension σ. Obviously, pores with radii smaller than $a < a^* = \Gamma/\sigma$ reseal, while for radii larger than a the area of the pores will increase to infinity. A critical radius $a^* = 5$ nm can be estimated from experimental values for similar lipids where a surface tension $\sigma = 2 \cdot 10^{-3}$ N/m and the edge energy $\Gamma = 10^{-11}$ N has been reported [4].

It is now widely accepted that electric fields are lowering the energy barrier (Eq. (1)) for pores in membranes. Recently, we induced the irreversible electrical breakdown with short charge pulses [4]. We concluded that the breakdown is mainly due to a negative contribution to the edge energy of the external field. A few μs after the pore area starts to increase, the electric potential difference across the membrane drops quickly and the contribution of the electric field to the kinetics will be determined only by the material properties of the membrane. As shown in our earlier study [4, 5], one can neglect the viscosity of the lipid film. The kinetics of the widening of the pore is driven by the finite surface tension and controlled by the inertia of the film. Balancing the decrease in elastic energy during the widening of the pore by inertia and the dissipation

yields the following time-dependency for the pore radius:

$$a(t) = a_0 + \sqrt{\frac{\Phi \cdot \sigma}{d\rho}} \cdot t = a_0 + \alpha \cdot t \; , \tag{2}$$

with Φ as a parameter depending on the unknown material flow as well as on dissipation effects. The lipid density is represented by ρ, and d is the membrane thickness. Knowledge of the pore radius allows the calculation of the pore conductance $G(t)$ as a function of time. Assuming a single pore in a thin membrane, we obtain:

$$G(t) = \frac{1}{R_{pore}} = 2 \cdot \kappa \cdot (a_0 + \alpha \cdot t) \; , \tag{3}$$

with κ as the specific conductivity of the aqueous solution.

During the irreversible membrane breakdown, the actual membrane conductance, $G(t)$, was calculated using the following equation:

$$G(t) = \frac{I(t)}{U(t)} = \frac{1}{U(t)} \cdot \frac{dQ(t)}{dt} = \frac{1}{U(t)} \cdot \frac{-d(C \cdot U(t))}{dt}$$
$$= \frac{-C}{U(t)} \cdot \frac{dU(t)}{dt} \; ; \tag{4}$$

$U(t)$ was the actual membrane voltage. Due to its considerable scatter the signal has to be averaged [4].

Black lipid bilayer membranes were formed from a 1% solution in n-decane of a mixture containing the specific lipid solution. During the first set of breakdown experiments, we formed the bilayer membranes with the lipid mixture described by Woodle and Lasic [6] to prepare sterically stabilized liposomes. This membrane forming solution contains 33 mol% cholesterol and 62 mol% EYPC (Avanti Polar Lipids, Inc., Alabaster, Alabama, USA), as well as the given mol% of the different Stealth (DSPE-PEG) lipids (Liposome Techn., Inc., Menlo Park, California, USA) in n-decane (Fluka AG, Buchs, Switzerland). Four different polymer moieties were available having molecular weights from 350, 750, 1900, and 5000 Da. In a second series, we used pure DPhPC (Avanti Polar Lipids, Inc.) with 0, 1 mol% and 5 mol% of the DSPE-PEG-1900, respectively. One dataset consists of at least 10 membranes. The membranes were spread across a circular hole with a diameter of about 1 mm^2 in a wall separating two aqueous phases in a teflon cell (see Fig. 1). The electrolyte, an unbuffered (pH around 6) 100 mM solution of potassium chloride (Merck, Darmstadt, FRG) in the case of the first series, and a solution of 100 mM potassium chloride buffered with 1 mM citric acid (Sigma Chemical Co., St. Louis, Missouri, USA) at pH 7.4 for the second series, was prepared using 18 MΩ/cm water (Millipore Super Q, Millipore Corp., Bedford, Massachusetts, USA). The conductivities of these solutions were

230
M. Winterhalter et al.
Electric field-induced breakdown in lipid membranes

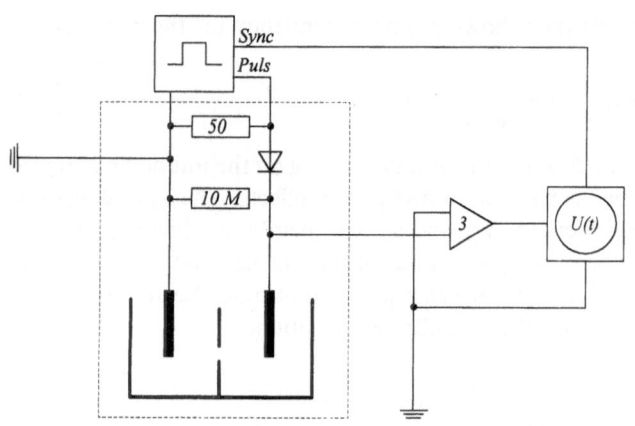

Fig. 1 Scheme of the charge-pulse instrumentation used for the measurement of electric field-induced irreversible breakdown of lipid bilayer membranes

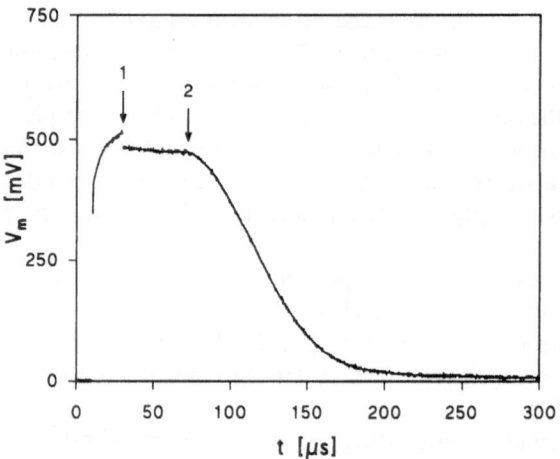

Fig. 2 Time-course of membrane voltage $V_m(t)$ during electric field-induced irreversible breakdown of a membrane formed of DPhPC and 1 mol% DSPE-PEG-1900. The length of the charge pulse was 20 μs. The aqueous phase contained 100 mM KCl and 1 mM citric acid buffered at pH 7.4; $T = 20\,°C$

$\kappa = 1.23$ S/m and $\kappa = 1.21$ S/m, respectively. The temperature was always kept at 20 °C. For temperature control the Teflon cell is fitted into a brass block which is streaked with cooling mains fastened to a heating stirrer (Haake Type FE) and a cooler (Lauda Type DLK 15). The detection system consists of extremely fast Ag/AgCl platinum black electrodes (Annex Instruments, Santa Anna, California, USA), one connected to a fast pulse generator (Hewlett Packard 214B) through a transistor used as a fast diode (reverse resistance $\geqslant 10^{11}\,\Omega$) and the other electrode grounded. The voltage between these two electrodes was measured with a high-input-resistance fast voltage amplifier (gain three fold) and a digital storage oscilloscope (Nicolet 4094 A). The voltage relaxations were analyzed with a PC 386/33 MHz computer. A short positive rectangular pulse with 20 μs duration charges the bilayer to an initial voltage of 10 mV through the fast diode. The value of the membrane capacitance was calculated from RC-time constant of the exponential discharge process of the membrane voltage across an external 10 MΩ resistor in parallel to the membrane.

To initiate the irreversible breakdown the membranes were charged to higher voltages in the range of 300 to 600 mV. In order to cause only single pores and not multipores, we carefully raised the applied voltage in small steps. At a critical voltage the electrical field led to a mechanical rupture of the membrane. A typical time-dependence of the voltage is shown in Fig. 2. The further analysis of the voltage vs. time plots of the breakdown experiments was performed according to Eqs. (3) and (4) and is shown in Fig. 3. In agreement with our former report [4, 5], we observe for all systems a linear increase of conductance with time. As earlier pointed out, viscosity of the lipid membrane would yield an exponential increase.

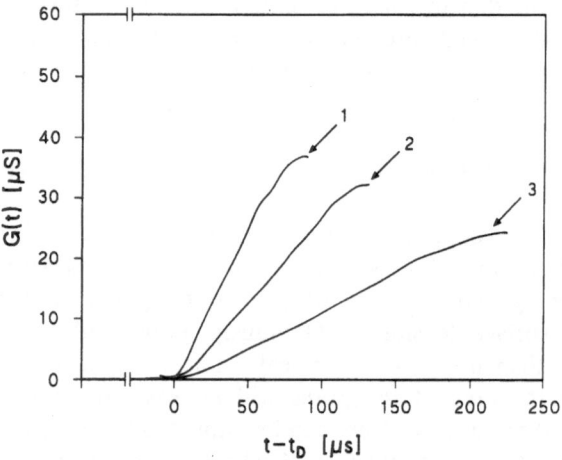

Fig. 3 Conductance $G(t)$ versus time curves during irreversible breakdown of membranes made of 1: DPhPC, curve slope 0.52 S/s (0.21 m/s); 2: DPhPC and 1 mol% DSPE-PEG-1900 added to the membrane forming solution, curve slope 0.29 S/s (0.12 m/s); 3: DPhPC and 5 mol% DSPE-PEG-1900 added to the membrane forming solution, curve slope 0.12 S/s (0.05 m/s). Note: the significant part we fitted covers the first 50 μs. $G(t)$ was calculated according to Eq. (4) by using the capacitance of the individual membranes and the velocity of the pore rim by using Eq. (3). The length of the charge pulses was 20 μs. Because of the considerable scatter of the signal the data have been averaged by standard methods [4]. The aqueous phase contained 100 mM KCl; $T = 20\,°C$

The linear increase could be related to inertia of the lipid film.

An increase of DSPE-PEG will increase the effective membrane thickness and therefore decreases the rupture velocity. As water is considered as a good solvent for PEG

Table 1 The data are given as mean ± SD of at least 30 experiments. For all systems 100 mM potassium chloride was used as electrolyte. The temperature was 20 °C. The EYPC concentration in the membrane forming solution was 10 mg/ml. The DSPE-PEG lipids were added in the mol percentage given in the table. The numbers in brackets below represent the values predicted from scaling theory. Velocity of pore increase during irreversible breakdown of lipid bilayer membranes made of pure and Stealth-endowed EYPC:cholesterol.

System	R_f	1 mol%	5 mol%	10 mol%
pure		0.10 ± 0.03		
DSPE-PEG-350	1.2	0.10 ± 0.02 (0.10)	0.08 ± 0.02 (0.09)	0.06 ± 0.01 (0.09)
DSPE-PEG-750	1.9	0.10 ± 0.02 (0.09)	0.09 ± 0.03 (0.08)	0.06 ± 0.02 (0.06)
DSPE-PEG-1900	3.4	0.10 ± 0.02 (0.07)	0.07 ± 0.02 (0.05)	
DSPE-PEG-5000	6.1	0.03 ± 0.01 (0.03)		

Table 2 The data are given as mean ± SD of n experiments. For all systems 100 mM potassium chloride was used as electrolyte, buffered with 1 mM citric acid at pH 7.4. The DPhPC concentration in the membrane-forming solution was 10 mg/ml. The mixed DPhPC/DSPE-PEG-1900 membranes were formed from a solution containing 10 mg/ml DPhPC and the mol percentage of Stealth-lipid given in the table. Influence of the protein adsorption on the rupture velocity of pure and stealth-endowed DPhPC membranes.

System	$\alpha/[m/s]$	n
DPhPC	0.21 ± 0.05	51
DPhPC 20 µg/ml Lipoprotein	0.17 ± 0.03	10
DPhPC 1 mol% DSPE-PEG-1900	0.12 ± 0.01	10
DPhPC 1 mol% DSPE-PEG-1900 20 µg/ml Lipoprotein	0.12 ± 0.03	14

the polymer is likely to be in a coiled state. According to Flory's theory [11], an average size of the coil can be estimated from scaling theory

$$R_f = l \cdot N^{3/5}, \tag{5}$$

with N being the number and l the length of the chain links which is in case of PEG about $l = 0.35$ nm. In Table 1 the calculated Flory radius for each polymer is presented. In a rough estimation, we assume that the effective increase in membrane thickness will be given by Flory's radius multiplied with the area fraction covered by the polymer coil. The obtained theoretical expectation for the rupture velocities α are given in brackets below the experimentally measured ones. Table 1 shows clearly the relation between increase of effective membrane thickness and rupture velocity. One reason for the observed deviation from this expectation is the unknown amount of interstitial water between the polymer coils, which is forced to follow the membrane movement. Such a movement may cause an additional dissipation of elastic energy.

Table 1 demonstrates also the actual limit in resolution of our method: the additional adsorbed or incorporated mass has to be more than 1 mol% PEG (which corresponds to 20% additional total weight or 150 g of adsorbed macromolecules per mol lipid). This has to be compared with other methods for protein detection [8] which, in contrast, are based on bulk properties. They allow to accumulate more adsorbed protein and are able to detect much lower values of adsorbed molecules per lipid.

In the following breakdown experiment we tested how the adsorption of the lipoprotein influenced the rupture velocity of a lipid membrane. In addition to the fact that

EYPC is a rather undefined mixture of lipids, the membranes appear more rigid in the presence of cholesterol which is known to inhibit protein adsorption [6]. In order to avoid these difficulties, we choose the pure and well defined DPhPC. Moreover, DPhPC bilayers have the advantage of higher interfacial tension, which results in a faster rupture velocity. Higher surface tension will emphasize effects caused by test molecules. The results in Table 2 indicate that at high concentrations (20 µg/ml) the lipoprotein sufficiently adsorbs to the membrane to detect a significant decrease in rupture velocity. Subsequently, we performed this experiment on Stealth-endowed DPhPC membranes. As shown in Table 2, the lipoprotein adsorbs on pure DPhPC membranes but the adsorption is inhibited if 5 mol% DSPE-PEG 1900 is present. According to Eq. (5), at such densities the polymer moiety very well covers the entire surface. Higher densities will cause an ordering of the PEG, called the brush state [11]. The dense polymer coil likely prevents proteins from coming close enough to overcome the steric barrier.

The experimental data suggest that the kinetics of defect formation may be used to investigate membrane interactions with macromolecules. This opens a wide field of applications to study the interaction of surface active molecules with membranes. In comparison to other methods to detect protein adsorption, this is a direct method to measure the integral amount of adsorbed proteins per lipid. It neither requires any specific fluorescence or radioactive labeling nor structural knowledge. Moreover, no separation into different molecular weights is needed. As we pointed out, this method may be applied as an easy and rapid test to improve the biological stability of drug carriers such as liposomes.

232
M. Winterhalter et al.
Electric field-induced breakdown in lipid membranes

The underlying physical processes of rupturing thin liquid films should be similar for films on solid supports. The presented technique could also be adopted for conducting solid support, and applied to study wetting properties [12].

Acknowledgements This work was supported by grants from the Deutsche Forschungsgemeinschaft (Graduiertenkolleg "Magnetische Kernresonanz" and project B7 of the Sonderforschungsgemeinschaft 176)

References

1. Borrebueck C, Hagen I (eds.) (1989) Electromanipulation in hybridoma technology. Stockton Press, NY
2. Zimm BH, Levence SD (1992) Quart Rev Biophys 25:171–204
3. Orlowski C, Mir LM (1993) Biochim Biophys Acta 1154:51–63
4. Wilhelm C, Winterhalter M, Zimmermann U, Benz R (1993) Biophys J 64:121–128
5. Klotz K-H, Winterhalter M, Benz R (1993) Biochim Biophys Acta 1147:161–164
6. Woodle CM, Lasic DD (1992) Biochim Biophys Acta 1113:171–199
7. Patel HM (1992) Critical Rev in Therapeutic Drug Carrier Systems 9:39–90
8. Chonn A, Semple SC, Cullis PR (1992) J Biol Chem 267:18759–18765
9. Gong EL, Nicholds AV (1979) Lipids 15:86–90
10. Winterhalter M, Helfrich W (1987) Phys Rev 36A:5874–5876
11. DeGennes PG (1985) Scaling Concepts in Polymer Physics, Cornell University Press, Ithaca
12. Redon C, Brochard-Wyart F, Rondelez F (1991) Phys Rev Lett 66:715–718

Progr Colloid Polym Sci (1995) 98:233–238
© Steinkopff Verlag 1995

J.L. Ortega-Vinuesa
M.J. Gálvez-Ruiz
R. Hidalgo-Alvarez

Aggregation behavior of F(ab')$_2$ fragments carrying polymer beads

Dr. J.L. Ortega-Vinuesa
M.J. Gálvez-Ruiz · R. Hidalgo-Alvarez
Biocolloid and Fluid Physics Group
Department of Applied Physics
University of Granada
18071 Granada, Spain

Dr. J.L. Ortega-Vinuesa (✉)
Department of Applied Physics
Science Faculty
Av. Fuentenueva S/N
18071 Granada, Spain

Abstract For quite some years ago, up to present, the high cost and complexity of diagnostic techniques in the immunochemistry field have prompted alternative approaches, such as the immunoprecipitation of colloidal particles carrying antibodies or antigens on their surfaces, which are less expensive and simpler.

It has been demonstrated that the use of antibody fragments (i.e., F(ab')$_2$), instead of the whole IgG, are useful in the development of certain biosensors, since the absence of the Fc moiety eliminates false positivity in diagnostic tests due to the presence of the Rheumatoid Factor. This is why we have used this protein in our studies. F(ab')$_2$ from polyclonal rabbit IgG was obtained by pepsin digestion.

As carriers of anti-CRP F(ab')$_2$, two different polymer beads were used: i) A conventional sample (RP) from Rhône Poulenc with an average size of 297 ± 7 nm and a surface charge density (σ_0) of $-6.9 \,\mu C/cm^2$, predominantly from sulfate groups. ii) A second sample (EU) obtained by emulsion copolymerization of methyl methacrylate (MMA) and butyl acrylate (BuA) with surfactants (SLS and Gafac RE610) using persulfate potassium as initiator. The diameter of the EU is 210 ± 20 nm and it has a $\sigma_0 = -7.2 \,\mu C/cm^2$, given by strong and weak acid groups.

The aim of this work was focused on the way in which the amount of adsorbed protein could affect both the stability of the system and its immunological response. Adsorption of F(ab')$_2$ under low ionic strength media at various pHs and with different coverage degrees was carried out. The more satisfactory results in the flocculation of the system by means of added CRP were found in those samples which had an optimum value of adsorbed protein that allowed to recognize antigen, but avoided the autocoagulation in the reactive medium.

Key words Immunoassays – protein fragments – polymer colloids

Introduction

The natural tendency of most proteins to adsorb onto different interfaces [1, 2] has led to discovering the possible applications in diverse fields of science. Since colloidal dispersions present a high value for the surface/ volume ratio, these could be excellent systems on which to perform adsorptions of different sorts of macromolecules using small bulks with an extremely high amount of area available. The increasing development in polymer science has allowed to create an extensive variety of colloidal polymer microspheres (latex) dispersed in aqueous medium, which may differ in size, chemical structure, surface charge density, etc. These particles can act as solid carriers onto which we can attach, physically or chemically, different proteins, finding in this way one of the most important applications of such systems: the detection of small concentrations of biological molecules, which has become essential in biological and medical research as well as in clinical medicine.

The adsorption of immuno gamma globulin molecules (IgG) onto polymer beads as an application for diagnostic test systems was first reported by Singer and Plotz [3]. This assay is based on the agglutination of antibody (or antigen)-coated carrier particles that are cross-linked and therefore, aggregated by complementary multivalent antigen (or antibody). The extent of particle aggregation is determined as a function of the agglutinator concentration. The sensitivity and accuracy of such immunotests depend on the optical technique used in the detection of the agglutination reaction [4–6], reaching sensitivities that are comparable with those obtained with radioimmunoassays.

The sensitization of polymer particles with IgG generally leads to a loss of stability; in addition, non-specific agglutination of the sensitized beads is often caused by substances which do not operate as antigens, resulting in erroneous diagnosis. In order to resolve these problems, the adsorption of antibody fragments, instead of the whole IgG molecule, has been suggested [7, 8]. The use of F(ab')$_2$ for antigen-binding sites, removing the Fc portion from IgG molecules, avoids false positives in immunotests due to the presence of the rheumatoid factor when anti-antibody is used in the final reagent [9], since this latter protein recognizes the Fc fragment.

Material and methods

In this work, two different latices have been used. The RP sample is a conventional free-surfactant polystyrene latex with sulfate groups on its surface, which was purchased from Rhône Poulenc (Estapor 030). The EU latex was synthesized in the Chemical Engineering Department of the Basque Country University [10]. Monomers used in the copolymerization reaction were methyl metacrylate (MMA, Fluka) and butyl acrylate (BuA, Fluka) and these were purified by low-pressure distillation (MMA, $t^a = 47\,°C$, $p = 100\,mmHg$; BuA, $t^a = 35\,°C$, $p = 10\,mmHg$). Potassium persulfate ($K_2S_2O_8$, Merck) was used as initiator. Sodium lauryl sulfate (SLS, Merck) and Gafac RE610 (GAF) were used as emulsifiers.

F(ab')$_2$ antibody fragments from rabbit polyclonal (anti C reactive protein)-IgG were kindly donated by Biokit S.A. (Spain). They were obtained by pepsin digestion of IgG, followed by a gel filtration chromatography

(Superose 12 HR 10/30 Pharmacia) and a Protein-A chromatography, HiPAc (ChromatoChem), to remove undigested IgG. Purity was checked by SDS-Page electrophoresis. The isoelectric point (i.e.p.) of F(ab')$_2$ molecules was determined by isoelectric focusing IEF, and the i.e.p. values are found in the range 4.6–6.0. The molecular weight is 102 kD.

C Reactive Protein (CRP) was used as antigen and it also was purified and donated by Biokit S.A.

The cleaning of latices was carried out by centrifugation, ion-exchange, and serum replacement; surface charge densities were determined by conductometric and potentiometric titration. Particles sizes were measured by transmission electron microscopy (TEM) and photocorrelation spectroscopy (PCS, 4700 SM Malvern Instruments), and the data provided by both techniques were in high concordance. The critical coagulation concentration (CCC) was determined by turbidimetric measurements using KBr as an electrolyte; the procedure was described in a previous paper [11].

Adsorption isotherms of aCRP-F(ab')$_2$ on both latices were carried out in a continuously shaken incubation bath for 14 h at 25 °C, adding 0.30 m^2 of polymeric surface from latex stocks to 10 ml of 55 and 230 μg/ml F(ab')$_2$ solutions in different buffers. Salt concentrations were calculated to get a final ionic strength of 0.002. F(ab')$_2$ concentrations before and after adsorption were determined by direct UV spectrophotometry at 280 nm ($\varepsilon = 1.48$ ml mg^{-1} cm^{-1}), using a Spectronic 601 spectrophotometer.

Immunoagglutination reactions were performed at pH 8 (borate 20 mM), NaCl 0.15 M, N$_3$Na 1 mg/ml used as a preservative and bovine serum albumin (BSA) 1 mg/ml to cover patches of latex free of aCRP-F(ab')$_2$ in order to avoid bridging coagulation of the complexes. Sensitized latices were centrifugated and the spun samples resuspended on the reaction buffer (BSA-saline), producing colloidal suspensions with an absorbance value (at $\lambda = 570$ nm) of around 0.5. On 950 μl of these solutions, 50 μl of CRP in BSA-saline at different concentrations among 0.2 and 40 μg/ml was added. It was rapidly and homogeneously mixed and its absorbance measured for 5 min. Initial particle concentrations in cuvette were 6.0 10^{11} part./ml for EU samples and 2.2 10^{10} part./ml for RP ones. The average size of F(ab')$_2$-latex complexes before starting the immunoagglutination was calculated by light scattering in order to determine the initial grade of aggregation of each sample.

Results and discussion

The main characteristics of the bare latices are shown in Table 1. In both cases, the polydispersity index (P.D.I.),

Progr Colloid Polym Sci (1995) 98:233–238
© Steinkopff Verlag 1995

Table 1 Some of the main characteristics of the bare latices are summarized in this table.

Latex	Diameter (nm)	PDI	σ_0 (μC/cm^2)	CCC (KBr)
EU	210 ± 20	1.040	-7.2 ± 0.1	800 ± 50 mM
RP	297 ± 7	1.005	-6.9 ± 0.1	200 ± 10 mM

which is defined as the ratio between the weight-average diameter and the number-average diameter, is very near the unity, indicating that these samples are highly monodisperse. It is considered that a colloidal particle dispersion is monodisperse if its P.D.I. is lower than 1.05 [12]. With regard to the surface charge density (σ_0), although the values are rather similar, the samples differ in the type of ionic group located on their surfaces. In RP particles, σ_0 comes from sulfate groups. However, for the EU sample, we can divide the total σ_0 value in two different contributions: a) strong-acid groups ($\sigma_{0s} = 1.4\ \mu$C/cm^2) that come from the SLS and the initiator used in its synthesis and, b) on the other hand, weak-acid groups ($\sigma_{0w} = 5.8\ \mu$C/cm^2) due to the presence of phosphate in Gafac RE610 molecules. In spite of having similar charge densities the stability of both dispersions are quite different as we can see in the CCC values obtained. Since there are adsorbed surfactants on the EU latex, the classical DLVO theory [13–15] cannot be used to explain this feature, and one must take into account the colloidal stability basis established for such systems by Mathai and Ottewill [16, 17] and Ottewill and Walker [18].

In Fig. 1 the adsorption isotherms of F(ab')$_2$ at two different coverage degrees and at different buffers are shown. The maximum amount of adsorbed protein appears around the i.e.p. of the F(ab')$_2$-latex complex [19] and this value is circa the theoretical F(ab')$_2$ monolayer (3.2 mg/m^2) considering a molecular area of 5400 Å^2 [20].

If the adsorption pH is higher than the i.e.p. of the F(ab')$_2$ molecules (around 5.4), the electrostatic repulsion between protein and interface increases, causing a decrease in the adsorbed amount at the same time that the pH moves away from such value. In those unfavorable conditions, at which protein and latex have the same charge sign, the amount of adsorbed protein on EU particles is lower than on the RP latex. The answer to the question this feature poses can be obtained if one takes into account the role played by the adsorbed emulsifiers on EU-latex/solution interface, which could prevent the approach of F(ab')$_2$ molecules towards the surface at such pHs and, consequently, its adsorption.

Figure 2 shows the dependence of the absorbance at 570 nm versus time, when different CPR concentrations were added to the sensitized EU latex. As is well known,

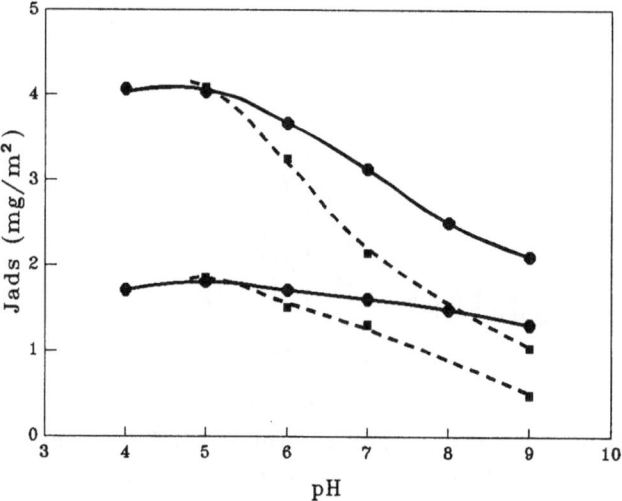

Fig. 1 Isotherm adsorptions at different pHs of aCRP-F(ab')$_2$ onto EU latex (■) and RP latex (●) at two different coverage degrees. They were carried out in a low ionic strength medium ($I = 0.002$)

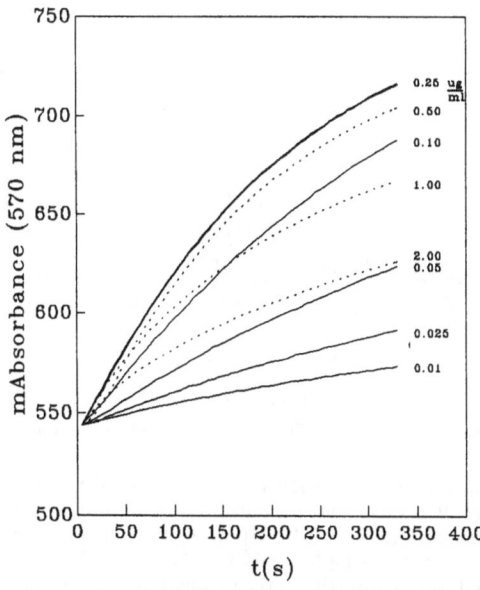

Fig. 2 Absorbance change, for different concentrations of CRP, caused by the immunoagglutination of the EU particles covered with 2.2 mg/m^2 of aCRP-F(ab')$_2$. The adsorption of this protein was performed at pH 7. Initial particle concentration was 6.0×10^{11} part./ml. All the immunoassays were carried out in BSA-saline

the sensitivity of this assay method, also called "latex immunoassay aggregation" (LIA), depends on the optical technique used in the detection of the aggregation reaction. We have chosen turbidimetry mainly because the set used in this technique, e.g., a spectrophotometer, is easily available in any laboratory, produces reproducible responses, has high sensitivity, and uses small sample volumes.

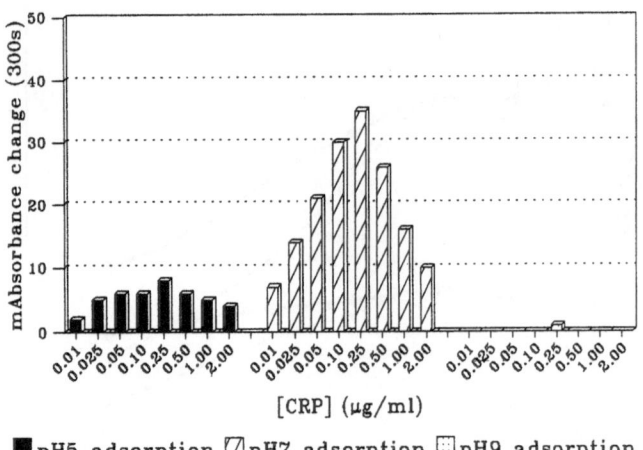

■pH5 adsorption ☑pH7 adsorption ▨pH9 adsorption

1030 ± 27 nm   275 ± 6 nm   211 ± 8 nm

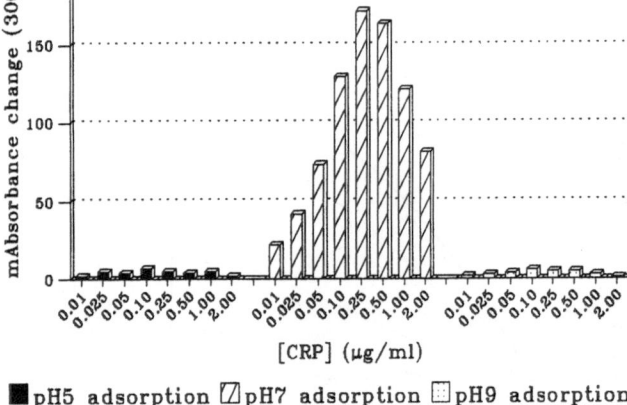

■pH5 adsorption ☑pH7 adsorption ▨pH9 adsorption

930 ± 25 nm   256 ± 6 nm   239 ± 11 nm

Fig. 3 Immunoaggregation of aCRP-F(ab')₂-EU complexes with different amounts of protein and the adsorptions of which were performed at three pHs (5, 7, and 9). Upper figure) The amounts of adsorbed F(ab')₂ were 1.8 mg/m² (pH 5), 1.3 mg/m² (pH 7) and 0.5 mg/m² (pH 9). Lower figure) shows the immunoresponse of these particles at other coverage degrees: 4.1 mg/m² (pH 5), 2.2 mg/m² (pH 7), and 1.0 mg/m² (pH 9). The initial average diameter of the samples is shown in these figures, indicating the extension of the particle aggregation before adding the antigen, since some of them were not stable in BSA-saline

Measuring the absorbance change over 5 min, similar increments for different CPR concentrations can be observed, just as would be expected considering the well known Heidelberger–Kendall curve for an immunoprecipitation reaction [21]. In spite of this, it is possible to differentiate regions of excess antigen and excess antibody-latex complex, comparing these curves, as Skoug and Pardue previously reported [22, 23]. Although the

equilibrium has still not been reached after 5 min, we can point out that in the region of excess antigen (dotted lines) this equilibrium value is obtained quicker than in those samples of excess antibody-latex particles, since the initial slopes for the former are higher than for the latter.

Figures 3a and 3b show the immunoresponse of the EU samples with different amounts of aCRP-F(ab')₂, where adsorptions of which were carried out at three different pHs (5, 7, 9). The average size of the colloidal particles, before starting the immunoagglutination, is also shown in this figure. Such values give an idea of the initial state of aggregation of each complex.

We should make a brief remark about the more important features that are required in these systems. The main characteristics of an immunosensor, such as ours, were able to be separated into two very different branches: i) the immunoresponse, which depends both on the immunological affinity and avidity of the antibody binding sites, and on the amount of aCRP-F(ab')₂ adsorbed; ii) the stability, in order to avoid inexpecific aggregations in the reactive medium. This feature also depends on the amount of adsorbed antibody, since the higher the value, the lower the stability will be. So, the use of F(ab')₂ fragments, instead of the IgG molecules, improves the stability of the system at the reaction pH, since in polyclonal IgG totally covered RP particles, the CCC value (given as mM of KBr) at such pH was < 10 mM [24], and for polyclonal F(ab')₂ RP complexes, at the same conditions, the CCC was around 50 mM [19]. In spite of this, those complexes having a F(ab')₂ full coverage surface are unstable in the reaction medium (BSA saline) and, therefore, totally useless. That is the reason why the samples adsorbed at pH 5, were coagulated before adding the antigen. However, the samples adsorbed at pH 7 have an amount of antibody in their surface quite sufficient to give a good immunoprecipitation, but not so much as to avoid autocoagulation. In contrast, we have the samples adsorbed at pH 9, where the lack of sufficiently adsorbed protein leads to rather stable complexes that are, however, devoid of immunoresponse. In addition, the orientation of the antibody molecules adsorbed on the particles could be affected by the adsorption pH, since, as they were approaching the polymer surface, the adsorption of short polypeptide chains or small patches of the whole macromolecule could be electrostatically favorable in a low ionic strength medium.

The immunoagglutination of the aCRP-F(ab')₂ RP complexes is more than slight. For those samples that have an amount of adsorbed protein higher than 2.2 mg/m² no increment of absorbance occurs. Only in the sample that has 2.0 mg/m², adsorption of which was carried out at pH 9, a small increase of no more than 10 miliabsorbance units takes place (figure not included). Similar results were

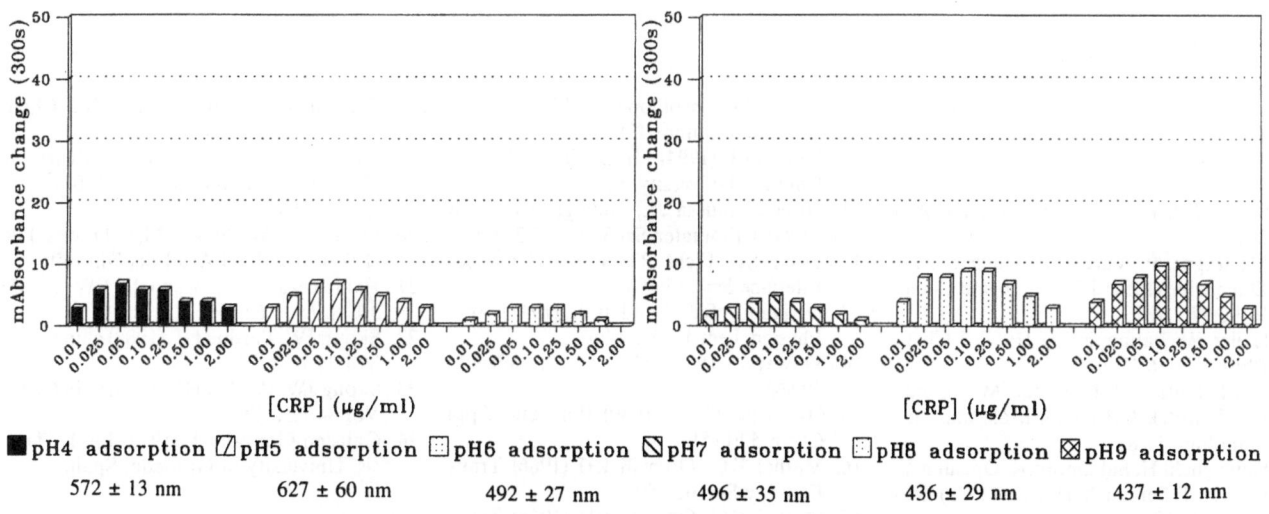

■pH4 adsorption ▨pH5 adsorption ⬚pH6 adsorption ◩pH7 adsorption ⬚pH8 adsorption ⊠pH9 adsorption

| 572 ± 13 nm | 627 ± 60 nm | 492 ± 27 nm | 496 ± 35 nm | 436 ± 29 nm | 437 ± 12 nm |

Fig. 4 Immunoaggregation of aCRP-F(ab')₂-RP complexes with different amounts of protein adsorbed at various pHs: 1.7 mg/m² (pH 4), 1.8 mg/m² (pH 5), 1.7 mg/m² (pH 6), 1.6 mg/m² (pH 7), 1.5 mg/m² (pH 8), and 1.3 mg/m² (pH 9). The initial average size of the complexes also appears in this figure. The particle concentration in these immunoassays were 2.2 10¹⁰ part./ml

obtained by those complexes with an amount of $F(ab')_2$ lower than 1.8 mg/m² (Fig. 4). As we claimed above, one must achieve microspheres with a certain amount of antibody adsorbed that optimizes the stability and the response of the immunosensor, which allows to reach highly stable complexes with the best immunological behavior. However, in RP latex it is almost impossible to get this optimum coverage, since the bare latex has a rather low CCC value itself (see Table 1) and, after only a slight amount of $F(ab')_2$ is adsorbed so the particles rapidly lose stability at the reaction medium.

The detection of the aggregation process in light-scattering immunoassays is more sensitive if the precipitation takes place among particles of a small size, compared with the wavelength used, compared to in those whose diameters are around the λ value [5]. This is another reason that justifies the better results achieved by EU samples in comparison with those obtained by RP ones.

Finally, after performing the adsorptions at pH 5, 7, and 9 in EU latex, and seeing its immunological behavior, we focused our aim on carrying out an adsorption at pH 6 in order to see if it was possible to achieve particles with the highest amount of $F(ab')_2$ adsorbed, while retaining the stability of the system. Figure 5 shows the absorbance increment caused by the immunoagglutination of these samples, which gives the best results for the two latices studied in the present work.

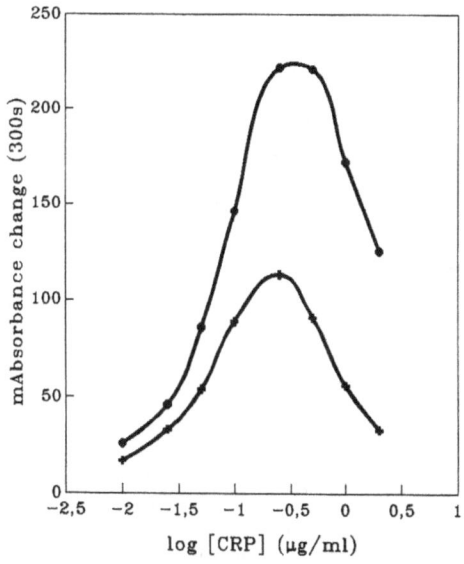

Fig. 5 Immunoresponse of EU particles covered with 1.5 mg/m² (+) and 3.0 mg/m² (●) adsorption, which was carried out at pH 6. The initial average size of these samples was 325 ± 11 nm for the former and 365 ± 16 nm for the latter

Acknowledgements We wish to express thanks to Elias Unzueta (Basque Country University) for preparing EU latex. We also wish to thank the CICYT (Spain) for providing funds for the MAT 94–0560 project.

References

1. Norde W (1986) Adv Colloid Interface Sci 25:267
2. Macritchie F (1978) Adv Protein Chem 32:283
3. Singer JM, Plotz CM (1956) Am J Med 21:888
4. Montagne P, Varcin P, Cuillère ML, Duheheille (1992) J Bioconjugate Chem, Vol 3 N°2:187
5. Heller W, Pangonis WJ (1957) J Chem Phys, Vol 26 N°3:498
6. von Schulthess GK, Giglio M, Cannell DS, Benedek GB (1980) Molecular Immunology 17:81
7. Kawaguchi H, Sakamoto K, Ohtsuka Y, Ohtake T, Sekiguchi H, Iri H (1989) Biomaterials 10:225
8. Muratsugu M, Kurosawa S, Kamo N (1991) J Colloid Interface Sci 147:378
9. Pereira AB, Theofilopoulos AN, Dixon FJ (1980) J Immunol 125:763
10. Unzueta E (1994) Ph. D. Thesis, Basque Country University, Spain
11. Ortega-Vinuesa JL, Hidalgo-Álvarez R (1994) J Biomater Sci Vol 6 N°3:269
12. Tsaur SL, Ficht RM (1987) J Colloid Interface Sci 115:450
13. Derjaguin BV, Landau L (1941) Acta Physicochem URSS 14:633
14. Verwey EJV (1942) Chem Weekbl 39:563
15. Overbeek JThG (1980) Pure and Appl Chem 52:1151
16. Mathai KG, Ottewill RH (1966) Trans Faraday Soc 62:750
17. Mathai KG, Ottewill RH (1966) Trans Faraday Soc 62:759
18. Ottewill RH, Walker T (1974) J Chem Soc Faraday Trans I 70:917
19. Ortega-Vinuesa JL, Hidalgo-Álvarez R (1993) Colloid and Surfaces B: Biointerfaces 1:365
20. Silverton EW, Navia MA, Davies DR (1977) Proc Natl Acad Sci 74:5140
21. Heidelberger M, Kendall FW (1935) J Exp Med 62:467
22. Skoug JW, Pardue HL (1988) Clin Chem Vol 34 N°2:300
23. Skoug JW, Pardue HL (1988) Clin Chem Vol 34 N°2:309
24. Galisteo-González F (1992) Ph. D. Thesis, University of Granada, Spain

Progr Colloid Polym Sci (1995) 98:239–242
© Steinkopff Verlag 1995

BIOCOLLOIDS

J. Krägel
D. Clark
P. Wilde
R. Miller

Studies of adsorption and surface shear rheology of mixed β-lactoglobulin/surfactant systems

J. Krägel
Universität Potsdam
Institut für Festkörperphysik
Potsdam, FRG

Dr. Clark · P. Wilde
Institute of Food Research
Norwich, United Kingdom

Dr. J. Krägel (✉) · R. Miller
Max-Planck-Institut für Kolloid-
und Grenzflächenforschung
Rudower Chaussee 5/Haus 9.9
12489 Berlin, FRG

Abstract The structure of mixed β-lactoglobulin/Tween 20 adsorption layers was studied by surface shear rheological and dynamic surface tension measurements. Surface shear rheology was carried out with a torsion pendulum set-up which provides information about the surface shear viscosity and elasticity in one single experiment.

The adsorption layer structure is controlled by the interactions between the protein molecules and the surfactant. The corresponding dynamic surface tension measure-ments confirm the peculiarities found by the shear rheology, which suggests dramatic changes of the mixed adsorption layer structure. With increasing surfactant at constant protein concentration the mixed surface film approaches, step-by-step, a state identical to a pure surfactant adsorption layer where the protein is completely removed.

Key words Protein surfactant interaction – adsorption layers – surface shear rheology – adsorption kinetics

Introduction

Shear rheological studies of adsorption layers along with other interfacial investigations are useful for the discussion of the structure of proteins and mixed protein/surfactant systems at liquid/gas and liquid/liquid inter-faces and are important for the understanding of various technological procedures, such as foaming [2–5], emulsifi-cation [6–9], or high-speed film coating [10, 11].

To avoid interfacial tension gradients during the gen-eration of a surface flow two different types of surface viscometers are typically used. The first type is based on the determination of interfacial velocity profiles (indirect methods) and the second on measurement of torsion stress values (direct methods).

The knife-edge surface viscometer of Brown et al. [12] as a classical direct method consists of a knife-edge bob suspended from a torsion wire such that the circular knife just touches the surface of a solution contained in a cylin-drical vessel. The measuring vessel is forced to rotate, and the torsion stress on the knife-edge is measured in order to determine the surface shear viscosity. Based on similar principles, other instruments were developed in the past [13–21].

Rheological investigations of adsorption layers at fluid interfaces require equipment which does not disturb their structure during the measuring procedure [22]. Therefore, in surface shear rheology the torsion pendulum is a prefer-red technique which allows experiments with very small mechanical deformations of the adsorption layer.

The present paper aims at the analysis of mixed β-lactoglobulin/Tween 20 adsorption layers by using a torsion pendulum apparatus [22] which is available as commercial set-up ISR1 from LAUDA, Germany. Simul-taneous dynamic surface tension measurements (ring tensiometer TD1/LAUDA; drop volume tensiometer TVT1/LAUDA and maximum bubble tensiometer

MPT1/LAUDA) are used to support the interpretation of rheological parameters as function of surfactant concentration and time.

Experimental

Material

β-Lactoglobulin (BLG) was purchased from Sigma Chemical Co. (from bovine milk, three times crystallised and lyophilised, LOT 51H7210), Tween 20 (polyoxyethylene 20 sorbitan monolaurate) from Pierce Chemical Co. (high purity, surfactant-amp grade). Both samples were used without further purification. The water was doubly distilled. All measurements were performed in a 0.01 mol/l sodium phosphate buffer solution (pH 7.0) at 20 °C.

Surface shear rheometer

The measuring principle of the surface shear rheometer ISRI from LAUDA, Germany, was described elsewhere in detail [22]. Briefly, it performs a simple pendulum experiment with a small deflection of 0.5 to 3 degrees so that a free oscillation of the ring, which touches the interface, results. The damped oscillation curve is recorded automatically by the instrument and the important quantities, the damping coefficient and the circular frequency of the torsion oscillation, are calculated from

$$y(t) = y_0 \cdot \exp\left(-\frac{t}{a}\right) \cdot \sin\left(\frac{2\pi}{T} \cdot (t - t_0)\right) + c , \qquad (1)$$

where a is the decay time $a = 1/\alpha$, T is the period of oscillation $T = 2\pi/\beta$, t_0 is the phase shift $t_0 = \varphi/\beta$, and c is the offset. The coefficients α and β in Eq. (1) are given by [23]

$$\alpha = \frac{F_r + \frac{\eta_s}{H_s}}{2 \cdot I_r}, \qquad \beta = \sqrt{\frac{E_r + \frac{G_s}{H_s}}{I_r} - \alpha^2} , \qquad (2)$$

where H_s is a geometric factor and I_r, E_r and F_r can be determined by calibration measurements. Knowing the values of α and β by fitting Eq. (1) to a set of experimental points the rheological parameters η_s and G_s of the adsorption layer can be calculated via Eq. (2).

The shear rheology measurements were performed in the following way. After the solution was filled into the vessel the surface layer was removed in order to start the experiments with a fresh surface. Then, each 10 min a pendulum experiment was performed and a time-dependence of the surface shear viscosity and elasticity was obtained.

Dynamic surface tension measurements

The dynamic surface tensions $\gamma(t)$ of the BLG, Tween and mixed BLG/Tween solutions were performed by using drop volume tensiometry (TVT1 from LAUDA) and ring tensiometry (TD1 from LAUDA). The drop volume experiments yield data in a time window from part of a second up to about 300 s while the ring tensiometry provides surface tension values from about 1 min to some hours. All measurements were performed at room temperature.

Results and discussion

The dynamic surface tension vs. time curves for BLG and three Tween 20 solutions are given in Fig. 1. At small adsorption time the BLG adsorbs faster than the surfactant Tween. At longer times the surface tensions do not decrease as much as for the Tween solutions, and after about 4 h values of only 53 mN/m are reached. The adsorption rate of the Tween solutions is of the same order of magnitude as BLG. This was expected from a transport-controlled adsorption model. The higher the concentration the faster the adsorption and the lower the $\gamma(t)$-values obtained after 4 h.

The surface tension vs. time curves of BLG/Tween 20 mixtures are shown in Fig. 2. The surface tension values at short times are close to those for pure BLG. With increasing time and concentration ratio of Tween: BLG the curves approach the $\gamma(t)$-dependences of the pure Tween solutions. This means that at short time the mixed adsorption layer contains significant BLG. This amount decreases with time and is more pronounced at higher Tween concentrations. Thus, at a ration of Tween: BLG of 1:5 almost no BLG remains at the surface after a sufficiently long adsorption time.

The surface shear viscosity as a function of time of BLG at different concentrations is shown in Fig. 3. At a concentration of $1 \cdot 10^{-6}$ mol/l the shear viscosity increases very fast with time and reaches a plateau value. At a lower concentration this viscosity increase is slower and the plateau value significantly lower. For higher BLG concentrations the viscosity decreases already in the time interval of the present studies. This decrease can be explained by the structure of the BLG adsorption layer. The higher the protein concentration, the faster the adsorption, and the first adsorbed molecules get less and less time to occupy a large interfacial area. Consequently, the molecules have to arrange an increasingly more compact conformation in the adsorption layer, which leads to the decreasing surface shear viscosity.

The surface shear viscosity as a function of time of BLG/Tween 20 mixtures of different mixing ratios is

Fig. 1 Dynamic surface tension as function of time of a $1 \cdot 10^{-6}$ mol/l β-lactoglobulin (⊠ ○) solution and three Tween 20 solutions at concentrations $1 \cdot 10^{-6}$ mol/l (▲), $2 \cdot 10^{-6}$ mol/l (♦) and $5 \cdot 10^{-6}$ mol/l (■)

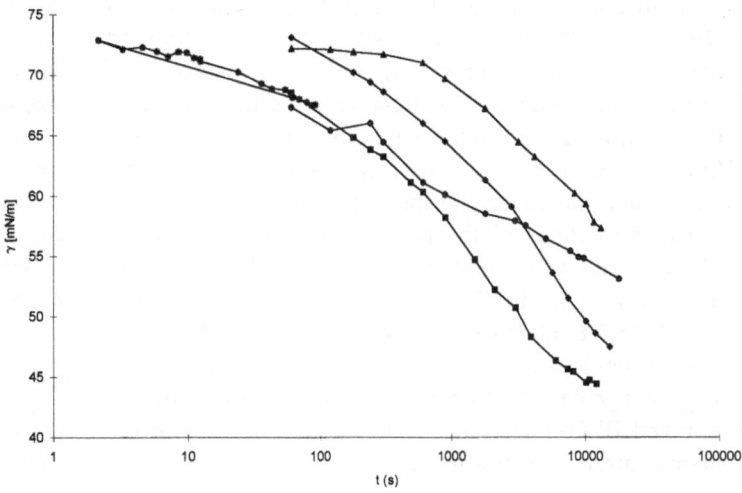

Fig. 2 Dynamic surface tension as function of time of β-lactoglobulin/Tween 20 mixtures at constant BLG concentration ($1 \cdot 10^{-6}$ mol/l) and different addition of Tween 20; 1:1 (◇), 1:2 (♦), 1:5 (■)

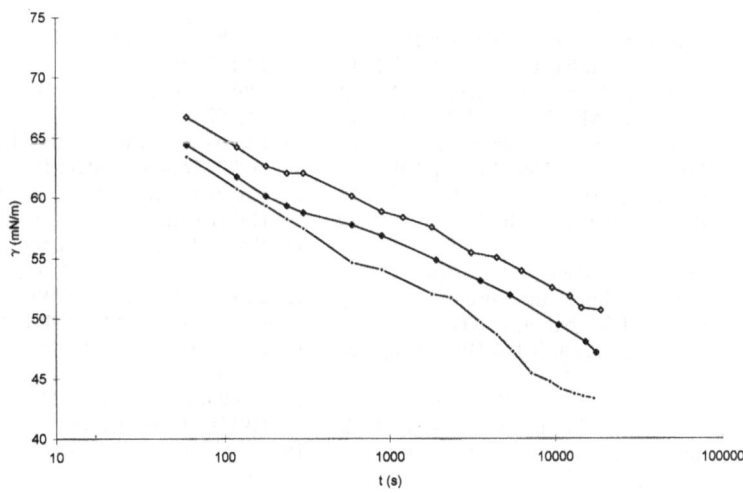

Fig. 3 Surface shear viscosity of BLG at different concentration; $1 \cdot 10^{-7}$ mol/l (*), $1 \cdot 10^{-6}$ mol/l (+), $2 \cdot 10^{-6}$ mol/l (×), $5 \cdot 10^{-6}$ mol/l (□), $1 \cdot 10^{-5}$ mol/l (■)

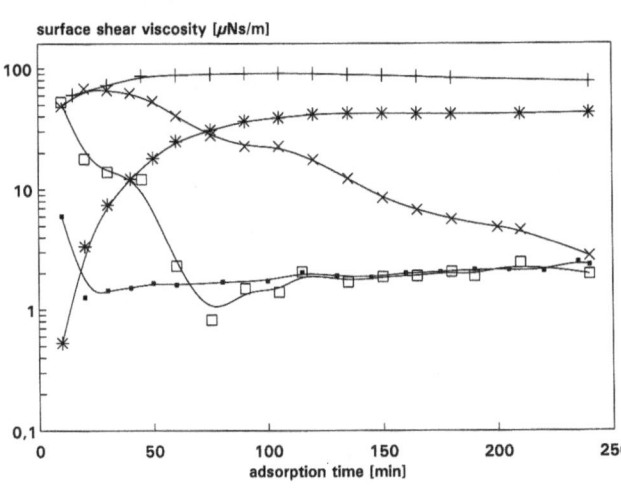

Fig. 4 Surface shear viscosity of BLG/Tween 20 mixture at constant BLG concentration ($1 \cdot 10^{-6}$ mol/l) and different addition of Tween 20; molar ratio (Tween 20/BLG): 0.2 (■), 0.5 (+), 0.8 (*), 1.0 (□), 1.2 (×), 2.0 (◇), 5.0 (△) and $1 \cdot 10^{-6}$ mol/l Tween 20 (X)

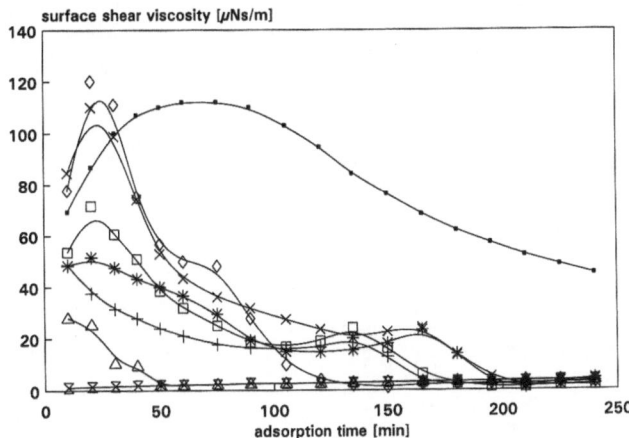

shown in Fig. 4. At the chosen BLG concentration of $1 \cdot 10^{-6}$ mol/l the highest shear viscosities were observed in absence of Tween. In presence of small quantities of Tween 20 the viscosities first increase, as those of pure BLG solutions go through a maximum, and then decrease. With increasing Tween concentration the decrease becomes faster and faster and the final values reach viscosities equal to those of the pure Tween solution. At the highest Tween concentration of $5 \cdot 10^{-6}$ mol/l a viscosity value comparable to that of a pure Tween solution is reached already after less than 1.

The time-dependence of the measured shear viscosities is in good agreement with the picture of the composition of the mixed BLG/Tween adsorption layer drawn from the dynamic surface tension measurements.

Conclusions

The experimental result shows that dynamic surface tension and surface shear viscosities of β-lactoglobulin/Tween 20 mixtures complement each other. At constant BLG concentration the composition of the mixed adsorption layer changes significantly with an increasing Tween 20 concentration. The Tween molecules replace the BLG in the layer so that at a mixing ratio of Tween:BLG of 5:1 the major part of the surface is occupied by the surfactant.

Acknowledgement This work was financially supported by a project of the European Community (HCM ERBCHRXT930322), the DFG (Wu 187/3-1), the Senate of Berlin (ERP 2659), and the Fonds der Chemischen Industrie (RM 400429).

References

1. Lucassen J (1981) In: Lucassen-Reynders EH (ed) Surfactant Sci Ser Vol 11. Marcel Dekker, New York, pp 217–265
2. Malhotra AK, Wasan DT (1988) In: Ivanov IB (ed) Surfactant Sci Ser Vol 29. Marcel Dekker, New York, p 829
3. Ronteltap AD, Damste BR, De Gee M, Prins A (1990) Colloids and Surfaces 47:269
4. Malysa K, Miller R, Lunkenheimer K (1991) Collioids and Surfaces 53:47–62
5. Wasan DT, Nikolov AD, Lobo LA, Koczo K, Edwards DA (1992) Prog Surf Sci 39:119–154
6. Tadros TF, Vincent B (1983) In: Becher P (ed) Encyclopedia of Emulsion Technology. Marcel Dekker, New York, Vol 1, pp 1–56
7. Dickinson E, Murray BS, Stainsby G (1988) J Chem Soc Faraday Trans 1 84:871–883
8. Doi M, Ohta T (1991) J Chem Phys 95:242
9. Lucassen-Reynders EH, Kuijpers KA (1992) Colloids and Surfaces 63:175–184
10. Fruhner H, Krägel J, Kretzschmar G (1991) J Inf Rec Mater 19:45–58
11. Kretzschmar G, Fruhner H, Krägel J (1993) Tenside 30:110–115
12. Brown AG, Thuman WC, McBain JW (1953) J Colloid Sci 8:491–507
13. Davies JT (1957) Proc Int Cong Surf Act 1:220
14. Lifschutz N, Hedge MG, Slattery JC (1971) J Colloid Interface Sci 37:73–79
15. Goodrich FC, Allen LH (1972) J Colloid Interface Sci 40:329–336
16. Briley PB, Deemer AR, Slattery JC (1976) J Colloid Interface Sci 56:1
17. Goodrich FC, Chatterjee AK (1970) J Colloid Interface Sci 31:36–42
18. Shail R (1978) J Engng Math 12:59
19. Oh SG, Slattery JC (1978) J Colloid Interface Sci 67:516–525
20. Davis AM, O'Neill ME (1979) Int J Multiphase Flow 5:413
21. Krägel J, Siegel S, Miller R, Born M, Emke B, Schano KH (1993) Prog Colloid Polym Sci 93:283
22. Krägel J, Siegel S, Miller R, Born M, Schano KH (1994) Colloids and Surfaces A: 91:169–180
23. Tschoegl NW (1961) Kolloid Z 181:19–29

Progr Colloid Polym Sci (1995) 98:243–247
© Steinkopff Verlag 1995

Structure of a layer of AOT adsorbed at the air/liquid interface at the critical micelle concentration determined by neutron reflection

Z.X. Li
J.R. Lu
R.K. Thomas
J. Penfold

Z.X. Li · J.R. Lu · R.K. Thomas (✉)
Physical Chemistry Laboratory
South Parks Road
Oxford OX1 3QZ, United Kingdom

J. Penfold
Rutherford-Appleton Laboratory
Chilton
Didcot, Oxon OX11 ORA, United Kingdom

Abstract Neutron reflectivity measurements have been used to study the monolayer of AOT adsorbed at the air-water interface at its critical micelle concentration (CMC = 2.45×10^{-3} M). Isotopic substitution has been used to label different parts of the AOT and to locate different groups within the surface region. For the saturated monolayer the thickness was found to be 18 ± 1 Å and the area per molecule 78 ± 2 Å^2. The width of the chain region was found to be 13 ± 1 Å and that of the head group region to be 15 ± 1 Å. The separations of the three distributions chain, head, and solvent, denoted by δ_{CS}, δ_{HS} and δ_{CH} were determined directly and found to be respectively 6.0 ± 1 Å, 4.5 ± 1 Å, and 2.0 ± 1 Å.

Key words Surfactant adsorption – aerosol OT – neutron reflection – structure studies

Introduction

AOT (sodium bis(2-ethylhexyl) sulfosuccinate; Fig. 1) is a surfactant with a small branched double tail. It has been widely used by many investigators because the hydrophilic and lipophilic properties are well balanced and therefore it is an ideal surfactant for forming microemulsions without cosurfactants [1–3]. The structure of any interfacial monolayer involving AOT is therefore of considerable interest. We report here the first such determination, using neutron reflectivity, of the structure of an adsorbed AOT monolayer at the air/water interface. Different isotopic AOT samples have been prepared in order to obtain details of the molecular conformation at the surface.

Experimental detail

Four isotopic species of AOT were used in the experiments, shown in Fig. 1 and designated dHdC, dHhC, hHdC and hHhC. The various partially deuterated compounds were prepared as described in [4–6] using D$_2$O, LiAlD$_4$, deuterated butanol and ethanol as initial sources of deuterium. Protonated AOT (hHhC) was obtained from BDH. All the samples were purified as described in [4]. The purity of samples was mainly assessed by surface tension measurements using the criterion of no minimum around the CMC. The values of the surface tension above CMC were all higher than commercial samples from Sigma or Fluka.

The neutron reflection experiments were carried out on the instrument CRISP at ISIS [7] and all the experiments were performed at 298 K.

Neutron reflection

The reflectivity is given approximately by

$$R(\kappa) = \frac{16\pi^2}{\kappa^2} |\rho(\kappa)|^2 , \tag{1}$$

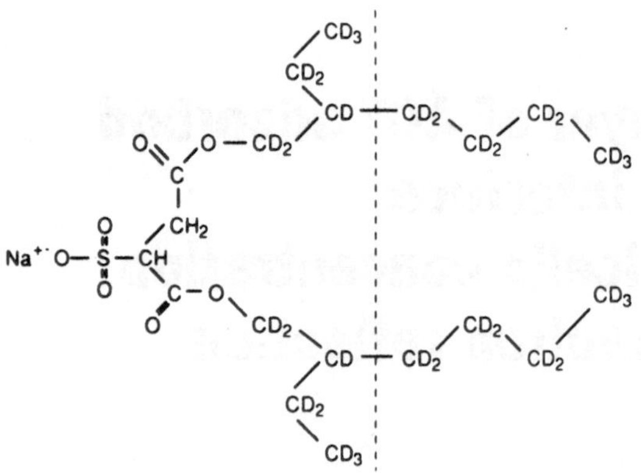

Fig. 1 The molecule structure of four partial isotopic AOT samples. The dashed lines are to indicate the separate chain and head groups

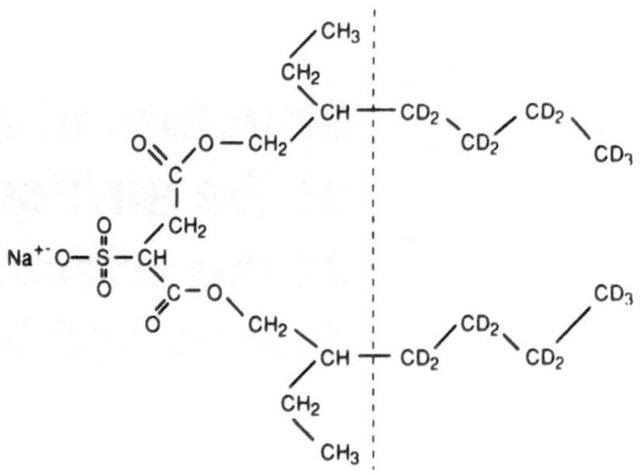

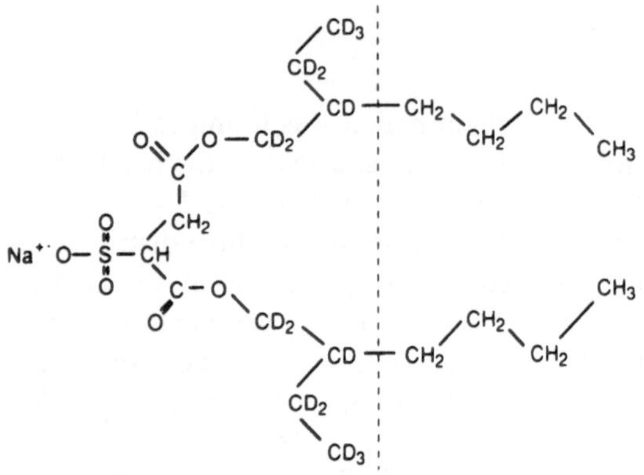

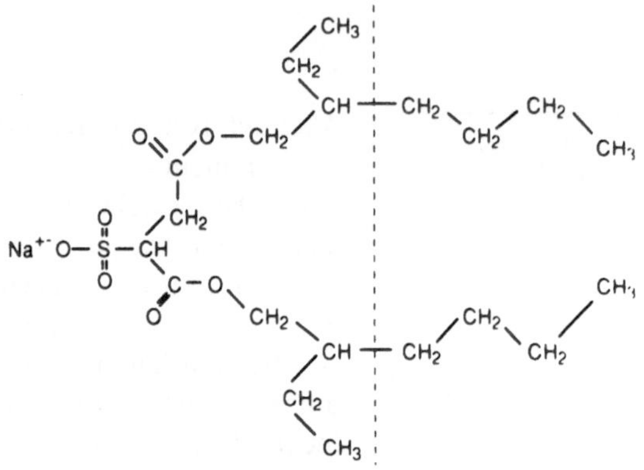

where $\rho(\kappa)$ is the one-dimensional Fourier transform of $\rho(z)$, the scattering length density profile normal to the interface [8], which is given by

$$\rho = \sum n_i b_i , \tag{2}$$

where n_i is the number density profile of species i and b_i is its scattering length.

The structure of the air/solution interface can be described by the distributions of the chain C, head H and solvent S, where the groups are as defined in Fig. 1. In terms of these labels the scattering length density can be written

$$\rho(z) = b_C n_C(z) + b_H n_H(z) + b_S n_S(z) . \tag{3}$$

The kinematic approximation for the reflectivity $R(\kappa)$ may be written in terms of the partial structure factors h_{ii}

$$R(\kappa) = \frac{16\pi^2}{\kappa^2} \{ b_C^2 h_{CC} + b_H^2 h_{HH} + b_S^2 h_{SS} + 2b_C b_H h_{CH}$$

$$+ 2b_C b_S h_{CS} + 2b_H b_S h_{HS} \} , \tag{4}$$

where h_{ij} are the partial structure factors given by

$$h_{ii}(\kappa) = |n_i(\kappa)|^2$$

$$h_{ji} = h_{ij}(\kappa) = Re\{ n_i(\kappa) n_j^*(\kappa) \} . \tag{5}$$

The $n_i(\kappa)$ are the one-dimensional Fourier transforms of $n_i(z)$, the average number density profile of atom, or group, i in the direction normal to the interface,

$$n_i(\kappa) = \int_{-\infty}^{\infty} \exp(i\kappa z) n_i(z) \, dz . \tag{6}$$

Equation (4) is only approximate, but the observed reflectivity can be corrected so that it satisfies the equation with sufficient accuracy for the present purpose [9].

It has been shown that a Gaussian distribution is the most accurate representation for the constituent parts of the surfactant. We use the Gaussian distribution defined by

$$n = n_i \exp(-4z^2/\sigma_i^2) \quad \text{for all } z , \tag{7}$$

where z is the distance in the direction normal to the interface and σ is the width parameter. The corresponding partial structure factor is

$$h_{ii} = \frac{1}{A^2} \exp\left(-\frac{\kappa^2 \sigma^2}{8}\right), \tag{8}$$

where A is the area per molecule.

The self partial structure factors, h_{ii}, contain information about the distribution of each labelled component, but not about the relative positions of the two components. The information about the relative positions is contained in the cross partial structure factor, h_{ij}. When the distributions are either perfectly even about their centres (this is exactly true for Gaussian distributions), or perfectly odd, the cross terms are

$$h_{CS} = \pm (h_{CC} h_{SS})^{1/2} \sin \kappa \, \delta_{CS} \tag{9}$$

$$h_{CH} = \pm (h_{CC} h_{HH})^{1/2} \cos \kappa \, \delta_{CH} . \tag{10}$$

From Eqs. (8), (9), and (10) h_{CH} is

$$h_{CH} = \frac{1}{A^2} \exp\{-\kappa^2(\sigma_C^2 + \sigma_H^2)/16\} \cos \kappa \, \delta_{CH} . \tag{11}$$

Since A, σ_C and σ_H are known, δ_{CH} can be determined directly from h_{CH}. A similar equation may be written for h_{CS} and h_{HS} using the expression for a "rough" solvent for h_{SS} [10], a Gaussian distribution for h_{CC} and h_{HH}, and Eq. (9),

$$h_{iS} = \frac{n_0}{A^2} \exp\{-\kappa^2(\sigma_i^2 + 8\tau_S^2)/16\} \sin \kappa \, \delta_{iS} , \tag{12}$$

where n_0 is the number density of bulk water and τ_S is the roughness parameter for the water surface, in the presence of surfactant.

Results and discussion

The reflectivity profiles from the adsorbed layer of three isotopic species of AOT in null reflecting water (n.r.w.) at the CMC are shown in Fig. 2, and those for a slightly different set of AOT in D_2O in Fig. 3. The continuous lines have been calculated using the exact optical matrix method using a model of one uniform layer for the n.r.w. profiles and a two-layer model for the D_2O profiles [11]. The surface excess of the surfactant, or area per molecule (A), can be obtained from the fitted parameters for the n.r.w. data, the area per molecule being given by

$$A = \frac{\sum n_i b_i}{\rho \tau} , \tag{13}$$

where n_i and b_i are as defined in Eq. (2). The structural parameters obtained from fitting this model are given in

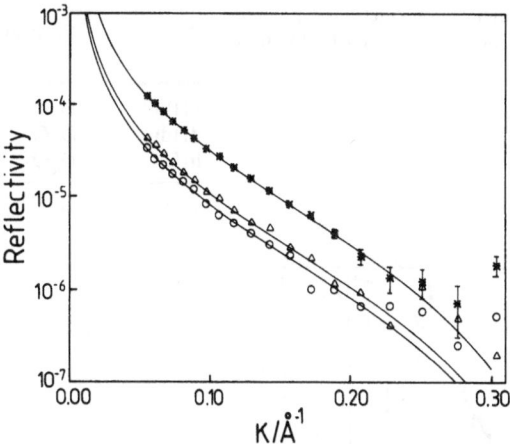

Fig. 2 Reflectivity profiles of fully deuterated AOT (dHdC) ($\times$), chain deuterated (hHdC) ($\circ$), and head deuterated (dHhC) ($\triangle$) in null reflecting water at the CMC. The continuous lines are the fits using a single uniform layer model with the structural parameters given in Table 1. The error bars are marked only for dHdC

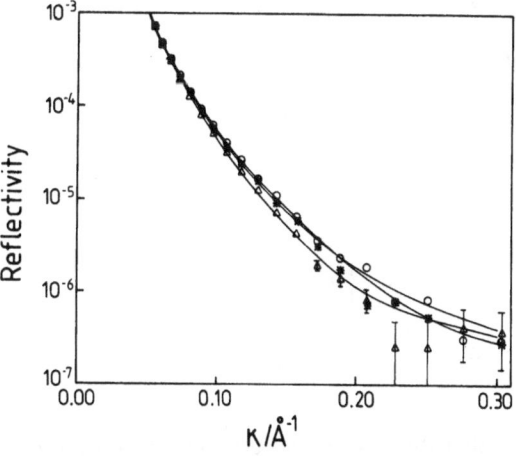

Fig. 3 Reflectivity profiles of fully protonated AOT (hHhC) ($\circ$), fully deuterated (dHdC) ($\times$), and chain deuterated (hHdC) ($\triangle$) in D_2O at the CMC. The continuous lines are the fits using the two uniform layer model and the structural parameters given in Table 2. The error bars are marked only for dHhC

Table 1 Parameters from one-layer model fitting for the profiles in n.r.w.

Surfactant	$A/\text{Å}^2$	$\rho \times 10^6/\text{Å}^{-2}$	$\tau/\text{Å}$
dCdH	78	2.80	18
hHdC	77	1.65	18
dHhC	79	1.45	18

Table 1. The area per molecule at the CMC was found to be 78 ± 2 Å. The thickness defined here as that of a uniform layer is 18 ± 1 Å.

In D_2O the two-uniform-layer model gives an outer layer of 6 ± 1 Å, and an inner layer of 12 ± 1 Å, with

246
Z.X. Li et al.
AOT adsorption at the air/liquid interface

Table 2 Parameters for calculated profiles using the two-layer optical matrix method

Species	Contrast	$A/Å^2$ ± 2	$\tau_c/Å$ ± 1	$\rho_c/Å^{-2}$ $\times 10^{-6}$	$\tau_h/Å$ ± 1	$\rho_c/Å^{-2}$ ± 1	n ± 1	f_c ± 0.1
hHhC	D_2O	78	6	-0.3	12	4.8	21	0
dHdC	D_2O	77	5	3.7	13	5.7	20	0
hHdC	D_2O	79	6	3.7	12	5.2	21	0

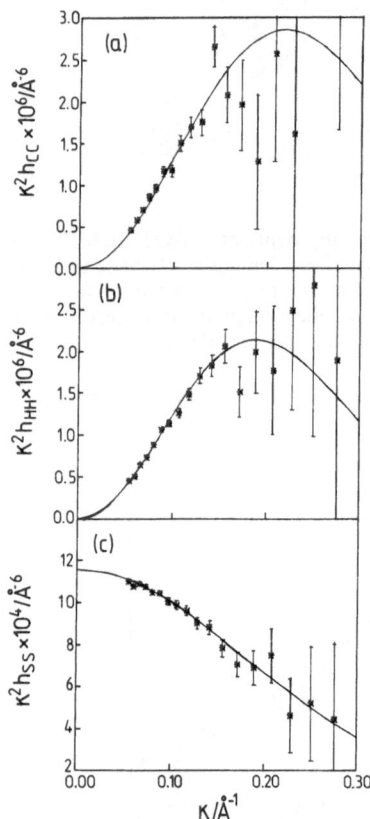

Fig. 4 The self-partial surface factors, (a) $\kappa^2 h_{CC}$, (b) $\kappa^2 h_{HH}$, (c) $\kappa^2 h_{SS}$. The continuous lines are calculated using Eq. (8) for $\kappa^2 h_{CC}$ ($\sigma_C = 13 \pm 1$ Å) and $\kappa^2 h_{HH}$ ($\sigma_H = 15 \pm 1$ Å), and a tanh distribution for the solvent ($\zeta_s = 4 \pm 1$ Å)

Fig. 5 The cross partial structure factors, (a) $\kappa^2 h_{CH}$, (b) $\kappa^2 h_{CS}$, and (c) $\kappa^2 h_{HS}$. The continuous lines are calculated using Eq. (11) for $\kappa^2 h_{CH}$ ($\delta_{CH} = 2 \pm 1$ Å) and Eq. (12) for $\kappa^2 h_{CS}$ ($\delta_{CS} = 6 \pm 1$ Å) and $\kappa^2 h_{HS}$ ($\delta_{HS} = 4.5 \pm 1$ Å)

a chain fraction f_c of zero (Table 2). This means that the butyl group of AOT is completely out of the water layer. The number of water molecules associated with each surfactant molecule in the head group region is 20 ± 2.

The kinematic approximation is a more direct method of analysing the reflectivity data. The six partial structure factors obtained by solving Eq. (4) are shown in Figs. 4 and 5. The continuous lines are the fits to h_{CC} and h_{HH} assuming Gaussian distributions using Eq. (7). The continuous line in Fig. 4(c) is the fit of a tanh distribution of water with a width parameter ζ_s [11] equal to 4 ± 1 Å. The cross partial structure factors, h_{ij}, were fitted using Eqs. (11) and (12) with the width parameters already obtained from the self-partial structure factors. The fits are shown as

continuous lines. The separation of the C and H fragments, δ_{CH}, was found to be 2 ± 1 Å. From the partial structure factors h_{CS} and h_{HS} the separations between chain and solvent distribution, δ_{CS}, and head and solvent, δ_{HS}, were found to be 6 ± 1 and 4.5 ± 1 Å respectively.

The width determined for any fragment of the surfactant depends on the average intrinsic dimensions of the molecule along the surface normal, l, and the roughness, w, and the relationship is [12]

$$\sigma^2 = l^2 + w^2 . \tag{14}$$

For a number of surfactant layers so far studied the roughness at the CMC seems to be greater than predicted by simple capillary wave models and is usually in the range

Progr Colloid Polym Sci (1995) 98:243–247
© Steinkopff Verlag 1995

Table 3 Structural parameters obtained from kinematic analysis

$l_C/\text{Å}$ ± 1	$\sigma_C/\text{Å}$ ± 1	$\sigma_H/\text{Å}$ ± 1	$A/\text{Å}^2$ ± 2	$\zeta_s/\text{Å}$ ± 1	$\delta_{CH}/\text{Å}$ ± 1	$\delta_{CS}/\text{Å}$ ± 1	$\delta_{HW}/\text{Å}$ ± 1
7	13	15	78	4	2	6	4.5

10–14 Å. Because the Gaussian thickness, σ, of the butyl fragment is 13 Å, w cannot be greater than 13 for AOT. However, given that the surface tension of the AOT at its CMC is substantially lower than for $C_n\text{TABs}$ [12], for which w is 12–14 Å, it is unlikely that w for AOT is less than 10 Å. If we take an arbitrary intermediate value of 11 Å for w, the intrinsic length along the surface normal of the butyl chain is 7 Å, and that of the remainder of the molecule 10 Å. A higher roughness of 12 Å would change these values to 5 and 9 Å respectively. If the butyl group were fully extended along the normal direction its length would be about 7 Å. It is unlikely that it is pointing along the normal direction because this would be totally inconsistent with the very small value of δ_{CH} of 2 Å. The latter, together with the large intrinsic dimension of the head group, suggests that the $O-CH_2-CH$ group (see Fig. 1) is oriented normal to the surface forcing the ethyl and butyl groups into orientations strongly inclined away from the surface normal. This would be consistent with the intrinsic length of the butyl group being less than the fully extended value of 7 Å, which suggests that the roughness is around 12 Å.

References

1. Shinoda K, Kunieda H, Arai T, Saito H (1987) J Phys Chem 88:5126
2. Aveyard R, Binks BP, Mead J (1985) J Chem Soc Faraday Trans 1, 81:2169
3. Williams EF, Woodberry NT, Dixon JK (1957) J Colloid Sci 12:452
4. Li ZX, Lu JR, Thomas RK (to be published)
5. Furniss BS, Hannaford AJ, Smith PWG, Tatchell AR (1989) Vogel's Textbook of Practical Organic Chemistry, 5th ed.
6. U.S. Patent 2,028,091
7. Lee EM, Thomas RK, Penford J, Ward RC (1989) J Phys Chem 93:381
8. Lu JR, Simister EA, Lee EM, Thomas RK, Rennie AR, Penford J (1992) Languir 8:1837
9. Crowley TL (1993) Physica A 195:354
10. Schwartz DK, Schlossman ML, Kawamoto EH, Kellogg GJ, Pershan PS, Ocko BM (1990) Phys Rev A 41:5687
11. Lu JR, Hromadova M, Thomas RK, Penfold J (1993) Langmuir 9:2417
12. Pleshanov NK, Pusenkov VM, Schebetov AF, Peskov BG, Shmelez GE, Siber EV, Soroko ZN (1994) Physica B 198:27

Progr Colloid Polym Sci (1995) 98:248–254
© Steinkopff Verlag 1995

J. Sánchez-González
M.J. Gálvez-Ruiz

Behavior and stability of lipid mixed monolayer at the air-water interface

J. Sánchez-González
M.J. Gálvez-Ruiz (✉)
Department of Applied Physics
University of Granada
C/Fuente Nueva s/n
18071 Granada, Spain

Abstract The study of mixed monolayers under different experimental conditions is of great interest, because it enables us to gain a much better understanding of the interactions between the monolayer compounds. In addition, superficial processes are strongly dependent on those interactions. Hence, understanding the behavior of mixed monolayers is a step towards the knowledge of the formation of Langmuir-Blodgett films onto solid surfaces.

Experimental isotherms, obtained by compressing monomolecular layers of lipids (distearoylphos phatidylcholine (DSPC) and Sphyngomyelin(Sph)) have been performed.

The monolayer properties of these compounds are influenced by the temperature, bringing about important changes in both interaction characteristics and the stability of the monolayers. These changes are mainly attributed to a decrease in the attractive forces of the Van der Waals type between the hydrophobic regions. Nevertheless, for mixed monolayers forming by DSPC and Sph, the mean contribution to the intermolecular interaction is the repulsive electrostatic force, between polar head groups, which tends to dominate those with attractive character due to the steric effect introduced by Sph molecules.

The behavior is largely influenced by the surface pressure values and, therefore, by the molecular orientation with respect to the water interface.

Key words Monolayers – lipids – miscibility – interaction – thermodynamic study

Introduction

It has been pointed out that the formation of different Langmuir-Blodgett (LB) multilayers is highly dependent on the previous experimental conditions of the monolayers at the air-water interface. For this reason, the aim of this work is to study the states, molecular interactions, and characteristics of the monolayers formed by lipids with large hydrocarbon chains. A similar study, but with lipids with shorter chains, has been carried out in order to compare the behavior of both molecule types [1–3].

In this paper, we have studied the behavior of two lipids with different structural characteristics, and also their respective mixtures in the air-water interface under diverse temperature conditions, focusing attention mainly on the miscibility among the different components involved.

Progr Colloid Polym Sci (1995) 98:248–254
© Steinkopff Verlag 1995

Investigating the properties of mixed monolayers is of great interest, because it makes it possible to gain knowledge about the interactions between the monolayer compounds. It is clear that all surface mechanisms, including the transfer to a solid surface, are influenced by these interactions. Hence, understanding the behavior of mixed monolayers can be considered an important contribution to a general study of the LB films formation. In particular, and following the thermodynamic formulation proposed by Goodrich [4], we can measure the excess free energies, entropies and enthalpies of mixing for our systems, under different experimental conditions, and arrive at conclusions concerning the extent and nature of the intermolecular forces involved.

Therefore, from the experimental isotherms surface pressure-molecular area (π-A), the excess free energy of mixing has been obtained as the following equation:

$$\Delta G^E = \int_0^\pi \Delta A^E \, d\pi \,, \tag{1}$$

where the excess area of mixing is:

$$\Delta A^E = N_A (A_{12} - x_1 A_1 - x_2 A_2) \,. \tag{2}$$

The equations used to calculate the others thermodynamic functions are:

$$\Delta S^E = -\left(\frac{\delta G^E}{\delta T}\right)_{\pi,\,x} - N_A \, \Delta A^E \frac{d\gamma_0}{dT} \tag{3}$$

$$\Delta H^E = \Delta G^E + T\Delta S^E \,. \tag{4}$$

Also, it is very useful to use the interaction parameter defined as:

$$\alpha = \frac{\Delta G^E}{RTx(1-x)} \tag{5}$$

which, as the rest of the thermodynamic quantities, quantify the interactions present in the monolayers [5].

Experimental

Materials

Water used in all the experiments was first twice distilled in an all-Pyrex apparatus and then passed through a Millipore Milli-Q Reagent Water System for further purification. Conductivity of the water obtained after this process was always $\leq 10^{-4} \, \Omega^{-1} \, cm^{-1}$ and its pH was 5.5–6.0.

The component monolayers, L-α-Phosphatidylcholine, Distearoyl (DSPC), and Sphyngomyelin (Sph) were supplied with analytical research grade by Sigma (USA).

The solvent used for spreading the monolayers formed by the above-mentioned compounds was a 4:1 (v/v) mixture of n-hexane and ethanol (Merck A.R. grade, Germany); a 0.05% of amylalcohol was added to improve spreading.

Experimental device and method

Surface pressure (π)-molecular area (A) isotherms were performed using the previously described [1] Langmuir method with a computer-controlled (80386, 33 MHz) "Lauda Filmwaage" balance.

In this work a new computer program was used in order to increase the possibilities of data processing, i.e., determination of phase transitions at the monolayers.

The main aspect that we point out is the elimination of the molecular area measurement calibration with respect to the original balance model, which allows us to work with a wide range of molecular sizes. (This program is available to any interested reader upon request).

Before the compression of the monolayers a 10-min lapse was estimated to be sufficient for evaporating the solvent without contaminating the film. Once this time had elapsed, the molecular films were compressed at a constant rate of $6.5 \times 10^{-4} \, m \, s^{-1}$. Previously, we checked that the shape of the isotherms was independent of the compression rate and in no experiment was a hysteresis phenomenon observed.

Finally, each experimental isotherm was recorded at different temperatures (298, 303, 308 and 313 K) controlled by a Haake F3K thermostatic bath to ± 0.1 K.

Results and discussion

Firstly, the study of the experimental surface pressure-molecular area isotherms is reported separately for two simple monolayers:

DSPC monolayers

The behavior of the DSPC monolayers under different temperature conditions is shown in Fig. 1. At low temperatures, the shape of the isotherms shows a liquid condensed (LC) state for the monolayers. These isotherms are displaced to higher areas when increasing temperature due to the diminution of the hydrophobic attractive forces between the hydrocarbon chains. Moreover, under the higher temperature a liquid extended (LE)-LC phase transition is observed for this compound.

Fig. 1 Experimental isotherms obtained for monolayers of DSPC at different temperature values (288, 293, 298, 303, 308, 313 K)

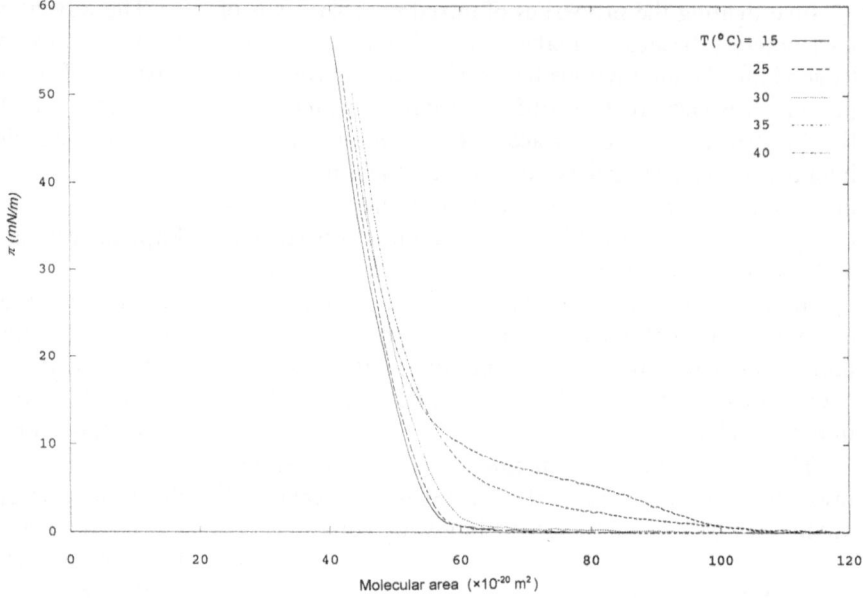

Fig. 2 Experimental isotherms obtained for monolayers of Sph at different temperature values (288, 298, 303, 308, 313 K)

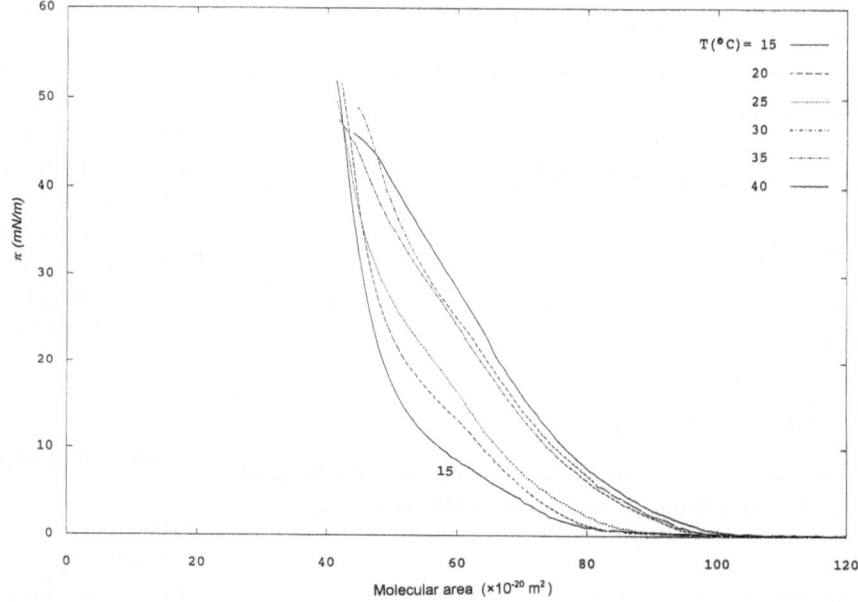

The collapse surface pressure decreases when temperature increases, demonstrating that the stability of the monolayers is modified as temperature changes.

Sph monolayers

In Fig. 2, the experimental isotherms corresponding to the Sph monolayers are plotted. In this case, the films are extended more when compared with DSPC monolayers due to the different structure of the hydrocarbon chains, asymmetric and with the presence of a double bound for the Sph molecules. For this reason, these films go through a LE-LC phase transition at low temperatures reaching LE state when temperature increases. This behavior, more extended with temperature, allow us to reject the presence of important hydration effects in the Sph molecules [6].

Mixed monolayers: DSPC-Sph system

In Fig. 3 (a and b), the experimental isotherms corresponding to the monolayers forming by different mixtures of DSPC and Sph, at 298 K, are shown.

Progr Colloid Polym Sci (1995) 98:248–254
© Steinkopff Verlag 1995

Fig. 3 (a and b) Experimental isotherms obtained for monolayers containing DSPC and Sph in different proportions at 298 K. Sph mole fraction is indicated

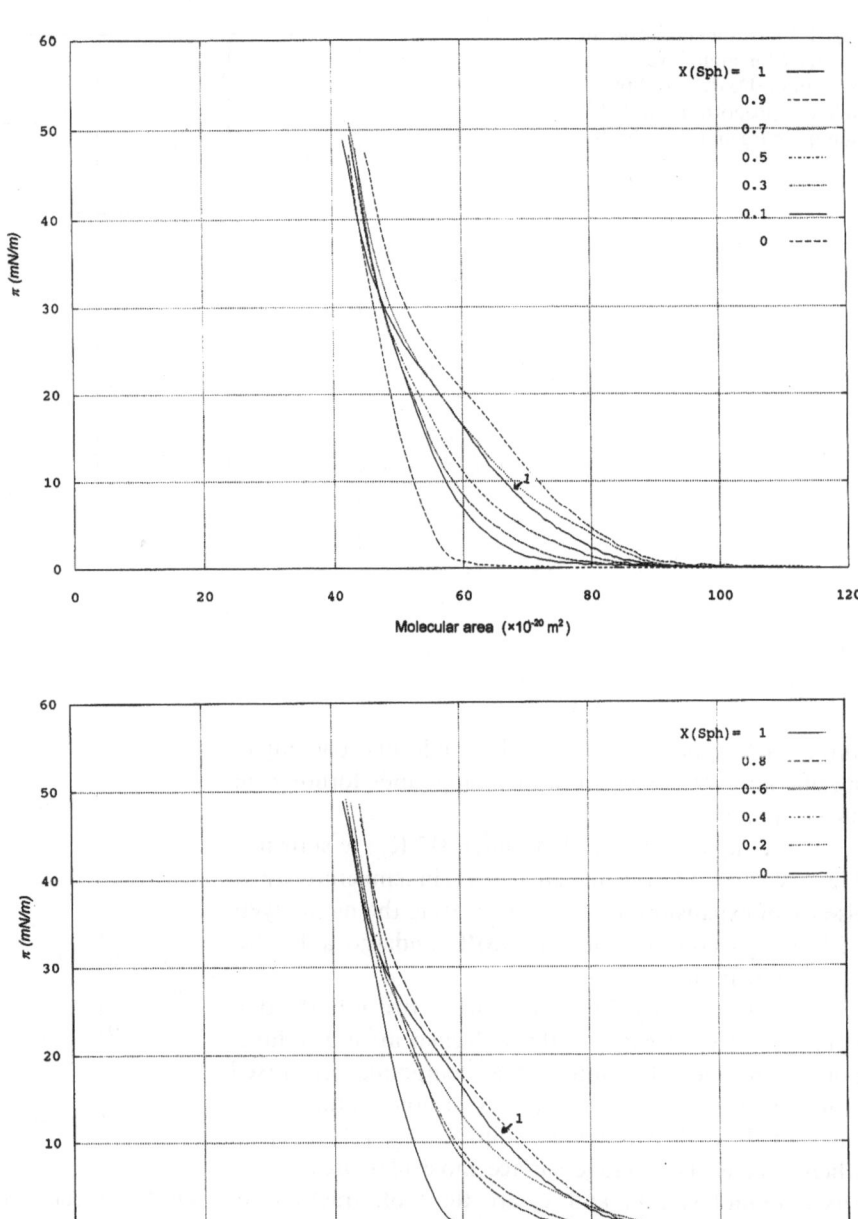

The different behavior of these compounds is due to the differences in the molecular structure. Both of the molecules have the same polar group, but in the case of the Sph molecules, the hydrocarbon chains are shorter and asymmetric, and the presence of a double bound provokes a structure more rigid than that corresponding to the DSPC molecules, which has two chains with 18 carbons and presents less steric effects in his bound with the polar group.

It can be observed that mixed monolayers of these compounds show a somewhat intermediate behavior compared with that shown by pure components, except for the mixtures $x_{Sph} = 0.7$, 0.8 and 0.9, which are more extended that the Sph pure monolayer.

The characteristic of the isotherms is a progressive condensation as the condensed structure of DSPC monolayers is approached. This is known as "condensing effect" of DSPC of Sph, and has been interpreted by a number of

Fig. 4 Experimental isotherms obtained for monolayers containing DSPC and Sph in different proportions at 313 K. Sph mole fraction is indicated

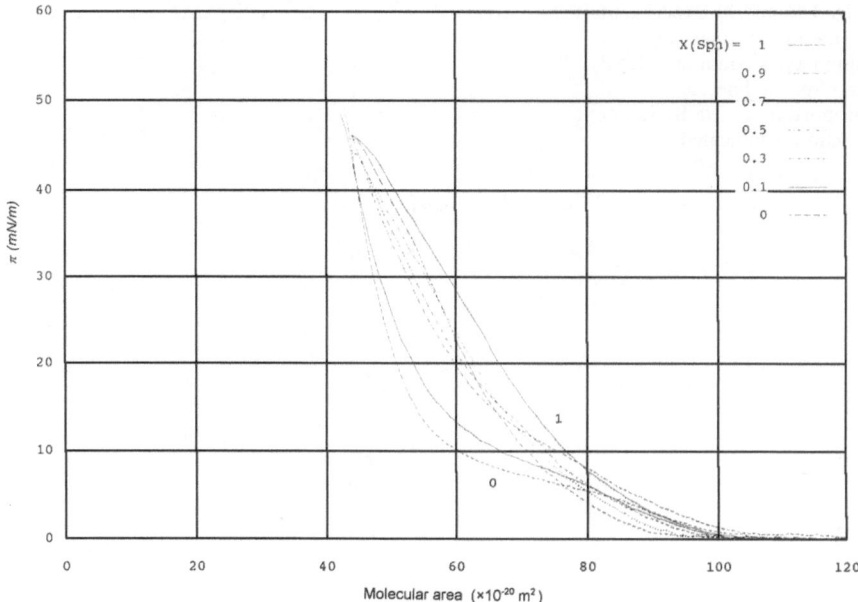

authors [7–9] on the basis of either molecular packing or specific interactions between the compounds forming the mixed monolayer.

The similar isotherms, but under 313 K, are shown in Fig. 4. We can observe that an increase in temperature give rise to an expansion in mixed films; even, the monolayers with a high concentration of DSPC undergo a LE-LC phase transition.

In order to study the interactions between both compounds, firstly, we examine the molecular areas as a function of the mole fraction of one compound for mixed monolayers (miscibility curves) at various surface-pressures for the additivity relationship: $A_{12} = x_1 A_1 + x_2 A_2$, where A_{12} is the average surface molecular area in the mixed monolayer, x_1 and x_2 are the mole fractions of components 1 and 2 in the monolayer and finally, A_1 and A_2 stand for the molecular areas of pure comonents (in simple monolayers) at the same surface pressure used to determine A_{12}, following a similar procedure to one chosen by different workers [1, 7, 10–12].

The miscibility curves for our system are shown in Fig. 5, where the surface-pressure values 5, 15, 25, and 40 mN m^{-1} have been choosen, because the two first are around the LE-LC phase transition surface-pressure values, the last is close to the collapse surface-pressure, and the other has an intermediate value.

When temperature is 298 K, a positive deviation from the linearity predicted by the previous relation is found. This means that the dominant interaction component is lightly repulsive, showing the maximum interaction at high concentrations of Sph. As was to be expected, this fact

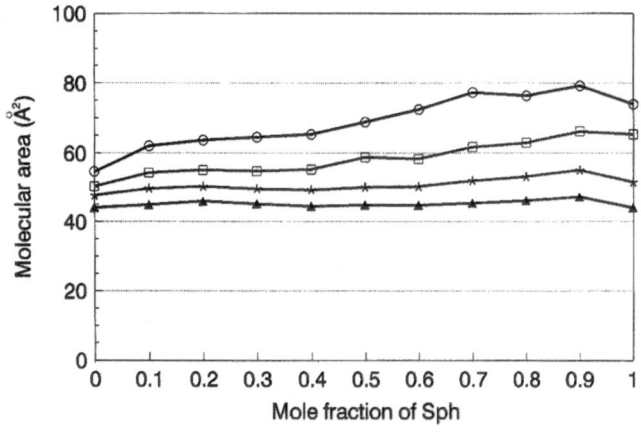

Fig. 5 Molecular area under constant surface pressure as a function of the Sph mole fraction in mixed DSPC-Sph monolayers: $\pi = 5$ mNm^{-1} (o), 15 mNm^{-1} (□), 25 mNm^{-1} (∗), 40 mNm^{-1} (▲)

disappears when surface-pressure increases, because at higher compression state, the repulsive interactions between polar groups are less important and are compensated by those of attractive character.

As temperature increases, the maximum of repulsive interaction is shifted towards higher concentrations of DSPC. It should be kept in mind that the presence of a double bound in the Sph molecules provokes a structure more rigid and then less susceptible to change by thermal effect with respect to the DSPC molecules.

In this work, a thermodynamic study of the interactions previously described has been introduced. Following the formulation due to Goodrich [4], the excess free energy, entropy, and enthalpy of mixing (ΔG^E, ΔS^E, ΔH^E) have been calculated for our system, at different conditions of temperature and for various surface-pressure values. Part of these results are shown in the next figures.

In Fig. 6, the excess Gibbs potential of mixing as a function of the mole fraction of DSPC, at 298 K, is represented. The behavior is analogous to that observed from the miscibility curves. In fact, the most outstanding results are the presence of a maximum of repulsive interaction at low concentrations of DSPC, when the temperature is 298 K, and also, the displacement of this maximum towards higher proportions of DSPC in the mixed films as temperature increases.

Moreover, we observe that the ΔG^E values increase with the surface-pressure, due to the diminution of the distance between molecules, increasing the interactions.

This fact can be seen more clearly in Table 1, which shows the interaction parameter values, at 298 K and various surface-pressures, and for different mixed monolayers forming by DSPC and Sph.

In Fig. 7 the ΔS^E values as a function of the mole fraction of DSPC, obtained at 298 K, are represented. The ΔH^E values present the same trend as the ΔS^E values and

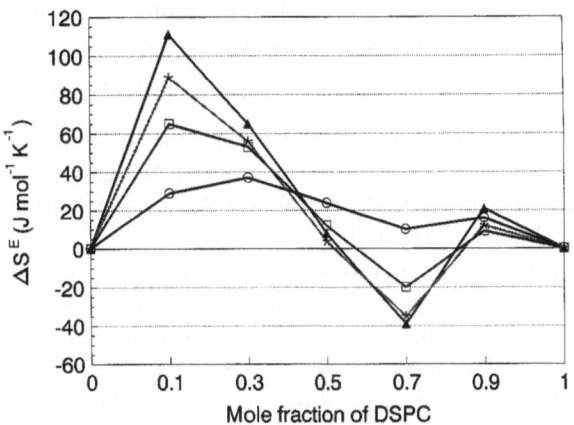

Fig. 7 Excess entropy of mixing under constant surface pressure as a function of the DSPC-Sph monolayers: $\pi = 5\,\text{mNm}^{-1}$ (o), $15\,\text{mMm}^{-1}$ (□), $25\,\text{mNm}^{-1}$ (∗), $40\,\text{mNm}^{-1}$ (▲)

for this reason are not shown. The contribution of the enthalpy to the excess free energy of mixing inform us about the stability of the mixed films with respect to those of pure components in relation with the molecular interactions, and from the dependence of the ΔS^E with the composition of the monolayer, we can say that high proportions of Sph disarrange the system; the opposite occurs with DSPC molecules. Again, the rigidity of the Sph molecules prevents the presence of interactions and reduces the possible configurations of the system.

Conclusions

From the study carried out on the interfacial behavior of monomolecular films forming by a lipid, DSPC or Sph, and by mixtures of both, we point out the following results:

- The films forming by Sph suffer an LE-LC phase transition when they are compressed in the temperature range between 288 and 298 K.
- The monolayers constituted by DSPC molecules are in the LC state under the temperature range 298–308 K, and suffer the LE-LC phase transition when the compression is carried out at 313 K.
- Both type of monolayers show an expansion by the effect of temperature.
- For both monolayers, the same collapse surface-pressure values have been observed, and then, under the studied experimental conditions, these compounds show the same stability at the air-water interface.
- From the miscibility analysis in the DSPC-Sph system, positive deviations from the linearity for the molecular areas have been observed in all the studied temperature range. At low temperatures the maximum of interaction

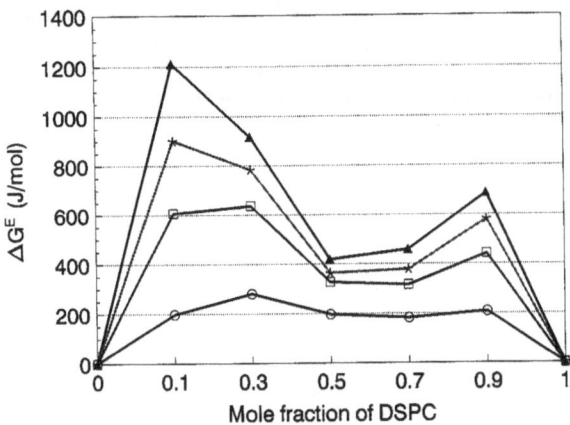

Fig. 6 Excess free energy of mixing under constant surface pressure as a function of the DSPC mole fraction in mixed DSPC-Sph monolayers: $\pi = 5\,\text{mNm}^{-1}$ (o), $15\,\text{mNm}^{-1}$ (□), $25\,\text{mNm}^{-1}$ (∗), $40\,\text{mNm}^{-1}$ (▲)

Table 1 Interaction parameter values for different DSPC-Sph mixed monolayers, at several surface-pressures.

π (mN/m)	$x_{\text{DSPC}} = 0.1$	0.3	0.5	0.7	0.9
5	0.9	0.5	0.3	0.4	0.9
15	2.7	1.2	0.5	0.6	2.0
25	4.0	1.5	0.6	0.7	2.6
40	5.4	1.8	0.7	0.9	3.1

is when the proportion of Sph in the monolayer is high. This result changes as temperature increases.

From these main results, we can conclude the following:

1) When the amphiphilic molecules contain symmetric hydrocarbon chains and a small polar group, the attractive interactions, of Van der Waals type, between the chains are favored.
2) The effect of temperature on the monolayers, forming by these lipids, provokes an increase of the mobility of the hydrocarbon chains, and therefore, a diminution of the Van der Waals interactions between them.
3) In the monolayers constituted by DSPC or Sph, i.e., by molecules with the phosphatidylcholine polar group, the presence of hydration forces is left out.
4) The lipid with high molecular asymmetry suffer a higher molecular reorganization when compressed

as monolayers, for this reason, the LE-LC phase transition is observed, mainly for monolayers forming by Sph.
5) Contrary to those previously observed in other studies with lipids containing short chains (< 14 C), when the hydrocarbon chains are larger the stability of the monolayers is independent of the polar group nature.
6) In the mixed monolayers formed by DSPC-Sph, the electrostatic repulsive interactions between polar groups dominate on those with attractive character between the hydrocarbon chains, due to the steric effects introduced by the Sph molecules.

Acknowledgement This work has been financially supported by the Comisión Interministerial de Ciencia y Tecnología (CICYT), project no. 94-0560.

References

1. Gálvez-Ruiz MJ, Cabrerizo-Vílchez MA (1991) Colloid Polym Sci 269:77–84
2. Gálvez-Ruiz MJ, Cabrerizo-Vílchez MA, Galisteo-González F, Hidalgo-Álvarez R (1991) Progr Colloid Polym Sci 84:494–501
3. Gálvez-Ruiz MJ, Cabrerizo-Vílchez MA (1992) In: Stroeve P, Balazs AC (eds) Macromolecular Assemblies in Polymeric Systems, ACS Symposium Series No. 493, USA, Chapter 13, pp 135–152
4. Goodrich FC (1957) In: Proc. 2nd Internat Congr Surface Activity 1:33–39
5. Queralto-Moreno A, Castro-Ruiz RM, Otero-Aenlle E (1980) An Quim 76:58
6. Bois AG (1985) J Colloid Interf Sci 105:24
7. Chifu E, Zsakó J, Tomoaia-Cotisel M (1983) J Colloid Interf Sci 95(2):346
8. Müller-Landau F, Cadenhead DA (1979) Chem Phys Lipids 5:315
9. Smaby JM, Baumann WJ, Brockman ML (1979) J Lipid Res 20:789
10. Kuramoto N, Sekito K, Motomura K, Nakamura M, Matuura R (1972) Mem Fac Sci Kysushu Univ C8(1):67–87
11. Costin IS, Barnes GT (1975) J Colloid Interf Sci 51(1):106–121
12. Puggelli M, Gabrielli G (1977) J Colloid Interf Sci 61(3):420–427

Progr Colloid Polym Sci (1995) 98:255–262
© Steinkopff Verlag 1995

G. Brezesinski
A. Dietrich
B. Dobner
H. Möhwald

Morphology and structures in double-, triple- and quadruple-chain phospholipid monolayers at the air/water interface

Dr. G. Brezesinski (✉)
A. Dietrich · H. Möhwald
Universität Mainz
Institut für Physikalische Chemie
Welder-Weg 11
55099 Mainz, FRG

B. Dobner
Universität Halle
Institut für Pharmazeutische Chemie
Weinbergweg 15
06120 Halle/S., FRG

H. Möhwald
Max-Planck-Institut für Kolloid-
und Grenzflächenforschung
Rudower Chaussee 5
12489 Berlin, FRG

Abstract The structure of double-, triple- and quadruple-chain phospholipid monolayers has been studied by Synchrotron x-ray diffraction. The double-chain mixed-linkage species exhibit an oblique structure at all pressures investigated. The triple-chain phospholipids show at lower lateral pressures a rectangular unit cell with a phase transition at higher pressures to a hexagonal packing of vertically arranged chains. The quadruple-chain lipid exhibits only the hexagonal phase structure. The position of the ether linkage and of the branched chain on the glycerol backbone has also a strong influence on the monolayer structures. Fluorescence microscopy shows different domain shapes for the different molecules investigated. The anisotropy of the domain shapes seems to be connected with the anisotropy of the structure. Obviously, the compound with the larger lattice anisotropy (larger chain tilt) exhibits dendritic domains.

Key words Branched-chain phospholipids – monolayers – fluorescence microscopy – grazing incidence x-ray diffraction

Introduction

Phospholipid monolayers at the air/water interface have revealed a number of novel physical properties using fluorescence microscopy as well as diffraction techniques in the last few years [1–3]. A variety of different phases has been observed. Visualizing phases in the coexistence region between gas-analoguous and liquid-expanded as well as between liquid-expanded and condensed states of phospholipids with fluorescence microscopy technique has led to detailed theoretical studies on the shapes and sizes of domain structures [4–6].

The domain shapes and sizes can be interpreted phenomenologically by a competition between line tension at the domain boundary and dipole-dipole electrostatic repulsion between molecules within and between domains [7, 8]. Line tension acts to minimize the length of the interface between different phases and thus results compact domain shapes while the electrostatic repulsions tend to maximize distances between molecules, favoring more elongated, noncircular domain shapes.

The chemical replacement of acyl groups by alkyl groups in phosphatidylcholine monolayers was described to influence the monolayer behavior of these molecules since the carbonyl dipoles determine the surface potential of these monolayers [9].

The present work concentrates on the investigation of domain shapes observed with fluorescence microscopy and the corresponding molecular structure detected by Synchrotron x-ray diffraction of double-, triple- and quadruple-chain phosphatidylcholines with ester- and

256

G. Brezesinski et al.
Morphology and Structure in Phospholipid monolayers

ether-linkages. Varying the position and the length of hydrophobic side branches, thus changing the relative size of hydrophilic and hydrophobic moieties will effect molecular interaction to a large extent. In case of triple- and quadruple-chain phospholipids two-dimensional ordering of the aliphatic chains is not influenced by any head group repulsion as is the case for the non-branched double chain system and, thus, the lateral density is enlarged, leading to a hexagonal molecular packing.

Materials and methods

The racemic double-chain phospholipids 1-palmitoyl-2-hexadecylphosphatidylcholine (1P-2H-PC) and 1-hexa-decyl-2-palmitoylphosphatidylcholine (1H-2P-PC), the racemic triple-chain lipids 1-hexadecyl-2-(2-tetradecyl-palmitoyl)-glycero-3-phosphocholine $(2-(2C_{14}-16:0)-1H-PC)$ and 1-(2-tetradecyl-palmitoyl)-2-hexadecyl-glycero-3-phosphocholine $(1-(2C_{14}-16:0))-2H-PC$ and the quadruple-chain phospholipid 1,2-di(2-tetradecyl-palmitoyl)-sn-glycero-3-phosphocholine $L-(1,2-Di(2C_{14}-16:0)-PC)$ were kindly synthesized by colleagues in the group of Prof. P. Nuhn in the Institute of Pharmaceutical Chemistry, Martin-Luther-Universität Halle, FRG. The chemical structures are presented in Fig. 1. The lipids were purified and characterized chromatographically. All lipids were spread from an approx. 1 mM p.a. grade chloroform (Merck, FRG) solution. The subphase was ultrapure water with specific resistance above 18 $M\Omega$/cm purified using a Millipore desktop unit. The pressure-area isotherms were recorded using a film balance with continuous Wilhelmy-type pressure measuring system.

The fluorescence microscopy measurements at the air/water interface were performed under argon atmosphere with a film balance described elsewhere [2] using the fluorescent dye probe L-1-hexadecanoyl-2-[N-(nitrobenz-2-oxa-1,3-diazol-4-yl)aminohexanoyl]-sn-glycero-3-phosphocholine (β-NBD-PPC) without further purification, purchased from Sigma.

The phospholipid monolayers doped with a small mole fraction of the fluorescent dye probe (1-2 mol-%) were spread onto the water surface and compressed isothermally to the liquid-expanded/condensed coexistence regime. With increasing the pressure above Π_c, nucleation occurs and solid domains are observed growing on slow compression. Since it is known that the fluorescent dye probe partitions into the liquid-expanded (LE) phase, the domains appear as dark patches surrounded by a bright fluid phase. The solid domains are solid in the sense that lateral diffusion within these domains is low, characteristic of the gel or crystalline phases of phospholipid bilayers [10].

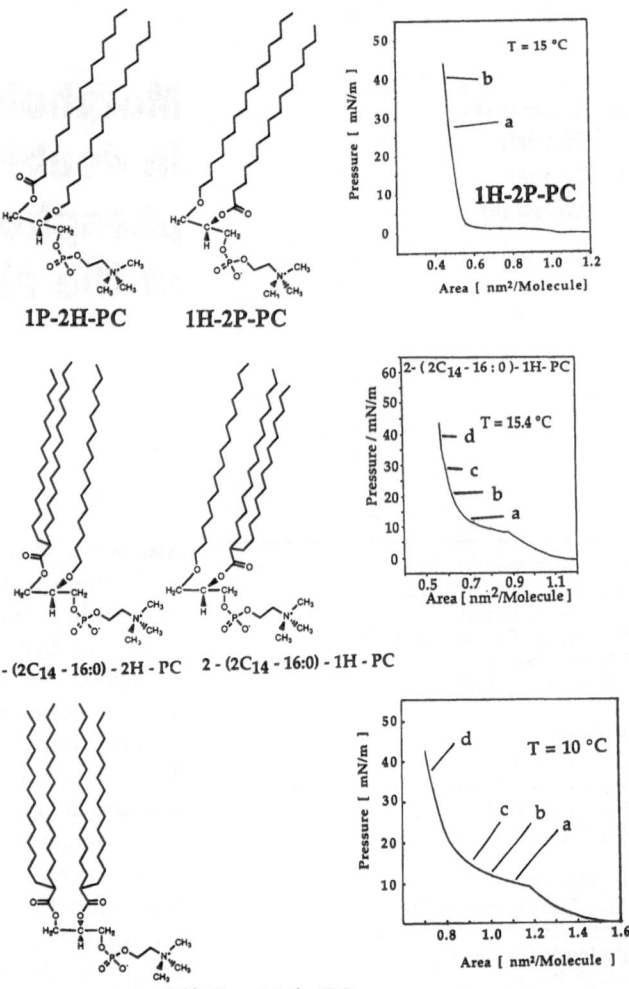

Fig. 1 Chemical structures of 1-palmitoyl-2-hexadecylphosphatidylcholine (1P-2H-PC), 1-hexadecyl-2-palmitoylphosphatidylcholine (1H-2P-PC), 1-hexadecyl-2-(2-tetradecyl-palmitoyl)-glycero-3-phosphocholine $(2-(2C_{14}-16:0)-1H-PC)$, 1-(2-tetradecyl-palmitoyl)-2-hexa-decyl-glycero-3-phosphocholine $(1-(2C_{14}-16:0)-2H-PC)$ and 1,2-di(2-tetradecyl-palmitoyl)-sn-glycero-3-phosphocholine $(1,2-Di(2C_{14}-16:0)-PC$ with typical pressure-area isotherms at 15°C and 10°C. The points indicate the positions at which diffraction data have been taken

Grazing incidence x-ray diffraction (GID) experiments were performed at 15 °C and 10 °C using the liquid surface diffractometer on the Synchrotron beam line BW1 at HASYLAB, DESY, Hamburg, Germany. The x-ray beam was made monochromatic by a Beryllium (002) crystal, and the angle of incidence was 0.85 α_c, where α_c is the critical angle for total external reflection. Detection was made by a linear position sensitive detector (PSD) mounted vertically behind a Soller collimator (consisting of many parallel vertical plates).

GID patterns from Langmuir monolayers arise from two-dimensional (2D) array of rods, called Bragg rods, which extend parallel to Q_z [11]. The scattered intensity measured by scanning different values of Q_{xy} and integrating over Q_z-intervals of the PSD yields Bragg peaks. The analysis of the in-plane diffraction data yields the lattice spacings according to $d_{hk} = 2\Pi/Q_{hk}$, where Q_{hk} is the horizontal component of the scattering vector at maximum intensity. In the general case of an oblique 2D crystal structure three reflection peaks of lowest order can be observed [11]. The lattice parameters a,b, angle γ between the a and b vectors can be calculated from the lattice spacings d_{hk} and from these the unit cell area A_{xy}. Simultaneously, the scattered intensity recorded in channels along the PSD, but integrated over Q_{xy}-intervals, produces Bragg rod profiles. These intensity profiles give information on the tilt angle t between the molecular axis and

the surface normal, and on the molecular orientation in the 2D crystals according to the general equation $\tan(t)\cos(\Psi_{hk}^*) = Q_z^{hk}/Q_{xy}^{hk}$, where Ψ_{hk}^* is the azimuthal angle between the tilt direction projected on the xy plane and the reciprocal lattice vector G_{hk}.

Results

All lipids investigated form stable monolayers on water. The pressure-area isotherms at 15 °C exhibit a two-phase region between a liquid-expanded and a condensed phase with a transition pressure Π_c. Figure 1 shows typical isotherms together with the structural formula of the compounds investigated. Π_c increases with increasing number

Fig. 2 Fluorescence micrographs of the double-chain phosphatidylcholines containing 2 mol% of the fluorescence dye. The lateral pressure is increased from a–e in the two-phase region between the liquid-expanded and condensed states

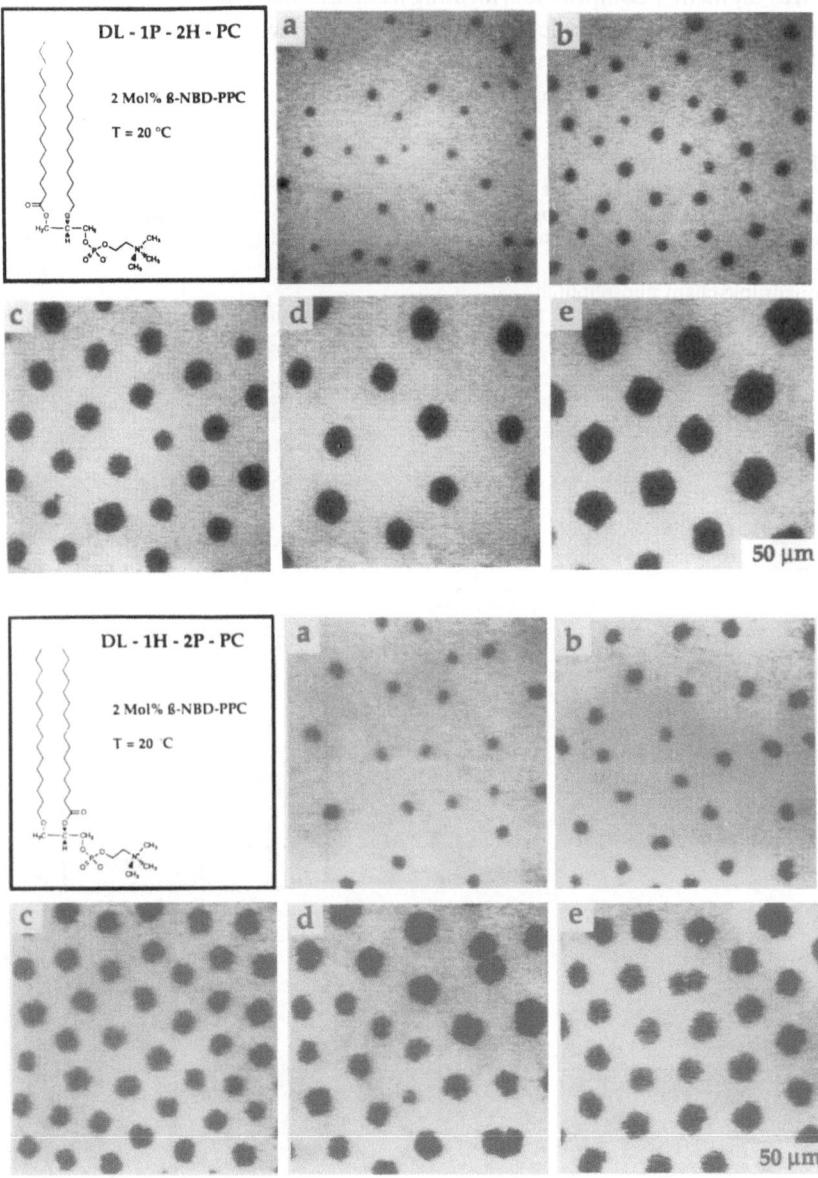

of chains. For the double-chain lipids a transition pressure between 2 and 3 mN/m was observed, while the triple-chain lipids exhibit a Π_c value between 7 and 8 mN/m, and the main transition of the quadruple-chain lipid occurs at 15 mN/m (for 15 °C measurements).

Figures 2 and 3 give the typical fluorescence micrographs for the investigated phospholipid systems at the air/water interface for various degrees of monolayer compression. For all monolayers compression was discontinuous and occurred rather slowly in order to ensure that the number of domains remains constant and that domain shapes do not change within times of measurement.

Increasing the pressure along the liquid-expanded/condensed phase coexistence region nucleation and growth of small solid domains is observed. Under the same growth conditions these domains increase uniformly in sizes on further compression, forming a superlattice due to interdomain electrostatic repulsions [12, 13]. Hence, an equilibrium size may be established from a competition between boundary energy, tending to decrease the domain size, and electrostatic energy, which tends to increase it.

Comparing the morphologies for the double chain lipids in Fig. 2, the domains seem to be rather similar and exhibit noncircular shapes with sharp edges; no chiral structure could be observed. Increasing the hydrophobic moiety with a third chain per molecule (Fig. 3, top), the monolayer of 1-(2C$_{14}$-16:0)-2H-PC (a) and 2-(2C$_{14}$-16:0)-1H-PC (b) clearly show different domain shapes depending on the position of the side branch at the C1 or C2 atom of the glycerol backbone [12]. In case of the quadruple-chain lipid in Fig. 3 (bottom) the domains exhibit smooth domain boundaries and undulating shapes.

Figure 4 presents in-plane diffraction scans for the double-chain lipid 1H-2P-PC and 1P-2H-PC with both an ester and an ether linkage at two different lateral pressures. Like the pure ester (DPPC) and the pure ether (DHPC)

Fig. 3 Fluorescence micrographs of the two triple-chain phospholipids triple-(*top*) a: 1-(2C$_{14}$-16:0)-2H-PC, b: 2-(2C$_{14}$-16:0)-1H-PC, and the quadruple-chain phospholipid (*bottom*)

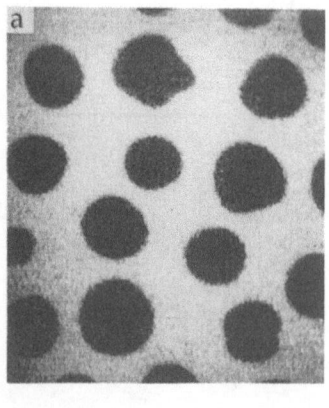
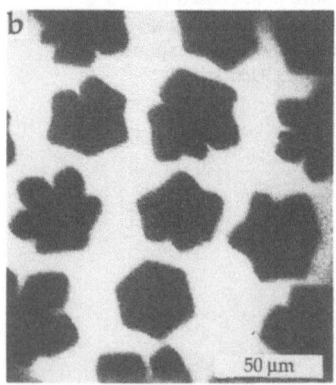

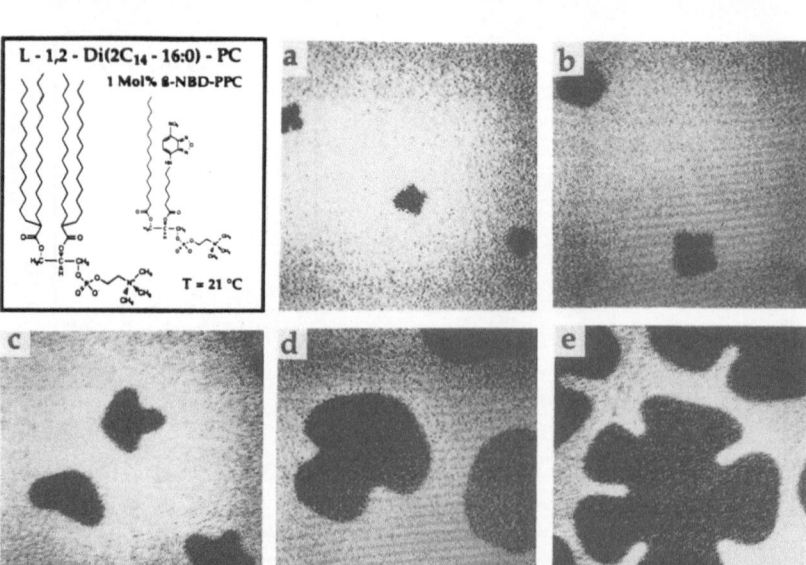

Progr Colloid Polym Sci (1995) 98:255–262
© Steinkopff Verlag 1995

Fig. 4 Q_z resolved in-plane diffraction scans for 1H-2P-PC (*left*) and 1P-2H-PC (*center*) at different surface pressures. Diffraction intensity as a function of out-of-plane scattering vector component Q_z for 1P-2H-PC integrated over small Q_{xy} intervals for a lateral pressure of 41 mN/m (*right*)

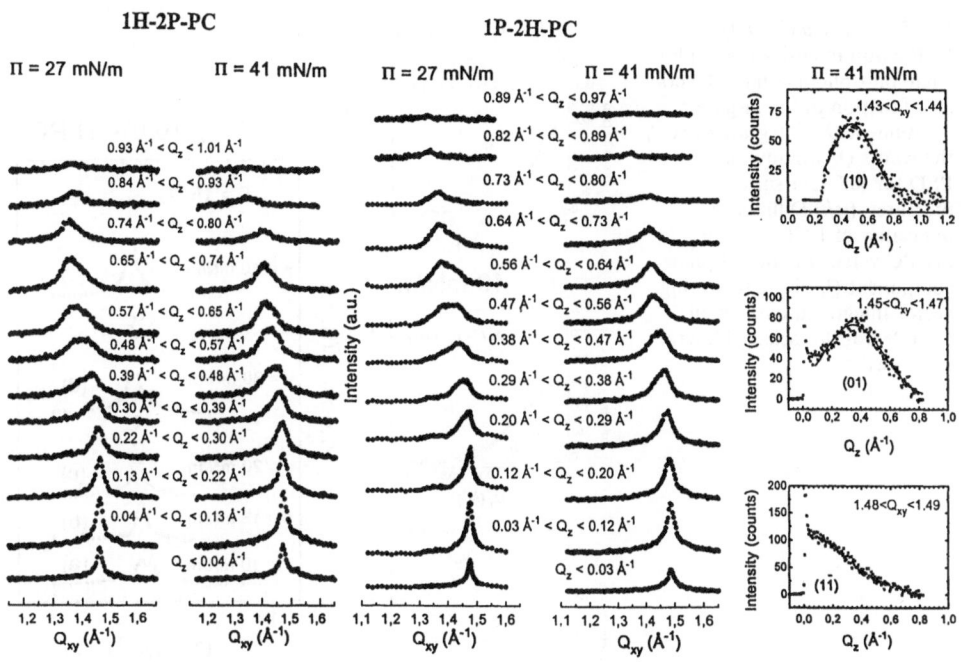

Table 1 Unit cell dimensions a, b, γ, projected area per chain A_{xy}, tilt angle t and chain cross-section A_0 for the double-, triple- and quadruple-chain lipids at 15 °C.

Compound	Π [mN/m]		a [Å]	b [Å]	γ [°]	A_{xy} [Å^2]	t [°]	A_0 [Å^2]
1H-2P-PC	27	DL	4.92	5.23	63.1	22.9	29	20.1
	41	DL	4.92	5.06	62.2	22.0	23	20.3
1P-2H-PC	27	DL	4.89	5.14	63.3	22.4	26	20.2
	41	DL	4.89	4.97	61.9	21.5	19	20.2
2-(2C$_{14}$-16:0)-1H-PC	13	DL	4.93	5.03	60.7	21.6	14	21.0
	20	DL	4.91	4.98	60.5	21.3	11	20.9
	30	DL	4.91	4.92	60.1	20.9	6	20.8
	40	DL	4.88	4.88	60.0	20.6	0	20.6
1-(2C$_{14}$-16:0)-2H-PC	10	DL	4.90	4.95	60.3	21.1	7	20.9
	13	DL	4.91	4.92	60.0	20.9	4	20.9
	20	DL	4.90	4.90	60.0	20.8	0	20.8
	30	DL	4.88	4.88	60.0	20.6	0	20.6
	40	DL	4.88	4.88	60.0	20.6	0	20.6
1,2-Di(2C$_{14}$-16:0)-PC	20	L	4.92	4.92	60.0	21.0	0	21.0
	33	L	4.91	4.91	60.0	20.9	0	20.9

phospholipids [14], both mixed-linkage species show three distinct peaks although two are barely resolved. The existence of three distinct diffraction peaks is clearly visible via the rod scans (right side of Fig. 4). The Miller indices (hk) are indicated for each peak. The intensity of the $(1\bar{1})$ Bragg rod is cut off at the horizon ($Q_z = 0$ Å^{-1}) by the liquid subphase. The unit cell parameters are summarized in Table 1.

Figure 5 compares the diffraction intensities as a function of the in-plane component Q_{xy} of the scattering vector obtained on compression of Langmuir monolayers of the

two triple-chain lipids. For both compounds, one observes two peaks at lower surface pressures and only one peak at higher pressures. Figure 5 shows also selected diffraction intensities along the plane normal z (Bragg rod) integrated over small Q_{xy} intervals of 1-(2C$_{14}$-16:0)-2H-PC. The two peaks at lower pressures have their maximum in Q_z above the horizon and $Q_z^{(11)} = (1/2)Q_z^{(02)}$. At higher pressures the Bragg rod has its maximum intensity at the horizon.

For the quadruple-chain lipid only one sharp diffraction peak is apparent at all pressures investigated above Π_c (Fig. 6). As a function of Q_z the maximum intensity

260
G. Brezesinski et al.
Morphology and Structure in Phospholipid monolayers

Fig. 5 (*Center and right*) Diffraction intensities of triple-chain phospholipid monolayers as a function of the in-plane component Q_{xy} integrated over the whole Q_z-window of the PSD for different surface pressures. (*left*) Diffraction intensities of 1-(2C$_{14}$-16:0)-2H-PC versus the out-of-plane component Q_z of the scattering vector integrated over small Q_{xy} intervals for selected lateral pressures

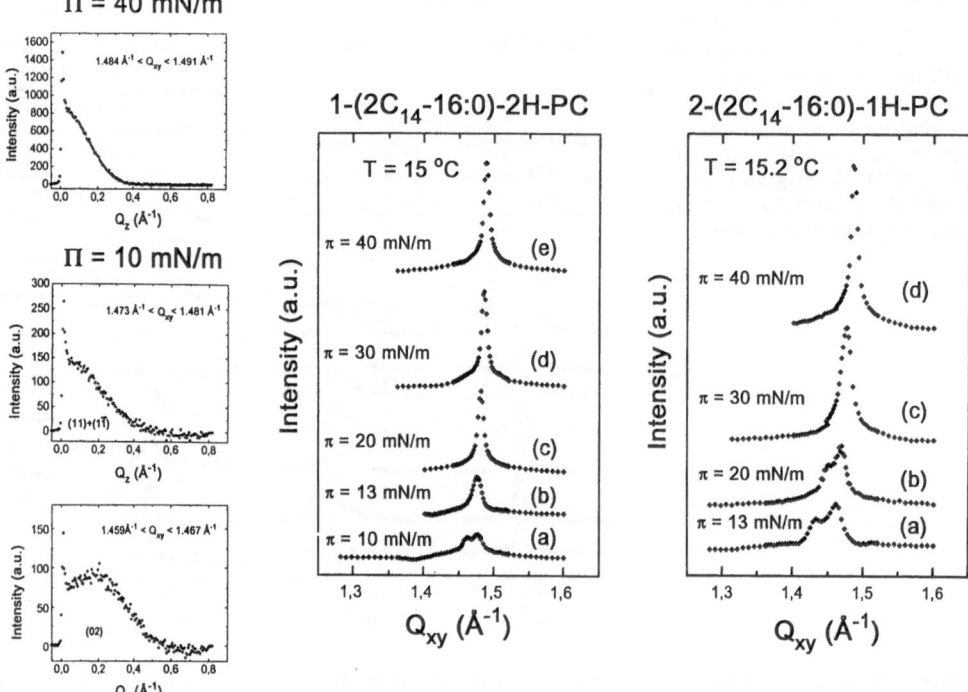

Fig. 6 Diffraction intensities of the quadruple-chain phospholipid monolayer as a function of the in-plane component Q_{xy} and the out-of-plane component Q_z for different surface pressures at 10 °C

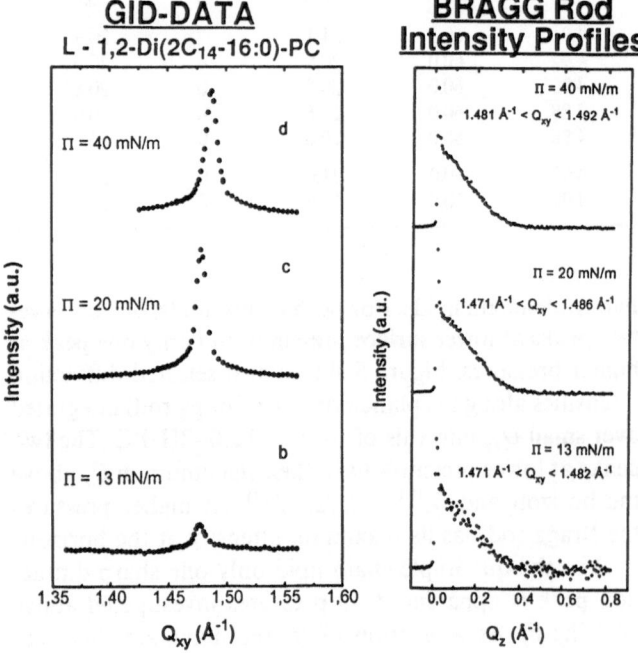

occurs in the plane of the water surface ($Q_z = 0$ Å^{-1}) and the intensity extends to $Q_z \sim 0.3$ Å^{-1} while the Q_{xy} peak position remains constant.

Discussion

Each molecule in a closely packed monolayer is surrounded by six nearest neighbors, forming a hexagon which can be distorted. A lattice of this type gives rise to three diffraction peaks of lowest order [11]. Therefore, the mixed-linkage double-chain lipids form an oblique lattice similar to that of the pure ester (DPPC) and pure ether (DHPC) lipids [14]. At a lateral pressure of 27 mN/m the construction of a distorted centered rectangular unit cell containing one molecule yields an angle of 93.5° indicating that the distortion from a rectangular unit cell is remarkable. While increasing the pressure the angle approaches 90°, but even at very high lateral pressures one observes an oblique unit cell. The tilt direction is always near to a next-neighbor (NN) tilt. Comparing the tilt angles t at the same surface pressures, one realizes that the lipid with the carbonyl group at the C1 chain exhibits the smaller t value (19° at 41 mN/m). Therefore, one might assume that the proposed nonequivalence of the two chains in lipid dispersions and crystals (the C2 chain is bend at the second carbon atom) [15, 16] is also valid for monolayer phases on water. The interaction of the carbonyl group at the C2 chain with the water leads to a force pulling into the subphase, and hence 1H-2P-PC exhibits the larger tilt angle (23° at 41 mN/m). Since the differences in tilt angle and cross-section area A_0 between the double-chain molecules at low pressures (where the fluorescence micrographs were taken) are not significant, thus the domain shapes are also not expected to be different (Fig. 2). The large tilt angle of the molecules seems to be responsible for the anisotropic shape of the domains.

The high-pressure phase of the triple-chain lipids shows only one diffraction peak indicating a hexagonal packing of the chains. The maximum intensity in Q_z occurs in the plane of the water surface ($Q_z = 0$ Å^{-1}) and therefore a vertical chain orientation can be deduced. The occurrence of two peaks at lower pressures indicates a (centered) rectangular unit cell. The observed intensity distribution along Q_z can be explained by a tilt of the chains towards their next-nearest neighbors (NNN) [17]. In the case of 1-(2C$_{14}$-16:0)-2H-PC the phase transition from the rectangular to the hexagonal packing occurs at a surface pressure between 13 mN/m and 20 mN/m. The low tilt angle just above the transition pressure Π_c (7° at 10 mN/m) indicates that the condensed phase is only slightly distorted from hexagonal packing. For this monolayer circular domain shapes occur. In contrast, the transition to the hexagonal phase of 2-(2C$_{14}$-16:0)-1H-PC could be observed only between 30 mN/m and 40 mN/m. At the lowest pressure investigated (13 mN/m) a tilt angle t of 14° was found, and at 30 mN/m t amounts still to 6°. These quantitative differences of the molecular structure may change domain morphology from circular to dendritic shapes. To understand the structural differences between the two isomers, one has to take into account also the possible conformational nonequivalence of the C1 and C2 chains [14], and the different chain linkage (ester or ether). Comparing the tilt angles with those of the mixed-linkage double-chain lipids, one notices that the triple-chain 1-(2C$_{14}$-16:0)-2H-PC with smaller tilt angle and lower transition pressure to the hexagonal lattice exhibits the ether linkage at the C2 position, too. Obviously, the compound with the larger lattice anisotropy (larger chain tilt) exhibits the dendritic domains.

The tilt angles in the rectangular phase just above the transition pressure were distinctly lower than those in the oblique phase, even at the highest lateral pressures.

The quadruple-chain lipid shows only one diffraction peak indicating a hexagonal packing. To our knowledge this is the first example of phospholipids where the transition from a liquid-expanded state leads directly to a hexagonal packing with vertical arrangement of the chains. According to the discussion about the connection between tilt angle and anisotropic domain shapes one can understand that the domain shapes of quadruple-chain lipids do not show any sharp edges or dendritic shapes.

The cross-section area A_0 of the mixed-linkage double-chain lipids amounts to (20.2 ± 0.2) Å^2, pointing to a free rotator phase [18]. The mismatch between two aliphatic tails linked via the glycerol backbone to a large head and this hydrated headgroup obviously prevents crystalline ordering of the tails as observed for single chain amphiphiles [18]. The triple- and quadruple-chain lipids exhibit cross-sections of (20.8 ± 0.2) Å^2. In this case the area per molecule is dominated by the ordering of the aliphatic chains which is not hindered by the headgroup (43–45 Å^2) as for the double-chain lipids. The covalent bonding of the side branches to the second carbon atom of the main chains is responsible for the larger area per chain compared to a lattice of single chain amphiphiles.

Acknowledgements This work was supported by the Bundesministerium für Forschung und Technologie (BMFT) and the Deutsche Forschungsgemeinschaft (DFG). We thank HASYLAB for providing excellent facilities and support. The collaboration with K. Kjaer and W.G. Bouwman is gratefully acknowledged.

References

1. von Tscharner V, McConnell HM (1981) Biophys J 36:409–419
2. Lösche M, Sackmann, E, Möhwald H (1983) Ber Bunsenges Phys Chem 87:848–852
3. Als-Nielsen J, Kjaer K (1989) In: Riste T, Sherrington D (eds) Phase Transitions in Soft Condensed Matter. NATO Series B, vol 211, Plenum Press, New York, pp 113–137
4. Vanderlick TK, Möhwald H (1990) J Phys Chem 94:886–890
5. Lee KYC, McConnell HM (1993) J Phys Chem 97:9532–9539
6. Muller P, Gallet F (1991) J Phys Chem 95:3257–3262
7. Helm CA, Möhwald H (1988) J Phys Chem 92:1262–1266
8. McConnell HM (1991) Annu Rev Phys Chem 42:171–195
9. Paltauf F, Hauser H, Phillips MC (1971) Biochim Biophys Acta 249:539–547
10. McConnell HM, Tamm LK, Weis RM (1984) Proc Natl Acad Sci USA 81:3249–3253
11. Jaquemain D, Wolf SG, Leveiller F, Deutsch M, Kjaer K, Als-Nielsen J, Lahav M, Leiserowitz L (1992) Angew Chem 104:134–158
12. Dietrich A, Möhwald H, Rettig W, Brezesinski G (1991) Langmuir 7:539–546
13. Möhwald H (1990) Annu Rev Phys Chem 41:441–476

14. Brezesinski G, Dietrich A, Struth B, Böhm C, Kjaer K, Möhwald H (1995) Chem Phys Lipids (in press)

15. Gaber BP, Yager P, Peticolas WL (1978) Biophys J 24:677–688

16. Pascher I, Sundell S, Harlos K, Ebil H (1987) Biochim Biophys Acta 896:77–88

17. Kjaer K (1994) Physica B 198:100–109

18. Kenn RM, Böhm C, Bibo AM, Peterson IR, Möhwald H, Kjaer K, Als-Nielsen J (1991) J Phys Chem 95:2092–2097

Progr Colloid Polym Sci (1995) 98:263–265
© Steinkopff Verlag 1995

L. Pauchard
J. Meunier

Experimental study of the breakage of a two-dimensional crystal

L. Pauchard · Prof. J. Meunier (✉)
Laboratoire de Physique Statistique
de l'ENS
URA 1306 du CNRS
associé aux Universités Paris VI et VII
24, rue Lhomond
75231 Paris Cedex 05, France

Abstract We study monolayers of an insoluble amphiphilic molecule at the free surface of water. During the compression of such a monolayer, we observe, by fluorescence microscopy, a phase transition between a two-dimensional liquid phase and a two-dimensional solid phase consisting of long rods (1 mm long and about 20 μm wide). Their shape is well adapted to micromechanical measurements. Studying the flexion of a rod in the plane of the water, the two-dimensional Young's modulus was measured. On the other hand, we have noticed that a domain (a monocrystal) under a bending stress, breaks after a delay time. Statistics show that this time-lag breakage is a well-defined function of the applied stress. For large deformations, a second mode of breaking appears and coexits with the previous one: a fraction of the rods breaks instantaneously while others break after a delay time. Different possibilities to explain the dependence of the breakage time on the stress are discussed.

Key words Amphiphilic molecule – monolayer – bidimensional solid – breakage

The subject of this paper is two-dimensional solids and, particularly, their mechanical properties at the air-water interface. We study monolayers of an insoluble amphiphilic molecule (12-NBD-stearic acid) at the free surface of water. The molecule is bolaform, which means it has two polar heads: an acid group (-COOH) and a fluorescent group (-NBD). If the monolayer is compressed, a first order liquid-solid phase transition [1] can be observed by fluorescence microscopy. The two-dimensional liquid phase (area per molecule ~ 83 Å^2) can be distinguished from the solid phase because, in this case, the two polar heads are in the water and, consequently, the fluorescence is quenched. In the two-dimensional solid phase (~ 30 Å^2/molecule) which occurs at a surface pressure $\Pi = 11$ mN/m, the molecules straighten up, lifting the NBD group out of the water, and an intense fluorescence can be observed. The domains that constitute the solid phase, in the absence of impurities which are due to the oxidation of the molecule, are long and rodlike [2] (1 mm long and about 20 μm wide) with two parallel sides, which makes them well suited for micromechanical measurements. Studying the flexion of such a rod in the plane of the water, when a force is applied in its middle, Bercegol et al. measured the two-dimensional Young's modulus [2]: its value (3660 ± 1300 mN/m) was found to be in agreement with the Kosterlitz–Thouless theory which gives a lower limit for the Young's modulus of a two-dimensional (2D) crystalline solid [5, 6]. Also, x-ray diffraction revealed a crystalline structure [3]. On the other hand, we have noted that a domain under a bending stress breaks after a delay time t_B (Fig. 1).

The time-lag breaking was studied measuring t_B for different applied strains τ (rate of extension of the external side of a rod under a bending stress) [7]. The different breaking times measured for a given value of τ are grouped, which allows for the definition of a mean value $\bar{t}_B$.

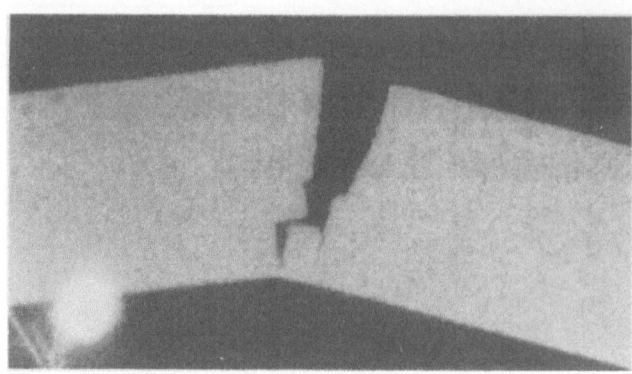

Fig. 1 Fracture of a rod (50 μm wide) after a three point flexion

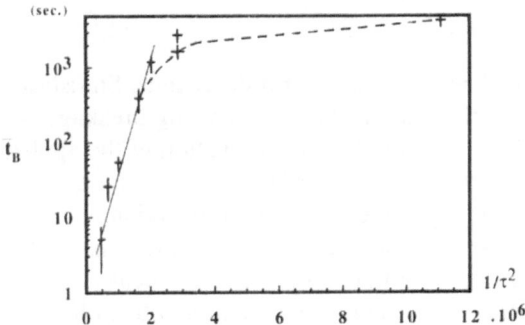

Fig. 2 Mean time-lag breaking $\bar{t}_B$ versus $1/\tau^2$ on a semilogarithmic scale. A point on this figure corresponds to a statistic where about 20 rods were broken. The dashed line indicates results affected by evaporation. For large t_B ($\sim 3 \times 10^3$ s), evaporation has not been avoided

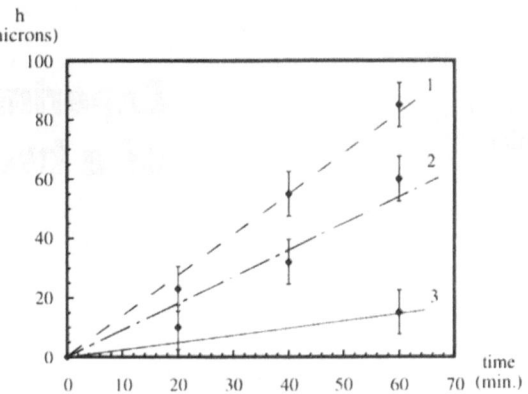

Fig. 3 h corresponds to the difference of level of water in the trough due to evaporation in the case of pure water (dashed line 1), water with monolayer (2), and after improvement (3)

The observed dispersion on the experimental values of t_B results from the low experimental accuracy of τ. These experiments have shown that the time-lag breaking is a clearly defined function of the applied stress. A plot of the breakage time t_B versus $1/\tau^2$, on a semi-logarithmic scale, is shown in Fig. 2, corresponding to five different values of the strain τ in the range $3 \times 10^{-4} < \tau < 14.5 \times 10^{-4}$. This plot is linear, excepted for large t_B where the dashed line suggests a finite value of t_B for a rod without strain ($\tau = 0$). This surprising result can be understood since, over long periods, it appears that the evaporation of water in the trough induces the sliding of the rod along the three glass fibers which are used for bending. This breaks some pieces of the solid domain around the glass fibers, inducing the fracture and, consequently, reducing the breaking time for low τ. Evaporation is measured focusing the microscope on the water surface; the lowering of the level of the monolayer is measured with an inaccuracy of about 5 to 10 μm (see Fig. 3). The change in water level due to the evaporation was estimated to be at least 60 μm per hour. To reduce the evaporation, we isolated the sample and placed a beaker of water at 30 °C in the sample chamber, thus increasing the relative humidity. With the isolated

system, $\tau = 5.9 \times 10^{-4}$ gives a mean value $\bar{t}_B \sim 46$ min instead of 29 min. The measurements for $\tau \geq 7.8 \times 10^{-4}$ do not change. For the lowest stresses, the measurement can also be affected by a small deformation of the water surface around each glass fiber. The stress induced by this deformation has not been taken into account.

At large τ ($\tau > 10^{-3}$), the time-lag behavior coexists with an instantaneous breaking which becomes more important at larger strains: a fraction of the crystals breaks instantaneously.

In each bending experiment, the flexion resulted from applying forces in three points on both sides of a rod. Flexion experiments with four pressure points (see Fig. 4a) have also been performed, and give the same results. In this case, both middle fibers are moved to apply the force; it is observed that most of the times (≈ 8 out of 10 times) the breaking occurs close to the opposite to one of these two fibers (Fig. 4b): this fact is probably due to an experimental assymmetry, the four points being not exactly on the same circle when a rod is bent.

The linear plot obtained (Fig. 2) is in agreement with a calculation of Pomeau which suggests that the time-lag breaking results from the time to nucleate a crack with an energy of activation E_{ac} so that $\bar{t}_B \approx \exp(-E_{ac}/kT)$. This nucleation time is proportional to $\exp(-c^n/\tau^n)$ (c being a molecular constant) [8], with $n = 2$ for a two-dimensional model and $n = 4$ for a three-dimensional model. The large exponent for three-dimensional crystals indicates that the range of strain giving observable times of nucleation is too small to be observed (the function $\bar{t}_B$ behaves like a step: $\bar{t}_B = 0$ for $\tau < c$ and $\bar{t}_B = \infty$ for $\tau > c$). In the case of a two-dimensional crystal $n = 2$, and the range of τ giving rise to observable times of nucleation is larger. However, the nucleation of a crack is due to a thermal process. Consequently, within Pomeau's theory, the mean value of the time for nucleation is $\bar{t}_B$, but the

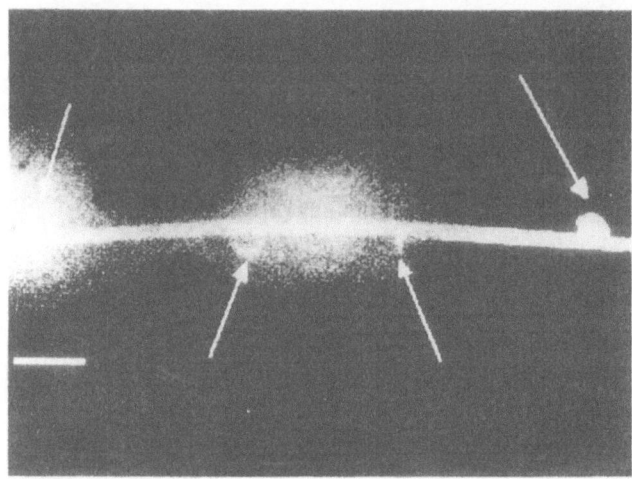

a)

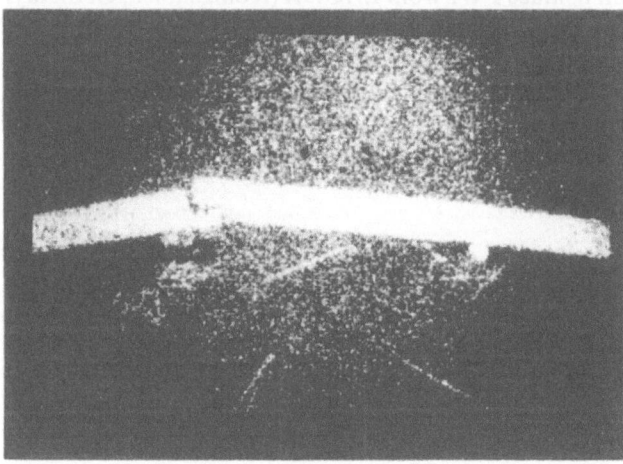

b)

Fig. 4 a) A two-dimensional rod bent by four glass fibers (marked by arrows) perpendicular to the water surface. The width of the rod is 20 μm and the two external glass fibers are 1mm apart. (bar $\sim$ 50 μm) b) The rod under stress has broken at time t_B

probability of nucleation is given by a Poisson law. We do not observe this law in our experiments: the breaking time is well-defined. This suggests that the breaking results from the percolation of a large number of cracks, N. According to the Poisson law, the distribution width for the time of nucleation of one crack is proportional to $\bar{t}_B$, however, a crack percolation model predicts a width proportional to $\bar{t}_B/\sqrt{N}$. Our experimental results are best explained by the latter, with $N \gg 100$.

As another breaking mechanism, one could consider the migration of defects towards the region of largest stress. Large reproducibility of these experiments indicates that this mechanism is independent of the density of defects existing in the crystal. Plastic deformations were not observed (within the 1 μm resolution of the microscope). But, releasing the stress briefly a short time before breakage, at t_1, and applying it again until breakage at t_2, it was observed that $t_1 + t_2 = \bar{t}_B$. The domains under stress keep the memory of the stress, which implies a plastic deformation.

Finally, another possibility can be suggested considering the very anisotropic growth and optical properties of the solid domains [1]. The mechanical properties are also probably very anisotropic. The appearance of a solid domain melting under the effect of a prolonged illumination suggests that a solid domain is made of long fibers, the molecular forces being very different in the direction of the length of a domain and in the perpendicular direction. Electron diffraction also suggests a fibrous structure at the molecular scale [3]. The process leading to the breaking being probably a succession of a large number of elementary processes: the elementary process could be the breaking of a fiber under stress, producing the progress of a crack from the edge of the domain. When the size of this crack reaches the critical size satisfying the Griffith critical condition [9], it propagates quickly.

References

1. Bercegol H, Gallet F, Langevin D, Meunier J (1989) J Phys (Paris) 50:2277
2. Bercegol H, Meunier J (1992) Nature (London) 356:226
3. Flament C, Graf K, Gallet F, Riegler H (1994) Thin Solid Films 243:411
4. Bercegol H (1992) J Phys Chem 96:3435
5. Nelson DR, Halperin BI (1979) Phys Rev B 19:2457
6. Nelson DR (1983) In: Dombs C, Lebowitz JL (eds) Phases transitions and critical phenomena vol 7, academic, London
7. Pauchard L, Meunier J (1993) Phys Rev Lett 70:3565
8. Pomeau Y (1992) CR Acad Sci (Paris) 314:553
9. Griffith AA (1920) Philos R Soc London A 221:163

Progr Colloid Polym Sci (1995) 98:266–268
© Steinkopff Verlag 1995

G. Emrich
D. Vollhardt
T. Gutberlet
B. Kling
J.-H. Fuhrhop

Morphology and growth of condensed phase structures in N-alkyl-aldonamide monolayers using Brewster angle microscopy

Introduction

The amphiphilic long-chain N-alkyl-aldonamides form a rich variety of three-dimensional characteristic ordered structures in aqueous solutions and gels, such as micellar fibers, helical bilayers, rods, ribbons, and platelets, dependent on their chiral carbohydrate head groups [1, 7].

The combination of a Langmuir trough with a Brewster angle microscope [4, 5] has been used to investigate two-dimensional phenomena by direct visualization of the monolayer organization. In particular, we compare the monolayer behavior of optically pure enantiomers with their corresponding racemic mixtures of N-dodecyl-gluconamide and N-dodecyl-mannonamide [2, 3].

Experimental

D-, L-enantiomers, and the racemic mixtures (1:1) of N-dodecyl-gluconamide and -mannonamide were prepared as described elsewhere [1]. Clear solutions (10^{-3} m; spreading solvent: chloroform/ethanol/water 7.0/2.9/0.1 v/v/v) were spread from a micro syringe onto a MilliQ water subphase. Surface pressure versus area per molecule isotherms were registered with an automatic Langmuir film balance FW2 from LAUDA (Königshofen, Germany). The Brewster angle microscope BAM1 from NFT (Göttingen, Germany) was mounted on the FW2 enclosed in a Plexiglas housing. Images of the monolayer morphology were stored by a video system. The lateral resolution of the BAM was about 4 μm [4].

Fig. 1 Surface pressure versus area isotherms compressed at 0.1 nm² molecule^{-1} min^{-1}: a) for the enantiomers and the racemic mixture (D:L = 1:1) of N-dodecyl-gluconamide at 10 °C; b) for the enantiomers and the racemic mixture (D:L = 1:1) of N-dodecyl-mannonamide at 20 °C

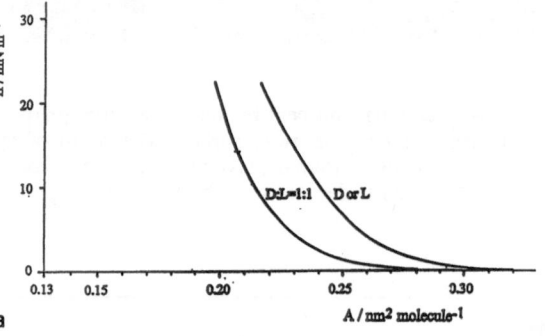

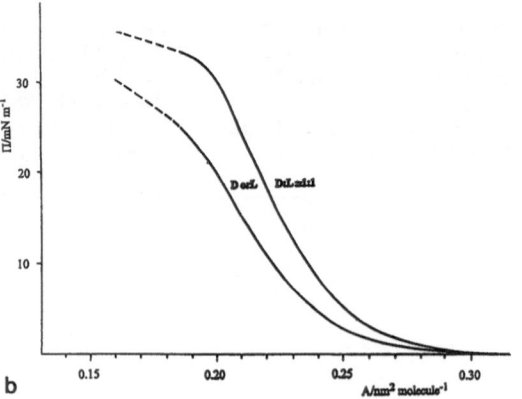

Dr. G. Emrich (✉) · D. Vollhardt · T. Gutberlet
Max-Planck-Institut
für Kolloid- und Grenzflächenforschung
Rudower Chaussee 5
12489 Berlin, Germany

B. Kling · J.-H. Fuhrhop
Institut für Organische Chemie
der Freien Universität Berlin
14195 Berlin, Germany

Progr Colloid Polym Sci (1995) 98:266–268
© Steinkopff Verlag 1995

Results and discussion

Isotherms of L-enantiomers were equivalent to those of D-enantiomers for each of the two aldonamides. The shape of the force-area curves was strongly influenced by temperature (range 10–40 °C) and compression rate. Surface pressure relaxation was observed for both the enantiomeric and the racemic monolayers. Compression-expansion cycles revealed hysteresis. In addition, the Π/A-plots were shifted to smaller area values per molecule at higher temperatures. A striking chiral discrimination can be observed comparing the pure enantiomers with their racemic mixtures (Figs. 1a and 1b). Isotherms showed the racemic N-dodecyl-mannonamide film is more expanded than the enantiomeric monolayer. This indicates that the mannonamide monolayer exhibits homochiral behavior. On the other hand, a more favorable D:L interaction suggests heterochiral association for N-dodecyl-gluconamide.

a   0 s

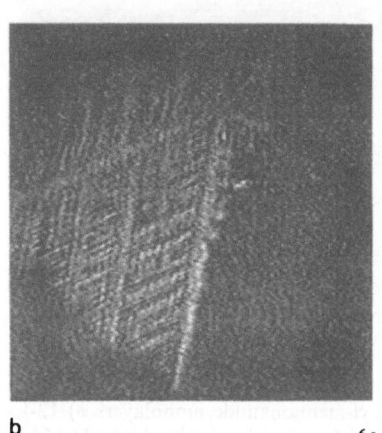

b   6 s

Fig. 3 Dendritic growth in N-dodecyl-gluconamide monolayers (20 °C; 4.3 mNm^{-1}; 0.27 nm^2 molecule^{-1}; time interval 6 s)

Fig. 2 a) Dendritic crystallization in a N-dodecyl-L-gluconamide monolayer (20 °C; 3.0 mNm^{-1}; 0.23 nm^2 molecule^{-1}); b) Isotropic aggregation for the racemic mixture of N-dodecyl-DL-gluconamide (20 °C; 17 mNm^{-1}; 0.16 nm^2 molecule^{-1})

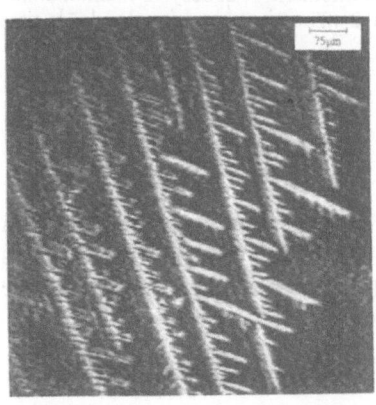

a

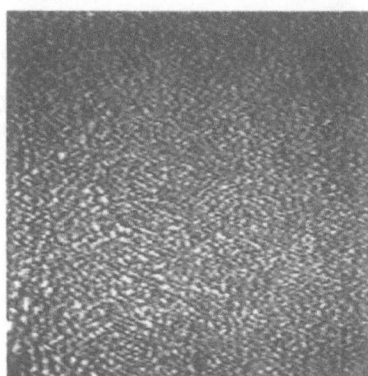

b

Important morphological information can be obtained by the recently developed microscopy at the Brewster angle. Condensed aggregates at the air-water interface have been formed even after spreading. In contrast to the dendritic crystallization of the pure enantiomeric films (Fig. 2a) the racemic N-dodecyl-gluconamide monolayer did not exhibit the growth of noticeable forms (Fig. 2b). This visualized evidence for chiral discrimination is supported by the measured Π/A-plots (Fig. 1a). Differences between the morphological patterns of the D- and L-enantiomers have not been found. It was possible to record dendritic growth under properly chosen experimental conditions (Fig. 3).

In the case of N-alkyl-mannonamide another morphology was observed. The L-enantiomer produces feather-like aggregates with a counterclockwise curvature (Fig. 4a). For the spread D-mannonamide similar crystals are developed with a clockwise curvature (Fig. 4b). Growth kinetics of L-enantiomeric monolayers are presented with a counterclockwise curvature, as seen in Fig. 5.

268

G. Emrich et al.
Structures in N-alkyl-aldonamide monolayers using BAM

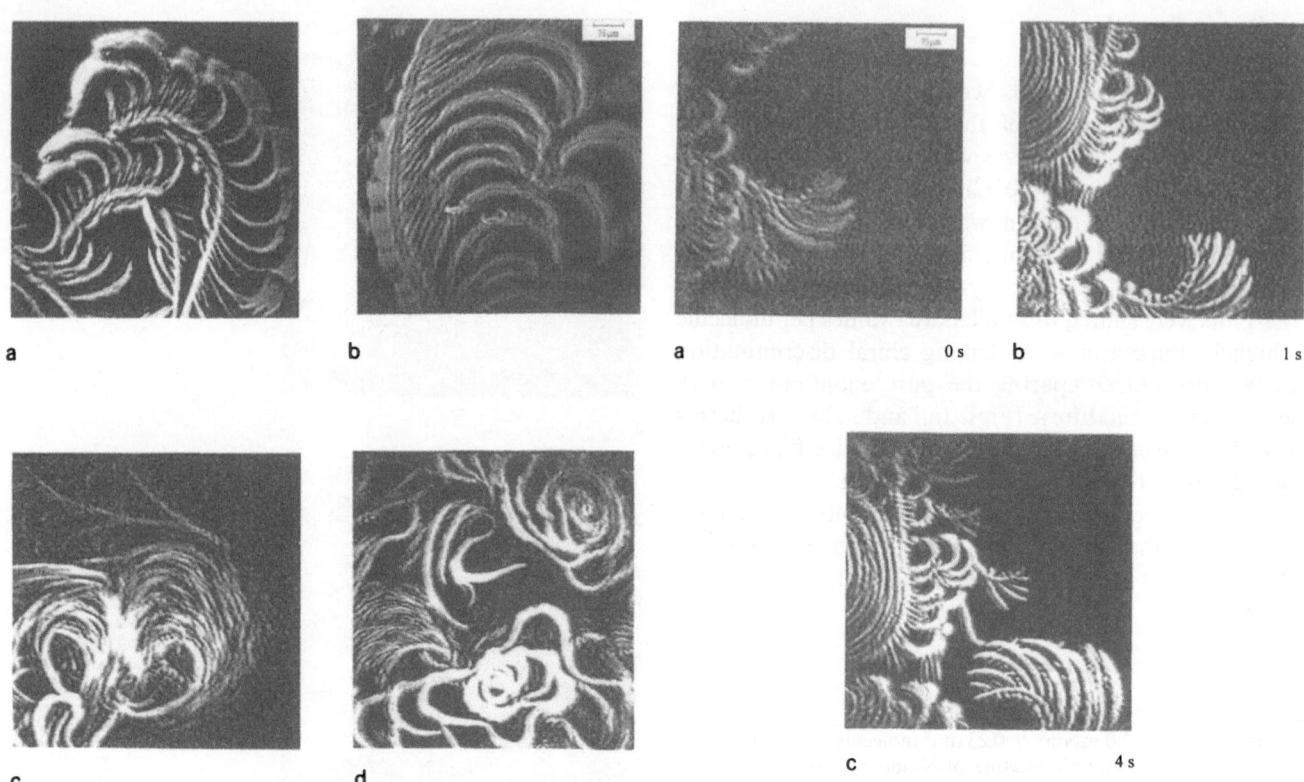

Fig. 4 Chiral discrimination in the morphology of condensed phase structures of N-dodecyl-mannonamide monolayers a) 12-L-MAN (30 °C; 0.2 mNm^{-1}; 1.5 nm^2 molecule^{-1}); b) 12-D-MAN (30 °C; 0.1 mNm^{-1}; 0.5 nm^2 molecule^{-1}); c) 12-DL-MAN (25 °C; 0.5 mNm^{-1}; 0.5 nm^2 molecule^{-1}); d) 12-DL-MAN (25 °C; 0.5 mNm^{-1}; 0.6 nm^2 molecule^{-1})

Fig. 5 Growth kinetics of enantiomeric N-dodecyl-L-mannonamide monolayers (5a–5c) 30 °C; 0.2 mNm^{-1}; 0.64 nm^2 molecule^{-1}; time intervals 1 s and 4 s

On the contrary, the racemic mixture shows curved aggregates without a favored direction under the same conditions for measurement (Figs. 4c and 4d). In addition to the thermodynamic chiral discrimination in N-alkyl-mannonamide monolayers indicated by the differences in the isotherms (Fig. 1b), the monolayers exhibit structures which depend on the chirality of the head groups in the monomolecular film [6, 8, 9].

References

1. Fuhrhop J-H, Boettcher Ch (1990) J Am Chem Soc 112:1768
2. Vollhardt D, Gutberlet T, Emrich G, Fuhrhop J-H, Langmuir (accepted)
3. Vollhardt D, Emrich G, Gutberlet T, Kling B, Fuhrhop J-H, J Am Chem Soc (submitted)
4. Hönig D, Möbius D (1991) J Phys Chem 95:4590
5. Hénon S, Meunier J (1991) Rev Sci Instr 62:936
6. Heath JG, Arnett EM (1992) J Am Chem Soc 114:4500
7. Fuhrhop J-H, Schnieder P, Boekema E, Helfrich W (1988) J Am Chem Soc 110:2861
8. Akamatsu S, Bouloussa O, To K, Rondelez F (1992) Phys Rev A 46:R4504
9. Stine KJ, Uang JY-J, Dingman SD (1993) Langmuir 9:2112

Progr Colloid Polym Sci (1995) 98:269–270
© Steinkopff Verlag 1995

S. Siegel

Non-equilibrium structures in monolayers of fatty alcohols

Key words Monolayers – Brewster
angle microscopy – tetradecanol –
hexadecanol

Experimental

Fatty alcohols with 14 to 18 carbon atoms (Sigma) were dissolved in heptane and spread on double distilled water. A computer-interfaced film balance with a Langmuir float (Lauda) and a Brewster angle microscope (BAM 1, NFT Göttingen) were used for our investigations. Because of the visual angle the images appear compressed in one direction, which was corrected by an image-processing system. Polarized laser light and a CCD camera are used in the microscope, which is sensitive to differences in density or orientation of the monolayer molecules. Its resolution is about 4 μm.

Results

1) A phase transition takes place in tetradecanol monolayers at 25 °C during compression. The condensed

S. Siegel
Technische Universität Berlin
Iwan-N.-Stranski-Institut
für Physikalische und Theoretische Chemie
Straße des 17. Juni 112
10623 Berlin

Dr. S. Siegel (⊠)
MPI für Kolloid- und
Grenzflächenforschung
Rudower Chaussee 5
12489 Berlin

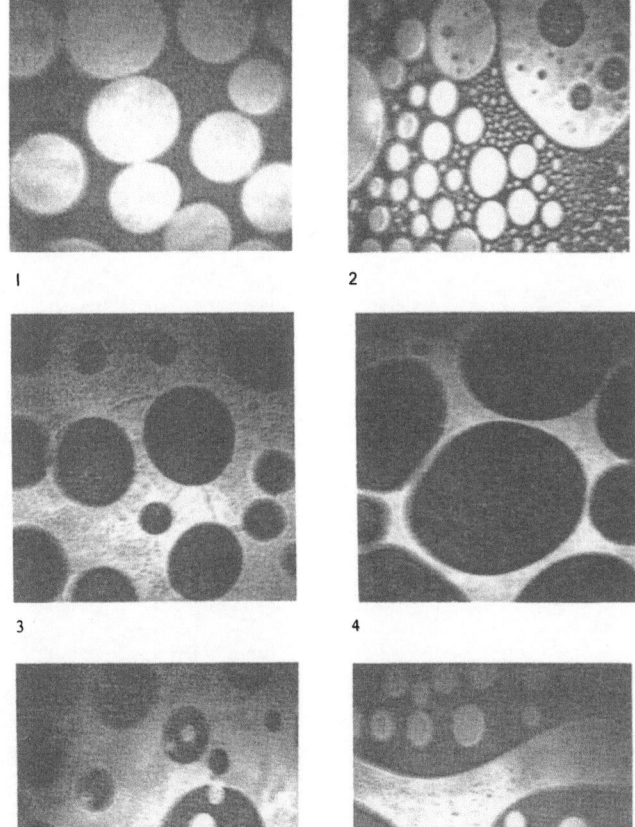

Figs. 1–6 Structures in monolayers of tetradecanol (Figs.1, 3–6) and hexadecanol (Fig. 2) at 25 °C. The condensed phase appears bright. The size of the Brewster angle micrographs is 700 × 700 μm

phase forms circular domains, starting at the beginning of the plateau region (Fig. 1).

2) Alcohols with 16 C atoms or more form inhomogeneous monolayers after spreading at room temperature. An expanded and a condensed phase coexist (Fig. 2, hexadecanol).

3) During decompression a foam structure can be produced. Although this foam exists over a long time, it is a quasi-equilibrium state stabilized by the line tension (Figs. 3, 4).

5) Repeated compression and expansion result in domains with different sizes or in domains within holes (Figs. 5, 6).

6) The nearly circular shape of the domains and of the holes in the monolayer and the fast process of foam rupture indicate a relatively high line tension between the condensed and the expanded phases of tetradecanol.

Progr Colloid Polym Sci (1995) 98:271–275
© Steinkopff Verlag 1995

M. Winterhalter

On the molecular origin of the bending elastic moduli

Dr. M. Winterhalter (✉)
Biozentrum
Biophysikalische Chemie
Klingenbergstraße 70
4056 Basel, Switzerland

Abstract In this work we outline the influence of the hydration pressure as well as of the polymer repulsive interaction on the bending elastic moduli of membranes. Starting from a continuum description for the molecular interaction, we calculate the curvature dependency of the local stresses due to the molecular forces. Applying Helfrich's momentum development yields the corresponding macroscopic material properties. Inserting experimental values for the hydration force into our model, we obtain fair agreement with measured elastic curvature moduli for many lipids. We conclude that the variation in hydration of the lipid head groups might account for observed differences in the curvature moduli of lipid films. Due to the lack of experimental data in the case of polymer interaction, we are only able to deliver theoretical estimations.

Key words Hydration – curvature elastic moduli – lipid membrane – pegylated lipids – polymer interaction

The behavior of flexible interfaces has become an important topic in widespread areas ranging from oil engineering to pharmacology [1]. Surfactants in aqueous systems are a well-known example. Surfactant molecules above the critical micellar concentration (CMC) spontaneously form micelles, lamellar, or more complex structures. The transition from one shape to another often requires only a few $k_B T$ (k_B is the Boltzmann factor and T the temperature) per aggregate. A slight change in temperature or composition causes different macroscopic features [1, 2].

Biological membranes have to be flexible and stable. The red blood cells exhibit astonishing material properties. During their life-time they change their shape drastically many times. It has been shown that at the early stage they are highly flexible and when later eliminated from the blood stream they appear to be more rigid [3]. Certain diseases as well as certain drugs show a strong correlation with the apparent flexibility of the erythrocytes [3]. The designing of artificial drug-delivery systems is a major goal in biotechnology. Prior to any practicable application of liposomal vehicles a very narrow size distribution, as well as thermal, mechanical, and chemical long-term stability is required. To avoid time-consuming measurements it is of great interest to predict material properties from molecular properties and composition [1].

A general theoretical model predicting the stability of membranes and the transition between various phases is still lacking. The most promising concept was suggested almost 20 years ago by Helfrich [4], who proposed a description of a flexible interface in terms of its elastic material properties. For membranes with vanishing surface tension he suggested, in analogy to smectic liquid crystals, that the behavior is determined by the bending energy. This model turned out to be a powerful tool [5–7] to describe a large number of shape transitions. Moreover, it is able to predict budding and vesiculation. The bending energy per unit area is [4]

$$g_c = \frac{1}{2} k_c (c_1 + c_2 - c_0)^2 + \bar{k} c_1 c_2, \tag{1}$$

which is quadratic in the principal curvatures c_1 and c_2. Within this model three macroscopic material properties are used to characterize the behavior of an interface. One is the spontaneous curvature c_0, which describes a possible asymmetry in composition between the inner and outer monolayer. As only the total energy of a sample can account for transitions between different states, the Gaussian module $\bar{k}$ will only be involved when the topology is changed. This approach is based on the assumption that the local molecular interaction which causes the bending rigidity is of short range, and therefore that the interaction can be developed in a rapidly converging series in curvature. If the interface in equilibrium is already highly curved, then additional coupling terms occur [8, 9].

The bending rigidity k_c for various lipids has been measured to be in the range of 5–50 $k_B T$ (see Table 1). However, the measurements of the bending rigidity differ substantially, sometimes even by more than one order of magnitude. Unfortunately, up to now no theoretical model is able to account for all observed features. Partly, this is due to different experimental methods, partly to so far unknown physical parameters.

Several models on the origin of the bending rigidity have been proposed. The influence of the hydrocarbon chains on the elastic moduli was already widely discussed [8–11]. Some years ago, several groups [9, 12] derived models which relate electrostatic interactions of charged head groups to the curvature elastic moduli. The free energies of a charged interface for planar, cylindrical and spherical geometries were expanded in power series of curvature c_1. A combination of the second-order terms yields the curvature elastic moduli. Within the Debye–Hückel approximation the contribution to the bending rigidity stemming from the surface charges is proportional to the third power of the Debye screening length. However, in most applications the Debye screening length is very small and the electrostatic contribution becomes neg-

ligible. Experimental limitations have so far prevented the proof or disproof of such a contribution [13–14].

In the first part of this work, we point out that hydration of the lipid headgroups might be the cause of the apparent curvature elasticity. In the second part, we derive similar equations for lipid mixtures where a variable fraction of the lipids carry covalently-bonded polymers on their headgroups.

Studies of the interaction of lamellar phases of lipid/water mixtures have shown a very strong exponentially-decaying repulsive force between the lipid layers for short intermembrane distances (less than 3 nm) [15–17]. Although there still seems to be some controversy on the origin of this repulsive force, it is now widely accepted that this behavior is associated with the orientation of the water.

Experimentally, the hydration pressure was measured by bringing two membranes in close contact. The repulsion shows an exponential decay as one goes further from the interface into the solution

$$\Pi(z) = \Pi_0 \exp(-z/\lambda) = -s(z) , \qquad (2)$$

where z is the distance from the water/lipid interface. λ is the measured decay length that is about the same as the molecular size of the solvent. Π_0 is the extrapolated hydration pressure at the water/lipid interface. Like in the case of electrostatic interaction of charged membranes, the hydration will cause local stresses $s(z)$. They are also present around single membranes and favour or unfavor bending of the interface. Our assumption is that each lipid molecule is surrounded by a fixed number of hydration water. Bending the interface will reduce the free volume, and will force the hydration pressure to decay more slowly than an exponential (in the case of the inner aqueous space and faster for the outer interface). A relatively small cylindrical curvature will cause a deviation from the simple

Table 1 Table of experimentally measured values for the hydration pressure and the bending rigidity for various lipids.

Lipid	λ [nm]	Π_0 [10^9 N/m²]	$4\Pi_0\lambda^3$ [10^{-19} J]	$k_c^{measured}$ [10^{-19} J]
SOPC [21]	0.2	3.2	1	0.9 [22]–1.8 [23]
DGDG [15]	0.17	2	0.4	0.1 [24]–0.4 [22]
DLPC [25]	0.2	4	1.3	0.3 [24]
DMPC [25]	0.22	3.2	1.3	0.6 [22], 2.4 [26]
DPPC [25]	0.21	10	3.7	2.0 [26]
DSPC [25]	0.13	790	70	1.8 [26]
EYPC [27]	0.21	4	1.5	0.2–2.3 [24, 28–33]
EYPC/Chol 1:1 [25]	0.11	6300	336	2.5 [22]
DPPC/Chol 1:1 [25]	0.15	31.6	4.3	
EYPC/Chol 1:1 [34]	0.19	8.2	2.2	2.5 [22]
DOPE/DOPC 3:1 [21]	0.18	1.6	0.4	0.4 [8]–1.6 [21]

exponential form of the hydration pressure for the outer layer:

$$\Pi(z) = \Pi_0 \frac{1}{1 + cz} \exp(-z/\lambda) = -s(z), \tag{3}$$

with c as the curvature of the interface. The pressure $\Pi(z)$ give rise to a negative local stress $s(z)$. The contribution to the interfacial tension is given by the zero moment of the stress

$$\gamma = \int_0^\infty s(z)dz = -\Pi_0\lambda, \tag{4}$$

assuming for a moment a planar monolayer. Switching on the hydration pressure will induce an area dilatation due to the additional interfacial tension γ. The change in area per molecule yields in a renormalization of the interaction [18–19]. However, the bare hydration pressure is experimentally not observable and the hydration pressure (Eq. 2) already represents the renormalized one.

Like in earlier approaches, we assume that the additional interfacial tension is balanced for each monolayer separately at the water/lipid boundary situated at half bilayer thickness $z_0 = \pm d/2$ [9]. This assumption is somehow arbitrary and the true neutral plane might be shifted inside the membrane. It should be noted that the apparent bending modulus is a function of the choosen plane. Knowledge of the exact deformation of the neutral plane allows the adjustment of the apparent bending modulus with represent to the other planes. [8]. Unfortunately, up to now the exact localization of the neutral plane is experimentally not yet accessible. Furthermore, a more complex model would require additional assumptions on the mechanical stress distribution inside the membrane and are beyond the scope of this work. It should be noted that if the interfacial tension is not completely balanced, coupling between the area dilatation and bending has to be taken into account [8].

Once the local stress distribution $s(z)$ for a cylinder is known the bending moduli can be readily calculated. As Helfrich [20] and Lekkerkerker [12] pointed out, the bending rigidity of a monolayer with its neutral plane fixed at $z_0 = d/2$ is given by

$$k_c^m = \int_0^\infty dz(z - z_0)\frac{\partial s(z)}{\partial c_{c=0}} = 2\Pi_0\lambda^3. \tag{5}$$

The second moment yields the Gaussian module,

$$\bar{k}_c^m = \int_0^\infty dz(z - z_0)^2 s(z)_{c=0} = -2\Pi_0\lambda^3. \tag{6}$$

For the sake of completeness, we also build the first moment

$$c_0 k_c = -\int_0^\infty dz(z - z_0)s(z)_{c=0} = \Pi_0\lambda^3. \tag{7}$$

From Eqs. (5) and (7), the spontaneous curvature of a monolayer due to its hydration is given by $c_0 = 1/\lambda$.

In order to compare our theoretical predictions with experimental data, we have to adapt them to the case of a bilayer. As pointed out earlier [8–9], the final result depends on how the actual bending is performed. Depending on the chosen neutral plane an eventual coupling to other elastic moduli occurs. For simplicity, we assume that the water/lipid interface for each layer is the neutral plane $z_0 = d/2$, and that the surface tension in this plane is balanced. Such a simplification will cause a doubling of the two bending moduli. Other models will involve terms containing powers of the quotient of membrane thickness and radius of curvature. However, no precise information on the bending processes is currently available on the molecular level.

In Table 1, we show the measured values for the hydration pressure as presented in the review by Rand and Parsegian [15]. In the third column, we give the calculated corresponding value based on our theory. These values have to be compared with the measured bending elasticities in the fourth column. Basically, it seems that the hydration based values are in a very good agreement with those from bending rigidity measurements. For such a conclusion, one has to take into consideration that bending rigidity measurements show a large scatter and often depend on the experimental method. The general feature of an increasing bending rigidity going from DGDG to SOPC, DPPC and lipid/cholesterol is well reproduced from the hydration data. However, for two lipids the two values differ significantly. In the case of DSPC this might be because the hydration pressure measurement has been performed at 25 °C and a phase transition at low water content could be observed, whereas the bending rigidity was measured at 56.5 °C. The second exception occurred in the EYPC/Chol mixture obtained by one group. In this case, the hydration data differ significantly from other similar compositions. Using their DPPC/Chol data again yields reasonable values for the bending rigidity. More recently, other measurements on EYPC/Chol mixture show better agreement with other hydration data. The calculated bending rigidity is again reasonably good.

It is interesting to note that the observed decay length λ is influenced by the solvent. One may speculate whether the large scatter in measured bending rigidity is due to the different solvent used in various groups.

In general that data seems to overestimate in general the bending rigidity. This could be due to the neglect of the effect of a possible variation in the neutral plane, as well as screening of the dipoles by their neighbors during bending. The presented data also depend on the kind of definition of the zero plane to which the experimental data of the hydration pressure has been fitted. A complete theory, however, has to take into account all possible interactions as pointed out in detail by several models [9, 18, 19].

The same principles as used above can be applied to consider the case where polymer is present and exerts a steric repulsion.

Surface adsorbed polymers are used widely to avoid coagulation of latex particles or of droplets [35]. The so-called Stealth lipids [1], which have a polyethylene glycol polymer covalently attached at the lipid head group, are a more spectacular application. Surprisingly, addition of a fraction of such lipids to the composition of liposomes gives a dramatic increase in blood circulation time as well as a very different final biodistribution after injection into the blood stream. Prior to any pharmaceutical application it is of primary importance to guarantee a very specific size distribution of the liposomes as well as a predictable mechanical stability. As pointed above, the curvature model is able to deliver certain criteria here.

The standard application involves ordinary lipids or lipid/cholesterol mixture including a few mol% of the polymer-bearing lipids. The lipid chains of the polymer-bearing lipids are present in the lipid membrane, whereas the polymer moiety is soluble in the aqueous part. At a very low molar concentration these polyethylene glycol chains will form mushrooms in the aqueous phase. We expect that such isolated polymer coils will scarcely influence the bending properties of the entire lipid membrane. In fact, a similar ansatz in this regime will yield a contribution of a few $k_B T$. The overlapping regime for higher polymer densities will be more important with respect to bending. Assuming a to be the monomer size, N the number of links, and σ the fraction of grafted sites, the polymer coils start to overlap [36–38] if $\sigma < N^{-6/5}$. According to DeGennes [36–38], a polymer in a good solvent is to be divided into individual parts whose size is given by the local correlation length ζ. Each of those is assumed to behave like an individual particle and exerts an osmotic pressure for a planar interface.

$$\Pi(z) = \frac{k_B T}{\zeta^3} = \frac{k_B T}{\zeta^3(z)}, \tag{8}$$

with $\zeta(z) = a[\rho(z)a^3]^{-3/4} = a\,\sigma^{-1/2}$ and $\rho(z)$ is the local density [36–38]. The presence of an additional local stress causes a stretching of the area per molecule which yields a renormalized effective pressure [18–19]. A complete model would involve additional assumptions on the area

stretching. We avoid these difficulties by replacing the theoretical grafting density σ by the corrected experimentally observed one. Furthermore, we assume that each monolayer is balanced at the water/lipid interface. Again, in the following step, we have to model how bending influences this pressure. A simple model is to assume that if the accessible volume increases due to bending, the density decreases. Successively, the correlation length ζ increases, which again reduces the pressure: One end of the polymer is fixed on a cylindrical surface of curvature c. The lipid density changes along the axis perpendicular to the surface $\rho(z) = \rho_0(1 - 2zc)$. Inserting this assumption into the expression for the pressure induced by the polymer repulsion yields, after truncation of second and higher-order terms,

$$\Pi(z) = \frac{k_B T}{D^3\left(1 + \dfrac{9}{2}cz\right)} \quad \text{with } D = \frac{a}{\sqrt{\sigma}}, \tag{9}$$

as the mean distance between the polymer grafting sites. Like in the first part, we obtain the bending module $k_c^{polymer}$ by differentiation with respect to the curvature c. In contrast to the hydration pressure the steric interaction has a natural cut-off given by the polymer length. As pointed out by DeGennes [36–38], the height in the overlapping regime scales with $h = Na\sigma^{1/3}$. A straightforward calculation of the first moment then yields the bending rigidity

$$k_c^{polymer} = \frac{3}{2} k_B T \sigma^{5/2} N^3. \tag{10}$$

Again, the bending rigidity has to be adapted to the bilayer. Similar assumptions as in the previous case yield for the bilayer twice the monolayer bending modulus.

We obtain 5/2 as the exponent for the grafting density, which differs only slightly from the value of 7/3 obtained by Milner and Witten [10] for the low-density regime. This is basically due to the difference in the assumption of what determines the radial variation of the polymer density. In our case this is independent of the specific grafting energy, whereas in theirs it is not. The Stealth lipid density in the low density regime is basically determined by the Stealth/lipid molar ratio, and can easily be varied within a certain range. Higher mole fractions could cause Stealth lipids to separate as micelles in the aqueous phase. On the other hand, the exponent for the dependency on the number of chain links N agrees well with that obtained by Milner and Witten [10].

These results have to be compared with experimental data. For low grafting densities, as well as for short chains, we expect that the apparent measured bending rigidity will be determined by factors other than polymer steric interactions. According to Eq. (10), higher densities should show a sudden onset of an increase in bending rigidity. Recently,

Progr Colloid Polym Sci (1995) 98:271–275
© Steinkopff Verlag 1995

we (Winterhalter and Meleard, in preparation) analyzed the fluctuation of large unilamellar vesicles by phase contrast light microscopy. In a first series of experiments, we used Stealth lipid DSPE-PEG2000 (around 47 links, Flory radius of about 3.4 nm) at various concentrations. Starting from pure DOPC (bending rigidity of about $k_c = 1.0$ 10^{-19}J) we found a 10% increase at about 4 mol% DSPE-PEG2000. However, higher concentrations gave rise to strongly connected vesicles and standard fluctu-

ation analysis became meaningless. We expect further experiments on single freely undulating vesicles to be able to distinguish between the theoretical predictions of Milner and [10] and that presented here.

Acknowledgments We thank W.M. Arnold and K. Gawrisch for constructive criticism of the manuscript. This work was supported by the Deutsche Forschungsgemeinschaft "Graduiertenkolleg Magnetische Kernresonanz."

References

1. Lasic DD (1993) Liposomes: From Physics to Application Elsevier, Amsterdam
2. Lipowsky R (1991) Nature 349:475–481
3. Fricke K et al (1986) Eur Biophys J 14:67–81
4. Helfrich W (1973) Z Naturforschung 28 C:693–703
5. Deuling HJ, Helfrich W (1976) Biophys J 16:861–868
6. Svetina S, Zeks B (1989) Europ Biophys J 17:101–111
7. Berndl K et al (1991) Europhys Lett 13:659–664
8. Kozlov MM, Winterhalter M (1991) J Phys (Paris) 2, II:1077–1084
9. Winterhalter M, Helfrich W (1992) J Phys Chem 96:327–330
10. Milner ST, Witten TA (1988) J Phys (Paris) 49:1951–1962
11. Szleifer I et al (1990) J Chem Phys 92:6800–6817
12. Lekkerkerker HNW (1990) Physica 167A:384–394
13. Song J, Waugh RE (1992) J Biomech Eng 112:235–240
14. Waugh RE et al (1992) Biophys J 61:974–982
15. Rand RP, Parsegian VA (1989) Biochim Biophys Acta 988:351–376
16. Israelachvili JN, Wennerstrom H (1990) Langmuir 6:873–876
17. McIntosh TJ et al (1992) Biochem 31:2020–2024
18. Cantor R (1981) Macromolecules 14:1186–1193
19. Leibler L (1988) Makromol Chem Macromol Symp 16:1–17
20. Helfrich WW (1980) In Physique des défauts, Les Houches XXXV, Balian R, Kléman M, Poirier J (eds), North Holland, Amsterdam 713–755
21. Rand RP et al (1990) Biochem 29:76–87
22. Evans EA, Rawicz W (1990) Phys Rev Lett 64:2094–2097
23. Bo L, Waugh RE (1989) Biophys J 55:509–517
24. Kummrow M, Helfrich W (1991) Phys Rev A44:8356–8360
25. Lis LJ et al (1982) Biophys J 37:657–666
26. Beblik G, Servuß RM, Helfrich W (1985) J Phys (Paris) 46:1773–1778
27. Rand RP et al (1988) Biochem 27:7711–7722
28. Servuß RM, Harbich W, Helfrich W (1976) Biochim Biophys Acta 436:900–903
29. Schneider MB, Jenkins JT, Webb WW (1984) Biophys J 45:891–899
30. Schneider MB, Jenkins JT, Webb WW (1984) J Phys (Paris) 45:1457–1472
31. Sakurei I, Kawamura Y (1983) Biochim Biophys Acta 735:189–192
32. Faucon JF et al (1989) J Phys (Paris) 50:2389–2414
33. Mutz M, Helfrich W (1990) J Phys (Paris) 51:991–1002
34. McIntosh TJ, Magid AD, Simon SA (1989) Biochem 28:17–25
35. Auroy P, Auvray L, Léger L (1991) Phys Rev Lett 66:719–722
36. DeGennes PG (1985) Scaling Concept in Polymer Physics; Cornell University Press, Ithaca, NY
37. DeGennes PG (1980) Macromolecules 13:1069–1075
38. DeGennes PG (1990) J Phys Chem 94:8407–8413

Progr Colloid Polym Sci (1995) 98:284–287
© Steinkopff Verlag 1995

Dilute and concentrated phases of vesicles at thermal equilibrium. Effect of bilayer elasticity

F. Auguste
A.M. Bellocq
D. Roux
F. Nallet
T. Gulik-Krzywicki

F. Auguste · Dr. A.M. Bellocq (✉)
D. Roux · F. Nallet
Centre de Recherche Paul Pascal
CNRS
Avenue A. Schweitzer
33600 Pessac, France

T. Gulik-Krzywicki
Centre de Génétique Moléculaire
CNRS
91190 Gif sur Yvette, France

Abstract It has been shown theoretically that phases of unilamellar and multilamellar vesicles are stable when the curvature energy of the bilayer, described by two elastic constants κ and $\bar{\kappa}$ associated respectively with the mean curvature and the Gaussian curvature, is small. We report here on a systematic study of the effect of the alcohol chain length on the stability and structure of vesicles phases made of brine, Sodium Dodecylsulfate (SDS) and alcohol. The alcohol chain length has been varied from C5 (pentanol) to C10 (decanol). Using different techniques we find that a phase of polydisperse vesicle exists in all the systems studied. In the dilute regime at very low membrane volume fraction ϕ_M, polydisperse unilamellar vesicles are present, their mean size increases with the alcohol chain length. As ϕ_M increases they coexist with oligolamellar vesicles with a small number of shells. The relative proportions of both types of vesicles depend upon the bilayer elasticity.

Key words Unilamellar and multilamellar vesicles – bending energy – light scattering – conductivity – freeze fracture electron microscopy

Recently, we reported experimental results on the system made of brine-sodium dodecylsulfate(SDS)-octanol, which demonstrate the existence of three different phases whose basic structural unit is a bilayer [1]. These phases include a lamellar phase L_α, a randomly connected sponge phase, L_3, and a phase of vesicles L_4. In this system the stability of the structures is controlled by the bilayer composition since the sequence L_4–L_α–L_3 is found at fixed SDS concentration by increasing the alcohol content. It has been proposed that thermodynamically unilamellar and multilamellar vesicles with a small number of shells may arise when the bending energy to create a sphere $4\pi(2\kappa + \bar{\kappa})$ is not too large compared to $k_B T$ [2], κ and $\bar{\kappa}$ are two elastic constants respectively associated with the mean and the Gaussian curvatures. κ controls the amplitude of the undulations of the bilayer and $\bar{\kappa}$ the bilayer topology; negative $\bar{\kappa}$ values favor vesicles or lamellar phases while positive $\bar{\kappa}$ values favor highly connected structures such as those found in the sponge phase. Moreover, one knows that the constant κ of SDS-alcohol bilayers is very sensitive to the alcohol chain-length with typical values increasing between 1.3 $k_B T$ and 13.0 $k_B T$ from hexanol to decanol systems [3]. Therefore, the mixtures brine-SDS-alcohol appear as good candidates to investigate the effect of the bilayer elasticity on the structure and stability of the L_4 phase. In this paper, we present a series of experimental results obtained with systems made of SDS-alcohol-brine; the alcohol chain length has been varied between pentanol (C_5) and decanol (C_{10}).

Phase diagrams

Six quaternary systems made of SDS-water-sodium chloride-alcohol have been studied. The following alcohols have been used: pentanol (C_5), hexanol (C_6), heptanol (C_7),

Progr Colloid Polym Sci (1995) 98:276–279
© Steinkopff Verlag 1995

octanol (C_8), nonanol (C_9) and decanol (C_{10}). We focus our study on the dilute part of the phase diagram corresponding to the brine corner. Figure 1 presents for the various alcohols, the phases obtained by varying the alcohol/surfactant ratio (A/S). As the alcohol content increases, the four following phases are identified: L_1, L_4, L_α, L_3 (L_1: micellar phase: L_4: vesicle phase; L_α: lamellar phase; L_3: sponge phase). All the diagrams exhibit the four phases, except the decanol system for which in the conditions of the study ($T = 25\,°C$, salinity 20 g/l NaCl) the L_α and L_3 phases do not exist. A similar behavior has also been observed in other surfactant-alcohol-water mixtures [4, 5].

The L_4 phases exist in a very narrow range of A/S ratios but they extend between 0 and 10% of surfactant. As in the case of octanol [1], all the phase boundaries are straight lines directed toward the brine corner, therefore corresponding to a constant A/S ratio. We may thus consider that any linear path pointing toward the brine corner corresponds to a dilution line at constant membrane composition. In the L_4 phases made with long alcohols ($C_7 - C_{10}$), the membrane contains less than one molecule of alcohol per surfactant molecule. Due to the partioning of hexanol and pentanol between the membrane and the solvent the L_4 phase is formed in these two cases for higher values of the A/S ratio. Dilute L_4 samples are isotropic, clear and low viscous; as the SDS concentration increases, they become weakly flow birefringent and very viscous. The change in behavior occurs around 5% of SDS in weight. All the L_4 samples strongly scatter light.

Like for octanol [1], the lamellar phase can be divided into two subregions: in region I, at low alcohol content, large multilamellar vesicles referred to as spherulites are dispersed in the lamellar phase; in region II the lamellar phase is free of these textural defects. The extent of region I increases with the chain length of the alcohol. The boundary of the L_4 phase at high alcohol concentration is difficult to determine because the separation between the L_4 and L_α phases occurs very slowly and can take several days or weeks. This limit has therefore been established by phase contrast optical microscopy which allows to easily detect the eventual presence of spherulites (L_α phase).

Freeze fracture electron microscopy

In order to achieve the best preservation of the sample structure upon cryofixation, we replaced water by a water-glycerol mixture (33% in volume). This substitution alters the phase diagram. In presence of glycerol, the dilute L_4 phases (SDS < 1%) prepared with hexanol and heptanol do not exist, and in the case of pentanol the entire L_4 domain disappears. Images of samples prepared at low SDS concentration (1% SDS) clearly show the existence of a polydisperse population of vesicles with sizes ranging from 200 to 2000 Å in the case of octanol and more than

Fig. 2 Freeze-fracture electron micrographs of L_4 phases (prepared with brine containing 33% glycerol). A and B illustrate the dilute regime and C and D the concentrated one. A) 1% SDS-octanol; B) 1% SDS-decanol; C) 9% SDS-hexanol; D) 9% SDS-octanol

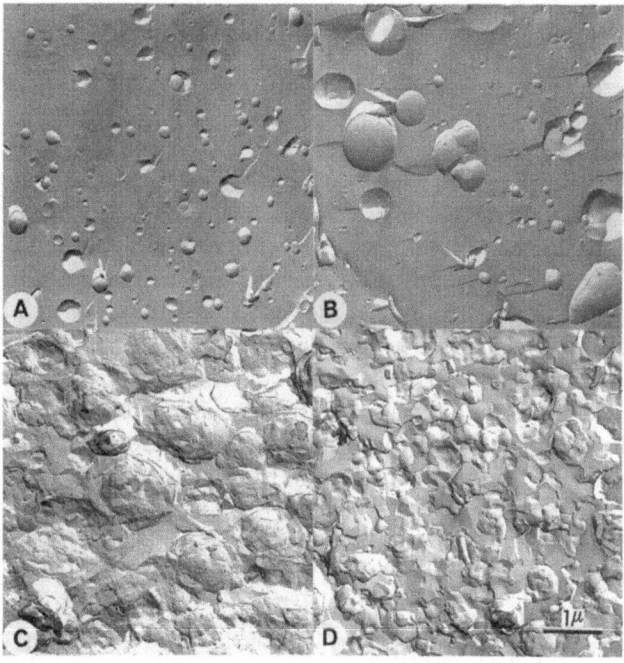

Fig. 1 Sequence of phases at $T = 25\,°C$ along the line at 5% SDS for the various systems SDS-alcohol-brine (20 g/l NaCl). (A/S is a molar ratio). L_1 micellar phase, L_4 vesicle phase, L_α lamellar phase (I and II are lamellar phases respectively with and without spherulites), L_3 sponge phase. Since in the dilute region (SDS < 10%) the limits of the phases are straight lines at constant A/S ratio, this figure totally describes the diagrams between 0 and 10% SDS

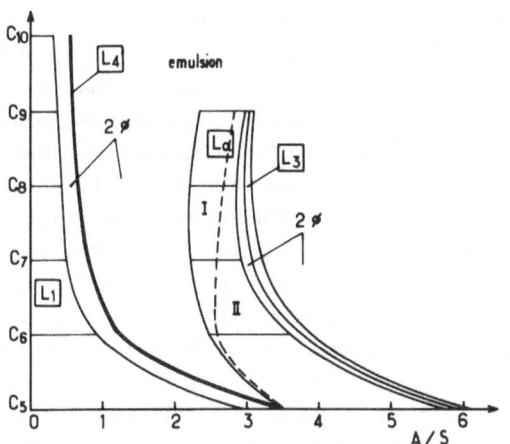

278
F. Auguste et al.
Effect of bilayer elasticity on vesicle stability

4000 Å for decanol (Fig. 2a). The dominant structure consists of unilamellar vesicles, only a few multilamellar aggregates are present with a higher proportion for decanol than octanol. At high SDS concentration (9% SDS), we see polydisperse vesicles almost closely packed for hexanol and heptanol, and dispersed in brine in the cases of longer alcohols (Fig. 2b). In these concentrated phases, the average size of the onions decreases as the alcohol chain length increases. In the intermediate concentration range ($\sim 5\%$ SDS) small unilamellar vesicles and multilamellar onions coexist.

Dynamic light scattering

Dynamic light-scattering experiments have been performed in order to determine the characteristic size of the vesicles at very dilute concentration. A monoexponential signal, characterized by a single frequency Γ, proportional to q^2, is observed at very low SDS concentration (SDS < 0, 5%). The proportionality constant is the translational diffusion coefficient D which is related through the Stockes–Einstein law ($R = \frac{k_B T}{6\pi\eta D}$) to the mean radius of the vesicles. At 0.1% SDS, we get $R = 150$ Å for hexanol, 645 Å for heptanol, 580 Å for octanol, 600 Å for nonanol, and 1260 Å for decanol.

Conductivity

Similar variations are observed in conductivity measurements for the various alcohols: as the volume fraction of bilayer ϕ_M (SDS + alcohol) increases, a very significant decrease, almost linear in ϕ_M, is first observed followed by a slower variation (Fig. 3). This observed decrease can be understood by taking into account two effects [3]: the first is the decrease of the conducting volume due to the encapsulation of brine inside the vesicles, and the second is an obstruction effect, due to the increase of the number of insulating spheres. At high ϕ_M, the ratio χ_0 between the conductivity of the solution σ and that of the solvent σ_0 increases with the alcohol chain length. In other words, the volume of brine encapsulated is larger for short alcohols than for longer ones. As previously discussed [1], the conductivity gives the volume fraction of the vesicles ϕ_V and, in the case of monodisperse unilamellar vesicles, their radius R ($\phi_V = \frac{R\phi_M}{3\delta}$):

$$\phi_V = \frac{1 - \chi_0}{1 + \chi_0/2} \tag{1}$$

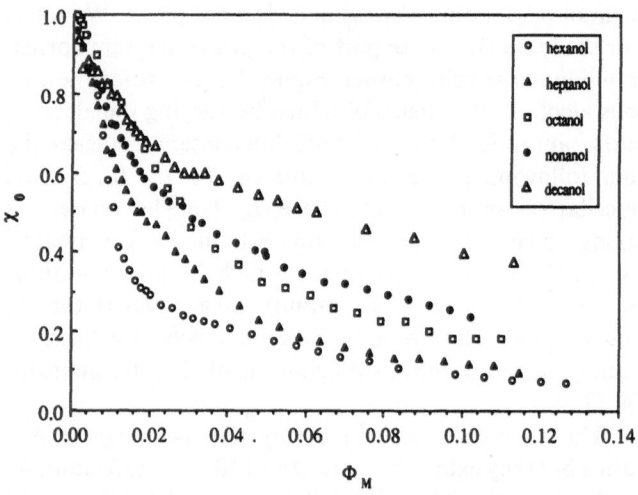

Fig. 3 Conductivity measurements in L_4 phases. Plot of the ratio χ_0 between the conductivity σ of the solutions and that of the solvent σ_0, versus the membrane volume fraction ϕ_M, for the five systems studied made of SDS-alcohol-brine (20 g/l NaCl)

$$R = \frac{3\delta}{\phi_M} \frac{1 - \chi_0}{1 + \chi_0/2}, \tag{2}$$

where δ is the bilayer thickness and has been measured by small-angle x-ray scattering in the lamellar phase [3] ($\delta = 21.5$ Å for hexanol, 22.5 Å for heptanol, 24.0 Å for octanol, 25.6 Å for nonanol, and 27.5 Å for decanol systems). From the conductivity data obtained in the dilute regime ($\phi_M \sim 0.02$) and Eq. (2), one gets the following results: $R = 250$ Å for hexanol, $R = 900$ Å for heptanol, $R = 1050$ Å for octanol, $R = 1150$ Å for nonanol, and $R = 1200$ Å for decanol.

Conclusion

Using different techniques, we have shown that a phase of vesicles at thermal equilibrium exists in all the systems studied. At low ϕ_M, the vesicles are unilamellar and polydisperse with typical sizes increasing between 150 Å for hexanol to 1250 Å for decanol. As ϕ_M increases, small sized multilayered vesicles are formed. At high ϕ_M ($\phi_M \sim 0.1$) freeze fracture electron micrographs and conductivity data indicate that these aggregates are close packed for short alcohols and dispersed in brine for the most long ones.

In all the systems investigated, the L_4 phase is very narrow, which means that a very fine tuning of $\bar\kappa$ is required to stabilize the L_4 phase. The elastic coefficient $E_0 = 2\kappa + \bar\kappa$ is related theoretically to the mean radius of the vesicles [1]. From the radii measurements, obtained by dynamic light-scattering experiments in the extremely dilute solutions, we found that E_0 varies between 2.7 $k_B T$ for hexanol and 3.7 $k_B T$ for decanol. Using κ values previously measured by NMR in the lamellar phase [3], one gets a rough estimate of $\bar{\kappa}$; $\bar{\kappa}$ decreases as the alcohol chain length increases from $-0.3\,k_B T$ for heptanol to $-22\,k_B T$ for decanol. A more complete analysis of these data will be presented elsewhere [6]. In conclusion, for all the L_4 phases studied here each constant κ or $\bar{\kappa}$ vary in a large range but their elastic energy remains small. Most of our findings are in agreement with Simons and Cates' theoretical predictions [2].

References

1. Hervé P, Roux D, Bellocq AM, Nallet F, Gulik-Krzywicki T (1993) J Phys II France 3:1255
2. Simons BD, Cates ME (1992) J Phys II France 2:1439
3. Auguste F, Barois P, Fredon L, Clin B, Dufourc EJ, Bellocq AM (1994) J Phys II France (in press)
4. Munkert U, Hoffmann H, Thunig C, Meyer HW, Richter W (1993) Prog Colloid Polym Sci 93:137
5. Porte G, Appell J, Bassereau P, Marignan J (1989) J Phys France, 50:1335
6. The results will be published elsewhere. Auguste F, Bellocq AM, Roux D, Nallet F, Gulik-Krywicki T, J Phys France (to appear)

Progr Colloid Polym Sci (1995) 98:288–290
© Steinkopff Verlag 1995

B. Pouligny
G. Martinot-Lagarde
M.I. Angelova

Encapsulation of solid microspheres by bilayers

Dr. B. Pouligny (✉) · G. Martinot-Lagarde
Centre de Recherche Paul
Pascal CNRS
Av. A. Schweitzer
33600 Pessac, France

M.I. Angelova
Institute of Biophysics
Bulgarian Academy of Sciences
1113 Sofia, Bulgaria

Abstract We analyze the problem of a bilayer membrane which is brought in contact with and adheres to a solid spherical particle. The membrane is that of a spherical giant lipid vesicle ($\sim 100\ \mu m$ in size), much larger than the solid particle. We discuss the specific case in which membrane phospholipids adhere to the particle surface in a "head-in" configuration, and take into account the fact that the membrane is made of two elastic surfaces which can slip relative to each other. This leads to non trivial results, such as a membrane topology transition driven by adhesion. We discuss the case of full particle encapsulation and derive criteria for experimental comparison. Finally, we briefly discuss experimental results obtained with charged polystyrene latex spheres.

Key words Microsphere – bilayer – phospholipid – adhesion – encapsulation

Introduction

Recently, Angelova et al. studied the interaction of charged latex microspheres with phospholipid vesicles [1]. Spheres were manipulated by means of an optical levitation trap [2] and were found to strongly adhere on vesicle bilayers. Most recent observations show that in all cases the solid spheres are not only adhered, but entirely coated by the membrane material [3]. These observations are important because they may help understand basic biological processes such as particle encapsulation and endocytosis [4] from a purely physical viewpoint.

At the present stage the molecular mechanism of phospholipid layer adhesion on the latex sphere surface is not known and is open to conjecture.

This short paper discusses one possible hypothesis, namely that of a "hydrophilic" particle, i.e., the case in which phospholipids in contact with the solid sphere surface are in a "head-in" configuration [5]. On this basis, we study the equilibrium configurations of the sphere-membrane complex. In our model, the fact that the membrane is made of two elastic surfaces that can slip relative to each other is essential, and leads to non trivial features such as an adhesion-induced transition of the membrane topology. This point is explained in the following section. The third section is dedicated to the equilibrium of fully encapsulated spheres. Our predictions about the influence of sphere size and the number of encapsulated spheres can be taken as criteria for experimental testing of the model. In the concluding section, we briefly discuss the case of latex spheres.

Bilayer topology transition driven by adhesion

The problem is illustrated in Fig. 1(a)(b)(c). The membrane of a spherical vesicle of radius R adheres to a solid sphere of radius r. Fig. 1(a) shows the intuitive configuration (hereafter denoted (a)) in which the membrane behaves as a single elastic surface. This study deals with large adhesion energies ($A \geq 10^{-2}\ \mathrm{J \cdot m^{-2}}$), so that membrane elasticity just involves a stretching term. We also suppose that the vesicle initially has no or little available excess area.

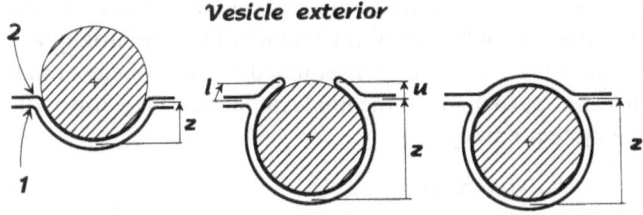

Fig. 1 Model scenario for coating ((a), (b)) and encapsulation ((c)) of a hydrophilic microsphere by a lipid bilayer. 1 and 2 are the two membrane monolayers. The membrane is that of a giant lipid vesicle, which is supposed to be much larger than the microsphere ($R \gg r$, see text). Lengths denoted z, u, l are normalized to sphere radius r. For clarity, the bilayer thickness has been considerably enlarged

Adhesion on the solid sphere consumes a finite surface area, which results in increasing the membrane tension. The equilibrium configuration is found by balancing adhesion forces and membrane tension.

We want to show that when A is large enough, because of the bilayer character of the membrane, configuration (a) is destabilized into the non trivial configuration (b) (Fig. 1(b)). We denote 1 and 2 as the inner and outer monolayers, respectively. In (b) monolayer 2 folds up and creeps along the solid sphere surface in a rolling motion. This process amounts to creating two defects, namely a dislocation loop and a pore.

The total energy involved in (b) is the sum of three terms, namely, stretching, adhesion and defects:

$$E = E_{st} + E_{ad} + E_{def} . \tag{1}$$

The system equilibrium is found by minimizing E. We assume $r \ll R$, which allows us to drop the condition of vesicle volume conservation. To calculate E_{st}, we suppose that each monolayer has a constant composition (no flip-flop) and a stretching modulus equal to $k_S/2$. Then:

$$E_{st} = \frac{1}{2}(k_S/2) \frac{(\Delta S_1)^2 + (\Delta S_2)^2}{S} . \tag{2}$$

Here, S is the initial membrane surface area and $\Delta S_{1,2}$ is the area increase of $1, 2$ in the final configuration. From simple geometry, we find:

$$\Delta S_1 = \pi r^2 (z^2 - \mu) \tag{3a}$$
$$\Delta S_2 = \pi r^2 (z^2 + 4u - \mu) , \tag{3b}$$

where z and u are the reduced distances defined in Fig. 1(b). Here, μ is a parameter, which is positive if the vesicle initially had a finite amount of excess area (quasi

spherical vesicle), or negative if the membrane was stretched. We will suppose $\mu \ll 1$ and will set out our results everywhere to first order in μ.

The adhesion energy is given by

$$E_{ad} = - A S_{ad} , \tag{4}$$

where S_{ad} is the adhered surface area:

$$S_{ad} = 2\pi r^2 (u + z) . \tag{5}$$

For well separated dislocation and pore boundary, the defect energy can be written in the form:

$$E_{def}^\infty = \gamma_d l_d + \gamma_p l_p . \tag{6}$$

Here, γ_p, l_p are the dislocation line tension [6] and length respectively, and γ_p, l_p are the equivalent quantities for the pore boundary. At the onset of the instability (small u), we expect E_{def} to increase continuously from 0 to E_{def}^∞ over a crossover distance presumably on the order of the bilayer thickness [7]. For a rough description, this can be set in the form:

$$E_{def} = E_{def}^\infty \frac{lr}{\xi} , \tag{7}$$

where ξ is a characteristic (molecular) length, and l is the reduced curvilinear length of monolayer 2 between the two defect lines (see Fig. 1(b)). Eq. (7) is valid for $l \leq \xi$ and E_{def}^∞ is defined for $l \gg \xi$.

$E(u, z)$ has a minimum at $u = \bar{u}$ and $z = \bar{z}$, with $\bar{u} \neq 0$ in general. A negative $\bar{u}$ is unphysical and means that configuration (a) is stable. This is always so at small adhesion energies. To find the stability limit of (a), we calculate the reduced height ($\bar{y}$) which is the value of z at equilibrium in Fig. 1(a). $\bar{y}$ is found by minimizing the energy in scenario (a), which reads:

$$E_{(a)}(y) = E(u = 0, z = y) \tag{8a}$$
$$= E_{st}(u = 0, z = y) + E_{ad}(u = 0, z = y) . \tag{8b}$$

This gives $\bar{y} = \chi + (1/3)\mu\chi^{-1}$, where χ is the undimensional parameter defined by $\chi = [2(A/k_s)(R/r)^2]^{1/3}$. The condition for (a) to be stable is found from the variation of $E(u, z)$ in the vicinity of $(u = 0, z = \bar{y})$.

We have:

$$\left(\frac{\partial E}{\partial z}\right)_{u=0, z=\bar{y}} = 0 . \tag{9a}$$

After some algebra, we find:

$$\left(\frac{\partial E}{\partial u}\right)_{u=0, z=\bar{y}} = 2\pi r^2 A \chi^{-3}(\chi - \chi_0)\left[-\chi^2 + \frac{1}{3}\mu(\chi + 1) \right] + 4\pi \frac{r^2}{\xi}\gamma , \tag{9b}$$

with $\gamma = 1/2\,(\gamma_d + \gamma_p)$ and $\chi_0 = 1 - (1/3)\mu$.

282

B. Pouligny et al.
Encapsulation of solid microspheres by bilayers

For $\chi \leq \chi_0$, or, equivalently, $A \leq A_0 = (1/2)k_S(r/R)^2(1-\mu)$, the right-hand side of Eq. (9) is positive whatever the value of γ, and then configuration (a) is stable. We find $\bar{y} = 1$ to first order in μ for $\chi = \chi_0$. This means that coating of the sphere by the membrane proceeds according to (a) at least up to the sphere equatorial plane. A finite γ stabilizes scenario (a) up to a limit $A_c > A_0$. The instability threshold is found by solving

$$\left(\frac{\partial E}{\partial u}\right)_{u=0, \, z=\bar{y}} \leq 0 \qquad (10)$$

for A. Equation (10) always has a solution $A \geq A_c$, but this solution is physical only if $A_c < A_e$, where $A_e = (4-\mu)k_S(r/R)^2$ is the minimum value of A to reach total encapsulation ($\bar{y} = 2$) in (a). Setting $A = A_e$ in (9b) and (10), we find the condition for the existence of configuration (b):

$$4\frac{\gamma}{\xi} < A_e. \qquad (11)$$

If (11) is fulfilled, Eq. (10) has a solution which is located between A_0 and A_e. To make an estimate, we may take $k_S \sim 0.2 \, \text{J} \cdot \text{m}^{-2}$ [8], which gives $A_0 \sim 0.1 \, (r/R)^2 \, \text{J} \cdot \text{m}^{-2}$ and $A_e \sim 0.8 \, (r/R)^2 \, \text{J} \cdot \text{m}^{-2}$. To set (11) in a rough numerical form, we estimate $\gamma \sim k_c/d$, where k_c is the bilayer bending elastic constant and d a molecular cut-off, say, the bilayer thickness. Equation (11) now reads:

$$\left(\frac{r}{R}\right)^2 \geq \frac{k_c}{k_s \, d \, \xi}, \qquad (12)$$

where ξ may be taken on the order of a bilayer boundary "fuzziness", say, $\xi \sim 1 \, \text{nm}$ [9]. With $k_c = 0.4 \times 10^{-19} \, \text{J}$ [10], $d = 4 \, \text{nm}$, we obtain $r/R \geq 0.2$. The value of A_e which corresponds to this lower boundary is about $0.04 \, \text{J} \cdot \text{m}^{-2}$ (with a surface area per phospholipid equal to $60 \, \text{Å}^2$ [11], this adhesion energy corresponds to about $6 \, k_B T$ per molecule). In summary, if $r/R > 0.2$ and if the adhesion energy is larger than $0.04 \, \text{J} \cdot \text{m}^{-2}$, scenario (b) is energetically more favourable than (a). The instability starts above the sphere equatorial plane ($z = 1$) and below the sphere top ($z = 2$).

Particle encapsulation

If A is large enough, adhesion of the membrane proceeds up to sphere top ($u + z = 2$), and pore fusion follows. The resulting geometry is shown in Fig. 1(c). This geometry is the direct consequence of pore fusion and can be viewed as general, whatever the coating process might have been, (a), (b), equilibrium or non equilibrium. In this section, we want to study this configuration at equilibrium, a task which amounts to calculating the position (z) and size of the dislocation line that runs around the sphere. Notice that the part of the membrane material which initially was in monolayer 2 and now is adhered on the sphere surface does not participate in the final equilibrium. We now write the system energy as:

$$E(z) = E_{\text{st}}(z) + E_{\text{def}}(z), \qquad (13)$$

where E_{st} is defined as in Eq. (2), with the same expression for ΔS_1 (Eq. (3a)). The expression for ΔS_2 now reads:

$$\Delta S_2 = \pi r^2 \left[(2-z)^2 - \mu\right] + \lambda 4\pi r^2. \qquad (14)$$

Here, $\lambda 4\pi r^2$ represents the amount of membrane material that has been taken from monolayer 2 for coating the sphere. λ is a number presumably on the order of unity. The defect energy is simply:

$$E_{\text{def}}(z) = 2\pi r \gamma_d (2z - z^2)^{1/2}. \qquad (15)$$

For $\lambda = 1$, $E_{\text{st}}(z)$ has a minimum at $z = \bar{z} = 1.39 + 0.01 \mu$ (to first order in μ). This means that in the limit of a small dislocation line tension the sphere appears as located across the vesicle membrane, with its center in the vesicle interior. This is the situation illustrated in Fig. (c). (Here, one may object that the constraint of vesicle volume conservation, which was ignored in our calculation, might change the value of $\bar{z}$. In fact it does, but only slightly. We find $\bar{z} \approx 1.39 \, [1 - 0.46 \, (r/R)]$ to first order in (r/R)).

The importance of the defect line tension in the equilibrium configuration can be estimated from the ratio $E_{\text{st}}(\bar{z})/E_{\text{def}}(\bar{z})$. This ratio scales as $k_s r^3/R^2 \gamma$, i.e. as $k_s r^3 d/k_c R^2$. For a numerical estimate, we give k_s, k_c, and d the same values as in the second section. With $R = 30 \, \mu\text{m}$ (giant vesicle), this results in $E_{\text{st}}/E_{\text{def}} \sim 1$ for $r \approx 0.4 \, \mu\text{m}$. We thus find that the defect energy is negligible for spheres several microns in size. If the membrane has little excess area ($\mu \ll 1$), the encapsulated sphere is across the vesicle contour. With small spheres ($r < 0.4 \, \mu\text{m}$) or large excess areas ($\mu > 1$), the dislocation loop shrinks out, which results in the sphere being completely inside the vesicle. In this description, the sphere keeps connected to the vesicle contour by a neck, or a tube if enough excess area is available. Notice that a true particle "endocytosis" [4] would require the rupture of this connection. But this point is beyond the scope of our analysis.

The fact that the sphere center is always inside the vesicle interior is due to the nature of the coating process that disconnects monolayer 2 into two parts. This results in creating more tension in 2 than in 1. If a second sphere is brought in contact with the same vesicle and encapsulated, this dissymmetry will be increased. The model thus predicts that two large ($r \gtrsim 0.4 \, \mu\text{m}$) identical and well separated encapsulated particles will be more inside the vesicle interior ($z > \bar{z}$) that one single particle ($z = \bar{z}$).

Progr Colloid Polym Sci (1995) 98:280–283
© Steinkopff Verlag 1995

Conclusion

We have described some features of how a "hydrophilic" solid spherical particle is coated by a bilayer. Our analysis shows that adhesion of the membrane on the particle surface can change the topology of the bilayer. We believe that this phenomenon is observable, for instance with fluorescent markers attached to the membrane.

The properties that we established about the penetration inside a vesicle of one or more encapsulated particles can be verified in experiments and used to test the adequacy of the model. These experiments should involve giant vesicles ($R \geq 20 \ \mu$m) and "hydrophilic" spheres a few microns in diameter. The spheres can be manipulated optically [1, 2]. An external control of the vesicle size or of the membrane tension [12] would be ideal.

In the Angelova et al. experiments [1, 2] spheres about 15 μm in diameter which were brought in contact with spherical vesicles (corresponding to $\mu \approx 0$) stay across the membrane after encapsulation. Similar experiments with flaccid vesicles ($\mu \geq 1$) lead to complete penetration of the solid particle in the vesicle interior. These features agree well with the above model. However, in some examples, encapsulated latex spheres are slightly outside of the vesicle contour [1, 2].

We believe that this observation cannot be reconciled with our prediction that the sphere center should be inside the vesicle contour, in spite of the many approximations that we made. This discrepancy raises questions about the molecular mechanism of adhesion of phospholipids on the surface of charged polystyrene latex beads. This mechanism might be more complex than or very different from the simple "head-in" anchoring that we took as our basic assumption. This point is currently under investigation.

Acknowledgements This work is supported by *Ultimatech* (CNRS) program, and is a part of a *Programme International de Coopération Scientifique* between CNRS and the Bulgarian Academy of Sciences.

References

1. Angelova MI, Pouligny B, Martinot-Lagarde G, Gréhan G, Gouesbet G (1994) Prog Colloid Polym Sci 97:293–297 (Proceedings of VII ECIS Conference, September 12–16 (1993), Bristol, UK
2. Angelova MI, Pouligny B (1993) Pure Appl Opt 2:261–276
3. Angelova MI, Martinot-Lagarde G, Pouligny B (1995) preprint submitted for publication
4. See e.g. Darnell J, Lodish H, Baltimore D (1986) Molecular Cell Biology. Scientific American Books, New York
5. See e.g. Hunter RJ (1989) Foundations of Colloid Science. Clarendon Press, Oxford, Chap 12
6. Kleman M (1977) Points, lignes, parois. Tome 1 Les Editions de Physique, Orsay
7. Dan N, Pincus P, Safran SA (1993) Langmuir 9:2768–2771
8. Kwok R, Evans E (1981) Biophys J 35:637–652
9. Egberts E, Marrink SJ, Berendsen HJC (1994) Eur Biophys J 22:423–436
10. Faucon JF, Mitov MD, Méléard P, Bivas I, Bothorel P (1989) J Phys France 50:2389-2414
11. Nagle JF, Wiener MC (1988) Biochim Biophys Acta 942:1–10
12. Evans E (1990) Colloids and Surfaces 43:327–347

Progr Colloid Polym Sci (1995) 98:291–294
© Steinkopff Verlag 1995

R.J. Pugh
M.W. Rutland
E. Manev
P.M. Claesson

A fundamental study of mica flotation in dodecylamine collector

Dr. R.J. Pugh (✉) · M.W. Rutland
E. Manev · P.M. Claesson
Institute for Surface Chemistry
Box 56 07
11486 Stockholm, Sweden

†A full account of this study will be published elsewhere in the future.

Abstract The batch flotation response of mica in dodecylamine solution was related to foam film experiments where the stability, thickness, and interfacial potentials at the air/dodecylamine solution interface was determined. These results were compared with surface force data (reported in an earlier publication) in which hydrophobic adhesion (pull-off force), adsorbed film thickness, and the interaction between molecularly smooth mica sheets in the amine collector solution was determined. The data covered a range of pH values. Maximum flotation occurred at pH 8 and correlated to a tightly packed hydrophobic collector monolayer giving maximum hydrophobicity to the mica surface. From extended DLVO theory, it was shown that heterocoagulation between the bubble and the mica could only occur providing there was a very long range hydrophobic interaction force to counterbalance the repulsive van der Waals and electrostatic forces.

Key words Flotation – mica – dodecylamine collector – surface force measurements – free aqueous foam films

Introduction

The solution properties of the dodecylamine chloride (DAHCl) surfactant system have been shown a change drastically with pH [1, 2]. In the literature the equilibrium constants are well defined and this has enabled the concentration of the different types of amphiphilic species (monomers, dimers) to be accurately mapped. Overall, an increase in pH will lead to a reduction of the number of charged species in solution and this can have an important influence on the intermolecular interactions between the various solution and precipitated species and on the adsorption of these species to the air/solution or mineral/solution interface. In the present study with dodecylamine surfactant, we have related the adsorption and interaction at the air/water interface and the mica/air interface to flotation of the mica.

Experimental

All the experimental studies were carried out within the temperature range 20°–22 °C. The chemical used in the experiments were Dodecylamine solution DAHCl (Eastman Kodak), sodium hydroxide and hydrochloric acid (Merck). Muscovite mica powder (MF60 grade, surface area 5 m²/g, Ernstrom Minerals) with the particle size specifications given as 100 w% < 250 μm, 70–90 w% < 150 μm, 20–40 w% < 75 μm and 5 to 20 w% < 53 μm. Smooth mica sheet supplied by Mica Supplies Ltd., UK,

Progr Colloid Polym Sci (1995) 98:284–287
© Steinkopff Verlag 1995

was used in the surface force measurements (the mica basal plane). For mica particles dispersed in aqueous solution, we have assumed that the basal plane plays a dominant role with regard to adsorption of amine and the interaction with air bubbles, since the surface area contribution from the faces of the relatively large sheet-like particles is considerably greater than the edges.

Flotation experiments

Flotation was carried out in a Clausson Cell (180 ml capacity) A suitable range of flotation data was obtained in DAHCl solution (initial concentration 10^{-3} M) with 4 gm of mica. The suspension was preliminary conditioned by slow stirring for 30 s followed by stirring at 100 rpm for 10 s which generated bubbles for flotation. The floated mica was scraped-off from the surface and weighed after drying.

Equilibrium film thickness measurements

Microscopic films of radius 0.1 mm were studied by the well-established microinterferometric method of Scheludko [3], which is shown in Fig. 1a. The horizontal film is formed between the tips of the menisci of a biconcave drop held in a vertical cylindrical glass tube of 2.2 mm radius by sucking the liquid out of the drop. The amount of liquid in the biconcave drop and the film radius are controlled by the Teflon microsyringe which contains the reservoir of solution and leads to contact with the drop through a capillary as indicated.

DLVO theory applied to thin aqueous films

The equilibrium thickness of an horizontal film is governed by the condition that the sum of the capillary pressure, P_c and the Π_i (e.g., the various components of the disjoining pressure), is zero.

$$P_c - \sum \Pi_i = 0 , \tag{1}$$

where $P_c = 2\gamma/R_c$ (where R_c is radius of curvature of the liquid surface). Also, the disjoining pressure between approaching interfaces is given by;

$$\sum \Pi_i = \Pi_{vdw} + \Pi_{el} , \tag{2}$$

where $\sum \Pi_i$ is the result of the balance of surface forces between Π_{vdw} the van der Waals attraction, and Π_{el} the electrostatic double layer forces,

$$\Pi_{el} = 64 \, C_{el} T \tanh^2 \phi_0 \exp(-\kappa h) , \tag{3}$$

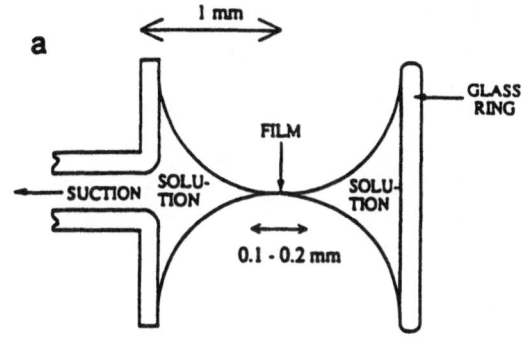

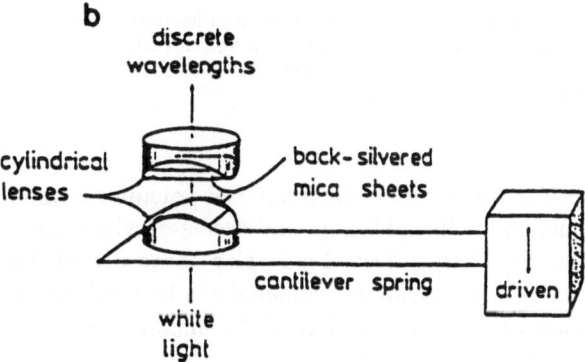

Fig. 1 (a) Measuring cell for investigating air/aqueous soln/air films and (b) surface force apparatus for studying mica/aqueous soln/mica films where optical interference fringes are used to measure the separation between the interfaces

where $\phi_0 = ze_0\psi_0/4kT$, z is ion valency, e_0 is elementary charge, and ψ_0 is potential of the diffuse electric layer. κ is the Debye–Huckel parameter and C_{el} is the electrolyte concentration. Only in thin films does Π_{vdw} become significant and can be calculated from the well known expression;

$$\Pi_{vdw} = -A(h)/6\pi h^3 , \tag{4}$$

where $A(h)$ is the "Hamaker function" of the molecular interactions weakly dependent on the film thickness.

Results and discussion

Thin aqueous equilibrium film thickness and interfacial potential at the air/aqueous solution interface

Equilibrium film thicknesses were determined in 10^{-4} M DAHCl solution as a function of pH carried out at constant ionic strength. From the film thickness values, the potential of the air/ solution interface can be calculated from Eq. (2) and the results are plotted in Fig. 2. In this

286

R.J. Pugh et al.
A fundamental study of mica flotation in dodecylamine collector

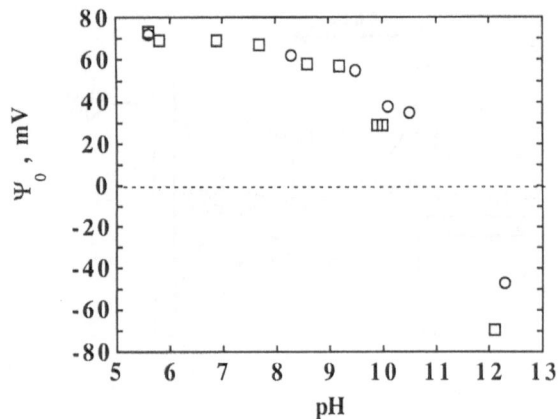

Fig. 2 Potential at the air/solution interface (determined from soap film studies) □ and the mica/solution interface (determined from surface force studies ○ versus pH for dodecylamine solution (10^{-4} M)

plot, it can be seen that the interfacial potential is about +70 mV in the low pH range and decreases with increasing pH with charge reversal occurring above pH 10. The gradual decrease in surface potential can be attributed to the decrease in concentration of the aminium ions and increase in the concentration of the free amine adsorbed at the interface or interfacial precipitated amine as the pH becomes alkaline. In the high pH region the surfactant is essentially insoluble and the restabilization of the film can be explained by a reversal of charge (positive to negative), possibly due to the preferential adsorption of OH^- or CO_3^{2-} species.

Summary of earlier reported surface force and contact angle data on the dodecylamine/mica system [4].

Details of the technique and the fitting procedures are explained fully in [4]. Briefly, the surface force technique permits the measurement of the force between two molecularly smooth mica surfaces as a function of their separation. The adhesion between the surfaces can also be obtained from the "pull-off" force – the force needed to separate the surfaces from contact. This can be used to estimate the hydrophobicity in aqueous systems since the large hydrocarbon water interfacial tension is reflected by a large pull-off force. From the nature of the forces between the surfaces in aqueous surfactant systems a great deal of information may be obtained, for example, the orientation and layer thickness of adsorbed molecules, the surface potential and the hydrophobicity of the surfaces.

In addition, the surface potential of the mica/amine solution interface determined from the curve fits could also be plotted versus pH, enabling a comparison to be made

with the charge at the air/solution interface as determined in the thin-film experiments (see Fig. 2). In this plot, it can be seen that the potential data at the air/solution and mica/solution interfaces determined at approximately the same ionic strengths appear to coincide.

Correlation of flotation with surface force
and contact angle data.

To rationalize the surface chemistry of the flotation process it is necessary to compare the data collected from the different experiments. Such a comparison is provided in Fig. 3, where the flotation efficiency, adhesion, (pull-off) force, the thickness of the dodecylamine surfactant layers adsorbed on the mica surface (as estimated from surface force measurements), and the contact angle data are also obtained from ref. [4]. From these curves it can be clearly seen that both the adhesion and the layer thickness undergo a sharp transition about pH 8.5 which is associated with the formation of a fairly thick film (equivalent to a bilayer), and this pH correlates with the region slightly beyond the maximum flotation response. In the pH range < 7, it appears difficult to correlate the hydrophobicity (determined from contact angles) with the adhesion. In the higher pH region (> 9), the thickness of the collector coating exceeds the maximum packing density and gradually builds up a bilayer, and this reduces the hydrophobicity and causes the adhesive forces between the mica to be drastically decreased. This event appears to correlate with reduced flotation performance. Also, although the

Fig. 3 Comparison of the experimental data. Flotation efficiency (–) adhesion (pull-off force) (– – – –). Dodecylamine collector layer thickness (—) and advancing equilibrium contact angle (– – –). The maximum value of the adhesion was 294 nN/m; the flotation efficiency was 75.3%; the advancing equilibrium contact angle was 98° and the dodecylamine collector layer thickness was 3.8 nm. The shaded region indicates the scatter in adhesion values

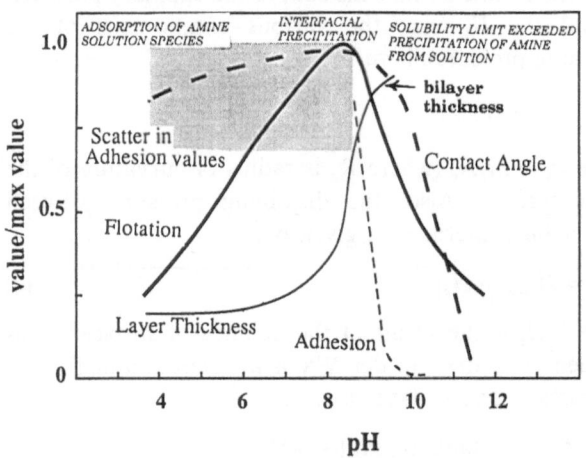

Progr Colloid Polym Sci (1995) 98:284–287
© Steinkopff Verlag 1995

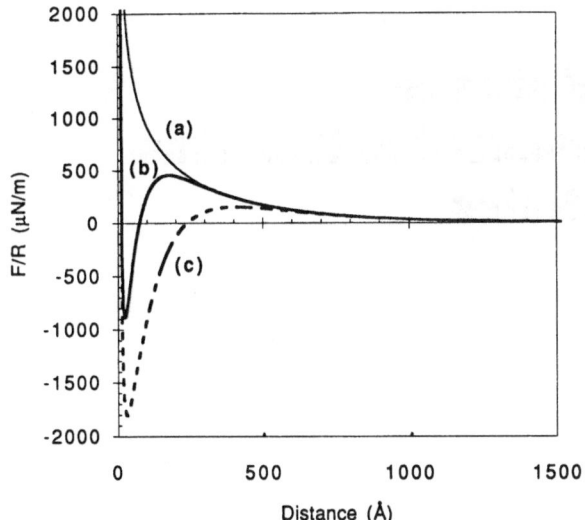

Fig. 4 Forces normalized by the particle radius between a hydrophobic mica particle ($R = 50\ \mu m$) and a flat bubble surface at pH 8. (a) $F_t = F_{vdw} + F_{el}$; no additional attractive force (b) $F_t = F_{vdw} + F_{el} + F_a$ where the additional attractive force $F_a = -C\exp(-h/D_0)$, $C = 5$ mN/m, $D_0 = 5$ nm (c) $F_t = F_{vdw} + F_{el} + F_a$, where $F_a/R = -C\exp(-h/D_0)$, $C = 5$ mN/m, $D_0 = 10$ nm, $\psi_0 = 65$ mV, $A_{123} = -2.4 \times 10^{-20}$ J and 10^{-4} 1–1 electrolyte

maximum value of the contact angle covers a wide range in pH, it can be seen that it shows a reasonable correlation with the maximum flotation response.

Application of DLVO theory to study the interaction between the mica/solution interface and the air/solution interface

From extended DLVO heterocoagulation theory [5] the total interaction force (F_t) between an air bubble approaching a hydrophobic mica particle in amine solution is given by

$$F_t = F_{vdw} + F_{el} + F_a ,\tag{6}$$

where F_{vdw} and F_{el} are the London–van der Waals (dispersion) and repulsive electrostatic interaction force, respectively. The additional attractive hydrophobic interaction force is designated (F_a). For the approach of interfaces, the van der Waal terms can be represented by the Hamaker equation and the electrostatic repulsive double-layer term

by the non-linear approximation of the Poisson–Boltzmann equation. Assuming this additional attractive force F_a is of the same nature as the so-called hydrophobic attractive interaction force, which has been shown (from surface force experiments) to occur between macroscopic hydrophobic surfaces [5], then for a flat bubble surface approaching a mica particle (radius R) it can be expressed by

$$F_a/R = -C\exp(-h/D_0) .\tag{7}$$

Equation (7) shows that the hydrophobic interaction decays exponentially with distance (h), separating the two surfaces and the two constants D_0 and C which have been defined as the exponential decay length and the pre-exponential hydrophobicity constant, respectively. C is negative since it is attractive. The greater the value of C, the more hydrophobic the surface and the larger the value of D_0 the longer the range of the hydrophobic interaction. Using a numerical program derived from Eq. (6), the magnitude of the total interaction force between a bubble and the hydrophobic mica could be computed against the separation distance. For each of the computation curves, the height of the force maximum (F_{max}) can be determined since this is the barrier which must be exceeded for film rupture. It is important to emphasise that the bubble/solution/particle interaction yields a negative value for the Hamaker interaction (-2.4×10^{-20} J) indicating a strong repulsive force. This is in addition to the strong electrostatic repulsion resulted from the overlap of the approaching double layers. For film rupture, the sum of these repulsive forces must be overcome by the attractive force.

For the computations, values of $+65$ mV were selected for the potential at the air/solution interface and mica/solution interface (which correspond to pH 8 or the region of maximum flotation response) in 10^{-4} 1:1 electrolyte and constant potential conditions were selected for the calculations. In the initial calculations, values of $C = 5$ mN/m and two D_0 values (5 and 10 nm) were selected. In Fig. 4, the results of the computations are presented in the form of the force versus distance curves. These results shows the influence of F_a/R on the repulsion forces.

In the case of $C = 5$ mN/m and $D_0 = 5$ nm then F_{max}/R is about 500 μN/m and occurs at about 15 nm, but on increasing D_0 to 10 nm then F_{max}/R is reduced to about 150 μN/m and occurs at a much greater separation distances (about 40 nm).

References

1. Somasundaran P, Ananthpadmanabhan (1979) Mittal KL (ed) In: Soln Chemistry of Surfactants Vol 2. Plenum Press pp 777–800

2. Pugh RJ (1986) Colloids and Surfaces. 18:19
3. Scheludko A (1967) Adv Colloid Interface Sci 1:391

4. Rutland M, Waltermo A, Claesson P (1992) Langmuir 8:176
5. Israelachvili JN, Intermolecular Forces Second Edit, Academic Press, 1992

Progr Colloid Polym Sci (1995) 98:295–298
© Steinkopff Verlag 1995

D. Papoutsi
P. Lianos

Photophysical studies of pyrene solubilized in thin films of glass substrates

D. Papoutsi · Dr. P. Lianos (✉)
School of Engineering
Physics Section
University of Patras
26500 Patras, Greece

Abstract Pyrene can be transferred from a cyclohexane solution into a titania film formed by the sol-gel method. Pyrene forms excimer in a dynamic manner. Excimer formation and other photophysical aspects of the fluorophore in the film are examined. Surfactant (Triton X-100 or AOT), used to control hydrolysis of titanium isopropoxide is also transferred on the film, possibly, by hydrophobic attraction.

Key words Sol-gel – titania – pyrene – excimers

Introduction

In this work, we show how a glass slide can be loaded with the excimer-forming hydrocarbon, pyrene, with the help of sol-gel-made TiO_2, in the presence of a non-ionic surfacant, Triton X-100 or an anionic surfactant AOT, both forming reversed micelles in cyclohexane. We have previously shown [1] that when titanium isopropoxide is dissolved in cyclohexane, containing Triton X-100 and water, the alkoxide is hydrolyzed and the ensuing hydroxide subsequently polymerized produces a clear gel. If a properly cleaned glass slide is dipped into the solution at an early stage of gelation, a thin, uniform, porous, transparent film forms. Its thickness is of the order of 100 nanometers. If pyrene is dissolved in the organic phase, it is carried along with the material attaching on the substrate, achieving high loading rates. Pyrene then forms excimers. The excimer formation capacity does not change with time and, as will be explained below, excimers are formed in a dynamic process.

Surfactants provide the environment necessary for the formation of a porous TiO_2 film. Other procedures of TiO_2 preparation, such as in a mixture of alcohol and water, do not succeed in incorporating pyrene in a thin film.

Materials and methods

Pyrene (Fluka), cyclohexane (spectroscopic grade, Aldrich), Triton X-100 (Aldrich), bis(2-ethylhexyl) sulfosuccinate sodium salt (AOT, Fluka) and titanium isopropoxide (Aldrich) were used as received. Millipore water was used in all experiments.

Titanium isopropoxide was added in a mixture of cyclohexane, surfactant and water. Optimum conditions were obtained when the final concentrations of surfactant, water and alkoxide in cyclohexane were 1:2:1 (molar ratio), respectively. A properly cleaned glass slide was dipped in the solution and withdrawn at a rate of 0.34 mm/s. Pyrene is introduced by solubilization in cyclohexane before alkoxide addition. Pyrene concentration in the original solution ranged from 1 to 16 mM.

Fluorescence measurements were made with a home-assembled spectroflourometer using Oriel parts and lifetime measurements with the photon-counting technique using a home-made hydrogen flash and ORTEC electronics.

Results and discussion

Figure 1 shows uncorrected pyrene fluorescence spectra. Figure 1A reveals that a substantial quantity of excimer is formed when pyrene concentration in the solution was 16 mM. The maximum of excimer emission appears in about the same position as the one observed with pyrene solution in pure solvents. This is an indication that excimer does not come from ground-state dimers which would emit at shorter wavelengths [2]. Pyrene in thin film finds itself in a very polar environment, as seen from the comparative spectra of Fig. 1B.

Fig. 1 Uncorrected fluorescence spectra of pyrene: (A): (——) in film, original concentration 16 mM; (- - -) in film, original concentration 1 mM; (...) in pure cyclohexane, 0.5 mM; (B): (——) in film, original concentration 1 mM; (- - -) in pure methanol, 10 μM; and (...) in pure cyclohexane, 10 μM

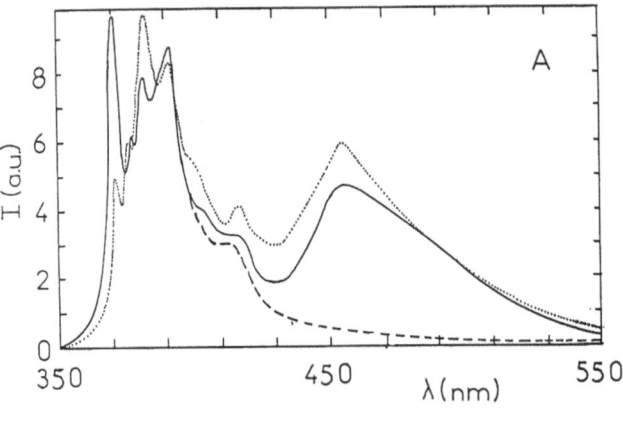

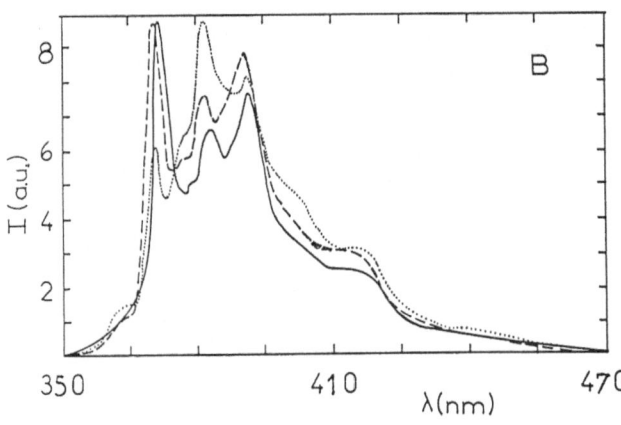

Comparison of absorption spectrum (not shown here) of pyrene in cyclohexane and of pyrene in the film indicates that there is no ground-state association, which would have been demonstrated with broadening of the film absorption [3].

Figure 2 shows the fluorescence decay profiles of monomer and excimer pyrene (normalized at maxima). The excimer profile, clearly demonstrates a rising part. This might indicate that excimer is formed in a dynamic process and it is not due to preassociated pyrene molecules. Note also that the excimer decay profile and the monomer decay profile are parallel at long times when pyrene concentration was 16 mM. This fact simply means that the excimer decay time is very short so that its fluorescence emission rate is controlled by the rate of excimer formation. Because excimer decays quickly, its dissociation rate is then negligible. Of course, the above finding also means that excimer is purely an excited-state and not a ground-state dimer, in confirmation of previous arguments.

The above results can be explained if we accept the following model: when a properly cleaned glass slide is dipped in the solution, the hydrolyzed groups adhere to the glass surface by forming –Ti–O–Si– bonds. Non-hydrolyzed isopropyl groups are exposed. Surfactant also adheres, possibly by hydrophobic interaction. Pyrene is carried along by confinement in the TiO_2 porous network.

Fig. 2 Fluorescence decay profiles of pyrene in the film: (——) monomer emission, original concentration 16 mM; (- - -) monomer emission, original concentration 1 mM; and (noisy curve) excimer emission, original concentration, 16 mM. The monomer curves have been pretreated to subtract scattered light and short-component contribution. They are convoluted with the excitation profile

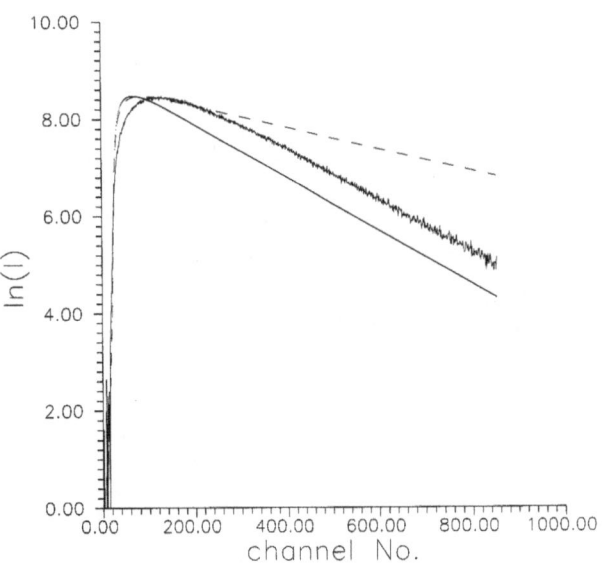

As soon as the mixture is exposed to ambient conditions, cyclohexane evaporates, while atmospheric humidity hydrolyzes the remaining non-hydrolyzed alkoxide groups. The titania film progressively becomes hydrophylic, as expected for the surface of most oxides exposed to ambient at low temperature [4, 5]. This hydrophobic to hydrophylic evolution possibly forces surfactant molecules to a self-organization in order to deal with the unfavourable conditions created by surface transformation. This subject is further studied in our laboratory.

The thinness and the porosity of the film limits the constraints imposed on pyrene molecules so that a substantial diffusion can occur allowing excimer formation in a dynamic manner.

References

1. Papoutsi D, Lianos P, Yianoulis P, Koutsoukos P (1994) Langmuir 10:1684–1689
2. L'Heureux GP, Fragata M (1989) J Photochem Photobiol B: Biology 3:53–63
3. Winnik FM (1993) Chem Rev 93:587–614
4. Ulman A (1991) In: An Introduction to Ultrathin Organic Films, from Langmuir-Blodgett to Self-Assebly; Academic Press: San Diego, CA
5. Pankasem S, Thomas JK (1991) J Phys Chem 95:7385–7393

Progr Colloid Polym Sci (1995) 98:299–302
© Steinkopff Verlag 1995

E.A. van der Zeeuw
G.J.M. Koper
D. Bedeaux

The optical properties of glass surfaces covered with latex particles

Dr. E.A. van der Zeeuw (✉)
G.J.M. Koper · D. Bedeaux
Department of Physical and
Macromolecular Chemistry
Leiden University
Gorlaeus Laboratories
P.O. Box 9502
2300 RA Leiden, The Netherlands

Abstract Scanning-angle reflectometry around the Brewster angle has been used to investigate the optical properties of positively charged polystyrene particles with radii ranging from 50 to 500 nm, deposited on a glass-water interface. A model based on Mie theory describes the data over the entire size range. This model ignores direct particle-substrate interactions, which is justified for the small differences in refractive indices in this system. Also the interaction between particles is ignored. The analysis yields three parameters: particle size, refractive index, and coverage.

Key words (scanning angle) reflectometry – adsorption – mie-scattering – latex particles – optical properties

Introduction

Many theoretical investigations have been done that are concerned with the optical properties of particles of various shapes and sizes that are in contact with a surface [1–7]. This situation is clearly more difficult than that of free particles, since both the interaction between particles and surface, and the interaction between particles themselves may play a significant role. For a film whose thickness is small compared to the wavelength considered, Vlieger and Bedeaux have developed a general description of the optical properties in terms of so-called constitutive coefficients [8]. Haarmans and Bedeaux [9, 10] have calculated these constitutive coefficients for small spheres on a substrate, taking the interactions mentioned above into account.

The results from theoretical investigations usually require calculations that are rather cumbersome. The experiments we report here allow for a simpler approach: because the differences between the various dielectric constants are small, the modification of the Mie-scattering function of the spheres due to interaction with the substrate can be neglected [3]. We also expect that for the low coverages observed (using light microscopy), it is permitted to neglect interactions between the particles. The strength of the interactions mentioned above is currently being studied [12].

We thus interpret the reflectivity as due to light reflected separately from the layer of spheres and the substrate. Using this simple model, we obtain information on the size of the particles, as well as on the coverage and their refractive index.

Theory

The latex particles are modeled as homogeneous spheres of radius a. We shall neglect size polydispersity, although it can easily be taken into account. The latex is non-absorbing at the wavelength considered, so that its refractive index is real. The particles are located at the glass/water interface. Experimentally, we observed that this interface behaved like a perfect Fresnel interface.

First, let us consider a free sphere. Given an incident plane wave, with electric field $\mathbf{E}_i = \exp\{i\mathbf{k}\cdot\mathbf{r}\}\mathbf{e}_i$, the scattered electric field is described by Mie theory [13, 14]. We are interested in light scattered in the plane determined by the wavevector $\mathbf{k}$ and the polarization vector $\mathbf{e}_i$. The scattered field is then given by:

$$E = \frac{e^{ik_2 r}}{ik_2 r} S_p(\theta) , \tag{1}$$

where k_2 is the wavenumber $k_2 = 2\pi n_2 / \lambda$, and n_2 the refractive index of the water. $S_p(\theta)$ is the scattering function for p-polarized light, and depends on the scattering angle θ, the radius of the sphere, and the ratio of the refractive indices of the sphere and water.

Next, let us consider an assembly of spheres in a plane. At sufficiently low surface coverages, the electromagnetic interactions between the spheres and the resulting modification of the scattering function can be neglected. For such a layer, light is scattered coherently in the direction with scattering angle $\pi - 2\theta_2$, where θ_2 is the angle between the incident light beam and the normal of the plane. The angle of reflection is equal to the angle of incidence, as with a homogeneous slab. The total field coherently scattered by the layer of spheres, neglecting the influence of the substrate, yields an amplitude reflection coefficient

$$r_l(\theta_2) = \frac{2\phi S_p(\pi - 2\theta_2)}{k_2^2 a^2 \cos \theta_2} , \tag{2}$$

where ϕ is the surface coverage [7].

For a pure glass/water interface, the amplitude reflection coefficient for p-polarized light is given by the Fresnel expression

$$r_{12} = \frac{n_2 \cos \theta_1 - n_1 \cos \theta_2}{n_2 \cos \theta_1 + n_1 \cos \theta_2} , \tag{3}$$

where n_1 and n_2 are the refractive indices of the glass and the water, respectively. The angle θ_2 on the waterside is related to the incidence angle θ_1 on the glass side by Snell's law: $n_1 \sin \theta_1 = n_2 \sin \theta_2$.

The total reflected field can now be approximated by considering these two reflections, and their interference. One then gets:

$$R \simeq r_{12}^2 + 2t_{12}t_{21}r_{12}\{\mathrm{Re}(r_l)\cos\Delta - \mathrm{Im}(r_l)\sin\Delta\}$$
$$+ t_{12}^2 t_{21}^2 |r_l|^2 , \tag{4}$$

with $\Delta = 2k_2 a \cos \theta_2$ the phase difference between the field reflected at the interface and the light coherently scattered from the spheres. Equality holds when interactions between spheres and substrate, and interactions between different spheres are completely negligible.

Experiment

Adsorption was performed onto the optically flat hypotenuse of a rectangular prism made of glass (refractive index

1.51509). Two such prisms were mounted in a holder, leaving a 3-mm space between the two hypotenuse faces. The method consists of measuring the reflection coefficient for p-waves at a number of angles around the Brewster angle (see Fig. 1). The light source (L) is a 5 mW HeNe laser operating at a wavelength $\lambda = 632.8$ nm. Light passes through a polarizer (P), aligned in the plane of incidence, perpendicular to the adsorption surface. Along the path to this surface, the light passes almost perpendicularly through the entrance face of the prism (C). After reflection it leaves the prism through its exit face, passes through a second polarizer (A) that again selects the p-wave, and impinges on the photodetector (D). The reflectometer is fully automated and computer-controlled. The adsorption surface was *chosen* to be horizontal (the reflectometer is oriented vertically, so one can really choose orientation of the sample by rotating the sample holder). The angle of incidence is selected by simultaneously rotating the laser and the detector supports.

The measured intensities are related to the reflectivity $R_p(\theta_1)$ by

$$I(\theta_1) = I_0 + AR_p(\theta_1) , \tag{5}$$

where I_0 is the residual intensity at the Brewster angle for a perfect interface and A is an instrument-dependent constant. The Brewster angle for the glass/water interface is about 41.3°. The amplitude reflection coefficient for a simple interface is given by the Fresnel expression (3).

At the beginning of each experiment, the reflected intensities I of the glass/water interface were measured for incidence angles θ_1 ranging from about 40° to 45°. This allowed us to determine the amplification of the signal, as well as the residual intensity at the Brewster angle. The latex suspension was then introduced into the cell and similar intensity curves were measured at about 10-min intervals. Typical curves are shown in Fig. 2. The reflected intensity evolved with time at a rate depending on the

Fig. 1 Schematic diagram of the reflectometer. (L) light source; (P), (A) polarizers aligned in the plane of incidence; (C) prism and sample cell, (D) photodetector

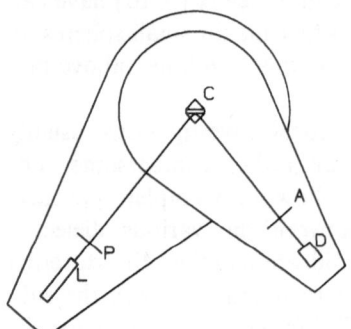

Progr Colloid Polym Sci (1995) 98:291–294
© Steinkopff Verlag 1995

Fig. 2 The reflectivity as a function of angle, before (solid symbols) and after (open symbols) injection of a solution of latex particles. Fits are performed using the Mie-based model. The radius of the particles was, according to the manufacturer, (a) 67 nm and (b) 239 nm

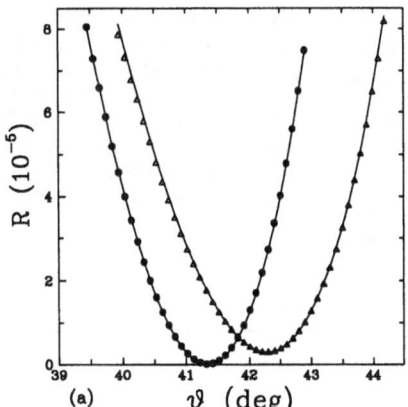 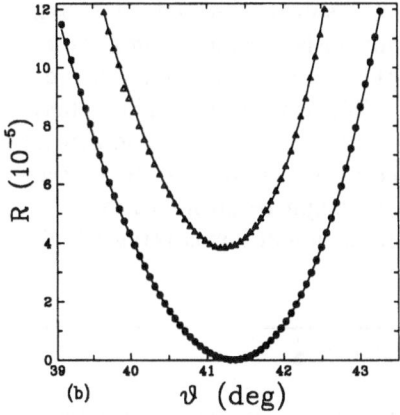

concentrations of the latex solution. The experiment ran, depending on the experiment, for 10 to 48 h. We flushed with water at the end of each experiment. Since the curves did not change significantly after that, we conclude that only adsorbed particles contributed to the evolution in reflected intensity. Data were stored in the computer for subsequent off-line analysis.

For the latex particles, we used polystyrene beads commercially available from Interfacial Dynamics Corporation (Portland, Oregon). Particles with positively charged amidine groups were chosen in order to ensure adsorption on the negatively charged surface. The particle solutions were diluted in pure water to concentrations ranging from 10^{-2} to 10^{-3}%. A range of different sizes, given in Table 1, was studied. The refractive index of the spheres is given as 1.591 (at 590 nm) by the manufacturer. The different elements of the cell were cleaned with a laboratory-use detergent (Hellmanex II; Hellma GmbH D-7840 Mullheim) and rinsed copiously with deionized superQ-Millipore water (used throughout the experiment), a dilute sulfuric acid solution, and again water.

Results and discussion

Examples of the angular dependence of the surface reflectivity are presented in Fig. 2. Similar measurements were performed on a series of latex particles with radii ranging from 50 to 500 nm (see Table 1). In the presence of the latex particles, the position of the minimum of the reflectivity may shift and also the value of this minimum may increase; one effect or the other can visually dominate depending on the particle size. The shape of the curve, most noticeably its width, also changes.

Our data were fitted to the previous discussed model, the results of which are shown in Table 1.

Table 1 The first two columns give the radius, polydispersity, and surface charge provided by the manufacturer; the remaining columns contain experimental values. The error estimates on a_{exp} refer to the variation in time and between experiments: They indicate the sensitivity of the fitting procedure to this parameter. Note that χ is normalized to the χ_f of the Fresnel curve.

a (nm)	$\sigma(\mu C/cm^2)$	a_{exp} (nm)	ϕ	χ/χ_f
58 ± 11.7%	8.1	53 ± 2	0.08	1.8
67 ± 10.9%	8.2	71 ± 1	0.09	1.6
96 ± 3.0%	4.3	99 ± 1	0.11	3.0
150 ± 3.3%	15.7	144 ± 1	0.11	1.8
239 ± 9.9%	11.0	240 ± 1	0.09	0.7
300 ± 6.3%	9.8	325 ± 1	0.07	1.9
345 ± 9.5%	15.3	373 ± 1	0.09	5.1
381 ± 3.1%	12.9	410 ± 1	0.05	2.8
477 ± 4.1%	21.4	524 ± 1	0.04	5.0

We also did the fit with an extra free parameter, introduced in Eq. (4):

$$R = r_{12}^2 + 2c_{12}t_{12}t_{21}r_{12}\{Re(r_l)\cos\Delta - Im(r_l)\sin\Delta\}$$
$$+ c_2t_{12}^2t_{21}^2|r_l|^2 . \tag{6}$$

This provided a test in using Eq. (4), because if particle interactions could be neglected, this would predict $c_1^2/c_2 = 1$. It was indeed found that this ratio was equal to one within 5%.

In general, we see good agreement between the experimental data and the fits using Eq. (4). The radius deduced from these fits, was essentially independent of time and surface coverage. The radius, as well as the final surface coverage, are given in Table 1. A measure of the quality of the fits is also given, by the average deviation χ between the experimental data and the fitted curves, normalized to the deviations observed for the Fresnel curves. For the smaller particles, determined radii are in fairly good agreement with the radii provided to us by the manufacturer. Deviations from the provided values get larger for the larger spheres. Theoretical work in progress must provide more insight in the strength of the interactions between the spheres, and furthermore, their dependence on size [12].

We also did static light-scattering experiments on a few lattices. The sizes we found were mostly in agreement with the values listed in Table 1, except for the largest ones. This might be an indication that for some particles, the interaction at low coverages is already appreciable.

Acknowledgements It is a pleasure to thank M.T. Haarmans, M.M. Wind and J. Vlieger for many illuminating discussions on the use of optical theory, A.E. Keulemans for doing the static light-scattering measurements, and G.H. Renes for help with the EM-pictures. We also thank J.P.M. van de Ploeg, G.H. van Veen, H. Verpoorten, and W.J. Jesse for technical support. The authors are grateful for financial support from the Netherlands' Technology Foundation (STW), and the EEC (contract SC1*-CT91-0696 (TSTS)).

References

1. Chauvaux R, Meessen A (1979) Thin Solid Films 62:1125
2. Ruppin R (1983) Surface Science 127:108
3. Wind MM, Vlieger J, Bedeaux D (1987) Physica A 141:33
4. Wind MM, Bobbert PA, Vlieger J, Bedeaux D (1987) Physica A 143:164
5. Bobbert PA, Vlieger J (1987) Physica A 147:115
6. Bobbert PA, Vlieger J (1986) Physica A 137:209
7. Bobbert PA, Vlieger J (1986) Physica A 137:243
8. Vlieger J, Bedeaux D (1980) Thin Solid Films 69:107
9. Haarmans MT, Bedeaux D (1993) Thin Solid Films 224:117
10. Haarmans MT, Bedeaux D (1994) Physica A 207:340
11. Bedeaux D, Koper GJM, van der Zeeuw EA, Vlieger J, Wind MM (1994) Physica A 207:285
12. Haarmans MT, personal communication
13. Van de Hulst HC (1981) Light Scattering by Small Particles, Dover, New York
14. Kerker M (1969) The scattering of light and other electromagnetic radiation, Ac. Press, New York

Progr Colloid Polym Sci (1995) 98:303–307
© Steinkopff Verlag 1995

T. Gisler
M. Borkovec
P. Schurtenberger
R. Klein

Poisson-Boltzmann cell model for charge renormalization in suspensions of interacting colloidal particles

T. Gisler (✉) · M. Borkovec
Institute of Terrestrial Ecology
Federal Institute of Technology (ETH)
Grabenstraße 3
8952 Schlieren, Switzerland

P. Schurtenberger
Polymer Institute
Federal Institute of Technology (ETH)
8092 Zürich, Switzerland

R. Klein
Faculty of Physics
University of Konstanz
78434 Konstanz, FRG

Abstract We present an investigation on the charging behavior of polystyrene latex particles in aqueous suspension. The particle bare charge is regulated by the dissociation equilibrium between protonated and ionized carboxylic surface groups. Effective charges are determined from the analysis of static light-scattering data on deionized suspensions of interacting particles. The experimental effective charges agree well with the predictions of the Poisson–Boltzmann cell model in the limit of infinite bare charge. This result is a strong indication that the renormalized particle charge is independent of the surface chemical details if the bare particle charge is sufficiently high.

Key words Light scattering – Poisson–Boltzmann equation – charge renormalization – polystyrene latex

In recent years, substantial progress has been made in the description of colloidal suspensions based on the theory of simple liquids (for recent reviews see [1, 2]). Colloidal suspensions of highly charged particles have, however, posed severe obstacles to a description on a primitive model level where the small ions and the macroions are correlated due to their mutual Coulomb interaction [3]. Thus, a contracted description for the sub-system of macroions is chosen for the interpretation of scattering experiments. In this one-component macrofluid description the presence of the small ions is accounted for by the screening of the bare Coulomb interaction between the macroions. The pair potential as a function of the particle separation r is given by the DLVO expression [4, 5]

$$u(r) = \frac{Z_{\text{eff}}^2 e_0^2}{4\pi\varepsilon\varepsilon_0} \left(\frac{e^{\kappa a}}{1 + \kappa a} \right)^2 \frac{e^{-\kappa r}}{r}, \tag{1}$$

where a is the particle radius, Z_{eff} is the effective charge of the macroion in units of the elementary charge e_0, and κ is the reciprocal Debye screening length; the solvent enters the interaction potential by its dielectric permittivity $\varepsilon\varepsilon_0$ only. Scattering experiments on micellar solutions [6, 7] and polystyrene latex suspensions [8] have, however, shown that the effective charge is usually smaller than the bare charge which is equal to the number of dissociated surface groups. For an understanding of the effective charge from the parameters of the composite system of macroions and small ions a renormalization scheme has to be applied which relates the observed effective charge to the bare charge and size of the particle, the volume fraction and the electrolyte concentration. In contrast to the rather involved renormalization schemes presented recently by Fushiki [9] and Löwen [10], the Poisson–Boltzmann cell model presented by Alexander et al. [11] replaces the system of many macroions interacting by the Yukawa potential (1) by a single particle placed in the center of a spherical cell. The size of this cell is determined by the particle volume fraction in the physical system. The Poisson–Boltzmann equation for the average electrostatic potential is solved within the cell for a given value of the bare particle charge. This solution is then matched with the solution of the linearized Poisson–Boltzmann equation at the cell boundary; the derivative of the linear potential at the particle surface is identified with the effective charge. For increasing bare charge the effective charge converges to a limit which corresponds to the saturation value (i.e., where additional counterions are said to condense on the particle surface, see Fig. 1). In aqueous solution, a common mechanism of charging a particle is the dissociation of acidic surface sites releasing protons. An understanding of bare and effective particle charges thus in principle requires the knowledge of the chemical equilibria at the particle surface; in this case the cell model has to be solved self-consistently with the chemical equilibria at the particle surface and in the electrolyte solution [12]. For the time being, we will, however, restrict our discussion to the case of a fixed bare charge without charge regulation.

In our experiments, we have used well-characterized polystyrene latex particles with an average radius of 50.7 nm and a polydispersity of 8.4% (see Table 1). The particles carry about $4 \cdot 10^4$ carboxylic surface groups with an average pK of 4.7 which was determined by titration experiments [13]. Structure factors $S(q)$ have then been measured by static light scattering at volume fractions between 10^{-5} and 10^{-4} on deionized suspensions where the excess salt had been removed by ion exchanger resin. On the other hand, $S(q)$ was calculated by solving the Ornstein–Zernike equation for the pair-correlation function with the Hypernetted-Chain (HNC) closure for a polydisperse mixture [14, 15]. The average particle radius and the polydispersity used in the calculation were

Fig. 2 Top: Measured structure factors (circles) and calculation with the polydisperse Hypernetted-Chain closure (full line) for volume fraction $5.8 \cdot 10^{-4}$ and effective charge $Z_{eff} = 550$. For the calculation, the distribution of the particle radii was replaced by an equivalent histogram distribution with three components. The mean particle radius $\bar{R} = 50.7$ nm and the polydispersity of 8.4% were taken from the experimentally determined size distribution. The relative error on each experimental point is approximately 2–3% for the whole range of scattering vectors investigated. Bottom: Same experimental data as above, but in comparison with the best-fit structure factor calculated with the monodisperse model (full line). Note that both models fit the data equally well, however, with different values for the effective charge

Fig. 1 The calculated effective charge Z_{eff} as a function of the bare charge in the spherical Poisson–Boltzmann cell model (thick line) with a volume fraction $\phi = 10^{-5}$ of particles with an average radius of 50.7 nm and a polydispersity of 8.4% at 25 °C. At low bare charges, Z_{eff} is equal to the bare charge; at high bare charges, Z_{eff} reaches a saturation value (thin line) which indicates the onset of counterion condensation. For the present system parameters, the saturation value is $Z_{eff} = 930$

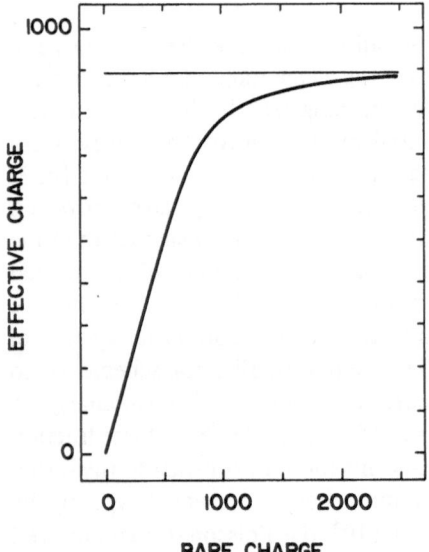

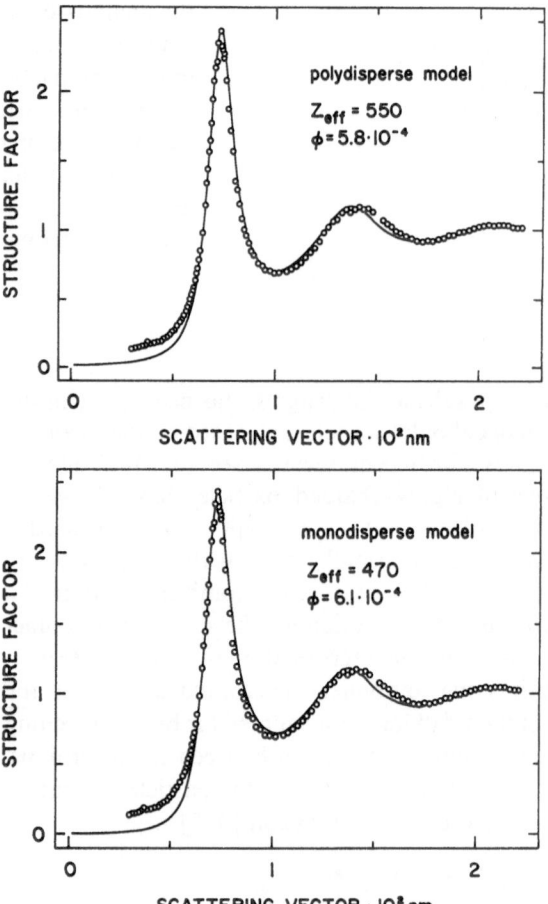

Progr Colloid Polym Sci (1995) 98:295–298
© Steinkopff Verlag 1995

Table 1 Characterization of the polystyrene latex particles. In the first column the number average particle radius $\bar{R}$, the polydispersity p_{TEM}, the radius of gyration $\overline{R_G}$: the hydrodynamic radius $\overline{R_H}$, and the specific surface are S are shown as they are directly calculated from the size distribution obtained from transmission electron microscopy. The second column contains the same calculated quantities, now corrected for shrinking under the electron beam, such that best agreement with the independently measured quantities from light scattering and N_2 gas adsorption experiments (third column) was reached.

	Uncorrected size distribution	Corrected size distribution	Independently measured
$\bar{R}$/nm	47.9 ± 0.3	50.7 ± 0.3	–
p_{TEM}/%	8.4 ± 0.5	8.4 ± 0.5	–
$\overline{R_G}$/nm	38.2	40.5	40.7 ± 0.6
$\overline{R_H}$/nm	49.3	52.0	51.8 ± 0.5
S/(m²/g)	58.8	56.1	56.8 ± 0.2

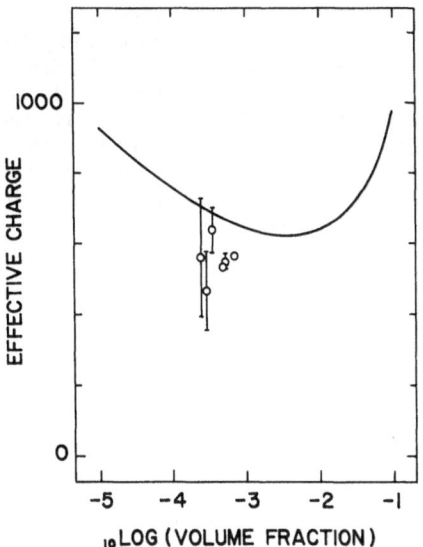

Fig. 3 Measured effective charges for the deionized carboxylate polystyrene latex (squares) and saturation values (thin line in Fig. 1) predicted by the Poisson–Boltzmann cell model in the limit of vanishing salt concentration (full line) as a function of the particle volume fraction. The system parameters are the ones used in Fig. 1. The errorbars on the experimental data indicate the reproducibility of the suspension preparation

taken from the size distribution measured by transmission electron microscopy. Effective charges and particle volume fractions have then been determined from fits of calculated structure factors to the experimental light scattering data (see Fig. 2); the effective charge is proportional to the height of the first peak of $S(q)$ and its position q_{max} increases with volume fraction. Satisfactory agreement between experiment and calculation was observed for scattering vectors around the first and second maximum in $S(q)$ for both the monodisperse and the polydisperse calculation. As the polydispersity of our particles is low, the calculated $S(q)$ for the monodisperse model and for the polydisperse model nearly coincide (except for low q), in contrast to the characteristic qualitative differences for higher polydispersity (broadening of the first peak and smearing of the oscillations). However, the effective charges from a monodisperse model are by about 15% smaller than when polydispersity is taken into account. In our experiments, we were limited to volume fractions between 10^{-5} and 10^{-4} since at lower volume fractions the peak in $S(q)$ appears at too low scattering angles; at higher volume fractions the particles started to crystallize and the signal was corrupted by multiple scattering. We observed effective charges of about 500 ± 50. These values are

slightly below the effective charges predicted from the low-salt Poisson–Boltzmann cell model in the limit of infinite bare charge (see Fig. 3). Explicit treatment of the cell model accounting for the surface chemical equilibrium between the protonated and the ionized surface groups yields effective charges which are lower by 10–20% which is within the range of the scatter of the experimental data. This small correction suggests that the detailed knowledge of the chemical equilibria at the particle surface is indeed not necessary for the understanding of charge renormalization within the Poisson–Boltzmann cell model if the bare particle charge is sufficiently high (i.e., higher than the saturation value given by the thick line in Fig. 3).

The authors are indebted to J.M. Méndez-Alcaraz and B. D'Aguanno for providing the HNC programs for the structure factor calculations. Financial support from the Swiss National Science Foundation is gratefully acknowledged.

References

1. Pusey PN (1980) In: Hansen JP, Levesque D, Zinn-Justin J (eds) Liquids, Freezing and Glass Transition. North-Holland, Amsterdam, pp 768–942
2. Nägele G (1994) Phys Reports (in press)
3. Belloni L (1993) J Chem Phys 98:8080–8095
4. Verwey EJ, Overbeek JTG (1948) Theory of the Stability of Lyophobic Colloids. Elsevier, New York
5. Medina-Noyola M, McQuarrie DA (1980) J Chem Phys 73:6279–6283
6. Chen SH, Sheu EY (1990) In: Chen SH, Rajagopalan R (eds) Micellar Solutions
and Microemulsions: Structure, Dynamics, and Statistical Thermodynamics. Springer, New York, pp 3–27
7. Bucci S, Fagotti C, Degiorgio V, Piazza R (1991) Langmuir 7:824–826
8. Versmold H, Wittig U, Härtl W (1991) J Phys Chem 95:9937–9940; Härtl H,

Versmold H, Wittig U (1992) Langmuir 8:2885–2888
9. Fushiki M (1992) J Chem Phys 97:6700–6713
10. Löwen H, Madden PA, Hansen JP (1992) Phys Rev Lett 68:1081–1084
    Löwen H, Kramposthuber G (1993) Europhys Lett 23:673–678
11. Alexander S, Chaikin PM, Grant P, Morales GJ, Pincus P, Hone D (1984) J Chem Phys 80:5776–5781
12. Gisler T, Schulz SF, Borkovec M, Schurtenberger P, D'Aguanno B, Klein R, Sticher H (1994) J Chem Phys 101:9924–9936
13. Schulz SF, Gisler T, Borkovec M, Sticher H (1994) J Colloid Interface Sci 164:88–98
14. D'Aguanno B, Klein R (1991) J Chem Soc Farad Trans 87:379–390
15. D'Aguanno B, Klein R (1992) Phys Rev A 46:7652–7656
16. Author to whom correspondence should be addressed: Thomas Gisler, Institute of Terrestrial Ecology, Federal Institute of Technology (ETH), Grabenstrasse 3, CH-8952 Schlieren, Switzerland

Progr Colloid Polym Sci (1995) 98:299–302
© Steinkopff Verlag 1995

J. Zajac
M. Chorro
C. Chorro

Calorimetric studies of micellization and adsorption of zwitterionic surfactants

Dr. J. Zajac (✉) · M. Chorro · C. Chorro
Laboratoire des Agrégats Moléculaires et
Matériaux Inorganiques
U.R.A. 79
Université Montpellier II
pl. E. Bataillon
34095 Montpellier, France

Abstract The energetics of the process of micellization of zwitterionic surfactants in aqueous solutions and their adsorption onto silica gel has been investigated. Differential molar enthalpies of dilution and adsorption were measured for (dodecyldimethyl-ammonio) ethanoate, (dodecyl-dimethylammonio) butanoate, and 3-(dodecyldimethylammonio)-1-propanesulfonate in pure water or in 0.1 M NaCl solution and at 298 K or 308 K. Endothermic molar enthalpies of micellization decrease with rising temperature, addition of electrolyte, and decreasing hydrophilicity of the zwitterionic headgroup, thus indicating the hydrophobic and electrostatic repulsive effects in the micelle formation. Adsorption of zwitterions at the silica/aqueous solution interface occurs by ion pairing between the quaternary nitrogen of the surfactant headgroup and negatively charged surface sites at very low surface coverages and by formation of surface micelles at higher coverages.

Key words Calorimetry – micellization – adsorption – zwitterionic surfactants – betaines

Introduction

Although zwitterionic surfactants do no represent a great percentage of the world production of surfactants, the market use is still increasing dramatically because of the unique properties of these materials [1–6]. Zwitterionic amphiphiles carry no formal net charge but the presence of both positively and negatively charged hydrophilic groups in the same molecule leads to the headgroup hydrophilicity intermediate between the ionic and conventional nonionic classes of surfactants. The properties of such molecules can be easily modified by varying the number of methylene groups separating the charged sites and by changing the anionic or cationic portion of the hydrophilic headgroup [3, 4].

The importance of investigation into the nature of the forces between all the components of colloidal systems has been recognized in colloid science for many years. The behavior of different zwitterionic surfactants in aqueous solutions and at interfaces has been widely investigated by means of a variety of experimental techniques [2–5], whereas calorimetry, which usually offers an opportunity for studying directly different types of interaction in the system, is applied very rarely.

This paper reports calorimetric data on dilution of micellar aqueous solutions for dodecylbetaines and dodecylsulfobetaines of varying intramolecular charge separation and on adsorption of these molecules from aqueous solutions onto silica gel. Differential molar enthalpies of dilution and adsorption are presented as functions of temperature, added electrolyte, tether length, and type of negatively charged center. These data are then interpreted to yield information on the energetics of micellization as well as on the mechanism of zwitterionic surfactant adsorption at the solid/liquid interface and its evolution with the surface coverage.

Materials and methods

Synthetic silica gel in the form of spherosil XOB015 was supplied by Rhône–Poulenc (France). This macroporous powdered substrate with spherical particles of diameter between 40 and 100 μm had a specific surface area of 25 m^2/g (the B.E.T method/nitrogen gas adsorption and the Harkins–Jura method/immersion in water). In a neutral aqueous environment this solid adsorbent is well above its point of zero charge and therefore carries a negative charge.

Carboxybetaines, $C_{12}H_{25}N^+(CH_3)_2(CH_2)_nCOO^-$, and sulfobetaines, $C_{12}H_{25}N^+(CH_3)_2(CH_2)_nSO_3^-$, with short interchange separation distances, $n = 1, 2, 3$, were synthesized in the laboratory; the methods used for their synthesis and purification are described in [6]. In neutral media, a polar headgroup of the carboxybetaines studied exists as a zwitterion [4]. The hydrophilicity of the headgroup increases with increasing tether length and reaches its maximum at n equal to 3 or 4 [4]. For similar structures, the carboxylate headgroup is more hydrophilic than the sulfonate [4].

All water for solution preparation was deionized and dust-free. The pH of solution was not adjusted during the experiments.

The enthalpy changes accompanying adsorption of zwitterions onto silica gel were measured by a microcalorimetric batch technique which allowed the adsorbing species to be introduced gradually into the calorimetric cell containing a homogeneous suspension of solid in solution [7, 8]. It was thus possible to follow the process step by step and detect enthalpy changes associated with subsequent steps. In order to cover the whole adsorption range, stock solution of molality 10 times greater than the corresponding CMC was injected into the calorimetric cell. A correction term arising from destruction of micelles and dilution of unmicellized species had to be subtract from the total enthalpic effect [7]. The number of moles of solute adsorbed on the surface at each step was determined with the aid of the adsorption isotherm. Enthalpies of dilution were measured with the same calorimeter [7, 8].

Results and discussion

Figure 1 shows the experimental enthalpic curves of dilution in different aqueous media and at different

temperatures for three zwitterionic surfactants: (dodecyldimethylammonio) ethanoate (abbreviated DDAET), (dodecyldimethylammonio) butanoate (DDABT), and 3-(dodecyldimethylammonio)-1-propanesulfonate (DDAPS). Five curves correspond to the following five systems: DDAET in pure water at 298 K, DDABT in pure water at 298 K, DDAPS in pure water at 298 K, DDAET in 0.1 M NaCl at 298 K, and DDABT in pure water at 308 K. The shape of each curve at low and moderate molalities (below and in the vicinity of the CMC) is consistent with an exothermic dilution of a micellar solution containing quite monodisperse micelles. At high molalities (much above the CMC), the effect of dilution becomes athermal, indicating that the resultant interactions between micelles are negligible from the energetic point of view. The first plateau on the curve (at low molalities) is thus a direct measure of the molar enthalpy of micelle formation. The experimental enthalpies are reported in Fig. 1. They all are endothermic; the driving free energy of micellization is principally entropic. It is well known that the major factor driving the surfactant molecules into aggregation in water at room temperature is a positive entropy change, presumably associated with breakdown of

Fig. 1 Differential molar enthalpies of dilution of three zwitterionic surfactants in different aqueous media and at different temperatures. The enthalpic values, taken with the opposite sign, are plotted against the molality of equilibrium bulk solution in the calorimetric cell. The numbers represent the corresponding molar enthalpies of micellization

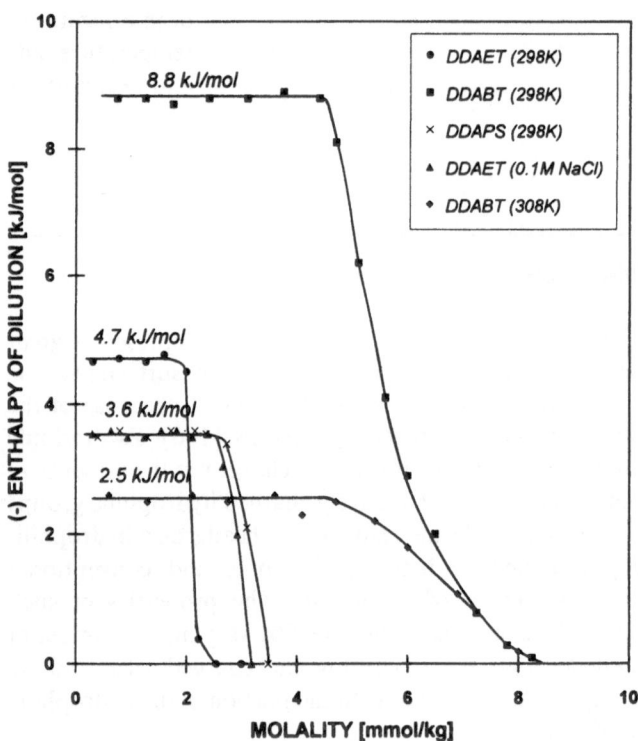

Progr Colloid Polym Sci (1995) 98:299–302
© Steinkopff Verlag 1995

the structured water which surrounds the hydrocarbon chain in the unassociated species [9]. A pronounced decrease in the endothermic enthalpies of micellization at a higher temperature and in the presence of electrolyte (whereas the corresponding CMC values change relatively weakly) suggests that the mechanism of micelle formation in the case of zwitterionic surfactants is also closely connected to the hydrophobic effect. At higher temperatures water loses most of its peculiar structural properties and the formation of structured water around the hydrocarbon becomes impossible. Relatively small ions, such as Na^+ and Cl^-, have net structure-making effect on water structure; surfactant monomers are salted out by electrolyte and micellization is energetically more favorable.

It can be noted that the effect of micellization becomes less endothermic with decreasing hydrophilicity of the zwitterionic headgroup (i.e., shorter intercharge bridge and sulfonate anionic group). This tendency shows that there exists another essential contribution to the energetics of micellization of zwitterionic surfactants; the contribution can be attributed to the nature of the surfactant headgroup. An increase in hydrophilic character, which improves the solubility of the molecule (lower Krafft point, greater CMC [4]), could signify stronger repulsive interaction between the different headgroups; it would be more difficult (effect more endothermic) to insert such a zwitterion into the micelle.

The analogous five enthalpic curves of zwitterionic surfactant adsorption onto silica gel are presented in Fig. 2. Two characteristic regions can be distinguished on each curve.

In the first region, of which the domain is limited to low surface coverage ratios, the decreasing tendency in the exothermic differential enthalpy with increasing amount of adsorption can be related to the direct solid-adsorbate interaction and is consistent with the picture of the more active sites being covered first [10, 11]. A proposed model is that individual zwitterions are localized on the negatively charged silica surface with the quaternary nitrogen as the surface binding site and the anionic group either away from the surface or extended on it. It can be seen that the enthalpy of adsorption is quite sensitive to the tether length and type of negatively charged center. The observation indicates that the polar headgroup may be oriented parallel to the surface, at least in the beginning of the process. The marked difference between enthalpic values corresponding to different temperatures provides evidence for the competitive character of this first adsorption step [10, 11]. Dewetting of a mineral surface is accomplished by the displacement of adsorbed water molecules by zwitterions of which the polar part interacts with ionic surface sites more strongly than with water. The endothermic contribution due to this effect decreases with a rise in

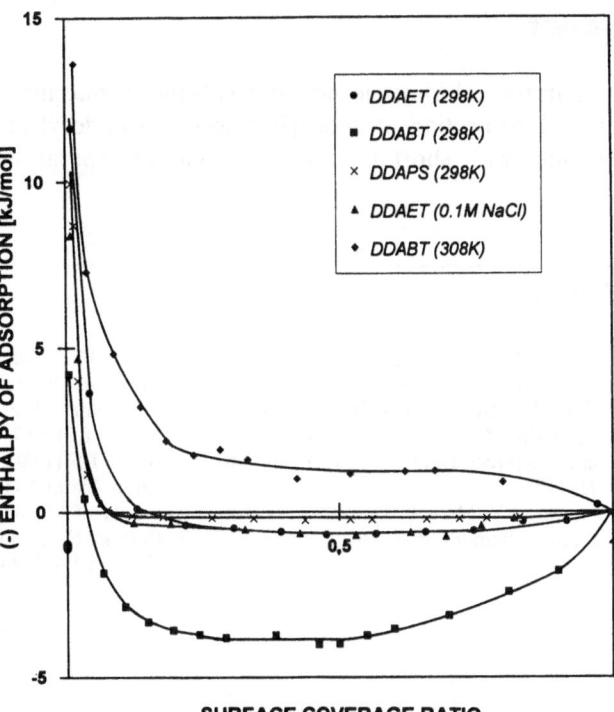

Fig. 2 Differential molar enthalpies of adsorption of three zwitterionic surfactants onto silica gel from different aqueous media and at different temperatures. The enthalpic values, taken with the opposite sign, are plotted against the ratio of the corresponding amount adsorbed to that at surface saturation

temperature, thus resulting in more exothermic enthalpies of adsorption.

In the second region, the enthalpy of surfactant adsorption is constant over a wide interval of surface coverage ratios. This enthalpic value, endothermic in most cases, exhibits almost the same dependence upon temperature and hydrophilic character of the headgroup as the enthalpy of micellization does. By analogy, adsorption in this region can be interpreted as the formation of three-dimensional-like micelles of surface. The fact that the enthalpic effect undergoes a regular change on transition from region 1 to region 2 permits to suppose that surface aggregates begin to form already at lower surface coverages and contribute to the total heterogeneity of the system. In the vicinity of the CMC, when the three-dimensional micellization starts in the bulk phase, the energetic difference between both phases vanishes and the enthalpy of adsorption tends to zero.

The addition of electrolyte slightly diminishes the exothermic enthalpy of adsorption in the first region and seems to have no effect in the second region. It is probable that the concentration of NaCl was too small to provoke a pronounced effect at the interface. On the other hand, the presence of monovalent ions in the vicinity of surface may cause an additional structuring of interfacial water and amplify the endothermic effect of its desorption [10, 11].

Conclusion

Comparative calorimetric measurements for the micellization and adsorption of dodecylbetaines and dodecylsulfobetaines with short intramolecular charge separation ($n = 1, 2, 3$) support a charge interaction mechanism of individual adsorption at the silica/aqueous solution interface at very low surface coverages and a three-dimensional-like micelle formation at higher values.

References

1. Rosen MJ (1978) Surfactants and Interfacial Phenomena. Wiley, New York
2. Ernst R, Miller EJ (1982) In: Bluestein BR, Hilton CL (eds) Amphoteric Surfactants. Marcel Dekker, New York, pp 71–173
3. Chevalier Y, Mélis F, Dalbiez JP (1992) J Phys Chem 96:8614–8619
4. Weers JG, Rathman JF, Axe FU, Crichlow CA, Foland LD, Scheuing DR, Wiersema RJ, Zielske AG (1991) Langmuir 7:854–867
5. Brode PF (1988) Langmuir 4:176–180
6. Amin-Alami A (1989) Ph.D. Thesis, University of Montpellier
7. Partyka S, Lindheimer M, Zaini S, Keh E, Brun B (1986) Langmuir 2:101–105
8. Partyka S, Keh E, Lindheimer M, Groszek A (1989) Colloids Surf 37:309–318
9. Tanford C (1973) The Hydrophobic Effect: Formation of Micelles and Biological Membranes, Wiley, New York
10. Trompette JL, Zajac J, Keh E, Partyka S (1994) Langmuir 10:812–818
11. Zajac J, Trompette JL, Partyka S (1994) J Thermal Anal 41:1277–1286

Progr Colloid Polym Sci (1995) 98:303–307
© Steinkopff Verlag 1995

J. Zajac
M. Lindheimer
S. Partyka

Calorimetric evidence for the similarity between the mechanisms of cationic and anionic surfactant adsorption on oppositely charged crystalline oxide surfaces

Dr. J. Zajac (✉) · M. Lindheimer
S. Partyka
Laboratoire des Agrégats Moléculaires et
Matériaux Inorganiques
U.R.A. 79
Université Montpellier II
pl. E. Bataillon
34095 Montpellier, France

Abstract Adsorption of one cationic surfactant, benzyldimethyldode-cylammonium bromide (BDDAB), and two anionic surfactants, sodium heptylbenzene sulfonate (SHBS) and sodium dodecyl sulfate (SDS), from aqueous solutions on the oppositely charged crystalline oxide surfaces, negatively charged quartz and positively charged zirconium dioxide, has been studied. Differential molar enthalpies of adsorption, supplemented by adsorption isotherms and electrophoretic data, are believed to reflect the same general scheme of the process, imposed by the particular geometry and properties of a crystalline solid surface. Some quantitative differences, due to the specific properties of every system, do not change this

impression. The following three sequential mechanisms can be proposed in each case: 1) electrostatic adsorption, involving ion-exchange with the pre-adsorbed counterions or ion pairing at oppositely charged surface sites, accompanied by the displacement of interfacial water molecules, 2) formation of the two-dimensional, monolayer aggregates, producing discrete and highly hydrophobic areas on the surface, and 3) process of growing of the transient monolayer aggregates into extended three-dimensional-like micelles of surface.

Key words Calorimetry – adsorption – cationic surfactants – anionic surfactants

Introduction

It is generally accepted that adsorption of ionic surfactants from aqueous solutions on mineral or inorganic substrates is governed mainly by electrostatic and hydrophobic inter-actions [1–9]. At low surface coverages, surfactant ions adsorb physically as individual ions on the oppositely charged surface sites; they displace water molecules and may exchange with other counterions present at the solid/water interface [5, 9]. The hydrophobic association of alkyl chains with the surface is less probable owing to the difficulty of disrupting the "icelike" region of water by

hydrophobic chains in the vicinity of a high concentration of surface hydroxyl groups [4]. The adsorbed surfactant ions with alkyl chains oriented toward the solution act as nucleation centers for surface aggregates formed through chain-chain association in the later stage of adsorption. Different structures of aggregates are usually discussed, such as hemimicelles, admicelles, or surface micelles [2, 3, 5–8]. Among several parameters of the adsorption system, which favor the formation of each of these struc-tures, the nature of a solid surface seems to be the most important.

Comparison of the various substrates for their abilities to adsorb a given surfactant molecule is obscured by the

304

J. Zajac et al.
Similarity of cationic and anionic surfactant adsorption

different particle size distributions or specific surface areas available for adsorbate, the different chemical pretreatments employed, and the crystalline structures involved. Thus, a study of adsorption on inorganic substrates of simple geometries and features can shed light on the interaction of ionic surfactants with many other substrates and will permit the development of models for the adsorption process.

In this paper, the mechanism of adsorption of cationic and anionic surfactants on oppositely charged crystalline oxide surfaces is examined by calorimetry. These results are supplemented by adsorption isotherms and electrophoretic data. The main emphasis is put on the similarity between the phenomena and its connection with the particular properties of a solid surface.

Materials and methods

Two types of solid sample were used: crystalline silica (quartz) and crystalline zirconium dioxide. Quartz, specified to contain 99.8% SiO_2, was supplied by Sifraco (France). Its crystalline structure was determined by x-ray diffraction analysis; a rhombohedral unit cell has the following parameters: $a = b = 0.49$ nm, $c = 0.54$ nm. Ground quartz crystals were sieved and sedimented to obtain a 1- to 10-μm-diameter fraction. This product was first leached repeatedly in boiling 2N HCl and thoroughly washed with deionized water until the filtrate showed no trace of chloride ion. The specific surface area, measured by nitrogen gas adsorption at 77 K (B.E.T. method, $a_m(N_2) = 16.2$ Å^2), was found to be 6.3 sq.m/g. Quartz particles, in the form of natural suspension in pure, deionized water (free pH), are negatively charged and are susceptible of adsorbing cations.

Zirconium dioxide, used in the study, was a commercial product supplied by T.S.K. Zirconia TZ-O (Japan). The zirconia crystals belong to the monoclinic system in which each zirconium atom is surrounded by the seven oxygen atoms at an estimated distance of about 2.04 to 2.26 Å. The mean particle size, observed in a scanning electron micrograph, was about 0.1 μm. This non-porous substrate with a B.E.T./N_2 specific surface area of 14 sq.m/g was used as received. In pure, deionized water, zirconia particles are positively charged and may adsorb anions.

The cationic surfactant, benzyldimethyldodecylammonium bromide (BDDAB), was a product of Fluka (France). It was purified by recrystallization from ethyl acetate and from water. The anionic surfactants used were sodium heptylbenzenesulfonate (SHBS), synthesized in the laboratory [11], and pure sodium dodecyl sulfate (SDS), purchased from Prolabo (France). The SHBS was successively purified by recrystallization from distilled water. Some basic parameters of these surfactant molecules are presented in Table1.

Table 1 Critical micelle concentration (CMC) and cross-sectional area of the polar headgroup at the water/air interface (A_o) for the ionic surfactants used in the study. Both parameters were determined experimentally, basing on the surface tension measurement [9, 11]

Surfactant	Temp. [K]	CMC [mmol/kg]	A_o [nm^2]
BDDAB	298	5.6	0.70
SHBS	298	22.6	0.36
SDS	308	8.37	0.49

Adsorption isotherms were measured by the solution depletion technique. Surfactant solutions were equilibrated with the powdered substrate for 24 h. The substrate was then removed by centrifugation, and the supernatants were analyzed by UV spectrophotometry (BDDAB, SHBS) or by differential refractometry (SDS).

Electrophoretic measurements were made on the Rank Brothers II type apparatus. After the adsorption equilibrium was attained, samples of the solid suspension in the supernatant were transferred to the thermostated microelectrophoresis cell and the average velocity of solid particles was determined at both stationary levels.

Enthalpies of adsorption were measured with a Montcal calorimeter [10]. The stock solution of molality 10 times greater than the corresponding CMC was injected into the calorimetric cell containing 16 g of solvent and 1 g of solid sample by steps of 0.9 mg/s. The dilution experiment had to be carried out in order to determine a correlation term arising from dilution of a micellar solution in the calorimetric cell [9, 10].

Results and discussion

The experimental adsorption isotherms, the electrophoretic curves, and the enthalpic curves have been collected in Figs. 1–3. Each figure presents a given type of data for three adsorption systems: BDDAB/water/quartz at 298 K, SHBS/water/zirconium dioxide at 298 K, and SDS/water/zirconium dioxide at 308 K. The shapes of the three curves of each type are believed to reflect the same, in a general sense, mechanism of the phenomenon. In particular, similarity between the calorimetric results provides a convincing argument for this hypothesis, since enthalpy of adsorption is the resultant macroscopic effect of intermolecular forces involved. Three characteristic regions can be distinguished on the enthalpic curve and they are attributed to three distinct modes of adsorption.

Progr Colloid Polym Sci (1995) 98:303–307
© Steinkopff Verlag 1995

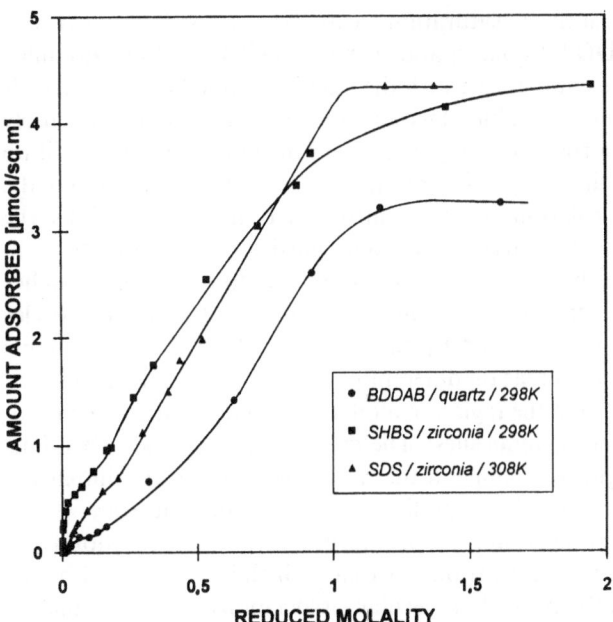

Fig. 1 Adsorption isotherms of three ionic surfactants from aqueous solutions on crystalline oxide surfaces. The amount adsorbed is plotted against the ratio of the equilibrium bulk molality to the corresponding CMC value

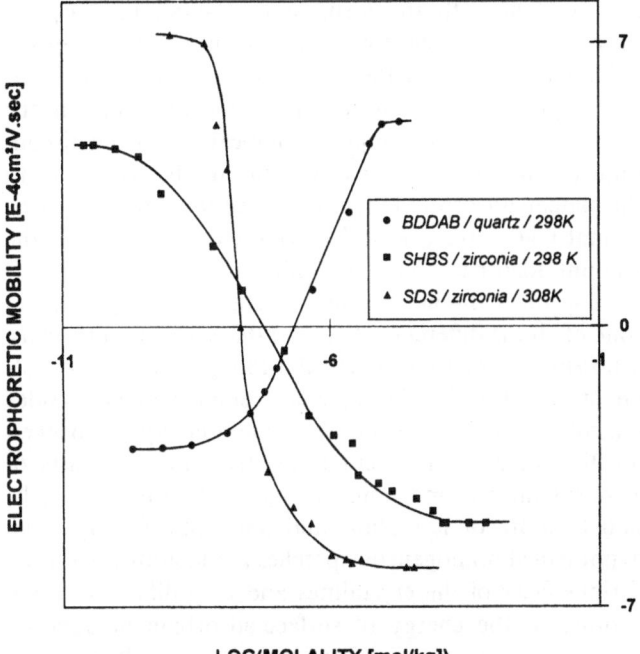

Fig. 2 Electrophoretic mobility of solid particles against the logarithm of molality for the adsorption of three ionic surfactants from aqueous solutions on crystalline oxide surfaces

At low surface coverage ratios, the exothermic enthalpy of adsorption decreases with increasing amount of adsorption. This tendency can be primarily related to the direct solid-adsorbate interaction: individual surfac-

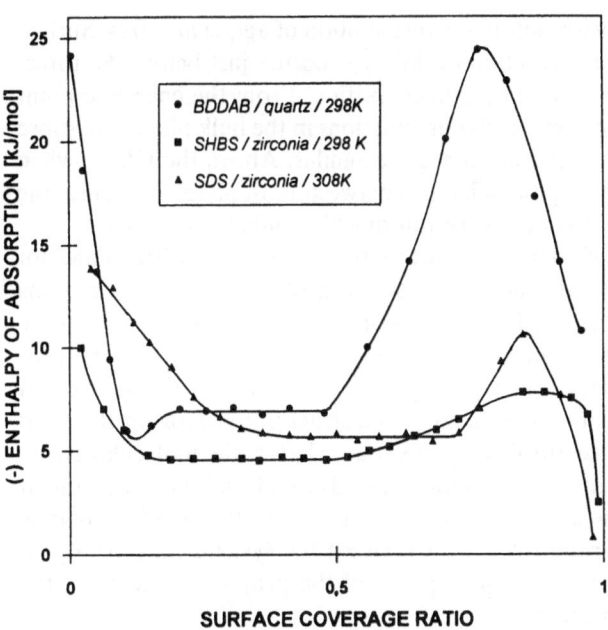

Fig. 3 Differential molar enthalpy of adsorption of three ionic surfactants from aqueous solutions on crystalline oxide surfaces. The enthalpic values, taken with the opposite sign, are plotted against the ratio of the corresponding amount adsorbed to that at surface saturation

tant ions are retained on the oppositely charged surface sites, the more active sites being covered first.

In a region of moderate surface coverages, the enthalpy of adsorption is constant. Since the beginning of this adsorption domain lies above the corresponding isoelectric point found in the electrophoretic measurements, the driving force of process may be chiefly ascribed to the combination of mutual attraction between hydrophobic groups of different surfactant molecules and their tendency to escape from an aqueous environment. The oncoming ions adsorb onto or adjacent to the previously adsorbed ones and this yields larger adsorbate structures having either monolayer (hemimicelles) or bilayer (admicelles) character.

The last adsorption region is characterized by an intricate peak on the enthalpic curve. In the initial subregion, the exothermic enthalpy of adsorption monotonously increases to a maximum value, which is situated not far from the CMC. There is great resemblance between this part of curve and that of dilution of a micellar surfactant solution in a dilute aqueous solution for concentrations around the CMC [9, 11]. The latter curve corresponds to the formation of micelles, of which sizes, number, and relative proportions are modified by the changing concentration in the neighborhood of the CMC. By analogy, adsorption in the third region can be interpreted as the formation of large, three-dimensional-like micelles of surface, involving a stepwise shift in the distribution of aggregate sizes. Surfactant aggregation of this type occurs just before the three-dimensional bulk micellization. From the energetic point of view, molecular associations in the bulk phase and those at the interface are quite similar. Above the CMC, when the energetic difference between both phases vanishes, the enthalpy of adsorption quickly tends to zero.

Of course, the three curves presented in Fig. 3 are not identical. They differ in width of the subsequent regions and in absolute values corresponding to the same region. These quantitative differences are due to the specific properties of every adsorption system, such as the nature and spatial arrangement of the structural groups on the solid surface, the density of surface charge, the molecular structure of the surfactant being adsorbed, and the environment of the aqueous phase. In some cases, the detailed analysis of certain differences between the systems may even provide evidence in support of the proposed scheme of the adsorption process.

The energy of individual adsorption of adsorbate, being a result of the balance between the affinity of surfactant ions to the surface and their affinity to the bulk solution, is strongly dependent upon the nature of both the adsorbate and the solid adsorbent. Three enthalpic curves show this general tendency in the first adsorption region.

Individual adsorption of surfactant ions has a competitive character: it involves replacement of counterions adsorbed onto the substrate from the solution by similarly charged surfactant ions or displacement of interfacial water molecules by the ionic headgroup, which interacts with oppositely charged surface sites more strongly than with water. It seems probable that a quasi-constant initial part of each electrophoretic curve supports the ion-exchange mechanism of adsorption at very low surface coverages. In each case, the total energetic effect includes such endothermic contributions as the energy of desorption of counterions and the energy of desorption of interfacial water. The latter depends mainly on the crystalline structure of surface. Crystalline solid surfaces with an orderly arrangement of the surface atoms have been found to exert an icelike structuring of adsorbed water molecules [12] and thus it is more difficult to remove water from them. A marked exothermic minimum on the enthalpic curve for the BDDAB/quartz system at low surface coverages may be explained by a greater endothermic contribution due to this local dewetting of a mineral surface; it is not surprising on account of certain similarity between open structures of water and quartz [13]. For the same surface structure and similar surfactant ionic heads, the above endothermic contribution decreases when the temperature is raised; this effect is also illustrated in Fig. 3.

Basing on the adsorption isotherms shown in Fig. 1, a limiting area, occupied per one adsorbed molecule at surface saturation, can be assessed at 0.51 nm^2 (BDDAB/quartz) and 0.38 nm^2 (SHBS or SDS/zirconia). The comparison with the values obtained at the water/air interface (Table 1) would argue in favor of either a densely adsorbed monolayer or a very incomplete bilayer built up on the surface. The results of qualitative observation of the flocculation do not confirm any of these conclusions: the flocs of quartz and zirconia particles were found to form even below the isoelectric point and to redisperse at higher surface coverages. Since flocculation of mineral particles may be justified by the existence of highly hydrophobic surface areas on different particles, surface aggregates produced in the region of moderate coverage ratios are monolayer (hemimicelles). The orientation of the adsorbed ions begins to change in the third adsorption stage, imparting increasing hydrophilic character to the substrate as adsorption continues. As a result, at surface saturation there is a certain fraction of surface which is not covered by the tightly packed molecular associations having a bilayer character. The condition of aggregate stabilization allows such fragmented bilayered structures to be regarded as the extended three-dimensional-like micelles of surface.

A slow increase in the enthalpy of adsorption of the SHBS onto zirconia in the region of high coverage ratios is consistent with the formation of small polydisperse micelles; the same effect was observed in the bulk solution [11]. The analysis of the SHBS/zirconia adsorption isotherm provides an additional argument: although the pseudo-phase separation model indicates that the discontinuity concentration corresponding to the transition to the plateau adsorption region must be equal to the experimental CMC, there is still a continuous increase in the amount adsorbed above this value.

The above discussed, uniform scheme of the process, in spite of all the differences between the adsorption systems, is possible owing to the particular topography of a crystalline surface. Powdered sample of crystalline mineral oxides is a collection of a large number of minute crystals packed together at random orientations. Its surface consists of a considerable proportions of edges and corners, in addition to more or less plane and unisorptic patches. The hypothetical homogeneous patches are identified with the various faces of the crystallites and they differ from one another in the energy of surface-adsorbate interaction. Such surface domains are filled with the vertically oriented surfactant ions in the sequence of decreasing energy of adsorption [2, 6, 8]. Strong lateral hydrophobic bonds result in the cooperative adsorption already in the first adsorption region. As adsorption continues, the number of hemimicellar structures and their average aggregation numbers increase, producing compact hydrophobic patches on the surface. Since the opposite surface charge has been screened in the first region, adsorption of surfac-

Progr Colloid Polym Sci (1995) 98:303–307
© Steinkopff Verlag 1995

tants in the next region (constant enthalpy) occurs on polar, non-ionized surface hydroxyl groups. It is possible because the affinity of surfactant headgroup to the surface is still greater than that of water. The maximum surface density of hydroxyl groups for a fully hydroxylated surface was shown to be about 5 $SiOH/nm^2$ [13] or 9 $ZrOH/nm^2$ [11]; in reality, only a fraction of the MeOH complexes are ionized and this is determined by the pH of solution. At the end of the second region, only submonolayer coverages can be achieved. Further monolayer adsorption would lead to a great excess of similar charge, accumulated on surface patches. The adsorbing ions, which now must additionally overcome electrostatic repulsion with the similarly charged adsorbed layer, are not able to displace interfacial water and to adsorb directly on the surface. In consequence, the primary aggregates do not fuse and begin to grow in a direction perpendicular to the surface.

Conclusion

The analysis of experimental data shows that the general mechanism of the ionic surfactant adsorption on the oppositely charged crystalline oxide surface is uniform in spite of all the differences between the adsorption systems. Such scheme of the process is determined by the patchwise-like topography of a heterogeneous solid surface.

References

1. Rosen MJ (1978) Surfactants and Interfacial Phenomena. Wiley New York
2. Scamehorn JF, Schechter RS, Wade WH (1982) J Colloid Interface Sci 85:463–478
3. Yaskie MA, Harwell JH (1988) J Phys Chem 92:2346–2352
4. Ingram BT, Ottewill RH (1991) In: Rubingh DN, Holland PM (eds) Cationic Surfactants. Marcel Dekker New York pp 87–140
5. Fuerstenau DW, Herrera-Urbina R (1991) In: Rubingh DN, Holland PM (eds) Cationic Surfactants. Marcel Dekker, New York, pp 408–447
6. Cases JM, Villieras F (1992) Langmuir 8:1251–1264
7. Böhmer MR, Koopal LK (1992) Langmuir 8:1594–1602 and 2649–2665
8. Lajtar L, Narkiewicz-Michalek J, Rudzinski S, Partyka S (1993) Langmuir 9:3174–3190
9. Trompette JL, Zajac J, Keh E, Partyka S (1994) Langmuir 10:812–818
10. Partyka S, Lindheimer M, Zaini S, Keh E, Brun B (1986) Langmuir 2:101–105
11. Grine N (1990) Ph.D. Thesis, University of Montpellier
12. Kavanau JL (1964) Water and Solute-Water Interactions, Holden-Day: San Francisco
13. Iler RK (1979) The Chemistry of Silica, Wiley: New York

Progr Colloid Polym Sci (1995) 98:311–312
© Steinkopff Verlag 1995

SUBJECT INDEX